I0065939

Power and Energy Engineering: Emerging Concepts and Applications

Power and Energy Engineering: Emerging Concepts and Applications

Editor: Linda Morand

CALLISTO REFERENCE

www.callistoreference.com

Callisto Reference,
118-35 Queens Blvd., Suite 400,
Forest Hills, NY 11375, USA

Visit us on the World Wide Web at:
www.callistoreference.com

© Callisto Reference, 2017

This book contains information obtained from authentic and highly regarded sources. Copyright for all individual chapters remain with the respective authors as indicated. All chapters are published with permission under the Creative Commons Attribution License or equivalent. A wide variety of references are listed. Permission and sources are indicated; for detailed attributions, please refer to the permissions page and list of contributors. Reasonable efforts have been made to publish reliable data and information, but the authors, editors and publisher cannot assume any responsibility for the validity of all materials or the consequences of their use.

ISBN: 978-1-63239-871-0 (Hardback)

The publisher's policy is to use permanent paper from mills that operate a sustainable forestry policy. Furthermore, the publisher ensures that the text paper and cover boards used have met acceptable environmental accreditation standards.

Trademark Notice: Registered trademark of products or corporate names are used only for explanation and identification without intent to infringe.

Printed in the United States of America.

Cataloging-in-Publication Data

Power and energy engineering : emerging concepts and applications / edited by Linda Morand.
  p. cm.
Includes bibliographical references and index.
ISBN 978-1-63239-871-0
1. Power (Mechanics). 2. Power resources. 3. Environmental engineering. I. Morand, Linda.
TJ163.9 .P69 2017
621.042--dc23

Table of Contents

Permissions

List of Contributors

Index

Preface

Every book is a source of knowledge and this one is no exception. The idea that led to the conceptualization of this book was the fact that the world is advancing rapidly; which makes it crucial to document the progress in every field. I am aware that a lot of data is already available, yet, there is a lot more to learn. Hence, I accepted the responsibility of editing this book and contributing my knowledge to the community.

Power engineering can be defined as the development of energy harnessing and distributing technologies that improve energy efficiency. This book on power and energy engineering discusses the design and maintenance of energy technologies. This book attempts to understand the multiple branches that fall under the discipline of power and energy engineering and how such concepts have practical applications. The various studies that are constantly contributing towards advancing technologies and evolution of this field are examined in detail. This text aims to advance the already existing knowledge on energy engineering and energy efficiency technologies. Students, researchers, experts and all associated with this field will benefit alike from this book. For all those who are interested in energy engineering, this book can prove to be an essential guide.

While editing this book, I had multiple visions for it. Then I finally narrowed down to make every chapter a sole standing text explaining a particular topic, so that they can be used independently. However, the umbrella subject sinews them into a common theme. This makes the book a unique platform of knowledge.

I would like to give the major credit of this book to the experts from every corner of the world, who took the time to share their expertise with us. Also, I owe the completion of this book to the never-ending support of my family, who supported me throughout the project.

Editor

Analysis of Nigeria-Sao Tome and Principe Joint Development and Suggestions for China

Huang Wen-bo

Institute of International Law, Wuhan University, Wuhan, China

Email address:

huang_19871021@163.com

Abstract: The Nigeria-Sao Tome and Principe Joint Development is a typical case in Africa. The two states are located in Gulf of Guinea, where delimitation disputes inside are serious and oil competitions among western developed countries outside are fierce, the Nigeria-Sao Tome and Principe Joint Development is carried out on this background. About the implement, several factors play a positive role: stable bilateral relations, adjustments of domestic policies and the establishment of Gulf of Guinea Commission; but oil steal and outside intervention are negative factors. About the content, many special provisions, such as Proportion, Management, Petroleum Development Contracts and Disputes Settlement are contained in joint development agreement. However, there are still some problems unsolved: unequal status of the two states, lack of information sharing mechanism and supervision mechanism, multinational corporations' monopoly. China should learn from it, deepen mutual understanding and mutual trust with countries nearby to promote joint development in South China Sea.

Keywords: Joint Development, Joint Development Agreement, International Law

1. Introduction

The Nigeria-Sao Tome and Principe Joint Development is a major attempt in Africa to develop offshore oil and gas resources. Nigeria and Sao Tome and Principe are located in Gulf of Guinea in West Africa, where oil and gas reserves are quite rich, attracting the world's attention. Oil carries substantial weight in these two countries. Nigeria is the world's second largest economy, oil exports, which contribute 98% of Nigeria's total export revenue and 83% of Nigeria's national income, are its most important economic source. Sao Tome and Principe also has rich offshore oil deposits, which are about 500 million barrels estimated by Exxon Mobil in 1999. At present, since China has been trying to promote joint development with countries in South China Sea, it is of great significance to analysis joint development cases, like Nigeria-Sao Tome and Principe Joint Development, to supply some useful lessons and experience.

2. Nigeria-Sao Tome and Principe Joint Development: Background and Factors

2.1. Background

The background of Nigeria-Sao Tome and Principe joint development project is mainly divided into two aspects:

External powers compete fiercely for oil in Gulf of Guinea. The Gulf of Guinea is one of the world's richest oil-producing areas,[1]and draws the world's attention because of high oil quality and convenient transportation. European and American developed countries try all means to have a share of oil in this rich land, which can be mainly manifested in two aspects:

Competition among traditional powers. The United States is the largest investor and beneficiary in Gulf of Guinea, and the advantages are still expanding. Gulf of Guinea, compared to Middle East, is nearer to the United States and free from terrorism, and of strategic importance for the United States. Nigeria, the largest oil producer in Gulf of Guinea, is an ideal oil supply country for the United States. After a series of diplomatic pushes and massive investments, many American oil companies like Chevron Texaco, Exxon Mobil and Esso have already established a foothold.[2]European countries represented by the United Kingdom and France don't want to be left behind. The United Kingdom expands business further in Nigeria, and also joins hands with France to compete with the United States to attain prospecting and mining rights.[3]France considers oil as a crucial factor when making diplomatic exchange and economic cooperation with West African countries.[4]These developed countries and oil giants mentioned above compete vigorously to carve oil resources.

Competition between traditional powers and emerging powers. Being aware of the tremendous business opportunities and the severe situation in Gulf of Guinea, newly emerging powers, such as China, South Korea and India, start competing with those traditional powers. It can be envisaged that, it is quite difficult for newly emerging powers to break the existing interests division mode. Due to traditional powers' continuous blocking, emerging powers have paid a huge price for growing. South Korea, for example, paid a record $310 million to win just No.323 deep-sea bloc in Nigerian oil bloc bidding, while the previous highest bid for single oil bloc was only $200 million.[5]In addition, competition also exists among emerging forces. South Korea and India produced dissatisfaction with each other because of claiming bidding bloc before.

Internal maritime delimitation disputes are prominent. Several countries located in Gulf of Guinea, which makes maritime delimitation very sensitive and complex. In the maximum concave, Nigeria, Cameroon, Gabon, Equatorial Guinea and Sao Tome and Principe, whose delimitation claims overlap, share a narrow sea. Any establishment or change of delimitation claims from one country will affect the others, which can be testified vividly in the case concerning Land and Maritime Boundary between Cameroon and Nigeria. Since Cameroon submitted the dispute to the International Court of Justice in March 1995, countries left began to assert their claims.

Equatorial Guinea's territorial water and exclusive economic zone boundary is delineated in accordance with international law and its domestic law. The boundary located in the northeast corner of Gulf of Guinea, overlaps with both Cameroon's and Nigeria's claims. Equatorial Guinea then sent letters to the Court respectively in June 1999 and in September 1999, requiring acting as a non-party intervener to participate in the Cameroon v. Nigeria to protect its legitimate interests. After that, Nigeria responded respectively in August 1999 and in September 1999, that Equatorial Guinea can be allowed to participate the case through Separation of the Proceedings at the premise of not affecting the trial process; Cameroon also took a stand respectively in August 1999 and in October 1999.[6]

Sao Tome and Principe didn't chose to intervene the Cameroon v. Nigeria, but its delimitation with neighboring countries is pending. Sao Tome and Principe reiterated its archipelagic baselines in 1998, and reached maritime delimitation treaties respectively with Equatorial Guinea in 1999 and with Gabon in 2001, but the treaties are not enforced finally. The tripartite boundaries among Sao Tome and Principe, northern Equatorial Guinea and southern Cameroon are undelineated.[7] In addition, Sao Tome and Principe has been pursuing negotiations with Nigeria because of overlapping EEZ claims.

2.2. Factors

There are some beneficial factors promoting Nigeria-Sao Tome and Principe joint development:

Nigeria and Sao Tome and Principe both adjust their domestic policies. How to attract foreign investment, especially international oil companies, becomes a key problem for Nigeria and Sao Tome and Principe. Therefore, Nigeria and Sao Tome and Principe have introduced various preferential policies to improve domestic investment environment to ensure smooth conduct of joint development.

For example, Nigeria polices, including foreign investors can obtain crude oil for refining from Nigerian National Petroleum Corporation, can enjoy the same price of crude oil with domestic refineries, can withdraw investments freely and so on, play an important role in attracting foreign investors.[8]

Sao Tome and Principe also enacted a new bill in 2004 to regulate petroleum fund management, aiming at establishing a national oil fund and ensuring transparency of national management of oil revenues. Oil contracts signed by government must be open, and at least 65% of oil revenues will be used for medical care, education and infrastructure construction annually. When the oil fund is converted into money, the President, the Prime Minister, the Central Bank President and other major financial officers have to sign; the oil sector, has to be audited twice a year by Accounting Council and International Accounting Firms.[9]The bill creates favorable investment environment in Sao Tome and Principe.

Nigeria and Sao Tome and Principe have stable bilateral relations. Although these two countries have disputes of offshore oil revenues, the bilateral relationship overall is stable, which with no doubt can promote the joint development.

On the one hand, relatively speaking, Nigeria and Sao Tome and Principe conduct joint development smoothly. From the formal signing of the Treaty in 2001, to the establishment of Joint Ministerial Council（JMC）in 2002, and to the first licensing round in 2003, the whole process is relatively short and the conduction goes successfully.

On the other hand, Nigeria and Sao Tome and Principe carry out some more mutually beneficial cooperation. For example, Nigeria intended to provide an interest-free loan of sixty-year repayment period to Sao Tome and Principe in July 2010. Before that, Nigerian government had already decided to provide a $30 million loan to Sao Tome and Principe.[10]In addition, a military coup happened in Sao Tome and Principe in July 2003, Nigeria strongly condemned the coup, and sent an envoy to Sao Tome and Principe to help Sao Tome and Principe to resolve the crisis.

The Gulf of Guinea Commission was established in 2006. Gulf of Guinea Commission was established in the summit meeting of Gulf of Guinea countries held in Libreville, capital of Gabon. The headquarter is based in Angola. The committee is responsible for consulting and mediating oil exploration and other aquatic resources disputes among Gulf of Guinea countries. Until now, the committee consists of eight member states: Angola, Cameroon, Congo (Brazzaville), Gabon, Equatorial Guinea, Nigeria, Congo (DRC) and Sao Tome and Principe.

Treaty establishing the Gulf of Guinea Commission builds a legal framework for the Gulf of Guinea Commission and provides guidance for oil exploration activities and disputes resolution in Gulf of Guinea. The treaty includes a preamble and 31 terms, the preamble emphasizes that member states should strengthen multilateral cooperation, enhance mutual trust, consolidate good-neighbor relations, and make full use of resources obtained to promote economic development and improve people's living standards. Member States express their willingness to develop social economy and obeying the treaty.[11]

The establishment of the Gulf of Guinea Commission provides a stable environment to carry on joint development activities for Nigeria-Sao Tome and Principe, and provides guarantee for African countries to manage and utilize energy with international organizations to develop social economy.

Nigeria-Sao Tome and Principe joint development is also influenced by some negative factors:

Oil losses are serious. Oil theft is rampant in Nigeria, causing huge losses to national oil industry. One Nigerian government report shows that, Nigeria loses nearly $35milliondaily due to oil theft, and explosions caused by excavations are common. In March 2014, Shell, working in Nigeria, said that Shell lost over $1billion caused by oil theft and infrastructure destruction in 2013. Therefore, when Shell found oil leakage in November 2014, the company shut down an important pipeline immediately[12] to minimize losses.

Besides, the waterways and forests in Niger Delta area are criss-cross, local folk armed groups, such as Delta People's Volunteer Army, taking advantage of this complex geographical condition, steal oil from underwater pipeline for a long time. They sell oil to buy weapons to fight against the government.[13] What is very worried is that the joint development zone is exactly located in the Niger Delta, thus huge oil losses and anti-government forces presence are with no doubt harmful to joint development.

Western penetration brings negative effect to joint development. Except giant oil companies, European countries and the United States also strengthen their military presence and economic penetration from a national strategic level. The United States is the most typical representative.

For example, in Nigeria, the United States sent former military officials to provide military training and assistance to improve Nigeria's combat capability,[14] to ensure safety oil transportation; in Sao Tome and Principe, in addition to huge investments, the United States has built a large-scale radar network to strengthen surveillance of Gulf of Guinea. Although those initiatives contribute to eliminating theft and ensuring maritime oil transportation, negative effects exist too:

Economically, the fate "oil curse" of Nigeria and Sao Tome and Principe is still inevitable. The US investments in Nigeria and Sao Tome and Principe are basically carried around oil, which although creates a lot of jobs for local people, these two countries are still very poor and the rich-poor division is still serious from a long-term point of view because of the absence of a stable economic system and an equal distribution. In addition, Sao Tome and Principe are excessively dependent on US investments,[15] lacking economic independence and creativity, and the government doesn't spend money on domestic infrastructure construction. Thus Nigeria and Sao Tome and Principe are impossible to get rid of "oil curse" in a short time.

Militarily, the strengthening of outside military exacerbates the two countries' instability. The United States provide military assistance in Nigeria to protect its own oil security. The intervene of the United States turns the former bilateral mode of Nigeria government-folk armed groups into a tripartite mode of the United states-Nigeria government-folk armed groups. What's more, local folk armed groups always kill American oil company's officials in order to revenge.

Politically, the decision-makers' personal role has been strengthened and corruption is rampant. The United States does not help Nigeria and Sao Tome and Principe governments to improve national management, but highlights the decision-makers' individual influence unprecedentedly, leading to frequent corruptions and scandals. To cooperate with Nigeria, the American officials inevitably have to transfer benefits to decision-makers, whose political principle is "cooperation with money givers" and who do not put long term development and political democracy as principal factors.

3. The Nigeria-Sao Tome and Principe Joint Development: Contents and Disadvantages

3.1. Contents

In February 2001, Nigeria and Sao Tome and Principe signed "Treaty between the Federal Republic of Nigeria and the Democratic Republic of Sao Tome and Principe on the Joint Development of Petroleum and other Resources, in respect of Areas of the Exclusive Economic Zone of the Two States", which contains mainly following aspects:

Proportion: Article 3(1) regulates that "Within the Zone, there shall be joint control by the States Parties of the exploration for and exploitation of resources, aimed at achieving optimum commercial utilization. The States Parties shall share, in the proportions Nigeria 60 percent, Sao Tome and Principe 40 percent, all benefits and obligations arising from development activities carried out in the Zone in accordance with this Treaty."

Management: Nigeria-Sao Tome and Principe joint development is a "two-tier" management structure: the Joint Ministerial Council and the Joint Authority. The first tier, the

Joint Ministerial Council, shall have overall responsibility for all matters relating to the exploration for and exploitation of the resources in the Zone, and such other functions as the States Parties may entrust to it. The Council shall comprise not less than two nor more than four Ministers or persons of equivalent rank appointed by the respective Heads of State of each State Party. The Council shall meet at least twice a year and as often as may be required, alternately in Nigeria and in Sao Tome and Principe. Meetings shall be chaired by a member nominated by the host State Party. All decisions of the Council shall be adopted by consensus. The second tier, the Joint Authority, shall have juridical personality in international law and under the law of each of the States Parties and such legal capacities under the law of both States Parties as are necessary for the exercise of its powers and the performance of its functions. In particular, the Authority shall have the capacity to contract, to acquire and dispose of movable and immovable property and to institute and be party to legal proceedings. The Authority shall be responsible to the Council. The Authority, subject to directions from the Council, shall be responsible for the management of activities relating to exploration for and exploitation of the resources in the Zone, in accordance with this Treaty.

Petroleum Development Contracts: No petroleum activities may be undertaken in the Zone other than pursuant to a petroleum development contract between the Authority and one or more contractors. Unless the Council otherwise decides, and in accordance with procedures laid down by the Council for tendering, the principle of holding licensing rounds must be followed prior to the signature of any petroleum development contract.

Settlement of Disputes: the Treaty concludes three kinds of disputes: disputes between the Authority and private interests, disputes arising in the work of the Authority or the Council, unresolved disputes between the States Parties. Disputes between the Authority and private interests, for example, concerning the interpretation or application of a development contract or operating agreement shall unless otherwise agreed between the parties thereto he subject to binding commercial arbitration pursuant to the terms of the relevant development contract or operating agreement.

3.2. Disadvantages

The Nigeria-Sao Tome and Principe joint development exists mainly the following disadvantages:

Sao Tome and Principe is of unequal status. Although Nigeria and Sao Tome and Principe both belong to the third world, there are still significant gaps between them, which decides the dominant position of Nigeria.

Sao Tome and Principe, being independent in 1975, was announced the world's least developed countries by the United Nations. In recent years, currency devalues, living standards decline, and people living under poverty line increase continuously, the government has to save the fragile economy and pay for high foreign debts through oil. Paradoxically, Sao Tome and Principe is a novice oil industry, it can only rely on Nigeria for guidance and experience.

In contrast, Nigeria is much stronger, which is the second largest economy and also the most populous country in Africa. Since 1970s, oil exports gradually become the country's main source of income. Nigeria, classified as aemerging entity, rapidly meets middle-income standard. As always, the Nigerian government has been working to promote and protect national oil industry and related infrastructures, oil exploration experience and technology are relatively well developed.

The treaty doesn't establish a specialized information sharing mechanism. The joint development agreement aims at promoting mutual cooperation and equal exchange, nevertheless, the lack of a specialized information sharing mechanism makes the two countries embarrassed to respond questionings to international community.

The joint development zone was conducted two bids since establishment. The first bid was conducted in April 2003 to develop 9 deep-sea blocs, which lasted for 6 months. Only Chevron-Texaco won the bid for only one bloc and other bidders for other 8 blocs withdrew because of substandard qualification. What is in stark contrast is the second bid in November 2004, which lasted for just 1 month to develop 5 deep-sea blocs. The second bid attracted many small, unknown, and relatively weak oil companies, who could actively participate in the bidding after only one month's preparation although they didn't meet technology, capital and experience standards before.[17]In this regard, the international community criticized and questioned seriously, but the two countries didn't explain and the reason is still far unknown.

The supervision mechanism is ineffective. According to the joint development agreement, the Joint Ministerial Council must have 2-4 members with the same level designated by each country's president, therefore the two presidents have been accorded enormous powers. Article 15.2 says that: "Unless otherwise expressly approved by the Council, no Executive Director, officer or other staff member of the Authority may have any direct or indirect financial interest in development activities in the Zone." It can be envisaged that the Council has great powers and it may be reduced to seek private interests, which can be proved by Fadriquede Menezes's remarks in 2004. Fadriquede Menezes condemned proposals of the Joint Authority, and declared that Nigeria national oil company ERHC, closely related to Nigerian President Olusegun Obasanjo, enjoyed 60% in bloc 4 and Filtim Huzod, belonging to former Nigerian politicians, enjoyed 85% in bloc 6, while these two companies did not have any deep-sea exploration experience before. Although Fadriquede Menezes dissatisfied with the phenomenon, he can do nothing about this. Therefore, power restriction and supervision are quite important.

Oil resources are monopolized by multinationals. Nigerian oil industry, for example, due to lacking self-exploration capabilities, mainly depends on cooperation with multinational oil companies. Therefore Nigerian oil industry is controlled by western oil companies especially multinational oil companies, 92% of oil production and 84%

of projects and services are undertaken by multinationals or foreign companies, whose oil daily production accounts for 95% of Nigerian oil production. Shell's daily production, for example, is 900,000 barrels of oil, accounting for 52% of Nigerian daily production. Nigerian oil economy is of high dependence, which will cause negative impact on Nigeria and Sao Tome and Principe economy.

4. Suggestions for China

China can draw lessons and experience from Nigeria-Sao Tome and Principe joint development from following aspects:

4.1. Promoting Equal Communication and Mutual Respect

No equity means no guarantee. Joint development countries should communicate and share benefits equally regardless of land size, economic development level and military power, etc.. Although Sao Tome and Principe makes a compromise on sharing (40%), it is hard to tell that Sao Tome and Principe stands to loss, because it has already gained a lot on its own terms. Comments should be made in an overall and comprehensive context. China should also have a clear understanding of this, and grasp overall situation when making decisions. Any behaviors absorbed only in immediate interests should be discarded and condemned.

It must be emphasized that, compared with states around South China Sea, China is in a dominant position in all aspects. To promote joint development, every country should respect others' sovereign rights within legitimate scope. China specially, should not be kidnapped by small countries using the concern "China is much stronger, and weak countries will be at a disadvantage in negotiations with Him", and should not make sacrifices and concessions blindly. As a big country, China also needs equality and respect.

4.2. Establishing Information Sharing Mechanism

Information transparency ensures joint development activities carried on equally, which is particularly important to countries whose economy is dependent heavily on oil. Nigeria and Sao Tome and Principe, both are African countries, but the gap between the two countries is still very large and Sao Tome and Principe has to rely on Nigeria. If Sao Tome and Principe is squeezed and suppressed by Nigeria, Sao Tome and Principe will suffer huge losses. China should learn from Sao Tome and Principe, establishes a information sharing mechanism to ensure information transparency, to maintain legitimate rights and interests at the most extent.

4.3. Establishing an Effective Supervision Mechanism

Oil Corruption will jeopardize national economy and security in long terms, thus establishing an effective supervision mechanism to restrict powers is a necessity in joint development agreement. An effective supervision mechanism should run through all processes. Supervision rights should not be totally centralized on heads of government, on the contrary, the subjects of supervision

rights should be diverse. There is no fixed pattern of supervision, how to design the mechanism depends on specific circumstances of each case. Countries should negotiate according to different situations.

4.4. Taking Advantage of National Oil Companies

Cooperation with multinational oil companies is a common practice. China needs to pay special attention that, China should take full advantage of its own national oil companies to accumulate deep-sea exploration and development experience and to seize the initiative in joint development activities. Highly dependent on cooperation with multinational oil companies goes against with national oil companies' growth.

4.5. Handling Relationship with Third States Properly

Article 45 of the joint development agreement says: "In the exercise of their rights and powers under this Treaty, the States Parties shall take into account the rights and freedoms of other States in respect of the Zone as provided under generally accepted principles of international law. If any third party claims rights inconsistent with those of the States Parties under this Treaty, then the States Parties shall consult through appropriate channels with a view to coordinating a response." Claims of maritime boundary delimitation between third states and the JDZ have not arisen.[18]However, China's situation is much more complex.

Claims overlap in South China Sea, and any change of claims will affect the whole situation in South China Sea. China has to face questions and objections from third states when carrying on joint development activities. Chinese government should design third party terms in an effective and comprehensive way, limiting disputes to bilateral negotiation occasions, deepening consensus and mutual trust, to avoid internationalization.

4.6. Strengthening Marine Environmental Protection

Sao Tome and Principe, although has great amount of oil resources, is still one of the world's least developed countries, which is closely related to vicious exploitation ignoring marine environmental protection. In Gulf of Guinea, many oil operating vessels pass through marine ecological zones in West Africa, oil spills cause damages to fishery and regional economies.[19]Maritime exploration, marine environmental protection and economic development are closely linked as a unified whole. China should learn lessons from Sao Tome and Principe, actively implement the article 194 and article 208 of United Nations Convention on the Law of the Sea, take measures to prevent, reduce and control marine pollution to promote sustainable development.

The paper is a stage achievement of the Key Projects of Philosophy and Social Sciences Research from Ministry of Education "Case Study and Practice of International Maritime Joint Development" (Project Grant No.13JZD039).

References

[1] See Tim Daniel, *Maritime Boundaries in the Gulf of Guinea*, Partner in DJ Freeman, London, available at http://www.gmat.unsw.edu.au/ablos/ABLOS01Folder/DANIE L.PDF, last visited on May. 2, 2014.

[2] See MO Ya, *Great Country was Busy Carving up Oil in Gulf of Guinea*, People, September 17, 2003, available at http://www.people.com.cn/GB/guoji/14549/2093180.html, last visited on Nov. 25, 2014.

[3] See YAO Gui-mei, *Western Powers' Strategy of Contending for African Oil Resources and its Consequences*, Africa Review, Vol.2, 2005.

[4] *Oil Scramble Next Step: Africa*, available at http://finance.eastday.com/eastday/finance/cjsp/userobject1ai2 78434.html, last visited on Nov. 25, 2014.

[5] See LIN Xiao-chun, DAI A-di, *Global Oil Giants Compete in Gulf of Guinea*, China Petrochemical News, Vol.8, Sept. 8, 2005.

[6] Available at http://www.icj-cij.org/docket/index.php?p1=3&p2=3&k=52& case=94&code=cn&p3=8, last visited on May. 1, 2014.

[7] See *Summaries of Judgments, Advisory Opinions and Orders of the International Court of Justice 1997-2002*, United Nations, p.241, available at http://legal.un.org/ICJsummaries/documents/english/ST-LEG-SER-F-1-Add2_E.pdf, last visited on Nov. 24, 2014.

[8] ZHANG Chang-bing, ZHANG Tai-qiu, *An Analysis on Autonomous Development of African Oil Economy and Its Main Influence*, Journal of International Trade, Vol.9, p.79, 2011.

[9] ZHENG Xiao-meng, *Sao Tome and Principe: West African Island Longing Black Gold Spewing*, China Petrochemical News, Vol.7, Nov. 5, 2010.

[10] Available at http://www.mofcom.gov.cn/article/i/jyjl/k/200905/200905062 84404.shtml, last visited on Nov. 24, 2014.

[11] See Gbenga Oduntan, *The Emergent Legal Regime for Exploration of Hydrocarbons in the Gulf of Guinea: Imperative Considerations for Participating States and Multinationals*, International and Comparative Law Quarterly, Vol 57, pp.253-258, 2008.

[12] See Chika Izuora, *Nigeria: Oil Export Suffers Setback As Shell Shuts Facility*, available at http://allafrica.com/stories/201411250221.html, last visited on Nov. 26, 2014.

[13] Japanese documentary film in 2005, *Nigerian Oil Battle*, available at http://share.renren.com/share/250086330/5633128640, last visited on Nov. 26, 2014.

[14] From Stratfor, *The Militarization of West Africa*, available at https://www.globalpolicy.org/the-dark-side-of-natural-resourc es-st/water-in-conflict/40165.html, last visited on Nov. 26, 2014.

[15] *US Naval Base to Protect Sao Tome Oil*, BBC News, Thursday, Aug. 22, 2002, available at http://news.bbc.co.uk/2/hi/business/2210571.stm, last visited on Nov. 26, 2014.

[16] Claire Woodside, *West Africa: America's Foreign Policy Post 911 and the "Resource Curse," A Head on Collision*, Journal of Military and Strategic Studies, Vol. 9, Issue 4, pp.26-27, 2007.

[17] Wendy N.Duong, *Chinese Law in the Global Context: Article: Following the Path of Oil: The Law of the Sea or Realpolitik-What Good Does Law Do in the South China Sea Territorial Conflicts ?* Fordham International Law Journal, Vol. 30, p.33, 2007.

[18] Nicholas Chinedu Eze, *Rethinking Maritime Delimitation and Promoting Joint Development of Petroleum: the Nigeria-Sao Tome and Principe Joint Development Model*, a thesis submitted in partial fulfillment of the requirements for the degree of Master of Laws in The Faculty of Graduate Studies, The University of British Columbia, September 2011, pp.172-173.

[19] Moreno, Carlos J., *Oil and Gas Exploration and Production in the Gulf of Guinea: Can the New Gulf Be Green?* Houston Journal of International Law, Vol. 31, No.2, p.429, 2009.

2

Multi Objective Dynamic Economic Dispatch with Cubic Cost Functions

Moses Peter Musau, Nicodemus Odero Abungu, Cyrus Wabuge Wekesa

Department of Electrical and Information Engineering, School of Engineering, The University of Nairobi, Nairobi, Kenya

Email address:
pemosmusa@gmail.com (M. P. Musau), abunguodero@gmail.com (N. O. Abungu), cyrus_wekesa@yahoo.com (C. W. Wekesa)

Abstract: The formulation and solution of the Dynamic Economic Dispatch (DED) problem is one of the key disciplines in modern power system operation, planning, operation and control. Past researches have considered DED on Quadratic Cost Functions (QCF), with only few works considering higher order cost functions which are more accurate. The Static Economic Dispatch (SED) has been widely tackled in past researches, however, it is the DED problem that represents a real life power system. There is need to review this problem and establish a more practical formulation of the same taking into consideration all the objectives and constraints possible. The methods used in the solution of DED problem have evolved from the traditional deterministic ones, to the pure heuristic, and finally to the state of the heart hybrids. The hybrids methods have been developed to exalt the strengths and improve the weaknesses of the base method. Such optimisation methods need to be reviewed and classified. This paper will do an in depth review of the DED problem on both quadratic and cubic cost functions. Further it will provide a detailed classification of the methods used to solve the problem as its complexity increases. Consequently, three method Hybrids is the way go as far as the solution of the cubic Multi Objective DED (MODED) with five objectives is concerned.

Keywords: Cubic Cost Functions (CCF), Dynamic Economic Dispatch (DED), Hybrid Methods,
Multi Objective DED (MODED)

1. Introduction

The fuel cost in in existing textbooks, for example *Optimization of Power System Operation* [1] is generally assumed to be a smooth Static ED (SED) modeled as a Quadratic Cost Function(QCF).According to Jizhong Zhu, pp85-88, (2006) [1], this definition however makes many assumptions which are impractical to real systems. These assumptions include ignoring emissions, uncertainties, reactive power dispatch, ramp rates, valve points and integration of Renewable Energy (RE) generators. These assumptions are impractical in the real time power system and can no longer be ignored. Consideration of such assumptions in ED formulation leads to the DED formulation and solution.

DED optimization problem is one of the most important issues which must be taken into consideration in power systems planning and operation. DED is aimed at planning the power output for each devoted generator unit in such a way that the operating cost is minimized and simultaneously,

matching real and reactive load demand, power operating limits and above all maintaining the system stability. Further, the security challenges and power trade issues need to be addressed in such a formulation. Based on convention, electrical power systems are operated based on minimizing operational cost while maintaining all the system constraints in place.

The integration of renewable energy into the grid and the need to account for the emission and transmission (real and reactive) losses in a more accurate sense has greatly affected the way in which the DED problem is formulated and solved. This is due to the use of different generators which utilize various fuels and the fact that their real and reactive power outputs are not deterministic but stochastic in such a case. Thus, there has been a desperate need to review the formulation and solution of the existing DED problem.

The Thermal, Renewable Energy (RE), Emissions and Transmission Loss cost functions when considered individually, results in a Single Objective DED (SODED). The SODED has been considered on a Quadratic Cost Function which is less accurate. Where more than two objectives are taken into consideration, a MODED problem results. In most researches, the thermal cost functions has been taken as the base cost function then the other functions and the corresponding constraints have been added depending on the interest of the researcher. However there is need to consider all these objectives and constraints simultaneously since they do affect the modern power system in such a practical manner. The Multi Objective DED(MODED) which is well suited for the modern smart grid need to be revisited to include wind ,solar, emission and transmission line losses in a more accurate scenario in which the five main objectives are handled simultaneously in a fully constrained environment.

Dealing with multiple and conflicting objectives in optimization engineering has always been a challenging undertaking. Many practical applications in optimization often involve many goals to be satisfied simultaneously. In power system planning, operation and control, MODED is a good example of such problems. As the MODED gets more complex, improved and higher order cost functions will continue to replace the Quadratic ones. This is because the higher order cost functions have proved to be more accurate and realistic than the traditional Quadratic ones. In addition, Hybrid methods which do accommodate the uncertain RE will replace the static ones which are restricted to one or two objectives. Also, the existing hybrids must also be tested on higher order cost functions for the MODED with more objectives. Further, three and four method hybrids will be embraced for more accurate realistic results.

Most of the existing security, reactive power and multi area problems of the economic dispatch has been considered in a static environment and using the classical methods. There is need to reconsider these problems in the MODED with the stochastic RE. This will lead to improved ED, increased system security, advanced voltage profile and better trade on electrical power.

This research paper aims at addressing the MODED problem by providing a detailed review of the existing DED works in terms of the problem formulation and the solution methods used. The paper has been divided into six main sections. There is an introduction to power system operation and planning, DED overview, MODED review, formulation of MODED and the existing sub problems, a classification of DED optimization methods and the review conclusion.

2. Modern Power System Operation and Planning

Power system operation and control problem include the Unit Commitment (UC), Optimal Power Flow (OPF), SED, Hydro-Thermal Scheduling (HTS), and DED which are integrated to each other. According to a review of the Science Direct databases, there is not as many studies of the DED

problem and it has not been as thoroughly investigated as other electric power system optimization areas of study [3].This is evident in Table 1. Similar to most of the real-world complex engineering optimization problems, the nonlinear and non-convex characteristics of the cubic cost functions(CCF) are more prevalent in the DED problem more so with the increasing number of objectives and constraints.

Survey of published works dealing with these optimization problems include N.P. Padhy,2004 [2] for the UC, J.A. Momoh et al ,1999 [4] and M.R. Al Rashid and M.E. El-Hawary,2009 [5] for the OPF and A. Mahor et al ,2009 [6] ,H. Altun and T. Yalcinoz ,2008 [7] for the SED.

Table 1. Power System Planning, Operation and Control [3].

Optimization Area	% Publications
SED	21
DED	4
OPF	27
HTS	14

3. Dynamic Economic Dispatch (DED)

Dynamic Economic Dispatch (DED) involves the formulation and solution of the thermal ED such that the ramp rates, valve points and all the possible constraints are accounted for. As evident in Table 1, the DED occupies a prominent place in a power system's planning, operation and control. The goal of DED is to determine the optimal power outputs of online generating units in order to meet the load demand subject to satisfying various operational equality and inequality constraints over finite dispatch periods. In practice, there are SED and DED problems where the latter considers additional practical constraints. Early research works addressing this aspect were published by T.E.Bechert and H.G. Kwatny, 1972 [8] and T.E. Bechert and Nanming Chen, 1977 [9].The DED problem has not been addressed and researched as thoroughly as other power optimization problems. Xia et al, 2010, [10] offered a review of DED highlighting the problem constraints and some available solution methods. The work presented in this research paper is different in that it categorizes earlier work based on the optimization algorithms, number of objective functions involved, number and type of constraints, type of coast function and published works with RE cost functions. Figure 1 illustrates annual published research on DED according to a review of the Science Direct databases [3].As shown in the figure, the number of publications on DED is on increase. This signifies the undeniable importance of investigating the DED problem.

From the published works sampled, the number of works integrating the Renewable Energy (RE) into the DED problem have been increasing significantly. This means that the inclusion of the RE into the ED problem can no longer be ignored since the security, reactive power and multi area aspects of the power system are affected . This is as shown in the Figure 2.

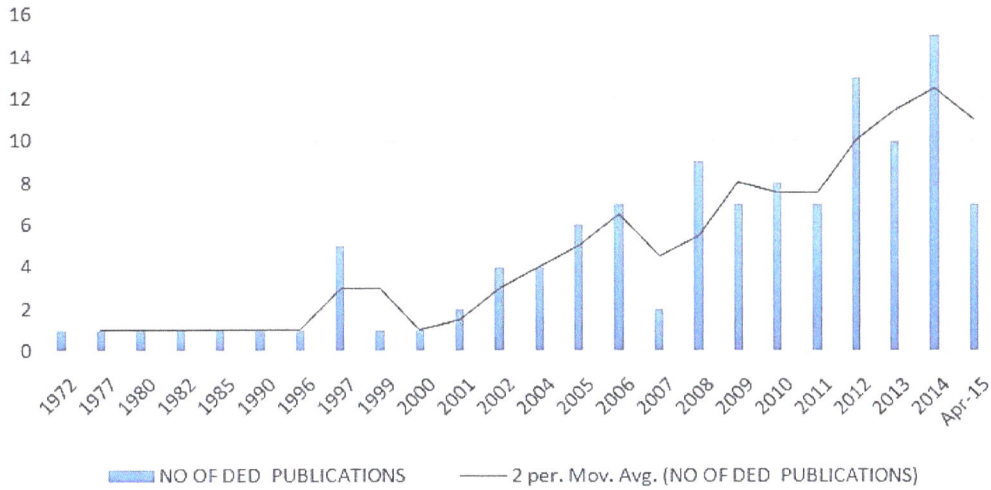

Figure 1. *DED Publications Since 1970s[3].*

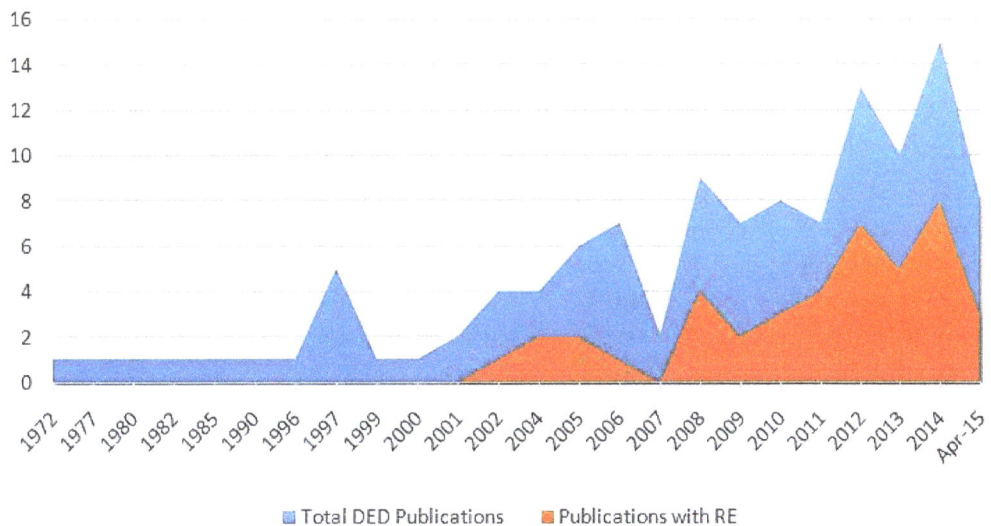

Figure 2. *DED Publications with RE Since 1970s[3].*

4. MODED Review

The history of the MODED problem dates back to early 2000 when transmission losses were considered as a cost function other than a constraint leading to the formulation of the first 2-objective MODED[12]

Table 2.2 summarizes the MODED problem so far. The summary includes the number and types of objectives, the type of method used and the number of constraints. From the table, it is clear that there is need to solve the MODED with all the five major objectives considered simultaneously on cubic cost functions. Further the problem need to be constrained further for more realistic results.

Table 2. *MODED Problem with Quadratic Cost Function (QCF).*

Researcher	n	Objectives	Solution Method(s)	Con	Type
Faisal .A. Mohamed ,2009[11]	4	T,W,S,E	Non Linear Constrained Method (NLCM)	2	Q
M.A Abibo, 2003 [12]	2	T,TL	Strength Pareto Evolutionary Algorithm (SPEA)	3	Q
Zhao Bo and Cao Yi-Jia,2005[13]	3	T,E,TL	Multi Objective Particle Swarm Optimization (MOPSO)	5	Q
M.A Abibo, 2006[14]	2	T,TL	SPEA and Strength Pareto Genetic Algorithm (SPGA)	4	Q
M. Basu, 2006 [15], Amitah Mahor ,2009[16],	2	T E	Particle Swarm Optimization (PSO)	3	Q
M .Basu, 2007 [17]	2	T,E	Evolutionary Programming (EP) and Fuzzy Logic (FL)-(EP-FL)	4	Q
C .Chen, 2007 [18]	2	T,W	Simulated Annealing (SA) and Direct Search Method (DSM)-(SA-DSM)	4	Q
M.Basu, 2008 [19]	2	T,E	Elitist Non-Dominated Sorting Genetic Algorithm-II (ENSGA-II)	3	Q
R.P Brent,2010 [20]	2	T,E	Preference-based Non-Dominated Sorting Genetic Algorithm (PNDSGA)	5	Q

Researcher	n	Objectives	Solution Method(s)	Con	Type
Zwe-Lee Gaing et al ,2009 [21]		T,E	Bacterial Foraging(BF)-PSO-Differential Evolution-(DE) (BF-PSO-DE)	4	Q
Alsumait et al, 2010 [22]	2	T,E	Pattern Search (PS)	3	Q
S.Subramanial et al, 2010 [23]	3	T,E,TL	Sequential Approach with Matrix Framework (SAMF)	6	Q
Chunghun et al,2013 [24]	3	T,W,S	Model Predictive Control (MPC)	5	Q
Haiwang Zhong et al,2013[25]	2	T,TL	Penalty-Based Algorithm (PBA)	4	Q
Azza A. ElDesouky, 2013 [26]	4	T,W,S,E	Weighted Aggregation (WA) and PSO(WAPSO)	7	Q
Bakirtzis et al ,2014 [27]	2	T,W	Variable Time Resolution (VTR) and Scheduling Horizon (SH)-(VTRSH)	4	Q
Soubache ID and Sudhakara Reddy,2014[28]	2	T,E	Multi Objective Computer Programming Based Method(MOCPBM)	5	Q

Key
Q-Quadratic Function,
C-Cubic Function,
n- Number of objective
T-Thermal cost objective function
W-Wind, cost objective function
S-Solar cost objective function
E-Emissions cost objective function
TL-Transmission Losses)
Con-Number of constraints

ED with cubic cost functions has been studied in the recent past . According Z.X Liang and J.D Glover ,1991[29],a very crucial issue in DED studies is to determine the order and approximate the coefficients of the polynomial used to model the cost function. This helps in reducing the error between the approximated polynomial along with its coefficients and the actual operating cost. According to Z.X Liang and J.D Glover,1992 [30] and A.Jiang and S.Ertem,1995 [31] to obtain accurate ED results, a third order polynomial is realistic in modelling the operating cost for a non-monotonically increasing cost curve. DED works using cubic cost functions include Bharathkumar S et al, 2013[32], Hari M.D et al, 2014[33], Deepak Mishra et al, 2006[34], and N.A.Amoli et al, 20 12[35]. Krishnamurthy, 2012 [36] used the static cubic function of the emissions dispatch in the Multi Objective Static ED(MOSED) using the Lagrange Method(LM).This provided better results as compared to the quadratic functions. In all these studies, the cubic cost function provided more accurate and practical results as compared to lower order cost functions. A summary of ED works using cubic cost functions is provided in Table 3.From the table, it is clear that cubic renewable and transmission losses cost functions have not been considered. Further, there is need to use more advanced hybrid methods for better results in these complex problems

Table 3. ED with Cubic Cost Functions.

Reference	Objectives	Nature of objective functions	Number of Constraints(n)	Method
Z.X Liang et al,1991[29]	T	Static	-	Gram-Schmidt, Least Squares
Z.X Liang et al,1992 [30]	T	Static	3	Dynamic programming(DP)
A.Jiang and S.Ertem,1995 [31]	T	Static	2	Newton Method(NM)
B.S et al ,2013[32]	T,E	DED with ramp rates and valve points	4	Fuzzy Logic (FL), BF and Nelder-Mead(NM) (FL-BF-NM)
Hari Mohan D.et al,2014[33]	T,E	Static	5	PSO-General Search Algorithm (PSO-GSA)
Yusuf Somez,2013[37]	T	General static	2	ABC
Deepak Mishra et al,2006 [34]	T	General static	2	OR-Hopefield Neural Network
N.A Amoli et al,2012 [35]	T	Static	2	Firefly Algorithm
Krishnamurthy .S et al ,2012[36]	T,E	Static	2	Langrange Method
T.Adhinarayananand M.Sydulu,2006[38]	T	Static	2	Lambda-logic based
T.Adhinarayananand M.Sydulu,2010 [39]	T	Static	2	Lambda-logic based
E.B Elanchezhian et al,2014[40]	T	Static	8	Teaching learning based optimization (TLBO)

Key
Q-Quadratic Function,
C-Cubic Function,
n- Number of objective
T-Thermal cost objective function
W-Wind, cost objective function
S-Solar cost objective function
E-Emissions cost objective function
TL-Transmission Losses)
Con-Number of constraints

5. MODED Formulation

DED considers change-related costs. The DED takes the ramp rate limits, valve points and prohibited operating zone of the generating units into consideration. The general form of DED was formulated by Yusuf Somez, 2013[136] as is given by

$$F(P_{ij}) = \{a_{0,i} + \sum_{j=1}^{L=n} a_{ji} P_{t,i}^j + r_i\} + |e_i \sin f_i(P_i^{min} - P_i)| \quad (1)$$

Where $a_{0,i}, a_{j,i}, e_i$ and f_i are the cost coefficients of the ith unit, P_i^{min} is the lower generation bound for it unit and r_i is the error associated with the ith equation. When L=3, the cubic form of the SODED results. This can be expressed as

$$F(P_{i,3}) = f_1(x_1) = \min F = a_{3,i} P_{t,i}^3 + a_{2,i} P_{t,i}^2 + a_{1,i} P_{t,i} + a_{o,i} + r_i + |e_i \sin f_i(P_i^{min} - P_i)| \quad (2)$$

Also the SODED can be written as

$$\min F = \sum_{t=1}^M \sum_{i=1}^N F_{t,i}(P_{i,m}) \quad (3)$$

where $P_{i,m}$ is the output power of the ith unit at time m, N is the number of generation units and M is the number of hours in the time horizon

To maximize the outcome of the Renewable energy(RE) systems, the implementation of ED combined Thermal and RE(TRE) systems has become of paramount importance. The formulations adopted in this paper are based on the work of Azza A. El Desouky,2013 [26]for the solar and John Hetzer et al ,2008 [41] for the wind cost functions. The operational cost objective function for wind power generation is formulated as

$$F(w_{ij}) = f_2(x_2) = F_{wi}(w_{ij}) + F_{p,wi}(w_{ij,av} - w_{ij}) + F_{r,wi}(w_{ij} - w_{ij,av}) \quad (4)$$

In this case, w_{ij} is the scheduled output of the ith wind generator in the jth hour $F_{wi}(w_{ij})$ is the weighted cost function representing the cost based on wind speed profile [26], $F_{p,wi}(w_{ij,av} - w_{ij})$ is the penalty cost for not using all the available wind power and $F_{r,wi}(w_{ij} - w_{ij,av})$ is the penalty reserve requirement cost which is due to the fact that that actual or available power is less than the scheduled wind power.

Similarly, the operational cost objective function for the PV power generation plant is formulated as [133]

$$F(PV_{ij}) = f_3(x_3) = F(PV_{ij}) + F_{p,PVi}(PV_{ij,av} - PV_{ij}) + F_{r,PVi}(PV_{ij} - PV_{ij,av})(PV_{ij}) \quad (5)$$

is the weighted cost function representing cost based on solar irradiance profile $F_{p,PVi}(PV_{ij,av} - PV_{ij})$ is the penalty cost for not using all the available solar power, and $F_{r,PVi}(PV_{ij} - PV_{ij,av})$ is the penalty reserve requirement cost due to the fact that the actual or available power is less than the scheduled PV power.

While attempting to find an optimum DED of generation, transmission loss is one important constraint since the generating centers and the connected load exist in geographically distributed scenario. Since the power stations are usually spread out geographically, transmission network losses must be taken into account to achieve true economic dispatch. Network loss is a function of power injection at each node [30]. Where the real power system transmission losses, PL_i is expressed using B- coefficients by the relation

$$P_L = \sum_{i=1}^n \sum_{j=1}^n P_i B_{ij} P_j + \sum_{i=1}^n P_i B_{0i} + B_{00} \quad (6)$$

Where i is the number of generators , j is the number of buses in the system, B_{ij} is the ij^{th} element of the loss coefficient Square matrix, B_{oi} is the ith element of the loss coefficient matrix and B_{00} is the constant loss coefficient .
When the loss expression is expanded for a particular number of generators and buses, the equation becomes

$$P_L = P_n^3 \sum_{i=1}^3 \left(B_{11} \frac{B_{1n}}{B_{11}} \frac{P_{1n}}{P_1}\right) + P_n^2 \sum_{i=1}^2 \left(B_{11} \frac{B_{1n}}{B_{11}} \frac{P_{1n}}{P_1}\right) + B_{0n} P_n + \frac{B_{00}}{n} \quad (7)$$

This can be expressed as a cubic function

$$P_L = \alpha_3 P_n^3 + \alpha_2 P_n^2 + \alpha_1 P_n + \alpha_0 \quad (8)$$

The cost of transmission line losses between plants are accounted with the actual fuel cost function by using a price factor g .This factor is defined as the ratio between the fuel cost at its maximum power output to the maximum power output .That is for this multi objective case

$$g_i = \frac{\sum_{i=1}^N F(P_{ij,max}) + \sum_{i=1}^M F(w_{ij,max}) + \sum_{i=1}^S F(PV_{ij,max})}{P_{i,max}} \quad (9)$$

Thus, the cost function for the losses at a particular time becomes

$$F(P_{L,i}) = f_4(x_4) = \min F = \sum_{i=1}^n g_i(\alpha_{3,i} P_{t,i}^3 + \alpha_{2,i} P_{t,i}^2 + \alpha_{1,i} P_{t,i} + \alpha_{0,i} P_{t,i}) \quad (10)$$

Several works on combined economic and emissions dispatch (CEED) have been considered in the past research. Some of these can be found in [11],[19],[35],[42],[43],[44],[23],[45],[46],[47] [48] and [26];just to list a few. All these works have considered emissions using the quadratic function which is less accurate[29]. Consequently, the more accurate equation for estimating the emissions cost is the third (cubic) form and it is the function which will be used in the proposed research. It is given by

$$E(P_{ij}) = \{\beta_{0,i} + \sum_{j=1}^{L=3} \beta_{ji} P_{t,i}^j + |\zeta_{j,i} \exp(\lambda_{j,i} P_i)|\} \quad (11)$$

$$E(P_{i,3}) = f_5(x_5) = \beta_{3,i} P_{t,i}^3 + \beta_{2,i} P_{t,i}^2 + \beta_{1,i} P_{t,i}^1 + \beta_{0,i} + \zeta_{3,i} \exp(\lambda_{3,i} P_i) + + \zeta_{2,i} \quad (12)$$

From equations (2), (4), (5), (10) and (12), the MODED

can be formulated as

$$\min F = \min\left[W_1\left\{\sum_{i=1}^{N} F(P_{ij}) + \sum_{i=1}^{M} F(w_{ij}) + \sum_{i=1}^{S} F(PV_{ij}) + F(P_{L,i})\right\} + W_2 E(P_{i,j})\right] \quad (13)$$

Or

$$\min F = \min[W_1 F + W_2 E] = WF + (1 - W) \quad (14)$$

Where

$$F = \sum_{i=1}^{N} F(P_{ij}) + \sum_{i=1}^{M} F(w_{ij}) + \sum_{i=1}^{S} F(PV_{ij}) + F(P_{L,i}) \ and \ E = \sum_{i=1}^{N,M,S} E(P_{ij}, PV_{ij}, w_{ij}) \quad (15)$$

Subject to

$$\sum_{i=1}^{N} P_{ij} + \sum_{i=1}^{M} Pw_{ij} + \sum_{i=1}^{S} PV_{ij} = P_{Dj} + P_{loss\,j} \quad (16a)$$

$$P_i^{min} \le P_i \le P_i^{max} \quad (16b)$$

$$0 \le w_i \le w_{ri} \quad (16c)$$

$$0 \le PV_i \le PV_{K_t\,max} \quad (16d)$$

$$P_{ij} - P_{ij-1} \le UR_i \quad (16e)$$

$$P_{ij-1} - P_{ij} \le DR_i \quad (16f)$$

$$-P_l^{max} \le P_{lj} \le P_l^{max} \ l = 1,2,3\dots.L \quad (16g)$$

$$P_i \le P^{PZ,LOW} \quad (16h)$$

$$P_i \ge P^{PZ,HIGH} \quad (16i)$$

$$P_r\left[\sum_i^{N} P_{im} + \Omega(w_i + PV_i)\right] \le P_{Dj} + P_{loss\,j} \le P_a \quad (16j)$$

W_1, and W_2 are non-negative weights used to make tradeoff (relative importance)between emission security and total fuel cost considering the three fuels such that W_1+W_2 =1.N, M and S are the number of thermal wind farm and PV power plants. W_1 Is the algebraic sum of the individual weight of the four objectives?

6. Applications of 5-Objective MODED

The MODED formulated can be used to investigate Multi Area Multi Objective Dynamic Economic Dispatch (MAMODED, Dynamic Reactive Power MODED (DRPMODED) and the Security Constrained MODED (SCMODED).These are discussed in this section.

The multi area single objective economic dispatch (MASOED) problem formulation is given by

$$\min F = \min \sum_{m=1}^{M} \sum_{n=1}^{N_M}(a_{mn}P_{mn}^2 + b_{mn}P_{mn} + C_{mn}) \quad (17)$$

Where P_{mn} Is the power output of generator n in area m, $a_{mn}, b_{mn},$ and $c_{mn,}$ are the fuel cost coefficients, M is the number   of are N_M is the number of online units for the area M .

According to Jizhong Zhu, pp. 211-248, (2006) [1], many approaches have been considered for static MAED. These include NFP with multi area wheeling (NFPMAW), ONN,

AHP, DP, Spatial DP(SDP), EP and NLCNFP. Other methods of solving the MAED problems are explained in [49]-[59]. In these methods however, only static MAED is considered. When a MAED problem is solved with spinning reserve constraints and RE cost functions the problem becomes further complicated. The power allocation to each unit is done in such a manner that after supplying the total load, some specified reserve is left for security reasons. This is the dynamic MAED(MAMODED) problem In the event that the power in these areas is changing significantly, the classical methods of MAED solution are no longer applicable in such stochastic environments .Thus more advanced hybrids must be applied to solved the MADED problem which has gained a lot of interest with the integration of renewable energy into the grid .

Thus, the multi area single objective DED (MASODED) with a cubic objective function that is of interest in the proposed research is defined by the relation

$$\min F_{S,M} = \min \sum_{m=1}^{M} \sum_{n=1}^{N_M}\left(a_{mn}\,P_{mn}^3 + a_{mn}P_{mn}^2 + a_{mn}P_{mn} + a_{mn} + r_{mn} + \left|e_{mn}\sin f_{mn}(P_{mn}^{min} - P_{mn})\right|\right) \quad (18)$$

The Multi Area Multi Objective DED (MAMODED) which incorporates the Renewable Energy, Transmission line and Emissions Cost functions can be represented by

$$\min F_{M,M} = \min \sum_{m=1}^{M} \sum_{n=1}^{N_k} F_{M,N}(P_{mn}) \quad (19)$$

Subject to Area power balance constraints (APBC)

$$APBC = \sum_{m=1}^{M} \sum_{n=1}^{N_M} P_{mn} - \sum_{m=1}^{M} \sum_{n=1}^{N_d} P_{Dmn} - P_l = 0 \quad (20a)$$

Generator capacity limit constraint

$$P_{mn,min} \le P_{mn} \le P_{mn,max} \quad (20b)$$

Tie-line power flow limit

$$|P_t| \le P_{t,max} \ t = 1,2,3\dots\dots.N_t \quad (20c)$$

Uncertainty constraint

$$P_r\left[\sum_i^{N} P_{im} + \Omega(w_i + PV_i)\right] \le P_{Dmnj} + P_{L\,j} \le P_a \quad (20d)$$

Generation constraint (GRC)

$$|P_{mn}| \le P_{mn,GRC} \quad (20e)$$

In this formulation, $(F_{M,M})$ is the MAMODED problem to be solved in the proposed research, $m = 1,2,3,\dots.M(areas)$, $n = 1,2,3,\dots.N(units)$ and $k = 1,2,3,\dots.K_n(fuels, P_{mn}$ is the power output of generator n in area m $a_{mn}, b_{mn}, c_{mn,}, d_{mn,}$ and $e_{mn,}$ are the fuel cost coefficients for area m and unit n, N_M is the number of online units for the area m , P_{Dmin} is the active load at node n in the area m, P_l is the total real power loss for multi area system, N_d is the number of loads in area m, N_t is the number of tie lines and P_t is the active power flow in the tie line t

The objectives of reactive power (VAR) optimization, which include RPED, are to improve the voltage profile, to minimize system active power losses, and to determine optimal VAR compensation placement under various operating conditions. To achieve these objectives, power system operators utilize control options such as adjusting generator excitation, transformer tap changing, shunt capacitors, and SVC. However, the size of power systems and prevailing constraints produce strenuous circumstances for system operators to correct voltage problems at any given time. In such cases, there is certainly a need for decision - making tools in predominantly fluctuating and uncertain computational environments. There has been a growing interest in VAR optimization problems over the last decade. Methods applied so far to solve the SORPED are as found in [60]-[65].Solving ORPED is gaining more importance due to their effectiveness in handling the inequality constraints and discrete values compared to that of conventional gradient-based methods. EAs generally perform unconstrained searches, and they require some additional mechanism to handle constraints. In the literature, various constraint handling techniques have been proposed. However, to solve ORPD the penalty function approach has been commonly used, while the other constraint handling methods remain untested. However, the excessive time consumption of EP and GAs will limit their applications in power systems, especially during real - time operation. Thus better method are needed to handle the more complex problems where stochastic reactive power from wind and solar generators are involved.

The fuel cost in terms of reactive power output can be expressed as

$$\text{F}(Q_{gi}) = a_{q,o} + \sum_{j=1}^{L=n} a_{qi} Q_{g,i}^j \qquad (21)$$

where $a_{q,o}$ and a_{qi} are the reactive power cost coefficients calculated using a curve fitting method Q_{gi} is the reactive power generated by generator i and n is the order of the fuel cost function

Subject to
Power balance constraints

$$P_i - V_i \sum_{j=1}^{N_B} V_j \left(G_{ij} \cos \theta_{ij} + B_{ij} \sin \theta_{ij} \right) = 0, i = \\ 1,2,3 \dots \dots N_B - 1 \qquad (22a)$$

$$Q_i - V_i \sum_{j=1}^{N_{PQ}} V_j \left(G_{ij} \sin \theta_{ij} - B_{ij} \cos \theta_{ij} \right) = 0, i = \\ 1,2,3 \dots \dots N_{PQ} \qquad (22b)$$

Continuous control variable (Generator Bus Voltage)

$$V_i^{min} \leq V_i \leq V_i^{max} \ i \in N_B \qquad (22c)$$

Discrete control variable (Transformer Tap Settings)

$$t_k^{min} \leq t_k \leq t_k^{max} \ i \in N_T \qquad (22d)$$

where t_k is the tap setting of transformer at branch k
State variables

$$Q_{Ci}^{min} \leq Q_{Ci} \leq Q_{Ci}^{max} \ i \in N_C \qquad (22e)$$

$$Q_{gi}^{min} \leq Q_{gi} \leq Q_{gi}^{max} \ i \in N_g \qquad (22f)$$

$$|S_i| \leq S_i^{max}, i \in N_i \qquad (22g)$$

Reactive power balance

$$\sum_{i=1}^{N_G} Q_{Gi} + \sum_{j=1}^{N_G} Q_{cj} = \sum_{k=1}^{N_D} Q_{dk} + Q_L \qquad (22h)$$

In this formulation Q_{Ci} is the reactive power generated by the ith capacitor bank, Q_{gi} is the reactive power generated at bus i, S_i is the apparent power flow through the ith branch, N_B is the total number of buses, N_T is the number of tap setting transformer branches, N_C is the number of capacitor banks and N_g is the number of generator buses,Q_{Gi} Reactive power generated by generator i, Q_{cj} Reactive power generated and absorbed by VAR compensation device j such as capacitors,SVC,Wind Based DFIGs, and PV generators, Q_{dk} Reactive power load at load bus k and Q_L Power system reactive power power loss and absorption .Further , Where V_i is the voltage magnitude at bus i , V_j is the voltage magnitude at bus j, P_i, Q_i is the real and reactive powers injected at bus i, G_{ij}, B_{ij} is the mutual conductance and suspectance between bus I and j , $N_B - 1$ is the total number of buses excluding the slack bus, N_{PQ} is the number of PQ buses and θ_{ij} is the voltage angle difference between bus I and bus j

The security - constrained economic dispatch (SCED) is one of the simplified optimal power flow (OPF) problems. It is widely used in power industry. The main objective of electric power dispatch is to provide electricity to the customers at low cost and high reliability. Transmission line failures constitute a great threat to the electric power system security.

The static single objective SCED (SSOSCED) problem is formulated as

$$F_{S,S} = \sum_{i \in N_G} F_{i(P_{Gi})} \qquad (23)$$

Where $F_{i(P_{Gi})} = a_{2,i} P_{t,i}^2 + a_{1,i} P_{t,i} + a_{o,i}$

The several approaches to solve the SCED from 1952 to 2014are presented in [1], [66], [67]-[71]. Single objective SCDED (SOSCDED) problem is formulated as

$$F_{S,S} = \sum_{i \in N_G} F_{i(P_{Gi})} \qquad (24)$$

where $F_{i(P_{Gi})} = a_{3,i} P_{t,i}^3 + a_{2,i} P_{t,i}^2 + a_{1,i} P_{t,i} + a_{o,i} + r_i + |e_i \sin f_i (P_i^{min} - P_i)|$

The SOSCDED in the proposed research uses the more accurate cubic function and incorporates the error ,valve points and ram rate terms. The multi objective SCED(MOSCDED) which investigates the effects of renewable energy ,power loss and emissions functions to the energy security of the power system is formulated as

$$F_{S,M} = \sum_{i \in N_G} F_{i(P_{Gi})} \qquad (25)$$

Where

$$F_{S,M} = [W_1\{\sum_{i=1}^{N} F(P_{ij}) + \sum_{i=1}^{M} F(w_{ij}) + \sum_{i=1}^{S} F(PV_{ij}) + F(P_{L,i})\} + W_2 E(P_{i,j})]$$

Subject to

$$\sum_{i \in N_G} P_{Gi} = \sum_{k \in N_{Dk}} P_{Dk} + P_l \qquad (26a)$$

$$|P_{ij}| \leq P_{ijmax} \ i \in N_T \qquad (26b)$$

$$P_{Gi\,min} \leq P_{Gi} \leq P_{Gi\,max} \qquad (26c)$$

$$P_r[\sum_i^N P_{im} + \Omega(w_i + PV_i)] \leq P_{Dj} + P_{loss\,j} \leq P_a \qquad (26d)$$

7. MODED Optimization Methods

The methods that have been to solve the DED Problem so far can be classified into three groups. These are, Deterministic, Heuristic and Hybrid methods. These methods have been discussed in the following sub sections.

7.1. Deterministic (D)

Unconstrained Methods (UCM) convert constrained problems into unconstrained form. This forms the basics of the formulation of the constrained optimization Algorithms. These methods include Gradient Search (GS),Line Search(LS),Lagrange Multiplier(LM),Newton Raphson(NR) ,Trust Region (TR) ,Quasi-Newton (QN),Double-dogleg(DD) and Conjugate Gradient(CG) Methods. In DED Only LM method has been used by W.G. Wood, 1982 [72] and W.R. Barcelo and P. Rastgoufard,1997 [73] and Maclaurin Series-Based Lagrangian Method (MSBLM) was later applied to the DED problem by Hemamalini et al, 2010 [74].

Linear Programming (LP) linearizes the nonlinear power system DED problem so that the objective function and the constraints have the linear form. It was Ahmed Farag et al, 1995 [75] and Y.H. Song and I. Yu, 1997 [76] and it proved reliable in terms of convergence .The method was found Quick in identifying infeasibility and thus able to accommodate large variety of power system operating limits and contingency constraints. However the method was found inaccurate in evaluating the power system losses and therefore had insufficient ability to find an exact solution as compared with more accurate Non Linear models. However the DED solutions generally met the requirements of Engineering precision when applied to the SCED and RPED optimization [1].

Non Linear Programming (NLP) is a first order method .Since DED problems are nonlinear, the method able to handle the nonlinear objectives and constraints .A search direction is chosen and the search is done using reduced gradient method. As applied by P.P.J. van den Bosch, 1985 [77] and an improved version of NLP called Non Linear Constrained Method (NLCM) developed by Faisal .A. Mohamed, 2009 [11],the method has several advantages as compared to the LP Method .These include higher accuracy and global convergence ,that is, convergence is guaranteed independent of the starting point. However the method has a

slow rate of convergence due to a zigzagging problem in the search direction. The method has also been used in SCED, MAED and RPED problems [1].

Other deterministic methods that have been used to handle DED problems include Economic Load Allocation and Supplementary Control Action (ELASCA), Bechert and Kwatny, 1972 [8] ,Muller Algorithm (MA), Chandram et al, 2006 [78],Brent Method (BM), K. Chandram et al, 2008 [79] and Model Predictive Control (MPC), X. Xia et al, 1999 [80], X. Xia et al,2011 [81] just to mention a few.

7.2. Non Quantity Approaches (N)

These methods are applied to problems with uncertainties in their parameter variations. Uncertainty due to insufficient information generate an uncertain region of decisions and therefore results from average values cannot be used to represent the uncertainty level. NQA methods are meant to account for the uncertainties in information and goals related to multiple and usually conflicting objectives in power system optimization .These methods find application in ED,HTS,UC,RPED and State Estimation problems with uncertainties. Compared to the deterministic methods, the NQA Methods are effective in handling uncertainties, that is, they compute the unavailable or uncertain data so that the MODED problem can be solved even when some data for the wind and solar objective cost functions are not available[1]. These methods include Optimization Neural Networks (ONN), Artificial Neural Network(ANN), Probabilistic /Characterization Theory (PT),Fuzzy Set Theory (FST),Analytical Hierarchical Process (AHP),Risk Management Tools(RMT) and Cost Benefit Analysis(CBA).These Methods have been used in the formulation of hybrids used in the solution of DED problems with uncertainties. Examples of such include,ANN [82],Heuristic Neural Network(HNN)[83-84],Trust Region (TR) and Goal Programming (GP) [85] , Chaotic Fuzzy(CF) and Variable Step Size (VSS) [86].

7.3. Heuristic (H)

Evolutionary Algorithms (EAs) are based on natural evolution .They are population based optimization processes. There is no need to differentiate the cost functions and the constraints in the ED formulation. The method can be categorized as GA, EP and DE. The three however use the same mutation, recombination, reproduction, cross over and selection operators [1]. DED using Genetic Algorithms (GA) and the derived methods include Elitist Genetic Algorithm (EGA),Fly et al, 1997 [87], Parallel Micro GA (PMGA), Ongsakul et al ,2002 [89] and Elitist non-dominated sorting genetic algorithm-II (ENSGA-II), Robert.T.F et al,2004[90].Evolutionary Programming (EP) include Constrained Evolution Programming (CEP),Shailti et al, 2005 [91]. EP, Joned et al, 2006 [92] and Quantum Evolutionary Algorithm (QEA), Babu et al, 2008 [93]. Differential Evolution (DE) is the most recent EA and the DED works done include Modified DE (MDE),Yuang et al ,2008 [4], K.Deb,2000 [94],Variable Scaling Hybrid Differential Evolution (VSHDE), J.Chiou,2009 [95] and Cellular Differential Evolution (CDE) Noman and Iba, 2011 [96].EAs methods have been used to solve simplified CED, DED SCED RPED and MAED problems. However these methods require all information to be included in the fitness function therefore it is difficult to consider all the MODED problem objectives and constraints in the objective function.

Particle Swarm Optimization (PSO) includes all the swarm intelligence algorithms inspired by the social dynamics and an emergent behavior that arises in socially organized colonies. These methods have been used in all hybrid algorithms where an accurate search is needed in DED problems. They include Particle Swarm Optimization (PSO) Z.L. Gaing, 2004 [97], Hardiansyah et al 2012[98], Zhao et al, 2004 [42] and Modified Adaptive PSO (MAPSO), Niknam et al, 2011[99].More specific PSO methods used in DED problems are BAT Algorithm, Anti Colony Optimization(ACO),Bee Colony Optimization(BCO) ,Artificial Bee Colony (ABC),Bacterial Foraging (BF),BAT Algorithm, Teacher Learning Algorithm(TLA) , Bandi Ramesh et al, 2013[44],Biogeography-Based Optimization (BBO), Divya Mathur, 2013,[100] .

Other Methods that have been used in DED include Simulated Annealing (SA), Panigrahi et al, 2006 [101],Pattern Search (PS),Alsumait et al, 2010 [22],Adaptive Look-Ahead (ALA), Han et al, 2001 [102],Feasibility-Based Selection Comparison (FBSC),K.Deb,2000 [94],Ongsakul et al, 2002 [103], X.Yuan et al,2009[104],Artificial Immune System (AIS), M.Basu, 2011[105] Hemmalini et al,2011 [106],Modified Teaching-Learning Algorithm (MTLA) Niknam, T et al ,2013 [107] and Optimality Condition Decomposition (OCD), Rabiee, A et al , 2014 [108]

7.4. Hybrids

The reason for increased use of hybrids is because they exalt the strengths and improve the weaknesses of the methods concerned. Deterministic,Heuristic and Non-Quantity Approaches are used in hybrid formation .There are two types of hybrids depending on the number and Type of methods used.

7.5. Two Method Hybrids

There are eight possible types of two-method hybrids. These include N-N, N-D, and N-H, D-D, D-H, D-N, H-D, H-H and H-H approaches.

Deterministic-Deterministic (D-D): These method involve two deterministic methods .They include Quadratic Programming (QP) and Linear Programming (LP) – (QP-LP), Somuah and Khunaizi, 1990 [109], QP-LP, Han et al, 2001 [110] and Interior Point (IP) and Quadratic Programming (QP) - (IP-QP), Lin and Chen, 2002[111]

Deterministic-Heuristic (D-H): These involve a deterministic method with a Heuristic method to fine tune it. These include SA and GA – (SA-GA), Ongsakul et al, 2001 [112], SQP-EP, Deterministically Guided (DG) PSO (DG-PSO), Victoire et al, 2005 [113-115] and SA and Direct Search Method (DSM)-(SA-DSM), C .Chen, 2007 [18]

Deterministic- Non Quantity Approaches (D-N) or Non Quantity Approaches-Deterministic (N-D): This involves a base deterministic method with a non-quantity method to deal with the uncertainities.The vice versa is also true. Examples in this context include Lambda-Iteration Technique (LIT) and Artificial Neural Networks (ANN)-(LIT-ANN), R.H Liang, 1999 [82] and Heuristic Neural Networks (HNN) and QP (HNN-QP), Abdul-Aziz et al, 2008 [83-84].

Heuristic -Deterministic (H-D): In this case a hybrid method is used as the base algorithm then its weaknesses are strengthened using a deterministic method. In DED ,examples include Genetic Algorithm (GA) and Gradient Search (GS)-(GA-GS), F.Li et al, 1997 [116],Relaxed GA (RGA) and Gradient Search (GS)-(RGA-GS) ,Li and Aggarwal,2000 [117],EP and Sequential QP (EP-SQP), P.A et al, 2002 [118],PSO-SQP, Victoire et al,2005 [113-115],Swarm Direction Technique (SDT) and Fast Evolutionary Programming (FEP) -(SDT-FEP), Zwe-Lee et al,2009 [119],Evolution (DE) and Local Random Search (LRS)-(DE-LRS), Y. Lu et al, 2009 [120],Improved Differential Evolutionary (IDE) and Shor's Algorithm (SHA)-(IDE-SHA), Yuan et al,2009 [104], F.Kappel and A.V Kuntsevich,2000 [121],PSO and Harmony Search (HS)-(PSO-HS), Ravikumar et al,2011 [122] ,Seeker Optimization Algorithm (SOA) and SQP- (SOA-SQP), S. Sivasurbramani et al,2010 [123],Hybrid Differential Evolution (HDE) and Sequential Quadratic Programming (SQP)-(HDE-SQP), A. M. Elaiw,2012 [124] ,Real Coded GA (RCGA) with Quasi-Simplex (QS)-(RCGA-QS), Zhang et al, 2006 [125]

Heuristic- Heuristic (H-H): In this type two heuristic methods are involved, on serves as the base and the other is used to improve the weaknesses of the first for accurate results. Examples include PSO and DE (PSO-DE), Dun –Wei Gong et al, 2010[126],Firefly Algorithm (FA) with GA Mutation (GAM)-(FA-GAM) , Niknam, T,2012 [127],Multi Objective PSO (MOPSO) and Brent method(BM)-(MOPSO-BM), H. Shayeghi and A. Ghasemi ,2012[128],Modified Artificial Bee Colony (MABC) with Differential Evolution (DE)-(MABC-DE), Hardiansyah ,2013[129] and PSO and Simulated Annealing (SA) –(PSO-SA),V.Karthikeyan et al,2013[130]

Non Quantity Approaches-Heuristic (N-H) 0r Heuristic-Non Quantity Approaches (H-N): In this case a non –quantity method is used to address an uncertainty before a heuristic method is applied. The vice versa is also true Examples include Fuzzy Optimization Technique (FOT) and Goal Satisfaction Concept (GSC)-(FOT-GSC) P.A et al, 2004 [131] and Chaotic Differential Evolution (CDE), Y.Lu et al,2011 [132].

Non Quantity Approaches-Non Quantity Approaches (N-N): Since non quantity methods are only used to address uncertainities,then a combination of two or more of such methods in the formulation of an hybrid is not feasible.

7.6. Three Method Hybrids

Recent trends in Hybrid formation involve the use of three methods. There are twenty seven possible three-method hybrids that can be used. In this section we discuss such methods that have been used in DED solution. These are summarized in Table 4

Heuristic algorithms seem to have shared the same dominance as deterministic algorithms. This is because heuristic algorithms, unlike deterministic ones, are derivative-free, and capable of solving optimization problems without requiring convexity.They are also independent of the initial solution, and have the ability to avoid being trapped in local optima.

On the other hand, heuristic algorithms have drawbacks such as being problem dependent, requiring parameter tuning, and unable to guarantee global solution attainment. A combination of a both methods into a hybrid has proved to solve this problem. This explains why publishing research work using hybrid methods has been becoming increasingly popular. This is as shown in Figure 3.

Also in the last decade integration of RE cost function has been of increasing interest. This is because RE is well utilized once in the national grid. Therefore, better methods for handling DED with wind and solar cost functions need to be developed.

Table 4. *Three Method Hybrids used in DED.*

Type	Examples in DED
H-H-D	EP, PSO, and SQP (EP-PSO-SQP), S. Titus and A.E Jeyakumar,2008 [133]
H-H-H	BF-PSO-DE, Praveena et al ,2010 [134]
H-D-H	GA and Local Search(LS) -Goal Programming(GP) -(GA-LS-GP), A.A.Mousa et al,2011 [135]
D-H-H	Mixed Integer Quadratic Programming (MIQP), the Warm Start Technique (WST) and the Range Restriction Scheme (RRS)-(MIQP-WST-RRS), Wang et al, 2014 [136]
N-N-H	Variable Step Size(VSS) Chaotic Fuzzy(CF) Quantum Genetic Algorithm, (VSS-CF-QGA), Wenxia Liu, et al,2014[86] Trust Region (TR),Goal Programming(GP) and PSO (TR-GP-PSO), Amhed et al,2013[48]

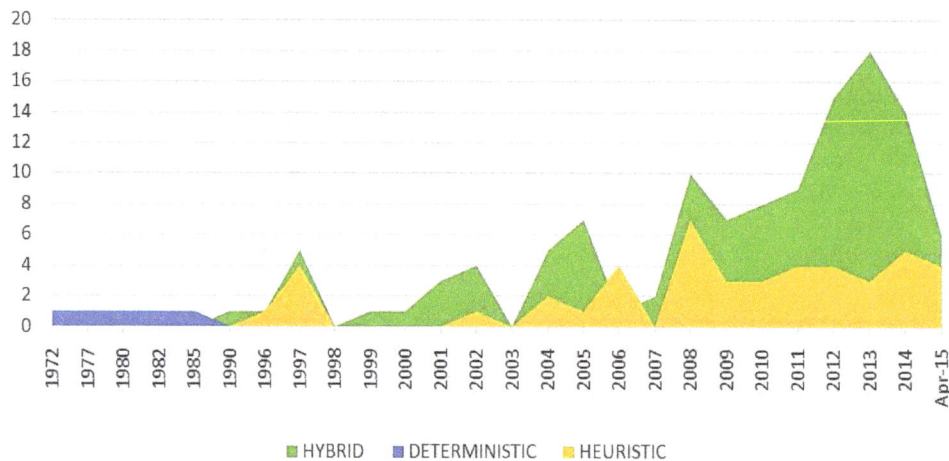

Figure 3. *Journals on DED Published Using Various Methods[3].*

8. Conclusion

This paper has psesented a detailed review of the MODED problem in terms of the formulation and the methods applied in the solution.It is apparent that cubic cost functions are found to be more accurate than quadratic ones. Further, fully constrained 5-Objective MODED with cubic functions to give more realistic results. However fourth order functions are predicted to be even better than the third order systems.

Dynamic trade in power is supposed be cheaper than the static one as the Improved Wheel Method (IWM) will be more realistic in modelling the areas and the tie lines. Dynamic reactive power (two-way) from the renewable sources will provide a far much improved voltage profile as compared to the static reactive power. In addition, there is foreseen better security due to the full constraining (with,

uncertainty) and two way reactive power.

Therefore ,there will be a more accurate MODED modelling ,new and better method for MODED Solution, Improved power system Security ,dynamic MAED and a new way for Reactive Power Dispatch.

A more recent trend for solving MODED is the three-method hybrids formulation in which all the weaknesses of the base methods are suppressed and the strengths exalted. This leads to increased accuracy and speed in handling higher order cost functions with more objectives.

References

[1] Jizhong Zhu (2009) *Optimization of Power System Operation,* New Jersey &Canada: John Wiley and Sons.

[2] N.P. Padhy,"Unit Commitment-a Bibliographical Survey," *Power Systems, IEEE Transactions on*, Vol. 19, No. 2, pp. 1196-1205, 2004.

[3] http://www.sciencedirect.com/science,April 2015.

[4] J.A. Momoh, M.E. El-Hawary and R. Adapa,"A Review of Selected Optimal Power Flow Literature to 1993. II. Newton, Linear Programming and Interior Point Methods," *Power Systems, IEEE Transactions on*, Vol. 14, No. 1, pp. 105-111, 1999.

[5] M.R. AlRashidi and M.E. El-Hawary, "Applications of Computational Intelligence Techniques for Solving the Revived Optimal Power Flow Problem, *"Electrical. Power Syst. Res.,* Vol. 79, No. 4, pp. 694-702, 2009.

[6] A. Mahor, V. Prasad and S. Rangnekar,"Economic Dispatch using Particle Swarm Optimization: A Review," *Renewable and Sustainable Energy Reviews*, Vol. 13, No. 8, pp. 2134-2141, 2009.

[7] H. Altun and T. Yalcinoz,"Implementing Soft Computing Techniques to Solve Economic Dispatch Problem in Power Systems," *Expert Systems Applications*, Vol. 35, No.4, pp. 1668-1678, 2008.

[8] T.E. Bechert and H.G. Kwatny, "On the Optimal Dynamic Dispatch of Real Power," *Power Apparatus and Systems, IEEE Transactions on*, Vol. PAS-91, no.3, pp. 889-898, 1972.

[9] T.E. Bechert and Nanming Chen,"Area Automatic Generation Control by Multi-Pass Dynamic Programming," *Power Apparatus and Systems, IEEE Transactions on*, Vol. 96, no. 5, pp. 1460-1469, 1977.

[10] X. Xia and A.M. Elaiw,"Optimal Dynamic Economic Dispatch of Generation: A Review," *Electrical Power Systems Res.*, Vol. In Press, Corrected Proof, 2010.

[11] Faisal A. Mohamed et al "Environmental/Economic Power Dispatch of Micro Grid Using Multiobjective Optimization" *International Conference on Renewable Energies and Power Quality* (ICREPQ'09) Valencia (Spain), 15th to 17th April, 2009.

[12] M.A. Abido and J.M. Bakhashwain " A Novel Multiobjective Evolutionary Algorithm for Optimal Reactive Power Dispatch Problem" in *Proceedings of the 2003 10th IEEE International Conference on Electronics, Circuits and Systems, 2003 (ICECS 2003)*, IEEE, Vol. 3, pp.1054--1057, December 2003.

[13] Zhao Bo and Cao Yi-Jia "Multiple objective particle swarm optimization technique for economic load dispatch" *Journal of Zhejiang University of Science and Technology* Vol.6 No.5 pp.420-427 2005.

[14] M.A. Abido." Multiobjective Optimal VAR Dispatch Using Strength Pareto Evolutionary Algorithm" in *2006 IEEE Congress on Evolutionary Computation (CEC'2006)*, pp. 2745--2751, IEEE, Vancouver, BC, Canada, July 2006.

[15] M. Basu, "Particle Swarm Optimization Based Goal-Attainment Method for Dynamic Economic Emission Dispatch," *Electric Power Components and Systems*, Vol. 34, No. 9, pp. 1015-1025, 2006.

[16] Amitah Mahor et al "Economic dispatch using particle swarm optimization: A review" *Renewable and Sustainable Energy Reviews*, Vol 11, pp. 2134-2141,2009.

[17] M. Basu, "Dynamic Economic Emission Dispatch using Evolutionary Programming and Fuzzy Satisfying Method," *International Journal of Emerging Electric Power Systems*, Vol. 8, No. 4, pp. 1-15, 2007.

[18] C.-. Chen, "Simulated Annealing-Based Optimal Wind-Thermal Coordination Scheduling," *Generation, Transmission & Distribution, IET*, Vol. 1, no. 3, pp. 447-455, 2007.

[19] M. Basu, "Dynamic Economic Emission Dispatch using Non dominated Sorting Genetic Algorithm-II," *International Journal of Electrical Power & Energy Systems*, Vol. 30, No. 2, pp. 140-149, 2008.

[20] R.P. Brent, Algorithms for Minimization without Derivatives, N.J.: Prentice-Hall,Englewood Cliffs, 1973.

[21] Zwe-Lee Gaing and Ting-Chia Ou, "Dynamic Economic Dispatch Solution using Fast Evolutionary Programming with Swarm Direction," *Industrial Electronics and Applications, 2009. ICIEA 2009. 4th IEEE Conference on*, pp. 1538-1544, 2009.

[22] J.S. Alsumait, M. Qasem, J.K. Sykulski and A.K. Al-Othman, "An Improved Pattern Search Based Algorithm to Solve the Dynamic Economic Dispatch Problem with Valve-Point Effect," *Energy Conversion and Management*, Vol. 51, No. 10, pp. 2062-2067, 2010.

[23] S Subramanian and S Ganesan. "A Simple Approach for Emission Constrained Economic Dispatch Problems" *International Journal of Computer Applications* Vol. 8, No. 11 pp39–45, October 2010.

[24] Chunghun Kim; Yonghao Gui; Chung Choo Chung; Yong-Cheol Kang, "Model predictive control in dynamic economic dispatch using Weibull distribution," *Power and Energy Society General Meeting (PES), 2013 IEEE* , Vol.10, pp.1,5, 21-25 July 2013.

[25] Haiwang Zhong; Qing Xia; Yang Wang; Chongqing Kang, "Dynamic Economic Dispatch Considering Transmission Losses Using Quadratically Constrained Quadratic Program Method," *Power Systems, IEEE Transactions on* , Vol.28, No.3, pp.2232-2241, Aug. 2013.

[26] Azza a. Eldesouky,"Security and stochastic economic dispatch of power system including wind and solar resources with environmental consideration" *International journal of renewable energy research* Vol.3, No.4 pp. 951-958, 2013.7.

[27] Bakirtzis, E.A; Ntomaris, AV.; Kardakos, E.G.; Simoglou, C.K.; Biskas, P.N.; Bakirtzis, AG., "A unified unit commitment — Economic dispatch model for short-term power system scheduling under high wind energy penetration," *European Energy Market (EEM), 2014 11th International Conference on the* , Vol.4, pp.1,6, 28-30 May 2014.

[28] Soubache ID and Sudhakara Reddy "solution Combined Economic and Emission Dispatch" *American Journal of Engineering Science and Research* Vol 1, No 1, pp 1-5 2014.

[29] Z.X Liang and J.D Glover "Improved cost functions for Economic Dispatch compensations *"Power Systems IEEE Transactions on* Vol 6.pp 821-829,1991.

[30] Z.X Liang and J.D Glover "A zoom feature for a Dynamic Programming Solution to Economic Dispatch including transmission Loses" *Power Systems IEEE Transactions* on Vol 7.pp 544-550, 1992.

[31] A.jiang and S.Ertem "Economic Dispatch with non-monotonically increasing incremental cost units and transmission system losses" *Power Systems IEEE Transactions on* Vol 10. pp 891-897, 1995.

[32] Bharathkumar.S et al "Multi Objective Economic Load Dispatch using Hybrid Fuzzy, Bacterial Foraging –Nelder-Mead Algorithm *"International Journal of Electrical Engineering and Technology*, Vol 4, Issue 3 pp. 43-52, May June 2013.

[33] Hari Mohan D.et al "A Fuzzy field improved hybrid PSO-GSA for Environmental /Economic power dispatch *"International Journal of Engineering Science and Technology*, Vol.6, No.4,pp.11-23,2014.

[34] Deepak Mishra et al "OR-Neuron Based Hopfield Neural Network for Solving Economic Load Dispatch Problem *"Letter and Reviews for Neural Information Processing* ,Vol.10,No.11 pp249-259,November 2006.

[35] N.A Amoli et al "Solving Economic Dispatch Problem with Cubic Fuel Cost Function by Firefly Algorithm" *Proceedings of the 8th International Conference on Technical and Physical Problems of Power Engineering,*ostfold University College Fredrikstad,Norway.pp 1-5,5-7th September 2012.

[36] Krishnamurthy, S.; Tzoneva, R., "Impact of price penalty factors on the solution of the combined economic emission dispatch problem using cubic criterion functions," *Power and Energy Society General Meeting, 2012 IEEE* , vol., no., pp.1,9, 22-26 July 2012.

[37] Yusuf Sonmez "Estimation of Fuel cost curve parameters for thermal power plants using the ABC Algorithm" ,*Turkish Journal Of Electrical Engineering and Computer Science* ,Vol .21 pp. 1827-1841,2013.

[38] T.Adhinarayanan and M.Sydulu " Fast and effective Algorithm for Economic Dispatch of Cubic Fuel Cost based thermal units " *First international conference on industrial and information systems ,ICIIS ,2006 ,Sirlanka ,8th -11th* August 2006.

[39] T.Adhinarayanan and M.Sydulu "An effective non-iterative $\lambda - Logic Based$ algorithm for Economic Dispatch of generators with cubic fuel cost function" *Electrical power and energy systems* ,vol 32,pp 539-542,2010.

[40] E.B Elanchezhian et al "Economic Dispatch with cubic cost models using Teaching learning Algorithm " *IET Generation,Transmission and Distribution* ,vol 8 issue 7 ,pp 1187-1202,2014.

[41] John Hetzer et al "An Economic Dispatch Model Incorporating Wind Power"*IEEE Transactions on Energy Conversion*, Vol .23, No.2 June 2008.

[42] Bo Zhao, Chuangxin Guo and Yijia Cao, "Dynamic Economic Dispatch in Electricity Market using Particle Swarm Optimization Algorithm," *Intelligent Control and Automation, 2004. WCICA 2004. Fifth World Congress on*, Vol. 6, pp. 5050-5054, 2004.

[43] M. Basu, "Particle Swarm Optimization Based Goal-Attainment Method for Dynamic Economic Emission Dispatch," *Electric Power Components and Systems,* Vol. 34, No. 9, pp. 1015-1025, 2006.

[44] Bandi ramesh et al "Application of BAT algorithm for Combined Economic Load and Emission Dispatch" *IJEETC* Vol 2 No1.pp 2512-2523, 2013.

[45] Zhao Bo and Cao Yi-Jia "Multiple objective particle swarm optimization technique for economic load dispatch" *Journal of Zhejiang University of Science and Technology* Vol.6 No.5 pp.420-427 2005.

[46] Amitah Mahor et al "Economic dispatch using particle swarm optimization: A review" *Renewable and Sustainable Energy Reviews*, Vol 11, pp. 2134-2141,2009.

[47] A. A. Mousa et al "A Hybrid Optimization Technique Coupling an Evolutionary and a Local Search Algorithm for Economic Emission Load Dispatch Problem "*Journal of Applied Mathematics,* Vol.2, pp. 90-898 ,2011.

[48] Ahmed Ahmed El-Sawy et al, "Reference Point Based TR-PSO for Multi-Objective Environmental/Economic Dispatch "Journal *Applied Mathematics*, 2013, Vol 4,pp 803-813 ,2013.

[49] Streiffert, D., "Multi-area economic dispatch with tie line constraints," *Power Systems, IEEE Transactions on*, Vol.10, No.4, pp.1946, 1951, Nov. 1995.

[50] T. Jayabarathi, G. Sadasivam and V. Ramachandran , "Evolutionary Programming-Based Multiarea Economic Dispatch with Tie Line Constraints", *Electric Machines & Power Systems*, pp 1165-1176,2000.

[51] P. s. manoharan et al , 2009, "A Novel EP Approach for Multi-area Economic Dispatch with Multiple Fuel Options "*Turk J Electrical Engineering & Computer Science* Vol.17, No.1, PP1-19, 2009.

[52] Prasanna.T.S and Somasundaram. P "Multi-Area Security Constrained Economic Dispatch by Fuzzy- Stochastic Algorithms" *Journal of Theoretical and Applied Information Technology*, pp 88-94, 2009.

[53] S.Chitra Selvi et al "Hybrid Evolutionary Programming Approach to Multi-Area Unit Commitment with Import and Export Constraints" *International Journal of Recent Trends in Engineering*, Vol.1,No. 3, 223-228, May 2009.

[54] Manisha Sharma et al, "Multi-area economic dispatch with tie-line constraints employing evolutionary approach" *International Journal of Engineering, Science and Technology* Vol. 2, No. 3, pp. 132-149, 2010.

[55] Manisha Sharma et al "Reserve Constrained Multi-Area Economic Dispatch Employing Evolutionary Approach *"International Journal of Applied Evolutionary Computation* Vol .1 Issue 3 pp. 49-69 July 2010.

[56] Sudhakar A.V.V et al "Multi Area Economic Dispatch using Secant Method" *Journal of Electrical Engineering Technology* Vol. 8, No. 4: 744-751,2013.

[57] Huynh This and Thanh Binh "Hybrid Particle Swarm Optimization for Solving Multi-Area Economic Dispatch Problem" *International Journal on Soft Computing (IJSC)* Vol.4, No.2, pp. 17-27 May 2013.

[58] M.Basu"Artificial bee colony optimization for multi-area economic dispatch "International Vol. 49, July 2013, pp. 181–187, 2013.

[59] De,Shankha Suvr et al "Artificial Immune System for Multi-Area Economic Dispatch" *International Journal of Emerging Electric Power Systems* Vol. 14, Issue 6, pp.581-590, Dec 2013.

[60] Lee, K.Y.; Park, Y.M.; Ortiz, J. L., "A United Approach to Optimal Real and Reactive Power Dispatch," *Power Apparatus and Systems, IEEE Transactions on* , Vol.PAS-104, No.5, pp.1147,1153, May 1985.

[61] Q.H. Wu, Y.J. Cao, J.Y. Wen"Optimal reactive power dispatch using an adaptive genetic algorithm" *International Journal of Electrical power and energy systems* Vol.20, Issue 8, Pages 563–569, November 1998.

[62] Serrano, B. R.; Vargas, A., "Active-reactive power economic dispatch of very short term in competitive electric markets," *Power Technology Proceedings, 2001 IEEE Porto*, Vol.1, No.10 pp.6-8, 2001.

[63] Worawat Nakawiro et al "A Novel Optimization Algorithm for Optimal Reactive Power Dispatch: A Comparative Study" *IEEE Transactions on Power Systems*, Vol.1 pp. 1155-1161, 2011.

[64] R. Mallipeddi et al "Efficient constraint handling for optimal reactive power dispatch problems" *Swarm and Evolutionary Computation* Vol. 5, Pages 28-36, August 2012.

[65] Lopez, J.C.; Munoz, J.I.; Contreras, J.; Mantovani, J. R S, "Optimal reactive power dispatch using stochastic chance-constrained programming," *Transmission and Distribution: Latin America Conference and Exposition (T&D-LA), 2012 Sixth IEEE/PES* , Vol.3, pp.1-7, 3-5 Sept. 2012.

[66] A. M. Elaiw, X. Xia,and A. M. Shehata "Dynamic Economic Dispatch Using Hybrid DE-SQP for Generating Units with Valve-Point Effects" *Hindawi Publishing Corporation Mathematical Problems in Engineering* Vol. 20,pp 1-10,2012.

[67] Kyoung-Shin Kim; Leen-Hark Jung; Lee, K.Y.; Un-Chul Moon, "Security Constrained Economic Dispatch Using Interior Point Method," *Power System Technology, 2006. PowerCon 2006. International Conference on*, Vol.12, pp.1-6, 22-26 Oct. 2006.

[68] Prasanna. T.S, Somasundaram. P "Multi-Area Security Constrained Economic Dispatch by Fuzzy- Stochastic Algorithms "*Journal of Theoretical and Applied Information Technology* pp. 88-94, 2009.

[69] Lizhi Wang and Nan Kong "Security Constrained Economic Dispatch: A Markov Decision Process Approach with Embedded Stochastic Programming" *Industrial and Manufacturing Systems Engineering Iowa State University 3016 Black Engineering, Ames, IA 50014, USA* PP 1-14,2010.

[70] Cvijic, S.; Jinjun Xiong, "Security constrained unit commitment and economic dispatch through benders decomposition: A comparative study," *Power and Energy Society General Meeting, 2011 IEEE*, Vol.10, pp.1-8, 24-29 July 2011.

[71] K.Vaisakh, P. Praveena, S. Rama Mohana Rao, Kala Meah, "Solving dynamic economic dispatch problem with security constraints using bacterial foraging PSO-DE algorithm" *International Journal of Electrical Power & Energy Systems*, Vol. 39, Issue 1, Pages 56-67, July 2012.

[72] W.G. Wood, "Spinning Reserve Constrained Static and Dynamic Economic Dispatch," *Power Apparatus and Systems, IEEE Transactions on*, Vol. PAS-101,no.2, pp. 381-388, 1982.

[73] W.R. Barcelo and P. Rastgoufard,"Dynamic Economic Dispatch using the Extended Security Constrained Economic Dispatch Algorithm," *Power Systems, IEEE Transactions on*, Vol. 12, no. 2, pp. 961-967, 1997.

[74] S. Hemamalini and S.P. Simon, "Dynamic Economic Dispatch using Maclaurin Series Based Lagrangian Method," *Energy Conversion and Management*, Vol. 51, no. 11, pp. 2212-2219, 2010.

[75] Ahmed Farag et al, "Economic load dispatch multiobjective optimization Procedures using linear programming techniques" *IEEE Transactions on Power Systems*, Vol. 10, No. 2, pp 731-738 May 1995.

[76] Y.H. Song and I. Yu,"Dynamic Load Dispatch with Voltage Security and Environmental Constraints," *Electrical Power Systems. Res.*, Vol. 43, no. 1, pp. 53-60, 1997.

[77] P.P.J. van den Bosch,"Optimal Dynamic Dispatch Owing to Spinning-Reserve and Power-Rate Limits," *Power Apparatus and Systems, IEEE Transactions on*, Vol. PAS-104, no. 12, pp. 3395-3401, 1985.

[78] K. Chandram, N. Subrahmanyam and M. Sydulu,"Dynamic Economic Dispatch by Equal Embedded Algorithm," *Electrical and Computer Engineering, 2006.ICECE '06. International Conference on*, pp. 21-24, 2006.

[79] K. Chandram, N. Subrahmanyam and M. Sydulu,"Brent Method for Dynamic Economic Dispatch with Transmission Losses," *Transmission and Distribution Conference and Exposition, 2008. T&D. IEEE/PES*, pp. 1-5, 2008.

[80] Xiaohua Xia, Jiangfeng Zhang and A. Elaiw, "A Model Predictive Control Approach to Dynamic Economic Dispatch Problem," *Power Tech, 2009 IEEE Bucharest*, pp. 1-7, 2009.

[81] X. Xia, J. Zhang and A. Elaiw, "An Application of Model Predictive Control to the Dynamic Economic Dispatch of Power Generation," *Control Eng. Pract.* Vol.19, no. 6, pp. 638-648, 2011.

[82] Ruey-Hsum Liang, "A Neural-Based Dispatch Approach to Dynamic Generation Allocation," *Power Systems, IEEE Transactions on*, Vol. 14, No. 4, pp.1388-1393, 1999

[83] A.Y. Abdelaziz, M.Z. Kamh, S.F. Mekhamer and M.A.L. Badr, "A Hybrid HNNQP Approach for Dynamic Economic Dispatch Problem," *Electrical Power Systems. Res.* Vol. 78, No. 10, pp. 1784-1788, 2008.

[84] A.Y. Abdelaziz, S.F. Mekhamer, M.Z. Kamh and M.A.L. Badr, "A Hybrid Hopfield Neural Network-Quadratic Programming Approach for Dynamic Economic Dispatch Problem," *Power System Conference, 2008. MEPCON 2008.12th International Middle-East*, pp. 565-570, 2008.

[85] Ahmed Ahmed El-Sawy et al, "Reference Point Based TR-PSO for Multi-Objective Environmental/Economic Dispatch "Journal *Applied Mathematics*, 2013, Vol 4,pp 803-813 ,2013.

[86] Wenxia Liu, Yuying Zhang, Bo Zeng Shuya Niu, Jianhua Zhang, and Yong Xiao "An Environmental-Economic Dispatch Method for Smart Microgrids Using VSS_QGA" *Hindawi Publishing Corporation Journal of Applied Mathematics* Vol.24, pp. 1-11 2014.

[87] F. Li, R. Morgan and D. Williams, "Towards More Cost Saving Under Stricter Ramping Rate Constraints of Dynamic Economic Dispatch Problems-a Genetic Based Approach," *Genetic Algorithms in Engineering Systems: Innovations and Applications, 1997. GALESIA 97. Second International Conference on (Conf.Publ. no. 446)*, pp. 221-225, 1997.

[88] W. Ongsakul and J. Tippayachai, "Parallel Micro Genetic Algorithm Based on Merit Order Loading Solutions for Constrained Dynamic Economic Dispatch," *Electrical Power Systems. Res.*, Vol. 61, no. 2, pp. 77-88, 2002.

[89] W. Ongsakul and J. Tippayachai, "Parallel Micro Genetic Algorithm Based on Merit Order Loading Solutions for Constrained Dynamic Economic Dispatch," *Electrical Power Systems. Res.*, Vol. 61, no. 2, pp. 77-88, 2002.

[90] Robert T. F. Ah King et al "Evolutionary Multi-Objective Environmental/Economic Dispatch: Stochastic vs. Deterministic Approach" *Department of Electrical and Electronic Engineering, Faculty of Engineering, University of Mauritius, Reduit, Mauritius*, pp. 1-15,2004.

[91] K. Shailti Swamp and A. Natarajan, "Constrained Optimization using Evolutionary Programming for Dynamic Economic Dispatch, *"Intelligent Sensing and Information Processing, 2005. Proceedings of 2005 International Conference on*, pp. 314-319, 2005.

[92] A.M.A.A. Joned, I. Musirin and Titik Khawa Abdul Rahman, "Solving Dynamic Economic Dispatch using Evolutionary Programming," *Power and Energy Conference, 2006. PECon '06. IEEE International*, pp. 144-149, 2006.

[93] G.S.S. Babu, D.B. Das and C. Patvardhan, "Dynamic Economic Dispatch Solution using an Enhanced Real-Quantum Evolutionary Algorithm," *Power System Technology and IEEE Power India Conference, 2008. POWERCON 2008. Joint International Conference on*, pp. 1-6, 2008.

[94] K. Deb, "An Efficient Constraint Handling Method for Genetic Algorithms, *"Computer Methods Applications in Mechanical Engineering*, Vol. 186, no. 2-4, pp. 311-338, 2000.

[95] J. Chiou, "A Variable Scaling Hybrid Differential Evolution for Solving Large-Scale Power Dispatch Problems," *Generation, Transmission & Distribution, IET*, Vol. 3, No. 2, pp. 154-163, 2009.

[96] Noman, N.; Iba, H.,"Solving dynamic economic dispatch problems using cellular differential evolution," *Evolutionary Computation (CEC), 2011 IEEE Congress on*, Vol.14. pp.2633,2640, 5-8 June 2011.

[97] Zwe-Lee Gaing, "Constrained Dynamic Economic Dispatch Solution using Particle Swarm Optimization," *Power Engineering Society General Meeting, 2004. IEEE*, pp. 153-158 Vol.1, 2004.

[98] Hardiansyah, Junaidi, and Yohannes MS"Solving Economic Load Dispatch Problem Using Particle Swarm Optimization Technique" *I.J. Intelligent Systems and Applications*, pp 12-18,2012.

[99] Niknam, T.; Golestane, F.; Bahmanifirouzi, B., "Modified adaptive PSO algorithm to solve dynamic economic dispatch," *Power Engineering and Automation Conference (PEAM), 2011 IEEE* , Vol.1, No.10, pp.108,111, 8-9 Sept. 2011.

[100] Divya Mathur "New Methodology for Solving Different Economic Dispatch Problems" *International Journal of Engineering Science and Innovative Technology (IJESIT)* Vol.2, No.1, pp.494-498 January 2013.

[101] C.K. Panigrahi, P.K. Chattopadhyay, R.N. Chakrabarti and M. Basu, "Simulated Annealing Technique for Dynamic Economic Dispatch," *Electric Power Components and Systems*, Vol. 34, No. 5, pp. 577-587, 2006.

[102] X.S. Han, H.B. Gooi and D.S. Kirschen, "Dynamic Economic Dispatch: Feasible and Optimal Solutions," *Power Engineering Society Summer Meeting, 2001.IEEE*, Vol.3, pp. 1704-1710, 2001.

[103] W. Ongsakul and J. Tippayachai, "Parallel Micro Genetic Algorithm Based on Merit Order Loading Solutions for Constrained Dynamic Economic Dispatch," *Electrical Power Systems. Res.*, Vol. 61, no. 2, pp. 77-88, 2002.

[104] X. Yuan, L. Wang, Y. Zhang and Y. Yuan, "A Hybrid Differential Evolution Method for Dynamic Economic Dispatch with Valve-Point Effects," *Expert Systems. Application*, Vol. 36, no. 2, Part 2, pp. 4042-4048, 2009.

[105] M. Basu, "Artificial Immune System for Dynamic Economic Dispatch," *International Journal of Electrical Power & Energy Systems*, Vol. 33, No. 1, pp. 131-136, 2011.

[106] S. Hemamalini and S.P. Simon, "Dynamic Economic Dispatch using Artificial Immune System for Units with Valve-Point Effect,"*International Journal of Electrical Power & Energy Systems*, Vol. In Press, Corrected Proof, pp. 1-7,2011.

[107] Niknam, T.; Azizipanah-Abarghooee, R.; Aghaei, J., "A new modified teaching-learning algorithm for reserve constrained dynamic economic dispatch," *Power Systems, IEEE Transactions on*, Vol.28, No.2, pp.749, 763, May 2013.

[108] Rabiee, A; Mohammadi-Ivatloo, B.; Moradi-Dalvand, M., "Fast Dynamic Economic Power Dispatch Problems Solution Via Optimality Condition Decomposition," *Power Systems, IEEE Transactions on* , Vol.29, No.2, pp.982,983, March 2014.

[109] C.B. Somuah and N. Khunaizi, "Application of Linear Programming Dispatch Technique to Dynamic Generation Allocation," *Power Systems, IEEE Transactions on*, Vol. 5, No. 1, pp. 20-26, 1990.

[110] X.S. Han, H.B. Gooi and D.S. Kirschen, "Dynamic Economic Dispatch: Feasible and Optimal Solutions," *Power Engineering Society Summer Meeting, 2001.IEEE*, Vol.3, pp. 1704-1710, 2001.

[111] W. Lin and S. Chen, "Bid-Based Dynamic Economic Dispatch with an Efficient Interior Point Algorithm," *International Journal of Electrical Power & Energy Systems*, Vol. 24, No. 1, pp. 51-57, 2002.

[112] W. Ongsakul and N. Ruangpayoongsak, "Constrained Dynamic Economic Dispatch by Simulated annealing/genetic Algorithms,"*Power Industry Computer Applications, 2001. PICA 2001. Innovative Computing for Power – Electric Energy Meets the Market. 22nd IEEE Power Engineering Society International Conference on*, pp. 207-212, 2001.

[113] T.A.A. Victoire and A.E. Jeyakumar, "Reserve Constrained Dynamic Dispatch of Units with Valve-Point Effects," *Power Systems, IEEE Transactions on*, Vol. 20, No. 3, pp. 1273-1282, 2005.

[114] T.A.A. Victoire and A.E. Jeyakumar, "A Modified Hybrid EP–SQP Approach for Dynamic Dispatch with Valve-Point Effect," *International Journal of Electrical Power & Energy Systems*, Vol. 27, No. 8, pp. 594-601, 2005.

[115] T.A.A. Victoire and A.E. Jeyakumar, "Deterministically Guided PSO for Dynamic Dispatch Considering Valve-Point Effect," *Electrical Power Systems. Res.*, Vol. 73, No. 3, pp. 313-322, 2005.

[116] F. Li, R. Morgan and D. Williams, "Hybrid Genetic Approaches to Ramping Rate Constrained Dynamic Economic Dispatch," *Electrical. Power Systems. Res.,* Vol. 43, No.2, pp. 97-103, 1997.

[117] F. Li and R.K. Aggarwal, "Fast and Accurate Power Dispatch using a Relaxed Genetic Algorithm and a Local Gradient Technique," *Expert Systems Application,* Vol. 19,No. 3, pp. 159-165, 2000.

[118] P. Attaviriyanupap, H. Kita, E. Tanaka and J. Hasegawa, "A Hybrid EP and SQP for Dynamic Economic Dispatch with Non smooth Fuel Cost Function," *Power Systems, IEEE Transactions on,* Vol. 17, No. 2, pp. 411-416, 2002.

[119] Zwe-Lee Gaing and Ting-Chia Ou, "Dynamic Economic Dispatch Solution using Fast Evolutionary Programming with Swarm Direction," *Industrial Electronics and Applications, 2009. ICIEA 2009. 4th IEEE Conference on,* pp. 1538-1544, 2009.

[120] Y. Lu, J. Zhou, H. Qin, Y. Li and Y. Zhang, "An Adaptive Hybrid Differential Evolution Algorithm for Dynamic Economic Dispatch with Valve-Point Effects, "*Expert Systems. Application,* Vol. In Press, Accepted Manuscript, 2009.

[121] F. Kappel and A.V. Kuntsevich, "An Implementation of Shor's r-Algorithm,"*Computational Optimization and Applications,* Vol. 15, no. 2, pp. 193-205, 2000.

[122] V. Ravi Kumar Pandi and B.K. Panigrahi, "Dynamic Economic Load Dispatch using Hybrid Swarm Intelligence Based Harmony Search Algorithm," *Expert Systems Applications,* Vol. In Press, Corrected Proof, pp. 1-6, 2011.

[123] S. Sivasubramani and K.S. Swarup, "Hybrid SOA–SQP Algorithm for Dynamic Economic Dispatch with Valve-Point Effects," *IEEE Transactions on Energy Conversion,* Vol. 35, No. 12, pp. 5031-5036, 2010.

[124] A. M. Elaiw, X. Xia,and A. M. Shehata "Dynamic Economic Dispatch Using Hybrid DE-SQP for Generating Units with Valve-Point Effects" *Hindawi Publishing Corporation Mathematical Problems in Engineering* Vol. 20,pp 1-10,2012.

[125] Guoli Zhang, Hai Yan Lu, Gengyin Li and Hong Xie, "A New Hybrid Real-Coded Genetic Algorithm and Application in Dynamic Economic Dispatch, "*Intelligent Control and Automation, 2006. WCICA 2006. The Sixth World Congess on,* Vol.1, pp. 3627-3632, 2006.

[126] Dun-wei Gong et al "Environmental/economic power dispatch using a hybrid multi-objective optimization algorithm" *IEEE Electrical Power and Energy Systems* Vol. 32 pp 607-614,2013.

[127] Niknam, T.; Azizipanah-Abarghooee, R.; Roosta, A, "Reserve Constrained Dynamic Economic Dispatch: A New Fast Self-Adaptive Modified Firefly Algorithm," *Systems Journal, IEEE,* Vol.6, No.4, pp.635, 646, Dec. 2012.

[128] H. Shayeghi and A. Ghasemi "Application of MOPSO for Economic Load Dispatch Solution with Transmission Losses*" International Journal on Technical and Physical Problems of Engineering (IJTPE),* Vol 4 No.1, Issue 10 pp. 27-34, 2012.

[129] Hardiansyah "Solving Economic Dispatch Problem with Valve-Point Effect using a Modified ABC Algorithm" *International Journal of Electrical and Computer Engineering (IJECE)* Vol. 3, No. 3, pp. 377-385 June 2013.

[130] V.Karthikeyan S.Senthilkumar and V.J.Vijayalakshmi "A New Approach to the Solution of Economic Dispatch using Particle Swarm Optimization with Simulated Annealing" *International Journal on Computational Sciences & Applications (IJCSA)* Vol.3, No.3,pp 37-49 June 2013.

[131] P. Attaviriyanupap, H. Kita, E. Tanaka and J. Hasegawa, "A Fuzzy-Optimization Approach to Dynamic Economic Dispatch Considering Uncertainties, "*Power Systems, IEEE Transactions on,* Vol. 19, No. 3, pp. 1299-1307, 2004.

[132] Y. Lu, J. Zhou, H. Qin, Y. Wang and Y. Zhang, "Chaotic Differential Evolution Methods for Dynamic Economic Dispatch with Valve-Point Effects," *Engineering Application of Artificial Intelligence ,* Vol. 24, No. 2, pp. 378-387, 2011.

[133] S. Titus and A.E. Jeyakumar, "A Hybrid EP-PSO-SQP Algorithm for Dynamic Dispatch Considering Prohibited Operating Zones," *Electric Power Components and Systems,* Vol. 36, No. 5, pp. 449-467, 2008.

[134] P. Praveena, K. Vaisakh and S. Rama Mohana Rao, "A Bacterial Foraging PSODE Algorithm for Solving Dynamic Economic Dispatch Problem with Security Constraints," *in Power Electronics, Drives and Energy Systems (PEDES) & 2010 Power India, 2010 Joint International Conference on,* pp. 1-7, 2010.

[135] A. A. Mousa et al "A Hybrid Optimization Technique Coupling an Evolutionary and a Local Search Algorithm for Economic Emission Load Dispatch Problem "*Journal of Applied Mathematics,* Vol.2, pp. 90-898 ,2011.

[136] Wang, M.Q.; Gooi, H.B.; Chen, S.X.; Lu, S., "A Mixed Integer Quadratic Programming for Dynamic Economic Dispatch With Valve Point Effect," *Power Systems, IEEE Transactions on ,* Vol.29, No.5, pp.2097,2106, Sept. 2014.

3

Thermo-Economic Analysis of Gas Turbines Power Plants with Cooled Air Intake

Rahim Jassim[1], Galal Zaki[2], Badr Habeebullah[2], Majed Alhazmy[2]

[1]Saudi Electric Services Polytechnic (SESP), Baish, Jazan Province, Kingdom of Saudi Arabia
[2]Mechanical Engineering, King Abdulaziz University, Jeddah, Saudi Arabia

Email address:
r_jassim@sesp.edu.sa (R. Jassim), bhabeeb@kau.edu.sa (B. Habeebullah), mhazmy@kau.edu.sa (M. Alhazmy)

Abstract: Gas turbine (GT) power plants operating in arid climates suffer from a decrease in power output during the hot summer months. Cooling the intake air enables the operators to mitigate this shortcoming. In this study, an energy analysis of a GT Brayton cycle coupled to a refrigeration cycle shows a promise of increasing the power output with a slight decrease in thermal efficiency. A thermo-economic algorithm is also developed and applied to the Hitachi MS700 GT open cycle plant at the industrial city of Yanbu, the Kingdom of Saudi Arabia (latitude 24°05" N and longitude 38° E). The results show that the power output enhancement depends on the degree of chilling the air intake to the compressor. Moreover, maximum power gain ratio is 15.46% whilst a slight decrease in thermal efficiency is of 12.25% for this case study. The study estimates the cost of the needed air cooling system. The cost function takes into consideration the time-dependent meteorological data, operation characteristics of the GT and air cooler, the operation and maintenance costs, interest rate, and lifetime. The study also evaluates the profit of adding the air cooling system for different electricity tariff.

Keywords: Gas Turbine, Power Boosting, Hot Climate, Air-Cooling, Mechanical Refrigeration

1. Introduction

High electricity demand during summer is a challenge for the local utilities in Saudi Arabia and neighboring countries. Air conditioning is a driving factor for electricity demand and operation schedules for these countries. In the Kingdom of Saudi Arabia (KSA), the utilities employ gas turbine (GT) power plants (present capacity 14 GW) to meet the peak load. Unfortunately, the power output and thermal efficiency of GT plants decrease in the summer because of the high inlet air temperature. The high air temperature at the GT intake decreases the air mass flow leading to the less power output generated [1]. For an ideal GT open cycle, the decrease in the net power output is approximately 0.4 % for every 1 K increase in the ambient air temperature [1-2]. Therefore, cooling the air intake to improve the GT performance receives considerable attention. Both direct methods (e.g., evaporative cooling) and indirect methods (e.g., mechanical vapor compression cooling) for cooling the air intake have been studied [1-10]. In the evaporative cooling method, the air intake is cooled off by contacting a cooling fluid such as atomized water sprays, fog, or their combination [2-3]. This method is suitable for GT power plants operating in dry and hot regions [1, 4-8]. The evaporative cooling leads to higher plan efficiency compared to vapor compression cooling when used in geographical regions having low ambient relative humidity and temperature [1]. The evaporative cooling provides low capital and operation cost and reliability, required moderate maintenance, and reduces the NOx content in the exhaust gases [9]. The main disadvantages of evaporative cooling method are low operation efficiency and large quantities of water consumption. Furthermore, the impact of the non-evaporated water droplets in the air stream could damage the compressor blades [10]. Replacing the water sprays by fogging system eliminates erosion problem [11-15].

The mechanical vapor compression [8, 16] and absorption refrigerator machines [17-20] are the two widely used approaches for indirect cooling. These cooling methods overcome the constraint set by the relative humidity of intake air. Direct cooling methods can reduce the air temperature below the ambient wet bulb temperature (WBT); however, extensive air chilling leads to ice formation as ice crystals or as solidified layer on compressor entrance [21]. The indirect cooling methods have gradually gained popularity over the evaporative cooling. For instance, 32 GT units have been outfitted with mechanical air chilling systems in Riyadh, Saudi Arabia. In general, application of the mechanical air-cooling increases the net power and reduces the thermal efficiency on the other hand [7].

New cooling methods are also available in the literature [22-27]. Farzaneh-Gord et al. [22] and Zaki et al. [23] proposed the use of a reversed Brayton refrigeration cycle for cooling the air intake to enhance the performance of GT. According to the results of the study [23], the air intake temperature could be lowered below the ISO standard with the power output increase up to 20% and decrease in the thermal efficiency of 6%. Jassim et al. [24] performed the exergy analysis of the system and showed the maximum improvement is 14.66% due to the components irreversibility. Khan et al. [25] proposed cooling the turbine exhaust gases and feeding them back to the compressor inlet with water harvested from the combustion products. Erickson [26-27] suggested a power fogger cycle that is a combination of waste heat driven absorption air-cooling and a water injection system.

Generally, even though thermal analyses of GT cooling are abundant in the literature, a few economic evaluations of implementing the air intake cooling methods have been considerably investigated. Such evaluations should account for the variations in the ambient conditions (temperature and relative humidity), the fluctuations in the fuel, electricity prices, and the interest rates. Therefore, the selection of a cooling technology (evaporative or refrigeration) and the sizing out of the equipment should base on not only the result of thermal analysis but also the cash flow. There are some outstanding studies focusing on the economic aspects of the cooling methods in literature. Gareta et al. [28] presented a computational algorithm to calculate the yearly additional power gain for combined cycle GT along with the economic feasibility for some cooling technologies. Chaker et al. [13] studied the economic potential of using evaporative cooling for GTs in USA. Yang et al. [17] did the same for combined GT in KSA while Hasnain et al. [29] and Shirazi et al. [30] examined the use of ice storage methods for GTs' air cooling in KSA. Investigations showed that the efficiency of inlet fogging was superior for the intake temperatures of 15-20oC, though it results in a smaller profit than inlet chilling.

This study presents a thermal and economic analysis of a GT system fitted with an external chilled water loop. The analysis accounts for the changes in the thermodynamics parameters as well as the economic variables (e.g., profitability, cash flow, and the lifetime of the system) for GT and the cooling components. The objective of this study is to assess the importance of using a coupled thermo-economic analysis in the selection of cooling system and operation parameters. The developed thermo-economic algorithm is then applied to the Hitachi MS700 GT, an open cycle plant, at the industrial city of Yanbu, KSA (latitude 24o 05" N and longitude 38o E). Finally, the cost analysis that is based on 10% interest rate and three-year payback period of the water chiller

is presented to determine the economic feasibility of using water chiller as an air cooler.

2. Thermodynamics Analysis

2.1. Gas Turbine Cycle Analysis

Figs.1a and b, respectively, show a schematic of a simple open GT Brayton cycle coupled to a refrigeration system and its T-S diagram. The power cycle consists of a compressor, a combustion chamber, and a turbine. The cooling system consists of a refrigerant compressor, air cooled condenser, throttle valve, and an evaporator. The chilled water from the evaporator passes through a cooling coil mounted at the air compressor entrance as shown in Fig. 1a. In this figure, the dotted line indicates a fraction of the electricity produced by the turbine and used to power the compressor and the pumps. Fig. 1c presents the T-S diagram of the refrigerant cycle by states a, b, c, and d.

As shown in Figs. 1a and b, the processes 1-2s and 3-4s are isentropic. Assuming air behaves as an ideal gas, the temperatures and pressures are related to the pressure ratio (PR) by

$$\frac{T_{2s}}{T_1} = \frac{T_3}{T_{4s}} = \left[\frac{P_2}{P_1}\right]^{\frac{k-1}{k}} = PR^{\frac{k-1}{k}} \qquad (1)$$

where k is the specific heat ratio.

The net power output of the GT with the mechanical cooling system is

$$\dot{W}_{net} = \dot{W}_t - (\dot{W}_{comp} + \dot{W}_{el,ch}) \qquad (2)$$

The first term of the RHS is the power produced by the turbine due to expansion of hot gases of mass flow rate $\dot{m}_t$ as

$$\dot{W}_t = \dot{m}_t c_{pg} \eta_t (T_3 - T_{4s}) \qquad (3)$$

In this equation, $\dot{m}_t$ is the total gas mass flow rate from the combustion chamber given in terms of the fuel air ratio $f = \dot{m}_f / \dot{m}_a$ and the air humidity ratio at the compressor intake ω_1 (kg$_w$/kg$_{dry\ air}$) at state 1(Fig. 1a) as

$$\dot{m}_t = \dot{m}_a + \dot{m}_v + \dot{m}_f = \dot{m}_a(1 + \omega_1 + f) \qquad (4)$$

The compression power for the humid air between states 1 and 2 is

$$\dot{W}_{comp} = \dot{m}_a c_{pa}(T_2 - T_1) + \dot{m}_v(h_{v2} - h_{v1}) \qquad (5)$$

where hv2 and hv1 are the enthalpies of saturated water vapor at the compressor exit and inlet states, respectively; $\dot{m}_v = \dot{m}_a \omega_1$ is the mass of water vapor.

The last term in Eq. 2 ($\dot{W}_{el,ch}$) is the power consumed by the cooling unit for driving the refrigeration machine electric motor, pumps, and auxiliaries. The thermal efficiency of a GT coupled to an air cooling system is

$$\eta_{cy} = \frac{\dot{W}_t - (\dot{W}_{comp} + \dot{W}_{el,ch})}{\dot{Q}_h} \qquad (6)$$

Substituting for T4s and $\dot{m}_t$ from Eqs. 1 and 4 into Eq. 3 yields

$$\dot{W}_t = \dot{m}_a (1 + \omega_1 + f) c_{pg} \eta_t T_3 \left(1 - \frac{1}{PR^{\frac{k-1}{k}}} \right) \qquad (7)$$

The turbine isentropic efficiency ηt can be estimated using the practical relation recommended by Alhazmy et al. [7] as

$$\eta_t = 1 - \left(0.03 + \frac{PR-1}{180} \right) \qquad (8)$$

Relating the compressor isentropic efficiency to the changes in temperature of the dry air and assuming that the compression of water vapor behaves as ideal gas, the actual compressor power becomes

$$\dot{W}_{comp_{air}} = \dot{m}_a \left[c_{pa} \frac{T_1}{\eta_c} \left(PR^{\frac{k-1}{k}} - 1 \right) + \omega_1 \left(h_{v2} - h_{v1} \right) \right] \qquad (9)$$

The compression efficiency η_c can be evaluated using the following empirical relation [7]:

$$\eta_c = 1 - \left(0.04 + \frac{PR-1}{150} \right) \qquad (10)$$

The heat balance in the combustion chamber (see Fig. 1a) gives the heat rate supplied to the gas turbine cycle as

$$\dot{Q}_h = \dot{m}_f NCV \eta_{comb} = \left(\dot{m}_a + \dot{m}_f \right) c_{pg} T_3 - \dot{m}_a c_{pa} T_2 + \dot{m}_v \left(h_{v3} - h_{v2} \right) \qquad (11)$$

Introducing the fuel air ratio $f = \dot{m}_f / \dot{m}_a$ and substituting for T2 in terms of T1 into Eq. 11 yields

$$\dot{Q}_h = \dot{m}_a T_1 \left[(1+f) c_{pg} \frac{T_3}{T_1} - c_{pa} \left(\frac{PR^{\frac{k-1}{k}} - 1}{\eta_c} + 1 \right) + \frac{\omega_1}{T_1} \left(h_{v3} - h_{v2} \right) \right] \qquad (12)$$

The simple expression for f is selected according to Alhazmy et al. [8] as

$$f = \frac{c_{pg} \left(T_3 - 298 \right) - c_{pa} \left(T_2 - 298 \right) + \omega_1 \left(h_{v3} - h_{v2} \right)}{NCV \eta_{comb} - c_{pg} \left(T_3 - 298 \right)} \qquad (13)$$

In this equation, hv2 and hv3 are the enthalpies of water vapor at the combustion chamber inlet and exit states, respectively, and can be calculated from Dossat [32]:

$$h_{vj} = 2501.3 + 1.8723 T_j; \quad j \text{ refers to state 1 or 3}. \qquad (14)$$

The four terms of the gas turbine net power and efficiency in Eq. 2 ($\dot{W}_t, \dot{W}_{comp}, \dot{W}_{el,ch}, \dot{Q}_h$) depend on the air temperature and relative humidity at the compressor inlet whose values are affected by the type and performance of the cooling system. The chillers' electric power $\dot{W}_{el,ch}$ is calculated in the following account.

2.2. Refrigeration Cooling System Analysis

For this analysis, the inlet air is cooled by using a cooling coil placed at the compressor inlet bell mouth. The chilled water from the refrigeration machine is the heat transport fluid as shown in Fig. 1a. The chiller's total electrical power can be expressed as the sum of the electric motor power ($\dot{W}_{motor}$), the pumps ($\dot{W}_P$), auxiliary power for fans, and control units ($\dot{W}_A$) as

$$\dot{W}_{el,ch} = \dot{W}_{motor} + \dot{W}_P + \dot{W}_A \qquad (15)$$

In this equation, $\dot{W}_A$ which is estimated to be from 5% to 10% of the compressor power is the input power to the auxiliary equipment, such as the condenser fans, control system, and so on. In this study, an air cooled condenser is used and 10% of the power required to drive the compressor motor is estimated for the cycle auxiliaries ($\dot{W}_A = 0.1 \dot{W}_{motor}$). The second term in Eq. 15 is the pumping power which is related to the chilled water flow rate and the pressure drop across the cooling coil as

$$\dot{W}_P = \dot{m}_{cw} v_f \left(\Delta P \right) / \eta_{pump} \qquad (16)$$

The isentropic compression process (a-bs) is the minimum energy utilized by the compressor, as depicted in Fig. 1c. The actual chiller power includes losses due to mechanical transmission, inefficiency in the drive motor converting electrical to mechanical energy, and the volumetric efficiency [32]. In general, the compressor electric motor work is related to the refrigerant enthalpy change as

$$\dot{W}_{motor} = \frac{\dot{m}_r \left(h_b - h_a \right)_r}{\eta_{eu}} \qquad (17)$$

The subscript r indicates refrigerant and η_{eu} is known as the energy used factor, $\eta_{eu} = \eta_m \eta_{el} \eta_{vo}$. The quantities on the right hand side are the compressor mechanical, electrical, and volumetric efficiencies, respectively. η_{eu} is usually determined by manufacturers and depends on the type of the compressor, the pressure ratio (P_b/P_a), and the motor power. For this analysis, η_{eu} is assumed as 85%.

Cleland et al. [31] developed a semi-empirical form of Eq. 17 to calculate the compressor's motor power usage in terms of the temperatures of the evaporator and condenser in the refrigeration cycle, T_e and T_c, respectively, as

$$\dot{W}_{motor} = \frac{\dot{m}_r (h_a - h_d)_r}{\dfrac{T_e}{(T_c - T_e)}(1 - \alpha x)^n \eta_{eu}} \qquad (18)$$

In this equation, α is an empirical constant that depends on the type of refrigerant and x is the quality at state d in Fig. 1c. The empirical constant is 0.77 for R-22 and 0.69 for R-134a [31]. The constant n depends on the number of the

compression stages and $n = 1$ for a simple refrigeration cycle with a single stage compressor. The nominator of Eq. 18 is the evaporator capacity $\dot{Q}_{e,r}$ and the first term of the denominator is the coefficient of performance of an ideal refrigeration cycle. Eqs. 2, 5, and 18 could be solved for the power usages by the different components of the coupled GT-refrigeration system and the increase in the power output as function of the air intake conditions. This thermodynamic performance analysis is coupled to a system economic analysis described later.

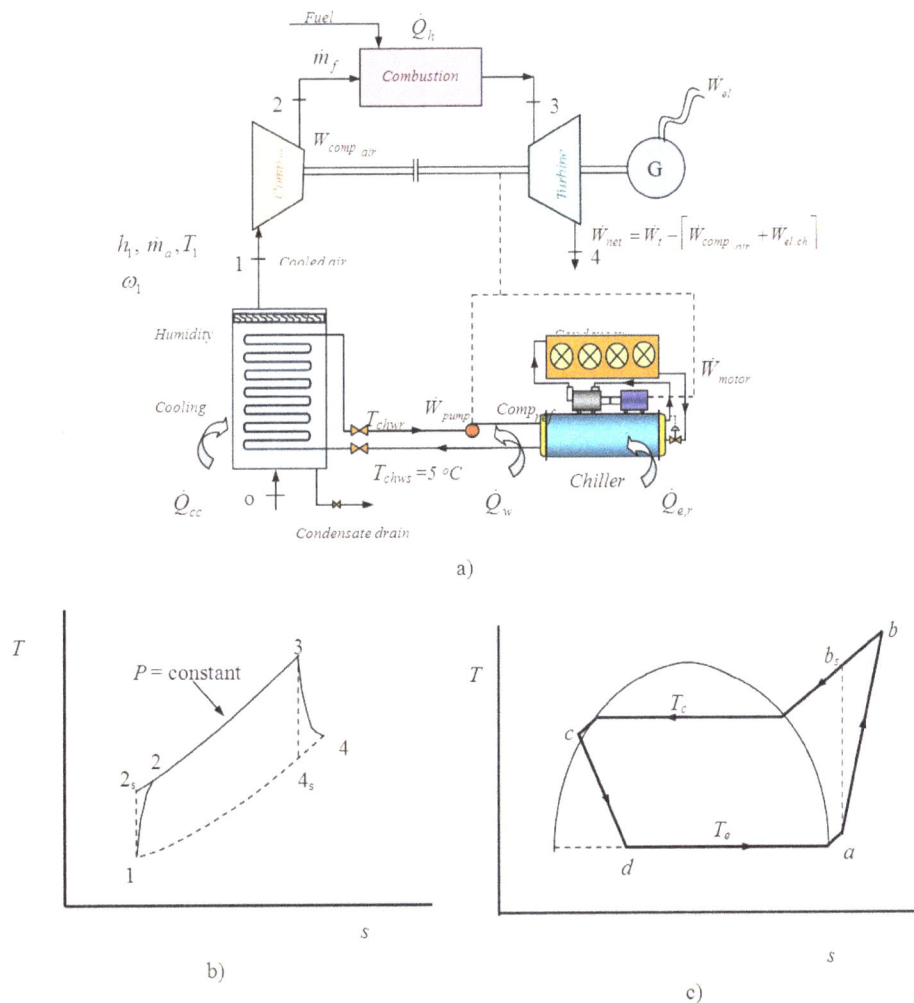

Fig. 1. a). Simple open type gas turbine with a chilled air-cooling unit; b) T-S diagram of an open type gas turbine cycle; c) T-S diagram for a refrigeration machine.

3. Economics Analysis

The increase in the power output will add to the revenue of the GT plant. However, this increase will partially be offset by the increase in capital cost associated with the installation of cooling system, personnel, and utility expenditures for the operation. For a cooling system including a water chiller, the increase in expenses involves the capital installments for the chiller (C_{ch}^c) and cooling coil (C_{cc}^c), the annual operational annual expenses. The latter is a function of the operation period top and the electricity rate. If the chiller consumes

electrical power $\dot{W}_{el,ch}$ and the electricity rate is Cel ($/kWh) then the total annual expenses are

$$C_{total} \ (\$ / y) = a^c \left[C_{ch}^c + C_{cc}^c \right] + \int_0^{t_{op}} C_{el} \dot{W}_{el,ch} \ dt \qquad (19)$$

where $a^c = i_r (1 + i_r)^n / \left((1 + i_r)^n - 1 \right)$ is the capital-recovery factor.

The chiller's purchase cost which is related to the chiller's capacity, $\dot{Q}_{e,r}$ (kW or ton/day), can be estimated from vendors or mechanical equipment cost index. For a particular chiller size and methods of construction and installation, manufacturers usually give the capital cost as

$$C_{ch}^c = \alpha_{ch} \times \dot{Q}_{e,r} \qquad (20)$$

where α_{ch} is a multiplication cost index in \$/kW. For simplicity, the maintenance expenses are assumed as a certain fraction α_m of the capital cost of the chiller; therefore, the total chiller capital cost is given as

$$C_{ch}^c (\$) = \alpha_{ch} \left(1 + \alpha_m\right) \dot{Q}_{e,r} \qquad (21)$$

Similarly, the capital cost of a particular cooling coil is given by manufacturers in terms of the cooling capacity that is directly proportional to the total heat transfer surface area (Accm2) Kotas [34] as

$$C_{cc}^c (\$) = \beta_{cc} \left(A_{cc}\right)^m \qquad (22)$$

In this equation, β_{cc} and m depend on the type of the cooling coil and material. For this study and the local KSA market, $\beta_{cc} = 30000$ and $m = 0.582$ are recommended. Substituting Eqs. 21 and 22 into Eq. 19 together with the assumption that the chiller power is an average constant value and the electricity rate is time independent for simplification, the annual total expenses for the cooling system become

$$C_{total} (\$/y) = a^c \left[\alpha_{ch} \left(1 + \alpha_m\right) \dot{Q}_{e,r} + \beta_{cc} \left(A_{cc}\right)^m \right] + t_{op} C_{el} \dot{W}_{el,ch} \qquad (23)$$

In Eq. 23, the heat transfer area Acc is used to evaluate the cost of the cooling coil. An energy balance for both the cooling coil and the refrigerant evaporator, taking into account the effectiveness factors for the evaporator $\varepsilon_{eff,er}$ and the cooling coil $\varepsilon_{eff,cc}$, gives

$$A_{cc} = \frac{\dot{Q}_{cc}}{U \Delta T_m F} = \frac{\dot{Q}_{e,r} \times \varepsilon_{eff,er} \times \varepsilon_{eff,cc}}{U \Delta T_m F} \qquad (24)$$

where U is the overall heat transfer coefficient for the chilled water-air tube bank heat exchanger. Gareta et al. [28] suggested a moderate value of 64 W/m2 K and recommended the factor F to be 0.98.

In Fig. 2, which shows the different temperatures in the combined refrigerant, water chiller, and air cooling system, the mean temperature difference for the cooling coil (air and chilled water fluids) is

$$\Delta T_m = \frac{\left(T_o - T_{chwr}\right) - \left(T_1 - T_{chws}\right)}{\ell n \left(\left(T_o - T_{chwr}\right) / \left(T_1 - T_{chws}\right)\right)} \qquad (25)$$

Equations 22 and 24 give the cooling coil cost as

$$C_{cc}^c = \beta_{cc} \left(\frac{\dot{Q}_{cc}}{U \Delta T_m F} \right)^m \qquad (26)$$

where $\dot{Q}_{cc}$ is the thermal capacity of the cooling coil. The atmospheric air enters at To and ω0 and leaves the cooling coil at T1 and ω1 before reaching the compressor, as shown in Fig. 1a. Both T1 and ω1 depend on the chilled water supply temperature Tchws and the chilled water mass flow rate $\dot{m}_{cw}$. When the outer surface temperature of the cooling coil falls below the dew point temperature (corresponding to the partial pressure of the water vapor), the water vapor condensates and leaves the air stream. This process may be treated as a cooling-dehumidification process as shown in Fig. 3.

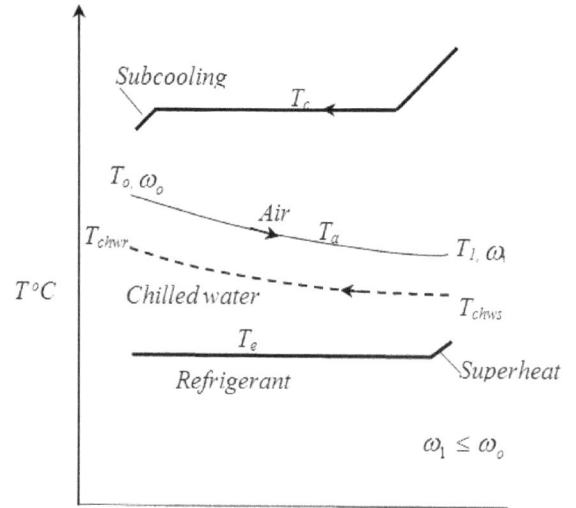

Fig. 2. *Temperature levels for the three working fluids, not to scale.*

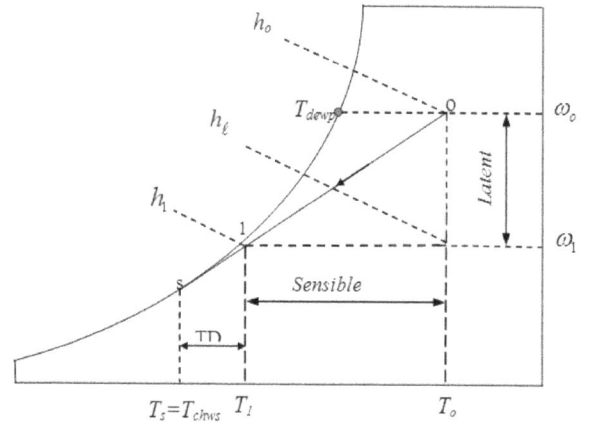

Fig. 3. *Moist air cooling process on the psychrometric chart.*

Steady state heat balance of the cooling coil gives $\dot{Q}_{cc}$ as

$$\dot{Q}_{cc} = \dot{m}_a \left(h_o - h_1\right) - \dot{m}_w h_w = \dot{m}_{cw} c_{pw} \varepsilon_{eff,cc} (T_{chwr} - T_{chws}) \qquad (27)$$

where $\dot{m}_{cw}$ is the chilled water mass flow rate, $\dot{m}_w$ is the rate of water extraction from the air, and $\dot{m}_w = \dot{m}_a \left(\omega_o - \omega_1\right)$. It is usually a small term when compared to the first and can be neglected [35].

In Eq. 27, the enthalpy and temperature of the air leaving the cooling coil (h_1 and T_1) can be calculated from

$$h_1 = h_o - CF(h_o - h_s) \qquad (28)$$

$$T_1 = T_o - CF(T_o - T_s) \qquad (29)$$

where CF is the contact factor of the cooling coil. It is defined as the ratio between the actual air temperature drop to the maximum at which the air theatrically leaves at coil surface temperature $T_s = T_{chws}$ and 100% relative humidity. Substituting h_1 from Eq. 28 into Eq. 27 gives

$$\dot{Q}_{cc} = \dot{m}_a \left[CF(h_o - h_{chws}) - (\omega_o - \omega_1)h_w \right] \qquad (30)$$

Eqs. 24 and 30 yield

$$\dot{Q}_{e,r} = \frac{\dot{m}_a \left[CF(h_o - h_{chws}) - (\omega_o - \omega_1)h_w \right]}{\varepsilon_{eff,er} \times \varepsilon_{eff,cc}} \qquad (31)$$

Eqs. 24, 30, and 31 give the cooling water flow rate, cooling coil capacity, and the evaporator capacity in terms of the air mass flow rate and properties.

Refrigeration Cooling System Analysis

$$C_{total} = \left[\frac{\dot{m}_a \left[CF(h_o - h_{chws}) - (\omega_o - \omega_1)h_w \right]}{\varepsilon_{eff,er} \times \varepsilon_{eff,cc}} \right] \times$$
$$\left\{ a^c \left[\alpha_{ch}(1+\alpha_m) + \beta_{cc} \left(\frac{\varepsilon_{eff,er} \times \varepsilon_{eff,cc}}{U \Delta T_m F} \right)^m \left(\frac{\dot{m}_a \left[CF(h_o - h_{chws}) - (\omega_o - \omega_1)h_w \right]}{\varepsilon_{eff,er} \times \varepsilon_{eff,cc}} \right)^{m-1} \right] + \right. $$
$$\left. t_{op} C_{el} \left[\left(\frac{1.1(T_c - T_e)}{(T_e)(1-\alpha x)^n \eta_{eu}} \right) + \left(\frac{\varepsilon_{eff,er} v_f(\Delta P)}{c_{p,w} \Delta T_{ch,w} \eta_p} \right) \right] \right\} \qquad (33)$$

4. Evaluation Criteria of GT-cooling System

In order to evaluate the feasibility of a cooling system coupled to a GT plant, the performance of the plant is examined with and without the cooling system. In general, the net power output of a complete system is

$$\dot{W}_{net} = \dot{W}_t - \left(\dot{W}_{comp} + \dot{W}_{el,ch} \right) \qquad (34)$$

The three terms in Eq. 34 are functions of the air properties at the compressor intake conditions (T1 and ω1), which in turn depend on the performance of the cooling system. This analysis considers the power gain ratio (PGR), a broad term suggested in [8] that takes into account the operation parameters of the GT and the associated cooling system

$$PGR = \frac{\dot{W}_{net,with\,cooling} - \dot{W}_{net,without\,cooling}}{\dot{W}_{net,without\,cooling}} \times 100\% \qquad (35)$$

For a stand-alone GT, PGR = 0. Thus, PGR gives the percentage enhancement in power generation by the coupled system. The thermal efficiency of the system is an important

Combining Eqs. 23 and 24 and substituting for the cooling coil surface area, pump, and auxiliary power give the cost function in terms of the evaporator capacity $\dot{Q}_{er}$. Total annual cost is given as

$$C_{total} = \left\{ a^c \left[\alpha_{ch}(1+\alpha_m)\dot{Q}_{er} + \beta_{cc} \left(\frac{\dot{Q}_{er} \times \varepsilon_{eff,er} \times \varepsilon_{eff,cc}}{U \Delta T_m F} \right)^m \right] + \right.$$
$$\left. t_{op} \dot{Q}_{er} C_{el} \left[\left(\frac{1.1(T_c - T_e)}{T_e(1-\alpha x)^n \eta_{eu}} \right) + \left(\frac{\varepsilon_{eff,er} v_f(\Delta P)}{c_{p,w} \Delta T_{ch,w} \eta_{pump}} \right) \right] \right\} \qquad (32)$$

The first term in Eq. 32 is the annual fixed charges of the refrigeration machine and the surface of air cooling coil, while the second term is the operation expenses that depend mainly on the electricity rate. The motor power has been increased by 10% to account for the auxiliaries' consumption. If the water pump's power is considered as infinitesimally small compared to the compressor power, the second term of the operation charges can be dropped. If the evaporator capacity $\dot{Q}_{er}$ is replaced by the expression in Eq. 31, the cost function, in terms of the primary parameters, becomes

parameter to describe the input-output relationship. The thermal efficiency change factor (TEC) proposed in [8] is defined as

$$TEC = \frac{\eta_{cy,with\,cooling} - \eta_{cy,without\,cooling}}{\eta_{cy,without\,cooling}} \times 100\% \qquad (36)$$

Both PGR and TEC can be easily employed to assess the changes in the system performance, but are not sufficient for a complete evaluation of the cooling method.

To investigate the economic feasibility of retrofitting a gas turbine plant with an intake cooling system, the total cost of the cooling system is determined (Eq. 32 or Eq. 33). The increase in the annual income cash flow from selling the additional electricity generation is also calculated. The annual energy electricity generation by the coupled power plant system is

$$E \,(\text{kWh}) = \int_0^{t_{op}} \dot{W}_{net} dt \qquad (37)$$

If the gas turbine's annual electricity generation without a cooling system is Ewithout cooling and the cooling system increases the power generation to Ewith cooling, and then the net increase in revenue due to the addition of the cooling system can be calculated from

$$Net\ revenue = \left(E_{with\ cooling} - E_{without\ coolig} \right) C_{els} \qquad (38)$$

The profitability due to the coupled power plant system is defined as an increase in revenues due to the increase in electricity generation after deducting the expenses for installing and operating the cooling system as

$$Profitability = (E_{with\ cooling} - E_{without\ cooling}) C_{els} - C_{total} \qquad (39)$$

The first term in Eq. 39 gives the increase in revenue whilst the second term gives the annual expenses of the cooling system. The profitability could be either positive or negative, which, respectively, means an economic insensitivity for adding the cooling system and an economic disadvantage, despite the increase in the electricity generation of the plant.

5. Results and Discussion

The analysis and economic feasibility are applied for the performance of the HITACH 700 model GT plant with a water chiller air cooling system. This plant, which has been already connected to the main electric grid, is located at the Industrial City of Yanbu (latitude 24o 05' N and longitude 38o E). The specifications of the GT plant are described in Table 1. The water chiller capacity is selected on basis of the maximum annual ambient temperature. On August 18th, 20xx, the dry bulb temperature (DBT) reached 50oC at 2:00 PM and the relative humidity was 84% at dawn time. The recorded hourly variations in the DBT (To) and RHo and the values are, respectively, shown in Fig. 4 and Table 2. The evaporator capacity of the water chiller (ton refrigeration) given in Eq. 30 is function of the DBT and RH. Fig. 5 shows that if the chiller is selected based on the maximum DBT = 50oC and RH = 18% (the data at 1:00 PM), its capacity would be 2200 tons. Another option is to select the chiller capacity based on the air maximum RH (RH = 0.83 and To = 28.5oC), which results in 3500 tons. It is more accurate, however, to determine the chiller capacity for the available climatic data of the selected day and to determine the maximum required capacity, as shown in Fig. 6. For the weather conditions at Yanbu city, a chiller capacity of 4200 tons is selected.

The hourly performance parameters of the GT plant, with and without cooling system (Eqs. 35 and 36), are calculated and compared. All thermo-physical properties are determined to the accuracy of the EES software [33]. The results show that the cooling system decreases the intake air temperature from To to T1 and increases the relative humidity to RH2 (Table 2). The chilled air temperature T1 is calculated from Eq. 29, assuming contact factor of 0.5 and a chilled water supply temperature of 5oC. Using the data in Table 2, the solution of Eqs. 35 and 36 gives the daily variation in the PGR and TEC (Fig. 7). There is certainly a potential benefit of adding the cooling system when there is an increase in the power output all the time; the calculated average for the design day is

12.25 %.

The PGR follows the same pattern of the ambient temperature, which simply means that the electric power of the GT plant increases during the hot hours of the day when electricity demand is high (10:00 AM to 6:00 PM). The increase in the output power of the GT plant reaches a maximum of 15.46% with a little change in the plant thermal efficiency. The practical illustrative application indicates that a maximum decrease in the thermal efficiency change of only 0.223% occurs at 13:00 PM when the air temperature is 45.2oC, and RH is 34%.

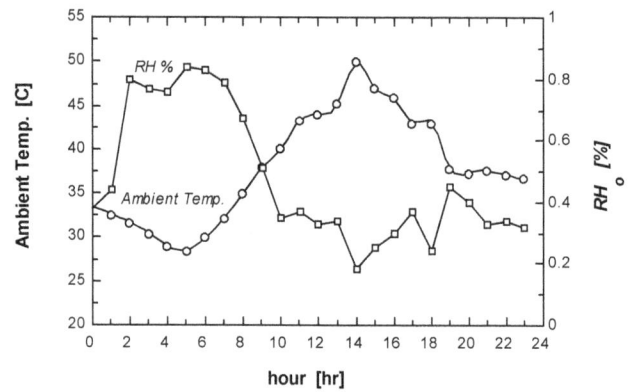

Fig. 4. *Ambient temperature and RH variations on August 18th at Yanbu Industrial City, KSA.*

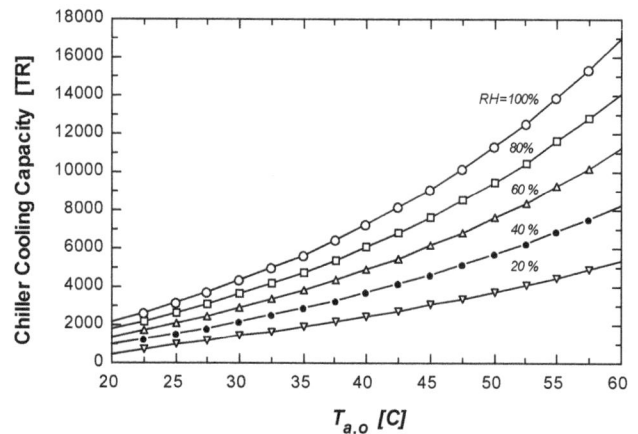

Fig. 5. *Dependence of chiller cooling capacity on the climatic conditions.*

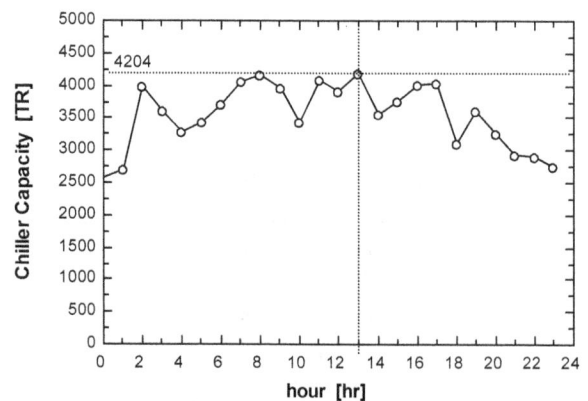

Fig. 6. *Chiller capacity with the variation of the climatic conditions (temperature and RH).*

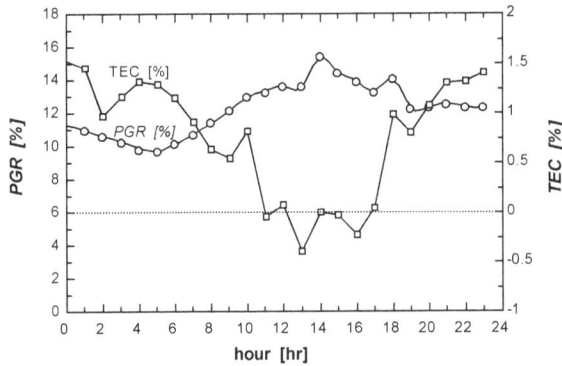

Fig. 7. *Variation of gas turbine PGR and TEC during August 18th operation.*

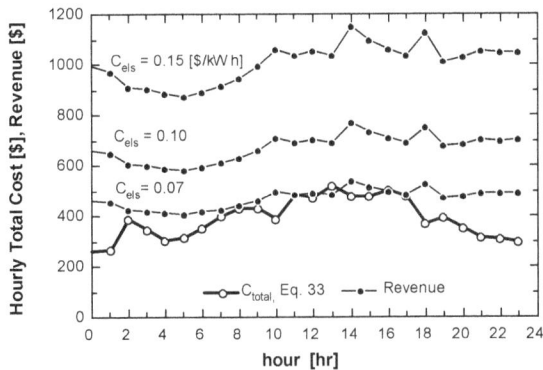

Fig. 8. *Variation of hourly total cost and excess revenue at different unit cost of selling electricity.*

Based on the daily variation of the ambient conditions on August 18th, assuming different values for selling the electricity (Cels), Eq. 39 gives the hourly revenues needed to payback the investment after a specified operation period (selected by 3 years). The different terms in both Eqs. 32 and 39 are calculated and presented in Fig. 8. Firstly, the effect of the climate changes is quite obvious on both the GT net power output as shown in Fig. 7 and on the total expenses as presented in Fig. 8. The variations in Ctotal are due to the changes in $\dot{Q}_{ev}$ in Eq. 32 that depends on T0, T1, ω0, and ω1. The revenues from selling additional electricity are also presented in the same figure, which shows clearly the potential of adding the cooling system. A profitability of the system, being the difference between the total cost and the revenues, is realized when the selling rate of the excess electricity generation is higher than the base rate of 0.07 $/kWh.

Fig. 8 shows that selling the electricity to the consumers at the same price ($C_{els} = C_{el}$ = 0.07 $/kWh) makes the cooling system barely non-profitable during the morning, night time, and hot hours of the day. This result encourages the utilities to consider adding a time-of-use tariff during the high demand periods, which is customary case in many courtiers. Should this become the case also in KSA, installing an air cooling system becomes economically feasible and profitable. Economics calculations for one year of 7240 operation hours with different electricity rates (C_{els}) and fixed electricity rate (C_{el} = 0.07 $/kWh) are summarized in Table 3.

Table 1. *Specifications of the GT plant.*

Parameters		Range
Ambient air	Ambient air temperature, T_o	28-50 °C
	Ambient air relative humidity, RH_o	18%-84%
	Pressure ratio, P_2/P_1	10
	Turbine inlet temperature T_3	1273.15 K
	Volumetric air flow rate	250 m³s⁻¹ at NPT
Gas turbine	Fuel net calorific value, NCV	46000 kJ kg⁻¹
	Turbine efficiency, η_t	0.88
	Air compressor efficiency, η_c	0.82
	Combustion efficiency, η_{comb}	0.85
Generator	Electrical efficiency	95%
	Mechanical efficiency	90%
	Refrigerant	R22
	Evaporating temperature, T_e	$T_{chws} - TD_e$ °C
	Superheat	10K
	Condensing temperature, T_c	$T_o + TD_c$ K
	Condenser design temperature difference TD_c	10 K
Water chiller	Evaporator design temperature difference TD_e	6 K
	Subcooling	3K
	Chilled water supply temperature, T_{chws}	5 °C
	Chiller evaporator effectiveness, $\varepsilon_{eff,er}$	85%
	Chiller compressor energy use efficiency, η_{eu}	85%
	α_{ch}	172 $/kW
Cooling coil	Cooling coil effectiveness, $\varepsilon_{eff,cc}$	85%
	Contact factor, CF	50%
	Interest rate i	10%
	Period of repayment (payback period), n	3 years
Economic analysis	The maintenance cost, α_m	10% of C_{ch}^c
	Electricity rate, C_{el}	0.07 $/kWh
	Cost of selling excess electricity, C_{els}	0.07-0.15 $/kWh
	Hours of operation per year, t_{op}	7240

Table 2. *The ambient conditions and the cooling coil outlet temperature and humidity during August 18th operation.*

Hour	$T_o\,^oC$	RH	$T_1\,^oC$	RH_1
0	33.4	0.38	19.2	0.64
1	32.6	0.44	18.8	0.70
2	31.7	0.8	18.35	0.99
3	30.5	0.77	17.75	0.98
4	29.0	0.76	17.0	0.99
5	28.5	0.84	16.75	0.97
6	30.0	0.83	17.5	0.99
7	32.2	0.79	18.6	0.96
8	35.1	0.67	20.05	0.99
9	38.0	0.51	21.5	0.84
10	40.2	0.35	22.6	0.64
11	43.3	0.37	24.15	0.69
12	44.0	0.33	24.5	0.64
13	45.2	0.34	25.1	0.66
14	50.0	0.18	27.5	0.43
15	47.0	0.25	26.0	0.53
16	45.9	0.30	25.45	0.61
17	43.0	0.37	24.0	0.69
18	43.0	0.24	24.0	0.50
19	37.9	0.45	21.45	0.76

Table 3. *Annual net profits out of retrofitting a cooling system to a GT, Hitachi MS700 GT at Yanbu for different product tariff and 3 years payback period.*

Electricity selling rate C_{els} (\$/kWh)	Annuity for chiller and maintenance (\$/y)	Annual operating cost (\$/y)	Annual net profit for the first 3 years (\$/y)	Annual net profit for the fourth year (\$/y)
0.07	1,154,780	1,835,038	-1,013,600	+141180
0.1	1,154,780	1,835,038	-166,821	+987,962
0.15	1,154,780	1,835,038	1,244,978	+2,399,758

The results in Table 3 show that there is always a net positive profit starting after the payback period for different energy selling prices. During the first 3 years of the cooling system life, there is a net profit when there is an increase in selling rate of the excess electricity generation to 0.15 S/kWh, which nearly doubles the base tariff.

6. Conclusions

There are various methods to improve the performance of gas turbine power plants operating under hot ambient temperatures far from the ISO standards. One proven approach is to reduce the compressor intake temperature by installing an external cooling system. In this paper, a simulation model that consists of thermal analysis of a GT coupled to refrigeration cooler and economics evaluation is developed. The performed analysis is based on coupling the thermodynamics parameters of the GT and cooler unit with the other variables such as the interest rate, life time, increased revenue, and profitability in a single cost function. The augmentation of the GT plant performance is characterized using the PGR and the TEC.

The developed model is applied to a GT power plant located at the city of Yanbu (20o 05" N latitude and 38o E longitude), KSA, where the maximum DBT has reached 50oC on August 18th, 20xx. The recorded climate conditions on that day are selected for sizing out the chiller and cooling coil capacities.

The performance analysis of the GT, for a pressure ratio of 10, rate of air intake of 250 m3/s, and 1000oC maximum cycle temperature shows that the intake air temperature decreases from 12 to 22 K, while the PGR increases to maximum of 15.46%. The average increase in the plant power output power is 12.25%, with slight change in plant thermal efficiency.

In this study, the profitability resulting from cooling the intake air is calculated for electricity rates between 0.07 and 0.15 \$/kWh and a payback period of 3 years. Cash flow analysis of the GT power plant in the city of Yanbu shows a potential for increasing the output power of the plant and increased revenues. The profitability is a result of adding the cooling system increase as the electricity rate increase during the peak demand periods, beyond the current base rate of 0.07 \$/kWh.

Acknowledgments

This project was funded by the National Plan for Science, Technology and Innovation (MAARIFAH)-King Abdulaziz City for Science and Technology-the kingdom of Saudi Arabia-award number (8-ENE 288-03). The authors also, acknowledge with thanks Science and Technology Unit, King Abdulaziz University for technical support.

Nomenclatures

A_{cc} cooling coil heat transfer area, m^2

C_{cc}^c capital cost of cooling coil ($)

C_{ch}^c capital cost of chiller ($)

C_{el} unit cost of electricity, $/kWh

c_p specific heat of gases, kJ/kg K

CF contact factor

E energy kWh

EES engineering equation solver

h_v specific enthalpy of water vapor in the air, kJ/kg

i_r interest rate on capital

k specific heat ratio

$\dot{m}$ mass flow rate, kg s^{-1}

$\dot{m}_a$ air mass flow rate, kg/s

$\dot{m}_{cw}$ chilled water mass flow rate, kg/s

$\dot{m}_r$ refrigerant mass flow rate, kg/s

$\dot{m}_w$ condensate water rate, kg/s

NCV net calorific value, kJ kg^{-1}

P pressure, kPa

PGR power gain ratio

P_o atmospheric pressure, kPa

PR pressure ratio = P_2/P_1

$\dot{Q}_h$ heat rate, kW

$\dot{Q}_{e,r}$ chiller evaporator cooling capacity, kW

$\dot{Q}_{cc}$ cooling coil thermal capacity, kW

T temperature, K

TEC thermal efficiency change factor

U overall heat transfer coefficient, kW/m^{2}K

x quality

$\dot{W}$ power, kW

Greek symbols

η efficiency

ε_{eff} effectiveness, according to subscripts

ω specific humidity (also, humidity ratio), according to subscripts, kg/kg$_{dry\ air}$

Subscripts

a dry air

cc cooling coil

ch chiller

$comb$ combustion

$comp$ compressor

el electricity

f fuel

g gas

o ambient

t turbine

v vapor

References

[1] Mohapatra AK, Sanjay. Comparative analysis of inlet air cooling techniques integrated to cooled gas turbine plant. J Energy Inst 2014; In press: 1-15.

[2] Cortes CPE, Williams D. Gas turbine inlet cooling techniques: An overview of current technology; Dec. 2004. Proc Power GEN, Las Vegas Nevada.

[3] Wang T, Li X, Pinniti V. Simulation of mist transport for gas turbine inlet air-cooling; Nov. 2009. ASME Int Mec Eng congress, Anaheim, Ca, USA.

[4] Ameri M, Nabati H. Keshtgar A. Gas turbine power augmentation using fog inlet cooling system; 2004. Proc ESDA04 7th Biennial Conf Eng Syst Des Anal, Manchester UK.

[5] Ameri M, Shahbazian HR, Nabizadeh M. Comparison of evaporative inlet air cooling systems to enhance the gas turbine generated power. Int J Energy Res 2007; 31: 483-503

[6] Jonsson M, Yan J. Humidified gas turbines - A review of proposed and implemented cycles. Energy 2005; 30: 1013-1078.

[7] Alhazmy MM. Najjar YS. Augmentation of gas turbine performance using air coolers. App Therm Eng 2004; 24: 415-429.

[8] Alhazmy MM, Jassim RK, Zaki GM. Performance enhancement of gas turbines by inlet air-cooling in hot and humid climates. Int J Energy Res 2006; 30:777-797.

[9] Sanaye S, Tahani M. Analysis of gas turbine operating parameters with inlet fogging and wet compression processes. Appl Therm Eng 2010; 30:234-244.

[10] Tillman TC, Blacklund DW, Penton JD. Analyzing the potential for condensate carry-over from a gas cooling turbine inlet cooling coil. ASHRAE Trans 2005; 111(Part 2) DE-05-6-3: 555-563.

[11] Chaker M, Meher-Homji CB, Mee M. Inlet fogging of gas turbine engines - Part B: Fog droplet sizing analysis, nozzle types, measurement and testing. ASME Proc Turbo Expo 2002; 4:429-442.

[12] Chaker M, Meher-Homji CB, Mee M. Inlet fogging of gas turbine engines - Part C: fog behavior in inlet ducts, cfd analysis and wind tunnel experiments. ASME Proc Turbo Expo 2002; 4:443-455.

[13] Chaker M, Meher-Homji CB, Mee M, Nicholson A. Inlet fogging of gas turbine engines detailed climatic analysis of gas turbine evaporation cooling potential in the USA. J Eng Gas Turbine Power 2003; 125:300-309.

[14] Homji-Meher BC, Mee T, Thomas R. Inlet fogging of gas turbine engines, Part B: Droplet sizing analysis nozzle types, measurement and testing; June 2002. Proc ASME Turbo Expo, Amsterdam, Netherlands.

[15] Gajjar H, Chaker M. Inlet fogging for a 655 MW combined cycle power plant-design, implementation and operating experience. ASME Proc of Turbo Expo 2003; 2:853-860.

[16] Elliot J. Chilled air takes weather out of equation. Diesel and gas turbine world wide; 2001, p. 49-96.

[17] Yang C, Yang Z, Cai R. Analytical method for evaluation of gas turbine inlet air cooling in combined cycle power plant. Appl Energy 2009; 86:848–856

[18] Ondryas IS, Wilson DA, Kawamoto N, Haub GL. Options in gas turbine power augmentation using inlet air chilling. Eng Gas Turbine and Power 1991;113: 203-211.

[19] Punwani D, Pierson T, Sanchez C, Ryan W. Combustion turbine inlet air cooling using absorption chillers some technical and economical analysis and case summaries. ASHRAE Annual Meeting; June 1999. Seattle, Washington.

[20] Kakarus E, Doukelis A, Karellas S. Compressor intake air cooling in gas turbine plants. Energy 2004; 29:2347-2358.

[21] Stewart W, Patrick A. Air temperature depression and potential icing at the inlet of stationary combustion turbines. ASHRAE Trans 2000; 106:318-327.

[22] Farzaneh-Gord M, Deymi-Dashtebayaz M. A new approach for enhancing performance of a gas turbine (case study: Khangiran refinery). Appl Energy 2009; 86: 2750-2759

[23] Zaki GM, Jassim RK, Alhazmy MM. Brayton refrigeration Cycle for gas turbine inlet air cooling. Int J Energy Res 2007; 31:1292-1306.

[24] Jassim RK, Zaki GM, Alhazmy MM. Energy and exergy analysis of reverse Brayton refrigerator for gas turbine power boosting. Int J Exergy 2009; 6:143-165.

[25] Khan JR, Lear WE, Sherif SA, Crittenden JF. Performance of a novel combined cooling and power gas turbine with water harvesting. ASME J Eng Gas Turbines Power; 2008; 130: 041702

[26] Erickson DC. Aqua absorption turbine inlet cooling; Nov. 2003. Proc IMEC 03, ASME Int Mech Eng Congress Exposition, Washington DC

[27] Erickson DC. Power fogger cycle. ASHRAE Transactions 2005; 111:551-554.

[28] Gareta R, Romeo LM, Gil A. Methodology for the economic evaluation of gas turbine air cooling systems in combined cycle applications. Energy 2004; 29:1805-1818.

[29] Hasnain SM, Alawaji SH, Al-Ibrahim AM, Smiai MS. Prospects of cool thermal storage utilization in Saudi Arabia. Energy Convers Manag 2000; 41:1829-1839.

[30] Shirazi A, Najafi B, Aminyavari M, Rinaldi F, Taylor RA. Thermal-economic-environmental analysis and multi-objective optimization of an ice thermal energy storage system for gas turbine cycle inlet air cooling. Energy 2014; 69:212-226

[31] Cleland AJ, Cleland DJ, White SD. Cost-Effective Refrigeration, Short course notes, Institute of Technology and Engineering, Massey University, New Zealand; 2000

[32] Dossat RJ. Principles of Refrigeration. New York: John Wiley and Sons; 1997.

[33] Klein KA, Alvarado FL. EES-Engineering Equation Solver, Version 6.648 ND, F-Chart Software, Middleton, WI; 2004.

[34] Kotas TJ. The exergy method of thermal plant analysis. Elsevier; 1995.

[35] McQuiston FC, Parker JD, Spilter JD. Heating, ventilating and air conditioning: Design and analysis. 6th ed. New Yorh: John Wily; 2005.

4

PMSG Based Wind Energy Conversion with Space Vector Modulation

M. Magesh Kumar[1], R. Sundareswaran[2]

[1]ME power system engineering, Chandy college of engineering, Tuticorin, India
[2]AP/EEE Department, Chandy college of engineering, Tuticorin, India

Email address:

makesh.engg@gmail.com (M. M. Kumar), sundarr877@gmail.com (R. Sundareswaran)

Abstract: Space Vector Modulation (SVM) based Direct Torque Control (DTC) scheme is used for a variable –speed direct drive Permanent Magnet Synchronous Generator (PMSG) in a wind power generation system. SVM Provides more flexibility for inverter voltage utilization, in order to compensate the torque and flux errors in a smoother way than conventional DTC. SVM based DTC maintains constant switching frequency and it can be applied using closed loop torque control, for minimization of torque ripple. Over the conventional Doubly Fed Induction Generator (DFIG), PMSG has several advantages. The PMSG based on Wind Energy Conversion System (WECS) can be connected to the turbine without gearbox. The gearbox causes in the cost of maintenance, and then it will decrease the weight of the nacelle. Additionally, the generators have a better efficiency and have been slightly cheaper. This provides maximum flexibility, enabling full real and reactive power control. The SVM-DTC regulated PMSG wind turbine can achieve fast torque response, relatively low torque ripples, and extracts the maximum power. The effectiveness of the SVM-DTC control scheme for PMSG has been demonstrated by MATLAB simulation.

Keywords: Space Vector Modulation, Direct Torque Control, Permanent Magnet Synchronous Generator,
Doubly Fed Induction Generator, Wind Energy Conversion Systems

1. Introduction

The issue of global warming has been raised due to the consumption of oil, coal and natural resource for the electricity producing process. Apart from that, the amounts of these fuels on earth are now decreasing day by day. Thus, the focus has been shifted to the green alternative energy which is not polluted and has no impact on the environment. The power of wind is now being explored which the researcher believes that it has all the qualifications to replace traditional fuel since it has less effect to global warming. During the past decade, the amounts of wind capacity have been installed every three years. Around 83% of wind capacities are located in these five countries, German, United States of America, Denmark, India and Spain [9].

Wind energy conversion system (WECS) consist of wind turbine, pitch angle control, drive train, generator and power converter [1]. Fig. 1 shows the typical Wind Energy Conversion System.

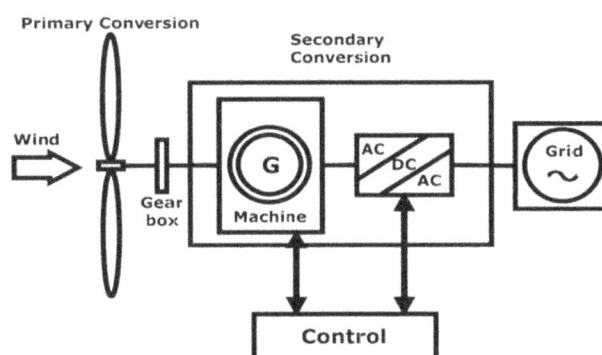

Fig. 1. Typical Wind Energy Conversion System.

There are various kinds of generators used in WECS such as induction generator (IG), doubly fed induction generator (DFIG) and permanent magnet synchronous generator (PMSG) [2][3]. The PMSG based on WECS can connect to the turbine without using gearbox, which significantly reduces the construction, operation, and maintenance costs, and then it will decrease the weight of nacelle [4][7]. These benefits have made the direct drive PMSG-based WTGs the most popular configuration in multi-MW offshore applications [4]. In the last few decades, a direct-torque control (DTC) method was proposed [6] and naturally introduced to the permanent-magnet synchronous motor (PMSM) drive systems [7], [8]. In the DTC, the inverter and the machine are conceived as a whole so that the stator voltage vectors can be selected directly according to the differences between the reference and actual torques and stator flux linkages. This method not only removes the need for complicated computation of reference frame transformation and high dependence of the decoupled current control on machine parameters but also achieves a faster torque response.

2. Direct Drive PMSG

The most important difference between the geared generator and the direct-drive generator is the rotation speed of the generator. The direct-drive generator rotates at low speed, because the rotor of generator is directly connected on the hub of rotor blades. This low speed makes it necessary to produce a very high rated torque in the direct-drive generator, since the size of this generator depends on the rated torque rather than the rated power. The direct-drive generator mostly used in the market can be classified to two concepts. They are (i) electrically excited synchronous generator (EESG) and (ii) permanent magnet synchronous generator (PMSG) [12] [13].

The advantages of the direct-drive generator compared to the geared generator are Simplified drive train by omitting the gearbox, high overall efficiency and reliability, high availability, low noise of the drive train.

A direct-drive generator is usually heavier than a geared generator because it has to make a high rated torque as mentioned above. To increase the efficiency, to reduce the weight of the active parts, and to keep the end winding losses small, the direct-drive generator is usually designed with a large diameter and small pole pitch. Fig 2.Shows Direct Drive PMSG for wind energy conversion system.

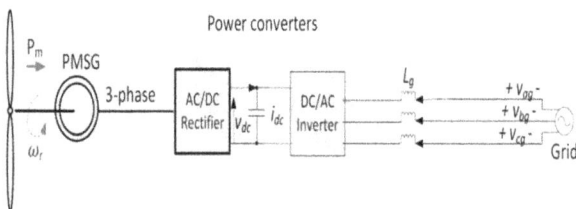

Fig. 2. Direct Drive Pmsg Wecs.

In the considerations of the energy yield and reliability, direct-drive systems seem to be more powerful than the geared systems, especially for offshore. Direct-drive systems, which are operated in low speed, also have disadvantages as the large diameter, large weight, and high cost compared to geared systems.

3. Direct-Drive Pmsg Wind Turbine System Modelling

The configuration of a direct-drive PMSG wind turbine is connected to the PMSG directly [18]. The electrical power generated by the PMSG is transmitted to a power grid and/or supplied to a load via a variable-frequency converter, which consists of a machine-side converter (MSC) and a grid-side converter connected back-to-back via a dc link [4][5][14].

A. Wind turbine aerodynamic model

The mechanical power that a wind turbine extracts from wind is given by [18],

$$p_m = \frac{1}{2}\rho A_r v_w^3 c_p(\lambda) = f(v_w, w_t) \tag{1}$$

Where,
ρ is the air density
A_r is the area swept by the blades
v_ω is the wind speed
C_P is the turbine power coefficient
ω_t is the turbine shaft or generator rotor speed
λ is the tip-speed ratio.
Tip-speed ratio is defined by [5],

$$\lambda = \frac{w_t}{v_w}R \tag{2}$$

Where, R is the radius of the wind turbine rotor plane.

B. Modeling of the PMSG

The dynamic equations of a three-phase PMSG can be written in a synchronously rotating dq reference frame as, [4][15]

$$\frac{d}{dt}\psi_q = v_{sq} - R_s i_{sq} - w_e \psi_d \tag{3}$$

$$\frac{d}{dt}\psi_d = v_{sd} - R_s i_{sd} - w_e \psi_q \tag{4}$$

Where,
v_{sq} and v_{sd} are the q- and d-axis stator terminal voltages, respectively,
i_{sq} and i_{sd} are the q- and d-axis stator currents, respectively,
R_s is the resistance of the stator windings,
ω_e is the electrical angular velocity of the rotor and,
ψ_q and ψ_d are the q- and d-axis flux linkages of the PMSG, respectively.
q-axis and d-axis flux linkages ψ_q and ψ_d of the PMSG are given by,

$$\psi_q = L_q i_{sq} \tag{5}$$

$$\psi_d = L_d i_{sd} + \psi_m \tag{6}$$

Where,
L_q and L_d are the q- and d-axis inductances of the PMSG, respectively and,
ψ_m is the flux linkage generated by the permanent magnets.
The electromagnetic torque T_e can be calculated by,

$$T_e = \frac{3}{4}\frac{P}{L_d}\frac{\psi_s}{L_q}\left[2|\psi_m|L_q \sin\delta + |\psi_s|\left(L_d - L_q\right)\sin 2\delta\right] \quad (7)$$

Where,

p is the number of pole pairs,

$|\psi_s|$ and $|\psi_m|$ are the magnitudes of the stator and rotor flux linkages, respectively,

δ is called the torque angle.

For a nonsalient-pole PMSG, $L_d = L_q = L_s$ (8)

Then, the electromagnetic torque can be simplified as,

$$T_e = \frac{3}{2}\frac{P}{L_s}|\psi_s||\psi_m|\sin\delta \quad (9)$$

C. Modeling of the shaft system

As the wind turbine is connected to the PMSG directly, the shaft system of the WTG can be represented by a two-mass model [16]. The motion equation is then given by,

$$2H\frac{dw_t}{d_t} = \frac{P_m}{w_t} + \frac{P_e}{w_t} - Dw_t \quad (10)$$

Where,

2H is the total inertia constant of the WTG, and D is the damping coefficient.

4. Space Vector PWM Technique

Space vector PWM refers to a special switching scheme of the six power semiconductor switches of a three phase power converter [7][8]. Space vector PWM (SVPWM) has become a popular PWM technique for three-phase voltage-source inverters in applications such as control of induction and permanent magnet synchronous motors. The drawbacks of the sinusoidal PWM and hysteresis-band current control are reduced using this technique. Instead of using a separate modulator for each of the three phases (as in the previous techniques), the complex reference voltage vectors processed as a whole. Therefore, the interaction between the three motor phases is considered. It has been shown, that SVPWM generates less harmonic distortion in both output voltage and current applied to the phases of an ac motor and provides a more efficient use of the supply voltage in comparison with sinusoidal modulation techniques. SVPWM provides a constant switching frequency and therefore the switching frequency can be adjusted easily. Although SVPWM is more complicated than sinusoidal PWM and hysteresis band current control, it may be implemented easily with modern DSP based control systems.

A. Principle of Space Vector PWM

Eight possible combinations of on and off patterns may be achieved. The on and off states of the lower switches are the inverted states of the upper ones [7]. The phase voltages corresponding to the eight combinations of switching patterns can be calculated and then converted into the stator two phase (αβ) reference frames. This transformation results in six non-zero voltage vectors and two zero vectors. The non-zero vectors form the axes of a hexagon containing six sectors (V1 − V6) as shown in Fig. 3.

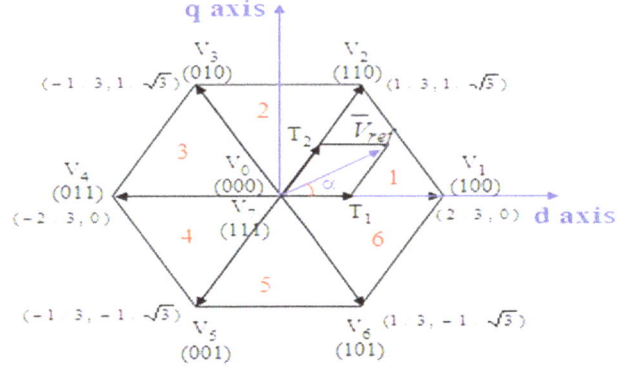

Fig. 3. *Non-zero vectors forming a hexagon and zero vectors.*

The angle between any adjacent two non-zero vectors is 60 electrical degrees. The zero vectors are at the origin and apply a zero voltage vector to the motor. The envelope of the hexagon formed by the non-zero vectors is the locus of the maximum output voltage [8].

The maximum output phase voltage and line-to-line voltage that can be achieved by applying SVPWM are:

$$V_{ph\,rms} = \frac{V_{dc}}{\sqrt{6}} \quad (11)$$

$$V_{11\,rms} = \frac{V_{dc}}{\sqrt{2}} \quad (12)$$

Therefore the dc voltage Vdc for a given motor R.M.S. voltage Vphrms is,

$$V_{dc} = \sqrt{6} * V_{ph\,rms} \quad (13)$$

Practically, only two adjacent non-zero voltage vectors Vx and Vx+60 and the zero vectors should be used. Depending on the reference voltages Vα and Vβ, the corresponding sector is firstly determined. The sector identification is carried out using the flow chart in Fig.4.

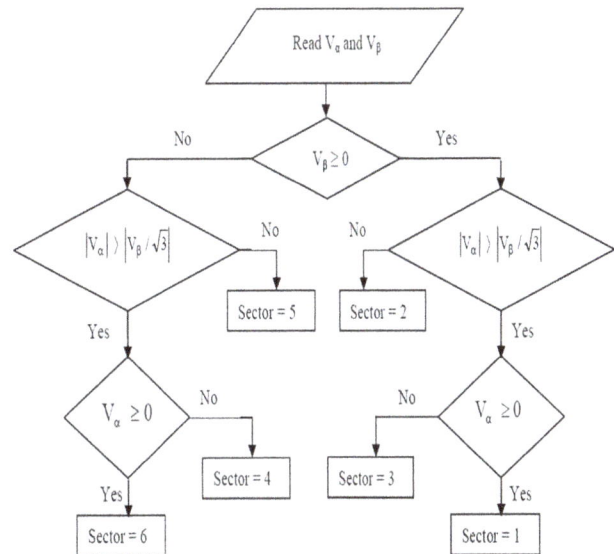

Fig. 4. *Flow chart for determining the sector.*

B. Comparison of Sine PWM and Space Vector PWM

- SVPWM generates less harmonic distortion in both output voltage and current.
- Provides a more efficient use of the supply voltage in comparison with sinusoidal modulation techniques.
- SVPWM provides a constant switching frequency and therefore it can be adjusted easily.
- SVPWM is more complicated than sinusoidal PWM and hysteresis band current control.
- It may be implemented easily with modern DSP based control systems.

Fig. 5 shows that the Voltage Utilization of SVPWM is $2/\sqrt{3}$ times more than Sine PWM

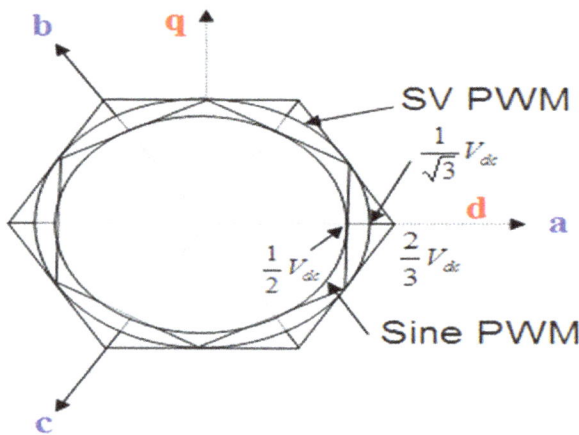

Fig. 5. *Voltage Utilization of SVPWM and SinePWM.*

C. Realization of Space Vector PWM
- Step 1. Determine V_d, V_q, V_{ref}, and angle (α).
- Step 2. Determine time duration T_1, T_2, T_0.
- Step 3. Determine the switching time of each transistor $(S_1$ to $S_6)$.

5. Space Vector Modulation-Direct Torque Control

In this new method disadvantages of the classical DTC are eliminated. Basically, the DTC-SVM strategies are the methods, which operate with constant switching Frequency. The DTC–SVM scheme combines advantages of FOC system (constant switching frequency, SVM and PI controllers) and DTC algorithm (simple structure without current control loops and coordinate transformation, direct torque and flux control) and recently has been successfully used in industrial drives.

Direct flux and torque control with space vector modulation (DTC-SVM) schemes are proposed in order to improve the classical DTC. The DTC-SVM strategies operate at a constant switching frequency. In the control structures, space vector modulation (SVM) algorithm is used. The type of DTC-SVM strategy depends on the applied flux and torque control algorithm. Basically, the controllers calculate the required stator voltage vector and then it is realized by space vector modulation technique.

A. DTC–SVM Principle
DTC–SVM can be implemented in various ways [6, 7, 8,

and 10]. The Fig.6 shows that the DTC-SVM structure. This control scheme does not have any differential algorithms, like in [6] or [7]. Instead of this in parallel structure of DTC–SVM reference voltage for space vector modulator is calculated by two linear PI regulators as two adjacent stator voltage components U_{sx} and U_{sy}. U_{sx} is calculated by stator flux PI controller, whereas U_{sy} is calculated by electromagnetic torque PI controller. Additionally, for high speed region, where stator flux magnitude has to be lowered, in parallel DTC–SVM structure field weakening algorithm can be applied easily.

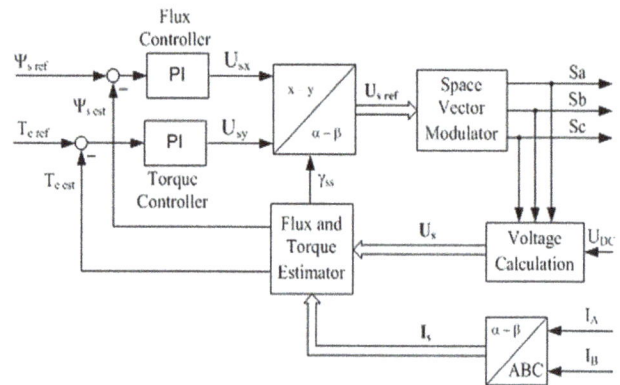

Fig. 6. *DTC–SVM structure.*

In DTC–SVM torque and flux are controlled directly. In traction torque control loop is used for speed regulation. Also field weakening algorithms are necessary for induction induction motor based traction drives. Therefore, presented algorithm is suitable for applications like: trams, trolleybuses, underground or fast city train.

B. SVPWM- DTC Modelling
The schematic of the proposed SVM-DTC for a nonsalient pole PMSG WTG is shown in Fig.7. $\psi s\alpha\beta = [\psi s\alpha, \psi s\beta]T$ and $us\alpha\beta = [us\alpha, us\beta]T$ are the estimated stator flux linkage vector and the resultant stator voltage space vector in the stationary $\alpha\beta$ reference frame, respectively and θre and ωt are the estimated electrical angular rotor position and rotor speed, respectively.

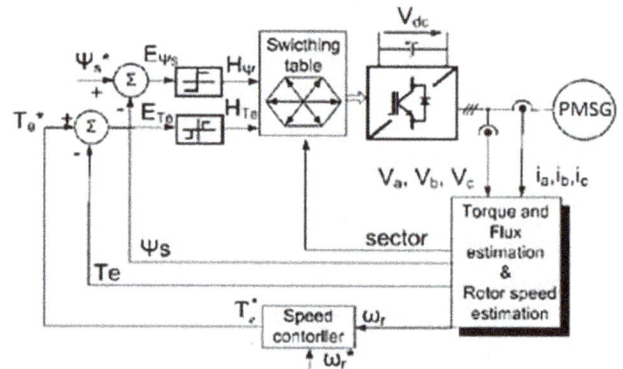

Fig. 7. *SVM-DTC for PMSG.*

SVM-DTC scheme retains the advantages of the conventional DTC, such as no coordinate transformation, no current control, etc. However, instead of adopting a switching table and hysteresis comparators, a reference voltage vector calculator (RVVC) is designed to determine the desired voltage vector $u\ \alpha\beta$. The detailed analysis is given below.

The reference torque angle of this system in a discrete time form can be written as,

$$\delta^*[k] = \sin^{-1}\left(\frac{2}{3}\frac{L_s}{P}\frac{T_e^*[k]}{\psi_s^*|k||\psi_m|}\right) \quad (14)$$

The reference stator flux angle $\theta*s[k]$ can be obtained from the following equation:

$$\theta_s^*|k| = \delta^*|k| + \theta_{re}|k| + w_e|k|T_s \quad (15)$$

Where the effect of the rotor speed is taken into consideration by adding the term $\omega e[k]Ts$ to compensate for the rotor position increment when the PMSG operates at a high speed. It is well known that the stator voltages of an electric machine can be expressed as,

$$u_{s\alpha\beta} = R_s i_{\alpha\beta} + \frac{d}{d_t}v_{s\alpha\beta} \quad (16)$$

Where,

$is\alpha\beta = [is\alpha, is\beta]T$ is the stator current vector in the stationary reference frame. Therefore, the desired stator voltage vector can be obtained in a discrete-time form with the information of the incremental flux linkage vector $d\psi s\alpha\beta[k]$. Thus,

$$u_{s\alpha\beta}|k| = \frac{\psi_{s\alpha\beta}^*|k| - \psi_{s\alpha\beta}|k|}{T_s} + R_s i_{s\alpha\beta}|k| \quad (17)$$

With knowledge of $us\alpha\beta[k]$, proper switching signals can be generated by the SVM module to achieve fast and accurate torque and flux linkage control. Since the reference torque angle for the next time interval is directly acquired, the calculated voltage vector can compensate for the torque and flux errors instantaneously. However, the maximum magnitude Vm of the voltage vector that can be provided by the three-phase bridge inverter is $Vdc/\sqrt{3}$, where Vdc is the dc-bus voltage. Once the magnitude of $us\alpha\beta[k]$ exceeds Vm, the values of $us\alpha\beta[k]$ should be modified as $us\alpha\beta[k]$ as follows,

$$u_{s\alpha}'|k| = \frac{v_m}{\sqrt{u_{s\alpha}|k|^2 + u_{s\beta}|k|^2}}u_{s\alpha}|k| \quad (18)$$

$$u_{s\beta}'|k| = \frac{v_m}{\sqrt{u_{s\alpha}|k|^2 + u_{s\beta}|k|^2}}u_{s\beta}|k| \quad (19)$$

In this way, sinusoidal output voltages can be obtained within the physical limitation of the inverter.

6. Simulations and Results

In order to investigate the effectiveness of the proposed models and control algorithms of the Closed Loop Vector Control and Space Vector PWM of a wind turbine, time-discrete dynamic simulations were implemented using Sim Power Systems of MATLAB/Simulink environment. In

order to have a fully functional and stable wind turbine system, both closed loop vector control and space vector PWM are required to operate as they are designed. On top of that, the results of the simulations are analyzed to ensure the output signals are implementable in real condition application. In order to operate a wind turbine system efficiently, grid-side controller is necessary because, the DC-link capacitor voltage varies with the wind. The SVPWM is designed to maintain the DC-link capacitor voltage at a constant magnitude by adjusting the power flow to the electrical grid.

Table 1 shows the range of parameters used to generate the power from wind. From this ranges we made the theoretical calculation of the power output of the generator.

Table 1. PMSG parameters.

PMSG	Ranges
No of poles	6
Stator phase resistance Rs	0.425
D-axis Inductances Ld(H)	0.0082
Q-axis Inductances Lq(H)	0.0082
Flux linkage established by magnets	0.433
Inertia	0.01197
Friction factor	0.001189
Rated speed	153 rad/sec
Rated current	12 amps
Rated power	6 kw

Therefore the power output from the generator using the table 1 is,

$$P_{avail} = \frac{1}{2} * 0.4 * 2.1519 * 12^3 * 1.24$$

$$P_{avail} = 922.18\ w$$

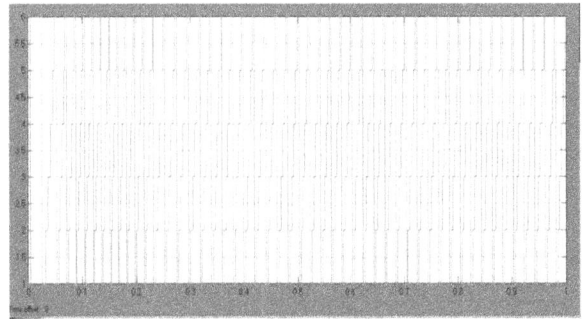

Fig. 8. *Gate signals of SVPWM.*

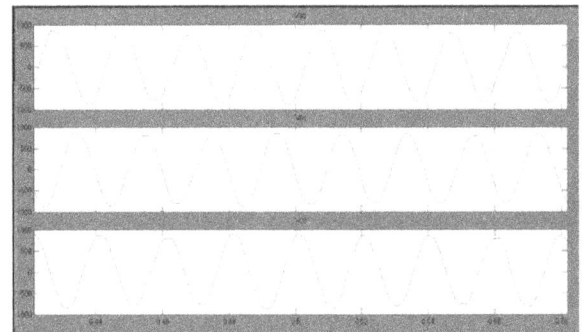

Fig. 9. *Phase voltage value of inverter output (V_a, V_b, V_c).*

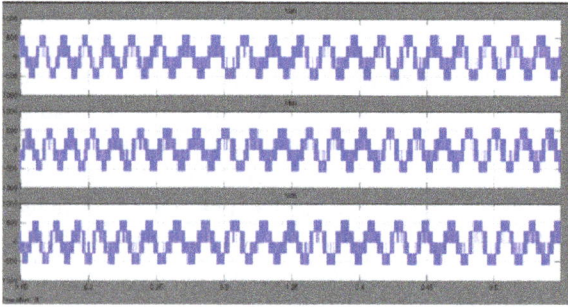

Fig. 10. *Line voltages value of inverter output (V_{an}, V_{bn}, V_{cn}).*

Harmonic reduction

PMSG, it is connected to the grid through power electronic converter fully, and even PMSG can omit the difficult gearbox system. Harmonic reduction method can be applied to any of the system to improve power quality [17]. The harmonic component for a typical six pulse rectifier can be calculated from,

$$Harmonic components = 6k \pm 1 \qquad (20)$$

Where k=1, 2, 3…

The harmonic components magnitude will decrease to a very low level when it reaches the 11th harmonic or higher. The 5th and 7th harmonic is the most important harmonics to reduce to a minimum because of their high magnitude. One way to reduce these harmonics is to use an inductor or a low pass filter [17]. Here THD stands for total harmonic distortion and it defines the harmonic distortion in terms of the fundamental current drawn by the load:

$$THD\% = \frac{\sqrt{\sum_{h=2}^{h=\infty} (M_h)^2}}{M_{fundamental}} * 100\% \qquad (21)$$

Where, M_h is the magnitude of either the voltage or current harmonic component and $M_{fundamental}$ is the magnitude of either the fundamental voltage or current.

Figure 11 shows the Harmonic Level or THD Level of the inverter output voltage.

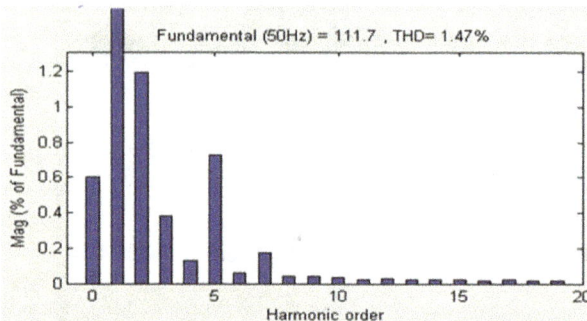

Fig. 11. *Harmonic Level or THD Level of the inverter output voltage.*

7. Conclusion

The paper is developed on the Closed Loop Vector Control and Space Vector PWM of a wind turbine. Direct Drive

Permanent Magnet Synchronous Generator is selected due to its high efficiency, high torque-to-size ratio, and low maintenance requirement. Thus the results show that the wind energy conversion using PMSG with SVM-DTC system can achieve excellent performance. And also it has a relatively good quality of output waveform with a low total harmonic distortion values.

References

[1]   Giuseppe S. Buja, and Marian P. Kazmierkowski, "Direct Torque Control of PWM Inverter-Fed AC Motors—A Survey", *IEEE Trans.on Industrial Electronics, Vol. 51, No. 4, August 2004.*

[2]   Md. Enamul Haque, *Member, IEEE*, Michael Negnevitsky, *Senior Member, IEEE*, A Novel Control Strategy for a Variable-Speed Wind Turbine With a Permanent-Magnet Synchronous Generator IEEE Trans. on Industry Applications, Vol. 46, No. 1, January/February 2010.

[3]   Wei Qiao, *Member, IEEE*, Liyan Qu, *Member, IEEE*, "Control of IPM Synchronous Generator for Maximum Wind Power Generation Considering Magnetic Saturation",IEEE Trans. on Industry Applications, Vol. 45, No. 3, May/June 2009.

[4]   S.VIJAYALAKSHMI, "Modeling and control of a Wind Turbine using Permanent Magnet Synchronous Generator", International Journal of Engineering Science and Technology (IJEST).

[5]   Satyam Kumar Upadhyay, Sudhanshu Tripathi, "Optimization and Control of a Variable Speed Wind Turbine with a Permanent Magnet Synchronous Generator", *International Journal of Engineering Trends and Technology (IJETT) - Volume4 Issue6- June 2013.*

[6]   Domenico Casadei, *Member, IEEE*, Francesco Profumo, *Senior Member, IEEE*, "FOC and DTC: Two Viable Schemes for Induction Motors Torque Contro", lIEEE Trans. on Power Electronics, Vol. 17, No. 5, September 2002.

[7]   Anjana Manuel, Jebin Francis, "Simulation of Direct Torque Controlled Induction Motor Drive by using Space Vector Pulse Width Modulation for Torque Ripple Reduction", International l Journal of Advanced Research in Electrical, Electronics and Instrumentation EngineeringVol. 2, Issue 9, September 2013.

[8]   K.Chikh, A.Saad, M.Khafallah and D.Yousfi," A Novel Drive Implementation for PMSM By using Direct Torque Control with Space Vector Modulation", Canadian Journal on Electrical and Electronics Engineering Vol. 2, No. 8, August 2011.

[9]   Ackerman T, editor, "Wind Power in Power Systems", West Sussex, England: 2005. Generators and Power Electronics for Wind Turbines; p 53-72.

[10]   Zhe Zhang, *Student Member, IEEE*, Yue Zhao, *Student Member, IEEE*, "A Space-Vector-Modulated Sensor less Direct-Torque Control for Direct-Drive PMSG Wind Turbines", IEEE Trans. on Industry Applications, Vol. 50, No. 4, July/August 2014.

[11]   Dr. M. Vijaya kumar," Modeling and Control of a Permanent Magnet Synchronous Generator Wind Turbine with Energy Storage ", International Journal of Engineering Research and Applications (IJERA) ISSN: 2248-9622 October 2012, pp.1900-1905.

[12] S. Engström, "NewGen a new direct drive wind turbine generator," *Nordic Wind Power Conference (NWPC'2006)*, Hanasaari, Espoo, Finland, 22-23 May 2006.

[13] T. Hartkopf, M. Hofmann and S. Jöckel, "Direct-drive generators for megawatt wind turbines", *Proc. European Wind Energy Conf.*, Dublin, Ireland, 1997, pp. 668-671

[14] Yicheng Chen, P. Pillay, and A. Khan, "PM wind generator topologies," *IEEE Trans. on Ind.Appl.*, vol. 41, no. 6, pp. 1619-1626, 2005.

[15] Johnson G.L. 2006, "Wind Turbine Power, Energy, and Torque," *In Wind Energy System*, Electrical Edition ed. Prentice-Hall Englewood Cliffs (NJ), pp. 4-1-4-54.

[16] Quang N.P. & Dittrich J.A. 2008, "Inverter control with space vector modulation," *In Vector Control of Three-Phase AC Machines*, First ed. Springer, pp. 17-59.

[17] Espinoza J.R. 2001, "Inverters," *In Power Electronics Handbook*, Rashid M.H., ed., Oxford University Press, pp. 353-404.

[18] Li S.H., Haskew T.A., Swatloski R.P., & Gathings W. 2012," Optimal and Direct-Current Vector Control of Direct-Driven PMSG Wind Turbines", *IEEE Transactions on power electronics*, 27, (5) 2325-2337.

5

Audit of Electricity Generation in University of Lagos, Nigeria

Oluwatosin M. Dada[1, *], **Ilesanmi A. Daniyan**[1], **Temitayo M. Azeez**[1], **Olalekan O. Adaramola**[2]

[1]Department of Mechanical and Mechatronics Engineering, Afe Babalola University, Ado Ekiti, Nigeria
[2]Department of Computer Engineering, Afe Babalola University, Ado Ekiti, Nigeria

Email address:
tossydada@yahoo.co.uk (O. M. Dada)
*Corresponding author

Abstract: Electricity is indispensable to a nation's economic development. In this paper, analysis of energy sources was carried out to ascertain their costs. The selected energy alternatives are PHCN 132 kV/33 kV/11 kV Substation and Captive Cummings Diesel generator. The case study-site was University of Lagos, Akoka located within Lagos, a commercial city in Nigeria. Practical surveys and data collection were carried out from the power house coupled with their load estimation to obtain the total cost of electricity consumed. The results indicate that, on a monthly basis, the university spends about $820,000 on diesel generator and $228,500 on PHCN.The results also show that other alternative energy sources should be included in the overall energy mix in order to minimise the total power cost of the university.

Keywords: Energy Consumption (kWh), Peak Load (kW), Base Load (kW), Cost Per kWh

1. Introduction

Electricity is indispensable to a nation's economic development as it drives industrialization, technology advancement and standard of living of the people. Over the past two decades, the stalled expansion of Nigeria's grid capacity, combined with the high cost of diesel and petrol, has crippled the growth of the country's productive and commercial industries. Nigeria's per capita electricity consumption is amongst the lowest in the world and far lower than many other African countries. Nigeria's per capita electricity consumption is just 7% of Brazil's and just 3% of South Africa's. Brazil has 100,000 MW of grid-based generating capacity for a population of 201million people. South Africa has 40,000 MW of grid-based generating capacity for a population of 50 million people. Nigeria, with a population of over 150 million people has a peak generation of only 3804 MW (The Presidency, 2010). Nigeria's electricity crisis significantly undermined the effort to achieve the sustained economic growth, competitiveness in regional and global markets, employment generation and poverty alleviation. The prolonged dismal electricity industry performance has been the most intractable infrastructural problem and policy challenge in the last half a century (Iwayemi, 2008). Electricity generation in Nigeria is characterized by excess capacity and inadequate supply. It has been observed that peak power generation is often about one-third of installed capacity due to the non-availability of spare parts and poor maintenance. A poorly-motivated workforce, vandalisation and theft of cables and other vital equipment, accidental destruction of distribution lines, illegal connections and resultant over-loading of distribution lines, are additional major problems of the electricity sector. These have been responsible for unannounced load shedding, prolonged and intermittent outages which most consumers of electricity in Nigeria have had to contend with over the years. The efficiency of the supply of electricity will not only influence returns on investment in existing enterprises, it also plays a major role in the creation of an economic environment which influences decisions on potential investments. The electric power sub-sector in Nigeria, dominated by the Power Holding Company of Nigeria (PHCN), has been unable to provide and maintain acceptable minimum standards of service reliability, accessibility and availability (Adesiji, 2009). However, the University of Lagos located in Akoka, Lagos also experiences shortage in electricity supply, of which the diesel generator is used to complement it. The main campus has various options for electricity generation; PHCN 132 kV/33kV/11 kV Substation, Captive Cummings Diesel generator, Solar cells and battery-powered inverter used in some faculties and departments (Unilag, Works and Physical Planning Department, 2012)). The first two are highly used to power most of load requirement on campus. The generator is used to power the electrical load and is sometimes complemented with PHCN.

2. Overview of Electricity Generation in Nigeria

Electricity plays a very important role in the socio-economic and technological development of every nation. The electricity demand in Nigeria far outstrips the supply and the supply is epileptic in nature. The country is faced with acute electricity problems, which is hindering its development. The electricity development in Nigeria can be traced back in 1898 when the first generating power plant was installed in Marina Lagos with a total capacity of 60 kW. The Electricity Corporation of Nigeria (ECN) was established by the Colonial Government under the ordinance no. 15 of 1950 to take over the various electricity supplies in the country (Awosope, 2014). In 1962, Act of Parliament established the Niger Dams Authority (NDA) for the development of Hydro Electric Power. However, a merger of the ECN and NDA was made in 1972 to form the National Electric power Authority (NEPA), which as a result of unbundling and the power reform process, was renamed Power holding Company of Nigeria (PHCN) in 2005 (Titus et al., 2012). The Electricity Power Sector Reform (EPSR) Act 2005 translated NEPA into the newly incorporated Power Holding Company of Nigeria (PHCN) comprising of 18 separate successor companies that took over the assets, liabilities and employees of NEPA, and responsible for the generation (6 companies), transmission (a company) and distribution (11 companies) (Okafor, 2005). The EPSR Act set the pace for power privatization in Nigeria. As part of the reform process, a regulatory agency, Nigerian Electricity Regulatory Commission (NERC) was inaugurated in November, 2005. The functions of the NERC were to regulate tariffs, provide quality service, and oversee the electricity industry effectively. This led to the new tariff regime that took effect through a Multi–Year Tariff Order (MYTO) in 2008. The National Electricity Power Policy (NEPP) and the Electric Power Sector Reform Act 2005 (EPSR) provide for the development of Nigeria electricity market (Makwe et al., 2012). Wholesale competition was recommended for Nigeria to assist in monopoly control and cost insensitivity. Thus, the reforms broke NEPA's monopoly and paved the way for the entry of independent

power producers (IPPs). Before the reform, tariffs in the Nigerian electricity industry were depressed by government order. The old NEPA was barred by decree from increasing tariffs, even when the cost of supply of electricity had increased. The result was underproduction of electricity and the absence of investment in the network. The EPSR Act 2005 isolated the NERC from the direct control of the government bureaucracy. Nigeria now has a cost reflective tariff, with voluminous traffic in foreign and local investment in the electricity market (Ajumogobia, 2015).

3. Electricity Supply Infrastructure in University of Lagos

The University of Lagos located in Akoka, Lagos experiences shortage in electricity supply, of which the diesel generator is used to complement it. The main campus has various options for electricity generation; PHCN 132 kV/33 kV/11 kV Substation, Captive Cummings Diesel generator, Solar cells and battery-powered inverter used in some faculties and departments (Unilag, Works and Physical Planning Department, 2012). The first two are highly used to power most of load requirement on campus. The generator is used to power the electrical load and is sometimes complemented with PHCN. The Power Holding Company of Nigeria (PHCN) supplies the University of Lagos a 11 kV voltage level distribution line from its 132/33/11 kV station located close to the University's control room panels. Power distribution on campus is effected by means of Ring main system arranged in four (4) rings comprising of eight (8) feeders. Each ring is on a standard cable of 150 mm 2x3 core High Tension Armoured cable which feeds the various sections of the University Community. The total High tension cable network is about 24 kilometers. There are 48 transformer substations with RMU (Ring Mains Units) from where the Low tension distributions are effected to some laboratories, faculties, residential buildings and commercial centers. Peak power requirement of the University is about 9.0 MW (Unilag, Works and Physical Planning Department, 2012). According to the Director of Works and Physical Planning (University of Lagos), Dr. A. E. Adeniran, PHCN charges the University $3.6 U.S Cents Per kWh of electricity used. The Nigerian Electricity Regulatory Commission (NERC) increased the electricity tariff in the country from $4.3 U.S Cents to $5.0 U.S Cents per kilowatt hour. This was as a result of its implementation of the schedule of the 2008/2013 regime of the Multi- Year Tariff Order (MYTO) (Uwe, 2011). Apart from the cost of power generation by PHCN, the cost of maintaining the generating station is quite expensive.

In March 2010, the University of Lagos and CUMMINS West Africa signed a contract agreement aimed at boosting electricity supply on the University Campuses through the supply and installation of 2x2000 kVA, 11000 Voltage Generators at the Power House, Main Campus, Akoka (Unilag, 2010). The two captive Cummings diesel generators with the same specification use diesel oil as fuel input. The diesel generator requires regular and preventive maintenance especially for every 250 hours in order to prevent sudden breakdown. The University spends huge amount on diesel monthly to prevent interruption in electricity supply. Strategic plan is currently developed by the University to increase its electrical supply facilities in order to ensure a steady, efficient, uninterrupted supply (Abolarin *et al.*, 2011).

4. Results and Discussion

To analyze the cost of generating electricity, data containing the daily consumption of electricity (kWh) of all the loads on campus was collated at the power plant for six months (1st September 2010 to 28th February 2011) in order to obtain the base load and the peak load and is shown in Figure (1).

The quantity of diesel in liters equivalent to the energy consumed by the loads (base load, peak load) is given as:

$$\text{Base load} = \frac{\text{Minimum Daily Energy Consumption(kWh)}}{\text{Duration (24)}} \quad (1)$$

$$\text{Peak load} = \frac{\text{Maximum Daily Energy Consumption (kWh)}}{\text{Duration (24)}} \quad (2)$$

$$\text{The area of the base of the diesel tank} = 121.8 \text{ cm x } 122.5 \text{ cm} = 14920.5 \text{ cm}^2 \quad (3)$$

Volume of the diesel oil used= Area of the base of the tank X Height of the tank
= 14920.5 X 19.1
= 284981.55 cm^3
= 284.981 liters
From Table (2), the energy supplied by the generator is 1201 kWh

$$\text{Thus 1201 kWh} = 284.981 \text{ liters} \quad (4)$$

1kWh = 284.981/1201= 0.237 liters
The mean of the liters per kWh used by the diesel generator at different time intervals
= (0.237 + 0.207+ 0.226 + 0.254 + 0.276)/ 5
= 0.240 liters per kWh
Thus, a diesel generator on campus produces

$$1/0.240 = 4.167 \text{ kWh per liter} \quad (5)$$

The lowest monthly base load for the University of Lagos, Akoka campus from September 2010 to February 2011 is 925.83 kW. Also, the highest monthly peak load for the same period is 8804.58 kW in January 2011 as shown in Table (1). Thus in order to calculate the quantity of diesel used per kWh, data was collated during the operation of the generator from the power house at the University of Lagos, Akoka. The results are shown in Table 2 & 3.

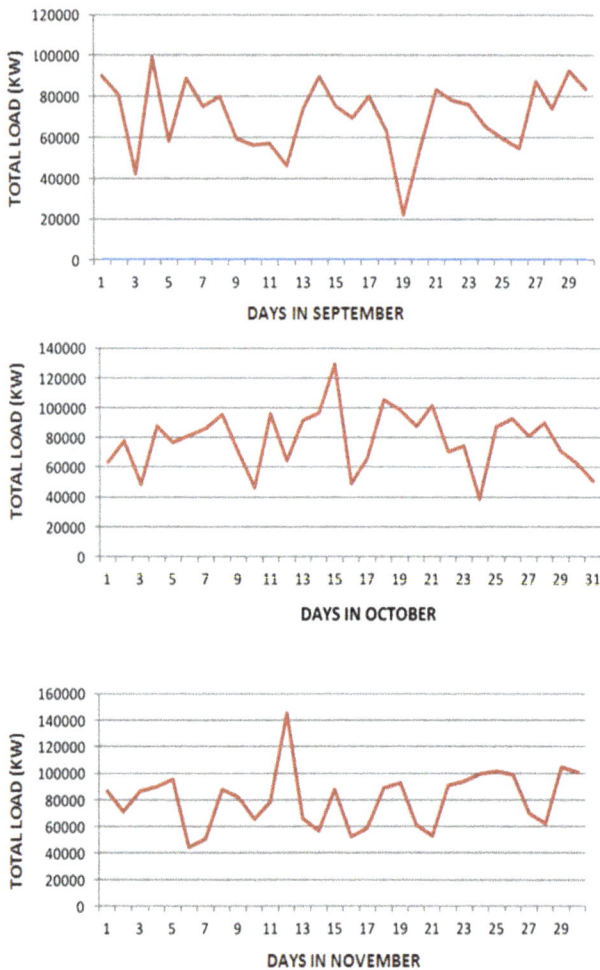

Figure 1. Load Duration Curve from 1ST September 2010 – 30TH November 2010.

Table 1. Minimum and Maximum daily energy consumption at the University of Lagos, Akoka Campus.

Month/year	Min. Daily Energy Consumption (kWh)	Quantity of diesel (liters)	Base load on campus (kW)	Max. Daily Energy Consumption (kWh)	Quantity of diesel (liters)	Peak load on campus (kW)
Sept.2010	22,220	5332.8	925.8	98,590	23661.6	4107.92
Oct.2010	38,660	9278.4	1610.83	128,720	30892.8	5363.33
Nov.2010	44,310	10634.4	1846.25	145,320	34876.8	6055
Dec.2010	30,240	7257.6	1260	96,330	23119.2	4013.75
Jan.2011	33,720	8092.8	1405	211,310	50714.4	8804.58
Feb.2011	29,690	7125.6	1237.08	155,070	37216.8	6461.25

Table 2. Data collected during operation of a diesel generator at the University of Lagos, Akoka Campus.

Date	kWh Reading (time)	kWh Reading (time)	kWh supplied by the generator	Difference in diesel level (cm)	Volume of diesel oil used (cm³)	Volume used in liters (cm³/1000)	Liters per kWh
31/5/2011	453549 (4pm)	454750 (5pm)	1201	19.1	284981.55	284.981	0.237
01/6/2011	528190 (1.28pm)	529105 (2:28pm)	915	12.7	189490.35	189.490	0.207
01/6/2011	529105 (2.28pm)	529911 (3.28pm)	806	12.2	182030.1	182.030	0.226
01/6/2011	530754 (3.28pm)	531507 (4.28pm)	753	12.8	190982.4	190.982	0.254
01/6/2011	531507 (4.28pm)	532188 (5.28pm)	681	12.6	187998.3	187.9998	0.276

Table 3. Estimate of the amount spent on diesel generator from September 2010 – February 2011.

Month/Year	Maximum Daily Energy Consumption (kWh)	Quantity of diesel in liters	Amount (USD) Per day	Amount (USD) Per Month
September 2010	98590	2366.6	19,218	576,540
October 2010	128720	30892.8	25,090	777,790
November 2010	145320	34876.8	28,326	849,780
December 2010	96330	23199.2	18,777	582,087
January 2011	211310	50714.4	41,189	1,235,670
February 2011	155070	37216.8	30,226	906,805

Table 4. The price of gas to power plants in cents and the corresponding cost per kWh of electricity.

Price of gas (cents)	64	70	80	100	120	140	160	180	200
Electricity cost/kWh (Naira ₦)	7.22	10.83	14.44	18.05	21.66	25.27	28.88	32.49	36.10

Table 5. Cost of electricity for a period (September 2010 – February 2011).

Month/ Year	Maximum Daily Energy Consumption (KWh)	Amount (USD) Per day	Amount (USD) Per Month
September 2010	98590	5419	162,570
October 2010	128720	7082	219,542
November 2010	145320	7988	239,640
December 2010	96330	5296	164,176
January 2011	211310	11627	360,437
February 2011	155070	8532	238,896

The cost per kWh of electricity provided by the Power Holding Company of Nigeria (PHCN) to the University of Lagos Akoka campus is $3.6 U.S Cents in 2011. When the price of gas goes up from 64 cents to $1.60, the expected cost per kWh of electricity provided by the PHCN will be $40 U.S Cents. If the cost of electricity from PHCN is $54 U.S Cents, the amount to be paid by the University to PHCN is analyzed using the data of the total energy consumed collated from the power house in Table 5. The amount of energy consumed in September increases from 98590 kWh to 145320 kWh in November 2010 and fell drastically to 96330 kWh in December 2010. The amount the University spends on electricity increases from $162,570 in September 2010 to $239,640 in November 2010 but fell towards the end of the year to $164,176 as a result of decline in energy consumption due to vacation of students and end of the year break. It was observed that the university spend huge amount on diesel generators in Table (3) there is need for alternative option for energy generation in the institution.

5. Conclusion

Electricity is the hub of both economic and technological advancement of any nation. The electricity industry in the developing countries has gone through quite a lot of metamorphosis in the recent past. Electricity generation in Nigeria is characterized by slow growth in generation capacity, market deregulation process interference by government, electrical transmission lines and distribution equipment vandalism, poor maintenance of existing electrical facilities and corruption. Nigeria should not be different in the vogue of global electricity market which focuses on building a cleaner, more diverse and more sustainable energy mix. The electricity market investment system should be quality, affordable and of proven security. However, University of Lagos is currently developing strategic plans to increase its physical infrastructure and utilities to ensure a steady, efficient, uninterrupted supply by adopting alternatives means for energy generation. Several ways to achieve this include; energy efficiency and conservation in energy audits, setting up committees on energy culture and energy budgeting, incorporation of renewable energy mixes into the present generating capacity. Developing cleaner energy should be part of the investment strategy with the focus on adopting cleaner fossil fuels based on renewable sources to meet the electricity demand in the institution.

References

[1] Abolarin, M. S., Gbadegesin, A. O., Shitta, B. M and Adegbenro, O. (2011). Energy Lighting Audit of four University of Lagos Hall of Residence. *Journal of Engineering Research, Vol. 16, no 2*, pp. 1-10.

[2] Adesiji, R. (2009). The Cost of Electricity in Nigeria. *The Energy Journal*, pp 15.

[3] Ajumogobia & Okeke (2015). *Nigerian Energy Sector: Legal and Regulatory Overview*, pp. 13-31.

[4] Awosope, C. A. (2014). *Nigeria Electricity Industry: Issues, Challenges and Solution. Public Lecture Series*, Vol. 3. No 2, pp. 5-6.

[5] Iwayemi, A. (2008). Investment in Electricity Generation and Transmission in Nigeria: Issues and Option. *The Energy Journal*, pp 37.

[6] Makwe, J. N., Akinwale, Y. O., and Atoyebi, M. K. (2012). *An economic assessment of the reform of Nigerian electricity market.* Journal of Energy and Power, Vol. 2. No 3, pp. 24-32.

[7] Okafor F. N. (2005). *Modelling the Ancillary Services in Deregulated Power Networks of Developing Economics*, 6th International Conference on Power System Operation and Planning (ICPSOP), pp. 222-227.

[8] The Presidency, (2010). Roadmap for Power Sector Reform. *A Customer-Driven Sector Wide Plan to achieve Stable Power Supply, pp. 16.*

[9] Titus, O. K, Abdul-Ganiyu A. J. and Phillips D. A. (2012). *The Current and Future Challenges of Electricity Market in Nigeria in the face of deregulation process*, EIE's 2nd Intl' Conf. Comp., Energy, Net., Robotics and Telecom, pp. 30-36.

[10] Unilag. (2010, April 9). *Unilag, Cummins sign contract agreement to boast power supply.* Retrieved July 31, 2012, from University of Lagos: www.unilag.edu.ng/newsdetailsphp?NewsID=407

[11] Unilag. (2012). *Works and Physical Planning Department.* Retrieved July 30, 2012, from University of Lagos: http//www.unilagworks.org/powersupply_n_Distribution.php

[12] Uwe, J. (2011). *High Electricity Tarriff: How PHCN Staff short change Nigerians through illegal electricty billing.* Retrieved July 31, 2012, from National Mirror: nationalmirroronline.net/index.php/insight/18689.html.

6

Power Optimization and Prioritization in an Island Supplied by a Rotating Machine Based Distributed Generator Using Artificial Bee Colony Algorithm

L. Mogaka[1], D. K. Murage[2], M. J. Saulo[1]

[1]Electrical and Electronics Department, Technical University of Mombasa, Mombasa, Kenya
[2]Electrical and Electronics Department, Jomo Kenyatta University of Agriculture and Technology, Nairobi, Kenya

Email address:
mogakaLucas@tum.ac.ke (L. Mogaka), dkmurage25@yahoo.com (D. K. Murage), michaelsaulo@tum.ac.ke (M. J. Saulo)

Abstract: Currently the greatest threat to the power systems reliability and security is the cascading of electric system failures thus causing power blackouts. For quite some time now, the world has been encountering many power blackouts as a result of these cascading failures. The cascading power failure instances pose great risks towards the integrity of power system network. This may finally lead to the splitting of the power system into various small unintentional islands. Hence, intentional or controlled islanding is then utilized as a preventive measure to mitigate the losses caused by unintentional islanding of the power system. Thus, by doing this, the entire power system is split into controlled island regions for the purposes of easy handling and control. In such situation, each islanded region should have sufficient generation to supply its connected loads in order to remain operative and stable. It should also be pointed out that intentional islanding is very important as it can prevent the entire power system from collapsing. The distributed generators supplying the loads in these islands may not be able to maintain the voltage and frequency within desired limits in the distribution system when it is islanded within the micro grid. There may be a power deficit within the island. This eventually leads to shedding of some loads within the island for the sake of stability of the system. Hence the main challenge here is to determine the appropriate and reliable method to optimize the power supply and the load demand in the island and thus maintain the voltage and frequency within the desired limit. In this study we focused on the determination of the minimum load amount for shedding within the islanded region and the prioritization of the buses for shedding so that electricity supply to customers could be maximized using ABC algorithm. From the results obtained, the ABC algorithm can be successfully applied for solving the optimization and prioritization problems within the island being supplied by a DG. The ABC algorithm has several merits over other algorithms which makes it suitable in this application. These advantages include; it is easily implemented, flexible, has few control parameters, easily hybridized with other optimization algorithms and can be modified very easily to suit any application. This system was simulated in MATLAB and SIMULINK using IEEE fourteen bus systems.

Keywords: ABC Algorithm, Islanding, Power Prioritization and Optimization

1. Introduction

An electric rotating machine can be defined as any form of apparatus which has a rotating member and generates, converts, transforms, or modifies electric power. Examples of these machines include motors and synchronous generators (SDGs). There are many types of rotating machines. The two basic and common rotating machine types are synchronous and induction generators [1].

Usually, the classical view of the power systems is characterized by a unidirectional flow of the power from one central generating point to the consumers through transmission and distribution systems. In this system, there is usually minimal amount of intelligent and automation functions involved. However, the ongoing deregulation of the power system that has given a new face to the power system by introducing the distributed generations (DGs) into distribution systems, leading to the bi-directional electric power flow.

There is also a current trend in the continued and increasing use of Distributed Generation (DG) in the distribution systems of the power grid to supplement the mains supply due to energy exhaustion and recent environmental issues [2]. This practice enables the collection of electrical energy from a variety of sources thus leading to the decreased environmental impacts and improved security of supply. These distributed generators are typically in the range of 1-10,000 kW and include wind farms, micro hydro turbines, photovoltaic (PV) system and other small generators which are supplied with biomass or geothermal fuel [3].

This continued and increased integration of DGs in the power system is due to its many advantages. These include: improved system reliability in the power supply, increased efficiency, avoidance of transmission capacity upgrades, improved power quality and reduced transmission line losses and environmental benefits (excluding diesel reciprocating engines often used as back-up distributed generators which tend to be the worst performers in terms of greenhouse gas emissions [4]) [1].

On the other hand, the incorporation of these DGs in the distribution system has one major drawback; unintentional islanding. An islanding condition occurs when the distributed generator continues to power a section of the grid system even after the connection to the rest of the system has been lost, either intentionally or unintentionally.

The unintentional islanding mode of operation is not desirable because of a number of reasons as it is stipulated by the IEEE standard 1547-2003 [5]. These include and not limited to the following; it poses a threat to the line workers' safety, the islanded system may not be properly grounded resulting in high voltage in the other phases when an earth fault occurs, and most importantly, the distributed generators may not be able to maintain the voltage and frequency within desired limits in the distribution system when it is islanded [2]. That is why it becomes necessary to determine the minimum load amount for shedding within the islanded region and the prioritization of the buses for shedding so that electricity supply to customers could be maximized using ABC algorithm.

The rest of the paper is organized as follows; section 2 discusses the concept of the artificial bee colony algorithm and its relevance in this study, then the methodology used to achieve the study objectives is highlighted in 3. The study results and discussions are elaborated in section 4 and the last section, 5, gives the conclusions that are drawn from the results of the study.

2. The Artificial Bee Colony Algorithm

2.1. The Nature of Bees

The Swarm Intelligence (SI) is a branch of Artificial Intelligence (AI) that has its basis on the collective characteristics of animals or certain unique phenomenon of natural setups such as bees, fish, ants and birds. In the process of searching for the food sources, the bee colony can move in several directions and over a distance of several kilometers.

This exercise of searching for new food sources commences by sending out a group of scout bees to search for flower patches at various bushes that contain a considerable amount of nectar and pollen. After this, the scout bees come back to their hive and then perform a special movement as others observe.

This dance is known as the waggle dance. This is shown in figure 1. This waggle dance is used by the employed bees to communicate to other bees in the hive to report three main types of information. This is with regards to the availability of flower patches, which are the direction of food sources location, their quality, quantity and distances from these food sources [6].

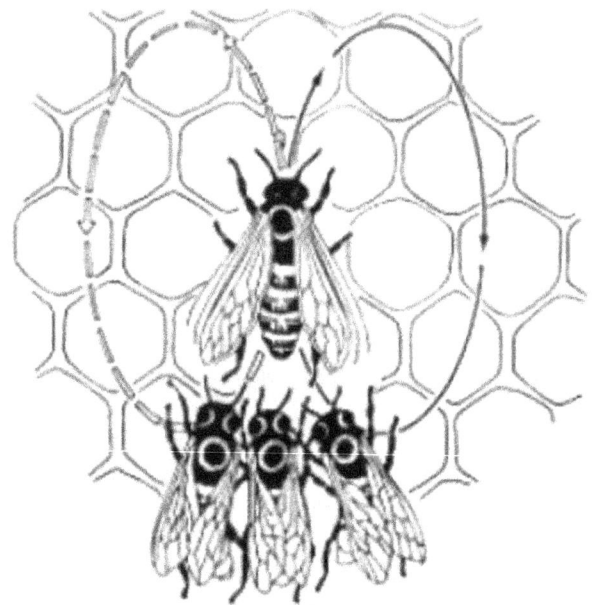

Figure 1. *The bees waggle dance.*

This information conveyed helps the other bees in the hive to travel towards the discovered flower patches more easily and precisely without the assistance from other bees. After the waggle dance, scout bees will fly back to the flower patches again with follower bees or worker bees [7].

The artificial bee colony algorithm consists of three important components in its operation. That includes the employed bees, unemployed foraging bees, and food sources.

- Employed bees: An employed artificial bee is often employed at one certain food source at a time which she exploits. She carries all important information about this particular food source and shares it with the rest of the bees waiting in the hive. Among other information she shares include the distance of the food source from the hive, its direction and how profitable it is.
- Unemployed bees: The group of forager bees that are looking for food sources to exploit are called unemployed bees. They can be either scout bees that search around the environment randomly or onlooker bees who try to find food sources by using the information given by the employed bees. The mean number of scouts is about percent.

• Food Sources: An artificial bee analyses a number of factors concerning a given food source before selecting that food source. These factors include the closeness of the food source to the hive, richness and quality of the energy, taste of its nectar, and the ease or difficulty of extracting this food from the source.

In short, the artificial foraging bees consist of a group of employed bees, onlookers and scout bees. Half of this colony comprise of the employed bees which forms the majority.

Every food source has an employed bee associated with it. Once a food source is depleted, the employed bee automatically becomes a scout. Thus the amount of nectar in a batch of flowers determines the fitness value of that solution, in this case the food position.

The basic mechanism search of ABC is well presented in figure 2 [8] where a) Initial situation, b) Final situation.

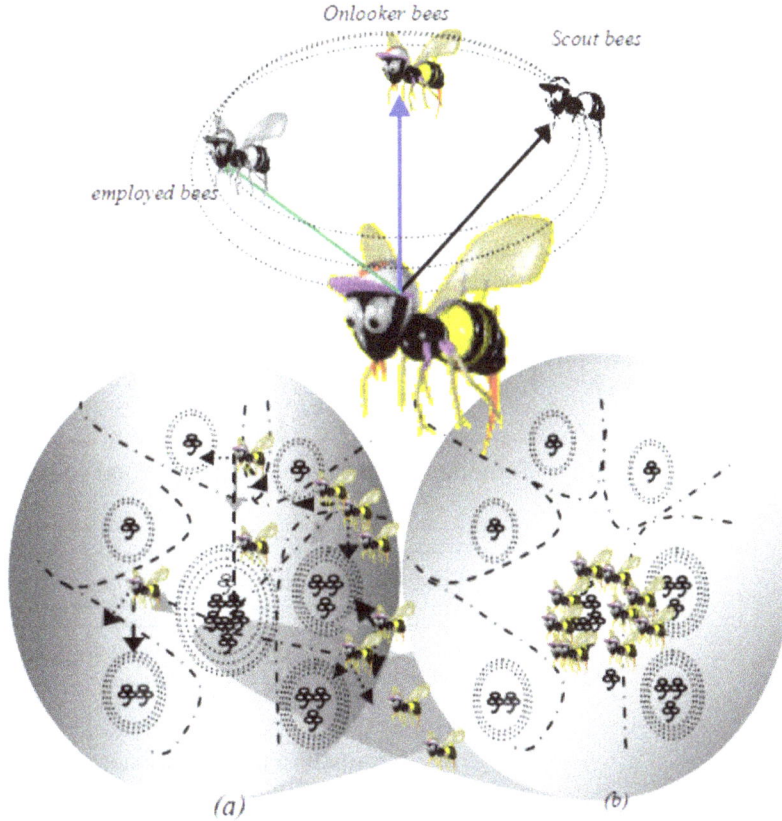

Figure 2. Basic mechanism search of ABC.

In the initialization stage of the ABC algorithm, it creates a randomly distributed initial population of solutions ($f = 1, 2...E_b$), where f signifies the size of population and E_b is the number of employed bees [9]. Each solution of the expression x_f is a D dimensional size vector, where D is the optimization parameters number. Throughout the optimization process, the artificial bees will memorize the new food position, that is, the modified solution, if the quantity of the new nectar position is higher than the previous nectar position.

Upon completion of each of the search process, the bees then share the nectar information they have found with onlooker bees in the beehive dance area. The onlooker bee will carefully observe these waggle dances and evaluate the information being conveyed and choose the food source with highest nectar quantity. The onlooker bees evaluate the nectar information and choose a food source depending on the probability value associated with that food source using the equation 1 below [10];

$$P_i = \frac{fit_i}{\sum_{j=1}^{n_e} fit_i} \qquad (1)$$

Where fit_i is the solution's fitness value i, which in turn is proportional to the amount of nectar of the source of food in the position i and n_e is the number of food sources which is equal to the number of employed bees in the colony [11].

On their turn, the onlooker bees also employ the same process of modification and selection of the food positions as the employed bees do. This can be demonstrated by the equation 2 below.

$$V_{ij} = X_{ij} + Q_{ij}\left(X_{ij} - X_{kj}\right) \qquad (2)$$

Where $k \in (1,2...n_e)$ and $k \in (1,2...D)$ are selected randomly. Although k is determined stochastically, it should not be equal to the value of i. Q_{ij} is a random number which should be between -1 and, +1. This controls the generation of the neighborhood food sources.

Once the new food position is determined as shown above, another cycle of the ABC algorithm begins. The same procedures are continuously repeated until the stopping criterion is met [9]. In nutshell, the ABC algorithm is a cycle which involve the following steps which are repeated until the

stopping criteria is achieved [12];
   Initialization Phase.
   REPEAT
   • Employed foragers Phase
   • Onlooker foragers Phase

   • Scout foragers Phase
   • Memorization of the best food solution achieved
   UNTIL (Cycle = Maximum number of Cycles)
   Generally the ABC algorithm steps can be summarized as shown in figure 3 [13]:

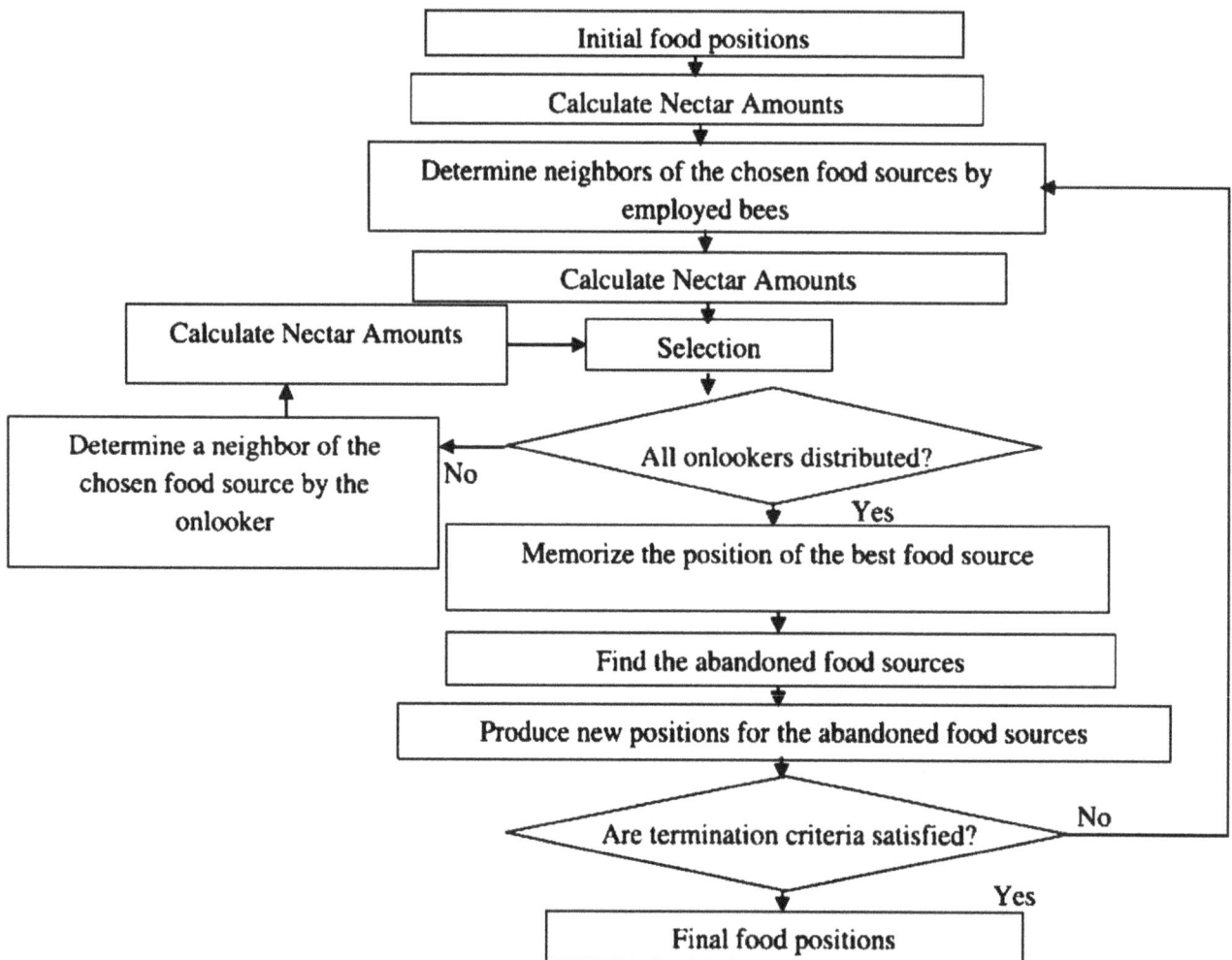

Figure 3. Artificial Bee Colony Algorithm flow chart.

2.2. Advantages of ABC Algorithm

The artificial bee colony algorithm system combines both the local search which is carried out by the employed and onlooker groups of bees, and also the global search which is managed by the onlookers and scout group of bees which attempts to balance the exploration and exploitation process [14]. The main advantages of the ABC algorithm over other optimization methods for solving optimization are [15] [16] [17] [18]:
   • It is simple to deploy
   • It has few control parameters
   • It is robust
   • It is highly flexible
   • Its ease of combination with other methods
   • Its ability to handle the objective with stochastic nature
   • Its fast convergence as it combines both exploration and exploitation processes.

2.3. Disadvantages of Artificial Bee Colony Algorithm

The artificial bee colony algorithm has some few weakness when it is put into practice.
   • First this method requires new fitness tests on every new algorithm parameters so as to improve its performance
   • It needs a high number of objective function evaluations
   • It slows down when used in sequential processing and the population of solutions increases the computational cost due to slowdown
   • It has many iterations and thus huge memory capacity required.

3. Methodology

The major aim of this study was to determine the minimum load amount for shedding within the islanded region so that we can maximize the electricity supply to customers in case the load surpasses the supply within that island.

To achieve this, artificial bee colony (ABC) algorithm was used to ensure there is optimum power supply and also perform power prioritization to determine the buses to be shed based on their priority index.

The IEEE fourteen bus was used in the analysis but with little modification on the bus data and line data. To start with, the distributed generator of 320W was connected at bus number two and a number of loads connected at different buses totaling to 362W. Then these parameters were varied separately while keeping the other constant and observations made.

The control parameters of ABC algorithm are assumed as follows:

- The number of colony size (employed bees and onlooker bees) is assumed to be 20.
- The number of food sources equals the half of the colony size.
- The limit is assumed to be 100. A food source which could not be improved through limit trials is abandoned by its employed bee.
- The number of cycles for foraging is assumed to be 100.

3.1. System Flow Chart

First, the frequency signals are sampled from the power line. In this study, current signals only were sampled and used in the analysis. Then features to be used in islanding detection were extracted by the use of discrete wavelet transform and the classification was done using fuzzy logic as either islanded or not islanded. In case of islanded condition, power optimization and prioritization within the island using ABC algorithm. Generally the system flow chart as shown in figure 4.

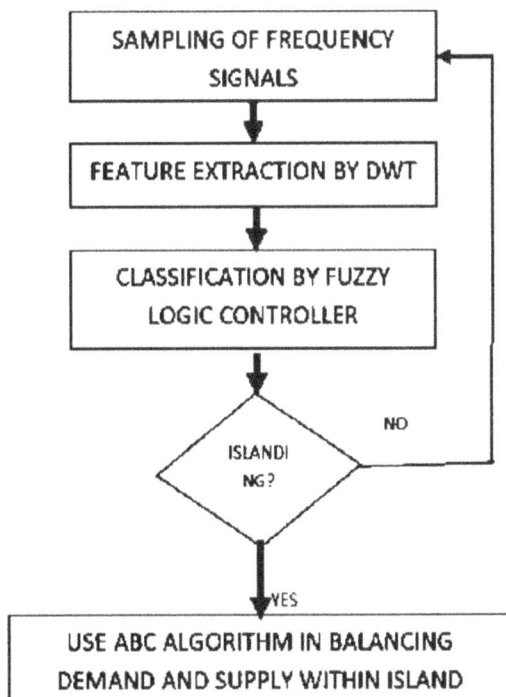

Figure 4. System flow chart.

3.2. IEEE 14-Bus Test System

After the occurrence of the islanding condition was successfully identified using DWT and FL, the load shedding of various buses was tested using an IEEE 14-bus test system. This system consists of five synchronous machines, including one synchronous compensator used only for reactive power support and four generators located at buses 1, 2, 6, and 8. In the system, there are twenty branches, fourteen buses and with 11 loads connected. This is shown in figure 5 below:

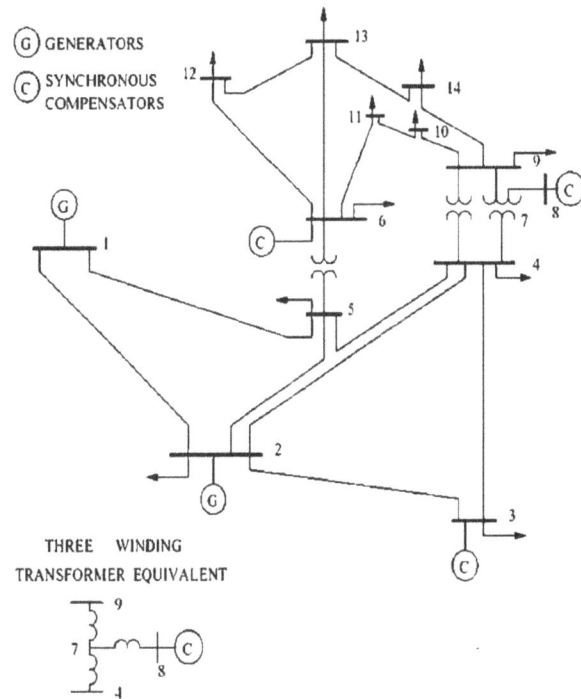

Figure 5. IEEE 14 bus.

The priorities for load shedding were set in the following decreasing order for Load buses 9, 10, 11, 12, 13 and 14. Out of these selected buses bus 14 is the one having highest sensitivity and therefore can be considered as the weakest bus for load shedding followed by bus 13.

4. Results and Discussion

The main aim of this study involved finding the optimal load to be shed and the selection of the buses to be shed using the ABC algorithm. Basically there are two main strategies of load shedding. The first is based on voltage which is called Under Voltage Load Shedding (UVLS) and the other one is based on frequency known as Under Frequency Load Shedding (UFLS).

The main objective of load shedding is to provide smooth load relief, in situations where the power system would otherwise go unstable. The buses for load shedding are selected based on the priority attached to those buses and the required amount of load to be shed.

On the part of power optimization and prioritization within the islanded region, the IEEE fourteen bus was used in the analysis but with little modification on the bus data and line data. The distributed generator of 320W was connected at bus number two and a number of loads connected at different buses totaling to 362W. Both the generator output and the connected load were varied and the signal variations observed.

Then using ABC algorithm and load flow, we were able to determine the power deficit and surplus in the island. In addition to this, the buses were given priorities and buses to be shed were picked based on the amount of deficit and the bus priority.

Table 1 is a snapshot of the system results obtained when a total of 362.5W load was connected to a 320W generation. From the simulation, the line losses were 32.805W and the total load to be shed should be 75.305W including line losses.

Table 1. Prioritization results.

Bus no.	Injection		Generation		Load	
	Mw	Mvar	Mw	Mvar	Mw	Mvar
1	75.305	0.000	75.305	0.000	0.000	0.000
2	298.300	0.000	320.000	0.000	21.700	0.000
3	-94.200	0.000	0.000	0.000	94.200	0.000
4	-47.800	0.000	0.000	0.000	47.800	0.000
5	-7.600	0.000	0.000	0.000	7.600	0.000
6	-11.200	0.000	0.000	0.000	11.200	0.000
7	-0.000	0.000	0.000	0.000	0.000	0.000
8	0.000	0.000	0.000	0.000	0.000	0.000
9	-30.000	0.000	0.000	0.000	30.000	0.000
10	-30.000	0.000	0.000	0.000	30.000	0.000
11	-30.000	0.000	0.000	0.000	30.000	0.000
12	-30.000	0.000	0.000	0.000	30.000	0.000
13	-30.000	0.000	0.000	0.000	30.000	0.000
14	-30.000	0.000	0.000	0.000	30.000	0.000
Total	32.805	0.000	395.305	0.000	362.500	0.000

Power generated = 320.000
Power demand = 362.500
Losses = 32.805
Load to shed = 75.305
Bus to shed = 14

Constant Demand and Constant Supply Characteristics.
The connected load was kept constant and the generator output varied and observations made. On the other hand the generator output was kept constant and the connected load varied and observations made. These are briefly shown in table 2 and 3.

Table 2. Constant demand with varying supply characteristics.

Simulation	Power generated	Power demand	Losses	Load to shed	Shed bus
1	362.5	362.5	33.817	-	-
2	330.0	362.5	32.983	65.483	14
3	300.0	362.5	32.563	95.063	14 & 13
4	270.0	365.5	32.486	124.986	14, 13 & 12

Table 3. Constant supply with varying demand characteristics.

Simulation	Power generated	Power demand	Losses	Excess generation
1	362.5	362.5	33.817	33.817
2	362.5	352.5	28.568	18.568
3	362.5	342.5	27.281	7.281
4	362.5	335.5	26.259	-3.741

5. Conclusion

In this study, the ABC algorithm was successfully applied for solving the optimization and prioritization problems in the island being supplied by the DG. The ABC algorithm is based on the foraging behavior of honey bees for finding global and local solution for optimization problems. The advantages of using this algorithm are its robustness, fast calculation of the error, flexibility, and few parameters to be set. However, the ABC algorithm suffers a drawback of the search space limited by initial solution. In fact, this drawback can be overcome using normal distribution sample in the initial step.

This proposed algorithm has been tested on a fourteen bus system and the obtained result for this system was analyzed and it was satisfactory to draw concrete conclusions.

In comparison with other methods of optimization, the proposed algorithm can obtain better optimal solution than many other methods with a fast computational manner, especially for large-scale systems. Therefore, the proposed ABC algorithm can be a favorable method for solving optimization and prioritization problems in power systems.

Acknowledgement

The authors would like to express the greatest gratitude to the Technical University of Mombasa for the continued support from time to time when required.

References

[1] Azakiah, K., Hussain, S., Erdal, B., & Tamer, K. (December 2013). A review of islanding dtection techniques for renewable distributed generation systems. *Renewable and sustainable energy reviews, 28*, 483-493.

[2] Belkacem, M., & Kamel, S. (2014). Solving Practical Economic Dispatch Problems Using Improved Artificial Bee Colony Method. *International Journal Intelligent Systems and Applications, 7*, 36-43.

[3] F, A., Mohamed, M., Elarini, M., & Ahmed, O. M. (2014). A new technique based on Artificial Bee Colony Algorithm for optimal sizing of stand-alone photovoltaic system. *Journal of Advanced Research, 5*, 397-408.

[4] Hardiansyah, Junaidi, & Yohannes, M. (2012). Application of soft computing methods for economic load dispatch problems. *International journal of computer applications, 58* (13).

[5] Hemamalini, S., & Sishai, P. (2010). Artificial bee colony algorithm for economic dispatch problem with non-smooth cost functions, electric power components and systems. *38* (7), 786-803.

[6] Karaboga, D., & Akay, B. (2009). A comparative study of artificial bee colony algorithm. *Applied Mathematics and Computation, 214* (1), 108-132.

[7] Karaboga, D., & Basturk, B. (2007). A powerful and efficient algorithm for numerical function optimization: artificial bee colony (ABC) algorithm. *Journal of Global Optimization, 39* (3), 459-471.

[8] Laghari, J., Mokhlis, H., Karimi, M., Bakar, A. H., & Hasmaini, M. (2014). Computational Intelligence based techniques for islanding detection of distributed generation in distribution network: A review. *Energy Conversion and Management, 88,* 139-152.

[9] Luong, D., Dieu, N., & Pandian, V. (2013). Artificial bee colony algorithm for solving optimal power flow algorithm. *The scientific world journal, 2013.*

[10] Mahani, Z. A. (2000). Malaysian economic recovery measures: A response to crisis management and for long-term economic sustainability. *ASEAN university network's conference on economic crisis in southeast Asia: Its social, political and cultural impacts.* Bangkok, Thailand.

[11] Martin, J. (2009). Distributed vs. centralized electricity generation: are we witnessing a change of paradigm? *paris.*

[12] Mogaka, L. O., Murage, D. K., & Saulo, M. J. (2015). Rotating Machine based DG islanding Detection Analysis using Wavelet Transform. *International Journal of Energy and Power Engineering, 4* (5), 257-267.

[13] Mogaka, L. O., Murage, D. K., & Saulo, M. J. (2015). Rotating Machine Based Islanding Detection Using Fuzzy Logic Method. *International Journal of Energy and Power Engineering, 4* (5), 311-316.

[14] Murthy, G., Sivanagaraju, S., Satyanarayana, S., & Rao, H. (2013). Optimal placement of DG in distribution system to mitigate power quality disturbances. *World academy of science, engineering and technology, 7,* 204-209.

[15] Pukar, M., Zhe, C., & Birgitte, B. J. (2008). Review on islanding operation of distribution system with distributed generation. *International Conference on Electric Utility Deregulation and Restructuring and Power Technologies.* Nanjing, china.

[16] Rao, R., Narasimham, S., & Ramalingaraju, M. (2008). Optimization of distribution network configuration for lossr eduction using artificial bee colony algorithm. *International Journal of Electrical Power and Energy Systems, 1* (2), 116-122.

[17] Singh, A. (2009). An artificial bee colony algorithm for the leaf constrained minimum spanning tree problem. *Applied Soft Computing Journal, 9* (2), 625-631.

[18] Sumpavakup, C., Srikun, I., & Chusanapiputt, S. (2010). A Solution to the Optimal Power Flow Using Artificial Bee Colony Algorithm. *International Conference on Power System Technology,* 1-5.

7

A Review of Particle Swarm Optimization (PSO) Algorithms for Optimal Distributed Generation Placement

Musa H., Ibrahim S. B.

Department of Electrical Engineering, Bayero University, Kano, Nigeria

Email address:

harunamusa2@yahoo.co.uk (Musa H.), hmusa.ele@buk.edu.ng (Musa H.), sabkibr@yahoo.com (Ibrahim S. B.)

Abstract: Particle Swarm Optimization (PSO) has became one of the most popular optimization methods in the domain of Swarm Intelligence. Many PSO algorithms have been proposed for distributed generations (DGs) deployed into grids for quality power delivery and reliability to consumers. These can only be achieved by placing the DG units at optimal locations. This made DG planning problem solution to be of two steps namely, finding the optimal placement bus in the distribution system as well as optimal sizing of the DG. This paper reviews some of the PSO and hybrids of PSO Algorithms formulated for DG placement being one of the meta-heuristic optimization methods that fits stochastic optimization problems. The review has shown that PSO Algorithms are very efficient in handling the DG placement and sizing problems.

Keywords: Distributed Generation, Power Losses, Optimization, Placement, Sizing, Objective Function, Power flow, Distribution Networks, Multi-objective

1. Introduction

Distributed generation placement and sizing is an important issue which requires special attention of both planners and system operators. DG installation at non-optimal places can lead to increase in system losses which imply increase in costs and hence having a negative impact opposite to the desired. The selection of the best location and size in large and complex system is a combinatorial optimization problem [1]. Researchers have employed various methods in addressing the placement problems. It has been observed that among all the methods reviewed so far analytical method is found to be the most accurate and more practical technique for placement. However, obtaining a truly optimal solution has presented a challenge as some computational methods do not yield global solution as many local solutions exists.

Due to this problem, deterministic algorithms such as Dynamic programming, NLP, LP, QP and SQP are considered to be the elegant options [2]. However, meta heusutic algorithms such as GA, PSO, EP, Tabu search (TS), simulated annealing (SA) seems to have shared the same dominance as deterministic methods [3]. This is due to the fact that meta heuristic are derivative free problems unlike

the deterministic methods and can be solved without need for convexcity. Apart from that the meta heuristic algorithms are independent of initial solution and can avoid local optima [4]. The techniques are robust and can provide near optimal solution for large and complex systems. The only drawback is the high computational efforts required for good solution. On the other hand, meta heuristic methods have their own drawbacks, such as use of trial and error process during parameter tunning and lack of guarantee of global solution attainment atimes. This is one of the reasons that made researchers to direct a lot of effort towards eliminating such problems by combining more than one algorithms to form a hydrid algorithm. The hybrid algorithms currently utilised are; such as LP – QP, EP – SQP GA – SA and PSO – SQP [5] e.t.c. These efforts have yielded results with a lot of enhancement over the individual algorithms when utilized alone.

Particle swarm optimization (PSO) algorithm is a population based optimization method that has gained much popularity among researchers since after its introduction. The algorithm mimics birds' behavior during flight in space. Each of the bird in the aggregation of the birds called swarm is represented as a particle. These particles that form the swarm searches for food based on their own experience and that of the other particles within the same swarm. The PSO have been studied by many researchers and several newer versions have been developed for applications in different real-world problems and are found to be robust and fast in solving nonlinear non-differentiable multi-modal problems.

Many survey papers with applications have been presented in literature reviews regarding these studies but it is still in its infancy stage requiring a lot of research work. Authors in [6] presented a method for optimal siting and sizing of multiple distributed generators (DGs) using PSO based approach. Similar multiple DGs placement was also presented in [7] with PSO as an optimization tool for variable power load with non-unity power factor. In another development an improved PSO algorithm was also proposed for optimal placement with an in built mechanism for better search that is capable of escaping local optima in [8]. As part of improvement on initial PSO algorithm, the concept of hybridized PSO was introduced by authors in [9]

in which Genetic Algorithm (GA) was combined with PSO. The GA is made to search for the DG site while the PSO optimizes the DG sizes which resulted in drastic reduction in system losses and improvement in voltage profile.

In this paper an introduction of PSO concepts and its algorithm is presented and survey of existing work follows based on objectives, methods and contributions of the work towards finding of optimal solutions to placement and sizing problems.

2. DG Placement and Sizing

The DG placement optimization problem has not been assessed thoroughly as done in many optimization problems. The review in this paper differs from previous reviews in the sense that all work done is going to be categorized based on optimization algorithms employed. Figure 1.0 shows the published research work done on placement and sizing based on IEEE Explore Digital library data base. The analysis shows significant improvement in papers published.

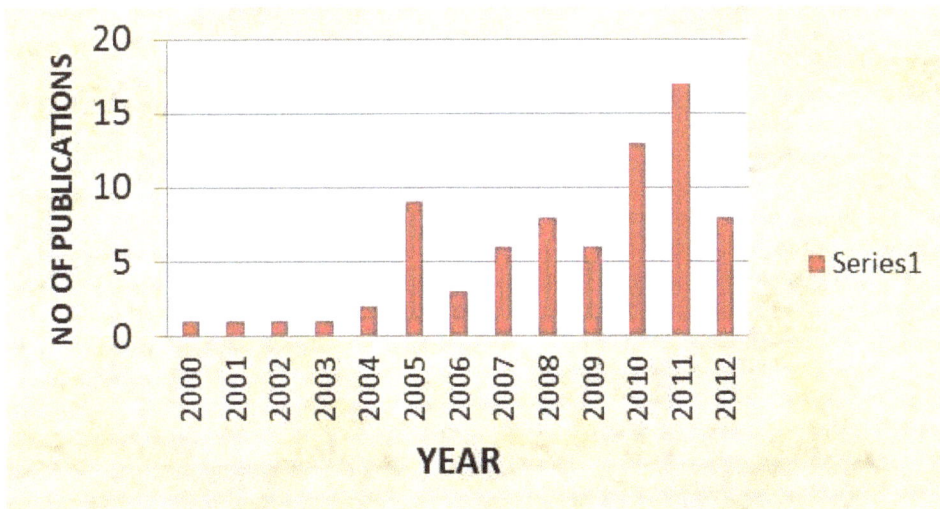

Fig. 1.0. Distributionof papers publsihed on DG placement and sizing

The gradual increase in published papers is a clear indication of growing interest of researchers willing to find solution to DG placement problems.

Many researchers have used various methods to tackle the placement and sizing problems. Methods of optimal placements and sizing of the DGs within networks always depend on the objectives and solution techniques employed. There are three basic models for DG optimization problems that are currently in use which are:

i)   Objective function model

ii) Constraints model

iii) Optimization algorithms model

Objective functions are optimized subject to operating constraints by using different techniques. The objective function can be single or multiple for the purpose of achieving maximum benefits of the DG without violating the equality and inequality constraints of the entire power system. The most popular objective is the minimization of

power losses [10]. Among other objectives considered by many researchers includes minimization of real and reactive power loss, maximization of DG capacity, maximization of profit and social welfare to mention a few [11-14]. Other objectives handled by many authors are the technical issues; environmental issues and voltage limit [15-16].

The constraints are state variable limits that are placed on operating conditions. Constraints are basically categorized into equality and inequality which must be satisfied by the objective functions either single or multi-objectives. The common constraints generally in DG placement includes but not limited to the following; line thermal limit, short circuit ratio limit, voltage step limit, phase angle limit, power generation limit, DG power generation limit, number of DG limit, power flow equality constraint, voltage profile limits, short circuit level limit, total line loss limit, substation transformer capacity limit, tap position limit, and power factor limit [12, 14, 17-23].

Many researchers proposed different methods such as analytic as well as deterministic and heuristic methods to solve placements problems. Out of these methods metaheuristic is the only method that has proven their effectiveness in solving optimization problems with appreciable feasible search space [4]. They can also be modified easily to become a hybrid of more than one algorithm to cope with different elements commonly used in most studies.

3. The Concept of PSO Algorithm

The search process is similar to the social behaviour of flying birds when searching for food. The individual bird called particles or swarm flies in the optimization problem hyperspace to search for optimal food location. The position and velocity of the particles is always changing and adjusted according to the cooperative communication among the particles and each individual's own experience simultaneously. Therefore the particle changes position by balancing its social and individual experience. Each particle is assigned a velocity V_i as well as position vector x_i

For a swarm of m-particle in hyperspace, the position and velocity vectors are [24];

$$S_i = [x_1^i, x_2^i - - - x_n^i] \quad i = 1,2 - -, m \quad (1)$$

$$V = [v_1, v_2 - - - - - - - - - - v_n] \quad (2)$$

where i is the particle index, V is the swarm velocity and n is the optimization problem dimension. The particle's new position is;

$$S_i^{(k+1)} = S_i^{(k)} + v_i^{(k+1)} \quad (3)$$

where

$S_i^{(k+1)}$ is particle i new position at iteration $k+1$

$S_i^{(k)}$ is particle i old position at iteration k

$v_i^{(k+1)}$ is particle i new velocity at iteration $k+1$

Equation (3) is the updated position equation. The updated velocity vector for particle i is

$$v_i^{(k+1)} = wv_i^{(k)} + c_1 r_1 \left(p_{besti}^{(k)} - s_i^{(k)}\right) + c_2 r_2 \left(g_{best}^{(k)} - s_i^{(k)}\right) \quad (4)$$

where

$V_i^{(k)}$ is the previous velocity of particle i

w is the inertia weight

c_1, c_2 are the individual and social acceleration positive constants.

r_1, r_2 are the random values in the range [0, 1], sampled from a uniform distribution

P_{besti} is the personal best position associated with particle i own experience

g_{besti} is the global best position associated with the whole neighborhood experience

3.1. Updated Velocity

The updated velocity as given by equation (4) has three major components consisting of the following;

1. The first component is related to particle's immediate previous velocity, and it consists of two variables. The variables are particle i last velocity $V_i^{(k)}$ and inertia weight w.
2. The second component is the cognitive component, which shows the individual's own experience.
3. The last term is the third components that represent the intelligent exchange of information between particle i and the swarm.

In the absence of the second and third terms of the velocity formula the particle will continue to fly in the same direction with a speed proportional to its inertia weight until it hits one of the solution space boundaries.

Therefore solution can never be obtained unless the solution lies in the same path of the previous velocity in such a case. The change in direction towards the solution is achievable with the help of second and third term of the equation. Those three components of the velocity update are responsible for the optimization process.

Versions of PSO algorithms have been proposed since after Kennedy and Eberhart which are the local best PSO and the global best PSO. The two models differ in the social component of the velocity update formula. In the case of the local best PSO the swarm is divided into several neighborhoods and the g_{best} of particle i is its neighborhood's global value. Whereas the global best model considers the swarm as one entity, and therefore the PSO particle's g_{best} is the best value for the whole swarm. Generally, the global model is more popular version since it needs less work to achieve result.

3.2. Previous Velocity Components

The component is given by the product of previous velocity of particle $V_i^{(k)}$ and the inertia weight w. This term connects the particle in the existing iteration with immediate past iteration which serves as the particle's memory. It is important as it prevented the particle from sudden change in its direction, and also allows the particle's own knowledge of its previous flight information to influence its newer course.

The first version of PSO has no inertia weight as was assumed to be unity. It was in subsequent versions that inertia weight was introduced in order to control the contribution of the particle's previous velocity in the current velocity decision making and this lead to a significant improvement in the PSO algorithm [25]. The implication of making its value too large is the broadening of the exploration mission of the particle, and if the value is small the exploration will be localized. Several dynamic inertia weights were proposed in literatures. The formulations of the literatures have expressed inertia weight as follows;

$$w^{(k)} = c_1 r_1^{(k)} + c_2 r_2^{(k)} \quad (5)$$

$$w^{(k)} = w^{(n_k)} + \left(w^{(1)} - w^{(n,d)}\right)\left(\frac{n_k - k}{n_k}\right) \quad (6)$$

$$w^{(k+1)} = \frac{w^{(k)} - w^{(n_k)}(n_k - k)}{n_k + w^{(n_k)}} \quad (7)$$

Where $w^{(k)}$ is the inertia weight value at iteration k
n_k is maximum number of iterations
$w^{(n}{}_k)$ is the inertia weight value at the last iteration n_k

Some authors in [26] have proposed using 0.9 and 0.4 as the initial and final weight values respectively.

Another factor introduced that is similar to inertia weight function is the constriction factor (λ) which is used for the balancing of the search mechanism between global and local exploration. Use of this factor improves convergence and the particles velocity is therefore constricted by a factor λ as expressed in the following equation;

$$v_i^{(k+1)} = \lambda\left(v_i^{(k)} + c_1 r_1\left(p_{best}^{(k)} - s_i^{(k)}\right) + c_2 r_2\left(g_{best}^{(k)} - s_i^{(k)}\right)\right) \quad (8)$$

where

$$\lambda = \frac{2}{\left|2 - \emptyset - \emptyset\sqrt{\emptyset^2 - 4\emptyset}\right|} \quad (9)$$

and

$$\emptyset = c_2 + c_2$$

$$\emptyset \geq 4$$

Hence, the constriction factor is a factor of individual and social acceleration positive constants C_1 and C_2 respectively. This factor is normally considered as a special case of inertia weight PSO algorithm because of the constraints imposed above. The factor λ controls the particle's velocity vector, whiles the inertia weight w controls the contribution of the particle's previous velocity towards calculating the new one velocity. Use of constriction factor eliminates velocity clamping and can safe guards the algorithm against explosion [27].

3.3. Cognitive Component

The cognitive component of the velocity update equation uses P_{best} which is referred to as the particle's best personal position that it has visited so far since the beginning of the PSO iterative process. Each particle in the swarm tries to evaluate its own performance by comparing its own fitness in the current PSO iteration with that evaluated in the proceeding one. The $p_{best_i}^{(k+1)}$ given that its $p_{best_i}^{(k)}$ is the best personal position so far, is defined as;

$$p_{best_i}^{(k+1)} = \begin{cases} p_{best_i}^{(k)} & if\ f(s_i^{(k+1)}) \geq f\left(p_{best_i}^{(k)}\right) \\ s_i^{(k+1)} & if\ f(s_i^{(k+1)}) \geq f\left(p_{best_i}^{(k)}\right) \end{cases} \quad (10)$$

Based on equation (5) each particle is suppose to remember its optimal personal best position achieved for use

in the update of velocity for future iteration. This component of velocity update equation diversifies searching process at the same time helps in avoiding possible stagnation.

3.4. Social Component

This represents the social behavior of the PSO particles. The g_{best} term in this component is referred to as the best position achieved among all the swarm particles. Whenever the best solution among the whole population of the swarm is achieved, all the particles are informed. The g_{best} fitness value is the optimal among all the particles during the current PSO iteration as;

$$g_{best}^{(k+1)} = min\left\{f(s_1^{(k)}), f(s_2^{(k)}), - - -, f(s_m^k)\right\} \quad (11)$$

where $f(S_i^k)$ is particle i fitness value at iteration k, and m is the swarm size.

3.5. Cognitive and Social Parameters

These parameters C_1 and C_2 are cognitive and social parameters respectively. They are factors that scaled the P_{best} and g_{best} in the updated velocity equation. The trust of the particle in itself is measured by C_1, while C_2 reflects the confidence it has on its neighbours. If C_1 is 0, the particle's own experience is eliminated during search process for a new solution, while if C_2 is 0 the search is localized and exchange of information between the particles is eliminate. The highest value recommended for the two in most literatures is 2.

r_1 and r_2 are two random numbers in the range of [0, 1] that are sampled from a uniform distribution. The stochastic exploration nature of PSO is due to these random numbers. A typical illustration for velocity and position update for a single PSO particle during iteration is shown in Fig. 2.0 during iteration.

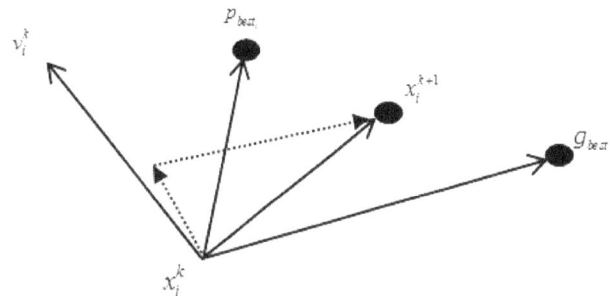

Fig. 2.0. *Velocity and position updates for a single particle during iteration k*

3.6. Pretty Features of PSO

The advantages associated with PSO are many just like other optimization algorithm. The main advantages are clearly distinct when compared to deterministic methods that are gradient based techniques and have no flexibility of dealing with objective functions that are not continuous or differentiable naturally. The search process for PSO does not involve use of derivative function; instead it uses the fitness function value as a guide for finding optimal solution in problem space. This concept of fitness function employed in PSO helps in eliminating the approximations and assumptions usually adapted on objective functions and constraints as in conventional optimization methods. For this reason PSO is considered as a stochastic optimization method and found to be very efficient in handling problems that their objective functions are time varying or stochastic in nature. Above all the solutions from PSO are not dependant on the initial solutions unlike the deterministic method.

4. Studies on PSO Algorithms for DG Placements and Sizing

This section deals with all the studies done on PSO and hybridized PSO Algorithms for DG placements and sizing. All the literature surveyed are summarized in table 4.1

Table 4.1. Some Published Works on PSO Algorithms

S/NO.	Ref.	Objectives	Optimization method	Contributions
1.	Haruna M. [28]	Minimization of power losses and voltage stability index (VSI)	PSO	Optimal DG Placement model for better search and improved (VSI)
2.	El-Zonkoly [7]	Multi-objective for short circuit level and other technical parameters	PSO	Determination of voltage collapse point for variable load
3	Nabavi, S.M.H[29]	Reduce network congestion and minimize locational marginal price (LMP)	PSO	congestion management and social welfare in placement/sizing
4	Amanifar, O. [30]	Minimization of investment cost of DGs and power losses	A PSO algorithm integrated in harmonic power flow algorithm	Investiment cost justifies DG placement for increase in power transfer capability
5	Dias, B.H [31]	Minimization of power losses	PSO in Nonlinear Optimal Power Flow (OPF)	Reduction in search space and improvement convergence
6	Gomez-Gonzadez [32]	Minimization of cost	PSO in Optimal power flow	ODGP model using discrete PSO and OPF
7	Wong, L.Y. [33]	Minimization of power losses	Pso in Newton –Raphson power flow	Effective solutions and improved voltage profile
8	Nguyen Cong Hien [34]	Maximizes reactive power flow	PSO	Enhancement of loadability of the primary distribution feeder

The PSO algorithms allow the system planner to find not only a single optimum point, but a family of near-optimum planning alternatives. This feature of PSO has become very useful in DG allocation because distribution network operators usually have little or no control on the DG integration and different planning alternatives can be necessary to face uncertainties and minimize risks.

Numerous publications and wide spread implementation of PSO algorithms has been conducted, a lot of barriers to proper implementation of these researchers by network operators is still lacking. Although the algorithm has many advantages over other meta-heuristic methods, the main challenge associated PSO is that of lack of solid mathematical background like many heuristic methods. It's solution method is problem dependent and for every solution parameters have to be tuned and adjusted for better solution. The conventional PSO introduced by Kennedy and Eberhart in 1995 and even those that had under gone modifications, are still dependent on a number of parameters that are externally set or arbitrary selected by the user [24]. This setting or selection by the user is a very delicate operation involving a lot of trials and errors before a reasonable tuning can be achieved especially in practical problems that require defining of inertia weight at each iteration step. Apart from that the usual initial assumption that inertia term is eliminated at an early stage of the optimization process, can cause the algorithm to be trapped at some local minimum. As a solution to this problem some authors have proposed some procedures of "re-seeding" the search by generating new particles at distinct places of the search space [35]. Other shortcoming of PSO is the random operation involved that a particle is initialized randomly and its location in the search space is updated during each iteration in the PSO algorithm. The problem is that when the particle initialization is not well done the issue of local minimum trapping can also arise or convergence can be prolonged as sharing of information between particles is not properly coordinated.

These challenges made researchers to direct a lot of effort towards eliminating such problems by combining more than one algorithms to form a hybrid algorithms. The efforts have yielded results with a lot of enhancement over the individual algorithms when utilized alone as indicated in table 4.2.

Table 4.2. Some Published Works on Hybrid PSO Algorithms

S/No.	Ref	Objective	Method	Contribution
1	Moradi [9]	Multi-objective with weights	Hybrid (GA &PSO)	Optimal DG Planning for placement using GA and sizing using PSO
2.	Sedighizadeh, M [36]	Minimization of power losses and voltage profile improvement	Hybrid (PSO and Clonal Selection)	Better quality of solutions and less number of iterations
3.	Musa, H. [37]	Multi-objective for power loss reduction and voltage stability index improvement	Hybrid (PSO & Evolutionary Programming)	Well-distributed Pareto optimal non-dominated solutions of DG sizes obtained
4.	Afzalan, M [38]	Minimization of power losses	Hybrid (PSO & HBMO)	Voltage profile improvement and branch current reduction
5.	Ziari, I [39]	Minimizes loss and improves system reliability.	Hybrid (Discrete PSO & GA)	Increase in the diversity optimization variables in DPSO not to be trapped in a local minimum
6	Soroudi, A. [40]	Multi-objective for technical constraint dissatisfaction, costs and environmental emissions	Hybrid (Binary PSO-based & Fuzzy)	Better timing of investment for both distributed generation (DG) units and network components obtained
7	Wen S. T.[41]	Multi-objective index-based approach for total real power losses, voltage profile, MVA intake by the grid, number of DG and greenhouse gases emission	Hybrid (PSO & Gravitational Search Algorithm)	Algorithm can provide efficient and robust solution to mixed integer nonlinear optimization problem
8	Haruna Musa[42]	power loss reduction (PLR) value	Hybrid (PSO & Ranked Evolutionary programming)	A simple and effective algorithm for power loss reduction and voltage profile improvement

5. Conclusions

Although there have been numerous publications in the area of the siting and sizing of DG, it is evident that, widespread implementation of the PSO optimizations has not been much. Development of the DG integration strategies and PSO optimization methods requires proper distribution system planning especially with the proliferation of electric vehicle in existing distribution networks. Further challenges that are yet to be tackled for better optimization are the modelling of the distribution networks with all the necessary details needed by network operators.

Acknowledgements

The authors acknowledged with gratitude the financial support offered by Bayero University Kano Nigeria and the provision of suitable research facilities.

References

[1] Borges, C., Falcao, D., "Optimal DG allocation for reliability, losses, and voltage improvement", International Journal of Electrical power & Energy systems, vol. 28 pp. 413-420. 2006

[2] T. Griffin, K. Tomsovic, D. Secrest, A. Law, " Placement of disperse generation systems for reduced losses", Proceedings of the 33rd Annual Hawaii International Conference on system Science, IEEE , pp. 164–164, Jan. 2000.

[3] M. Gandomkar, M. Vakilian, and M. Ehsan, "A genetic-based tabu search algorithm for optimal DG allocation in distribution networks," *Elect. Power Compon. Syst.*, vol. 33, no. 12, pp. 1351–1362, Aug. 2005

[4] Haruna Musa "A Review of Distributed Generation Resource Types and their Mathematical Models for Power Flow Analysis" International Journal of Science, Technology and Society Doi: 10.11648/j.ijsts.20150304.21, 2015; 3(4): 204-212

[5] G. P. Harrison, A. Piccolo, P. Siano, and A. R. Wallace, "Hybrid GA and OPF evaluation of network capacity for distributed generation connections," *Elect. Power Syst. Res.*, vol. 78, no. 3, pp. 392–398, Mar. 2008

[6] Jain Naveen, Singh SN, Srivastava SC. Particle swarm optimization based method for optimal siting and sizing of multiple distributed generators. In Proc 16th national power systems conference, 15–17th December, 2010. p. 669–74.

[7] A. M. El-Zonkoly, "Optimal placement of multi-distributed generation units including different load models using particle swarm optimisation," *IET Gener., Transm., Distrib.*, vol. 5, no. 7, pp. 760–771, Jul. 2011.

[8] W. Prommee and W. Ongsakul, "Optimal multiple distributed generation placement in microgrid system by improved reinitialized social structures particle swarm optimization," *Euro. Trans. Electr. Power*, vol. 21, no. 1, pp. 489–504, Jan. 2011.

[9] M. H. Moradi and M. Abedini, "A combination of genetic algorithm and particle swarm optimization for optimal DG location and sizing in distribution systems," *Int. J. Electr. Power Energy Syst.*, vol. 34, no. 1, pp. 66–74, Jan. 2012.

[10] D. Q. Hung and N. Mithulananthan, and R.C. Bansal "An analytical expression for DG allocation in primary distribution networks," *IEEE Trans. Energy Convers.*, vol. 25, no. 3, pp. 814–820, Sept. 2010.

[11] Rau N. S. and Wan Y.-H., "Optimum location of resources in distributed planning". *IEEE Trans. Power Syst.*, vol. 9, pp. 2014– 2020. 1994

[12] A. Keane and M. O'Malley, "Optimal allocation of embedded generation on distribution networks," *IEEE Trans. Power Syst.*, vol. 20, no. 3, pp. 1640–1646, Aug. 2005.

[13] D. Gautam and N. Mithulananthan, "Optimal DG placement in deregulated electricity market," *Elect. Power Syst. Res.*, vol. 77, no. 12, pp. 1627–1636, Oct. 2007.

[14] C. J. Dent, L. F. Ochoa, and G. P. Harrison, "Network distributed generation capacity analysis using OPF with voltage step constraints," *IEEE Trans. Power Syst.*, vol. 25, no. 1, pp. 296–304, Feb. 2010.

[15] E. Haesen, J. Driesen, and R. Belmans, "Robust planning methodology for integration of stochastic generators in distribution grids," *IET Renew. Power Gener.*, vol. 1, no. 1, pp. 25–32, Mar. 2007.

[16] Y. M. Atwa and E. F. El-Saadany, "Probabilistic approach for optimal allocation of wind-based distributed generation in distribution systems," *IET Renew. PowerGener.*, vol. 5, no. 1, pp. 79–88, Jan. 2011.

[17] A. Kumar and W. Gao, "Optimal distributed generation location using mixed integer non-linear programming in hybrid electricity markets," *IET Gener. Transm. Distrib.*, vol. 4, no. 2, pp. 281–298, Feb. 2010.

[18] Algarni Ayed A. S. and Bhattacharya Kankar, 2009. Disco operation considering DG units and their goodness factors. IEEE Trans. Power System, vol. 24, no. 4, pp.1831-1840.

[19] D. Singh, D. Singh, and K.-S. Verma, "Multiobjective optimization for DG planning with load models," IEEE Trans. Power Syst., vol. 24, no. 1, pp. 427–436, Feb. 2009.

[20] D. H. Popovic, J. A. Greatbanks, M. Begovic, and A. Pregelj, "Placement of distributed generators and reclosers for distribution network security and reliability," Int. J. Elect. Power Energy Syst., vol. 27, no. 5, pp. 398–408, Jun. 2005.

[21] W. El-Khattam, K. Bhattacharya, Y. Hegazy, and M. M. A. Salama, "Optimal investment planning for distributed generation in a competitive electricity market," *IEEE Trans. Power Syst.*, vol. 19, no. 3, pp. 1674–1684, Aug. 2004.

[22] P. N. Vovos, G. P. Harrison, A. R.Wallace, and J.W. Bialek, "Optimal power flow as a tool for fault level-constrained network capacity analysis," *IEEE Trans. Power Syst.*, vol. 20, no. 2, pp. 734–741, May 2005.

[23] R. A. Jabr and B. C. Pal, "Ordinal optimisation approach for locating and sizing of distributed generation," *IET Gener., Transm., Distrib.*, vol. 3, no. 8, pp. 713–723, Aug. 2009.

[24] Eberhart, R.C., Kennedy, J. " A new optimizer using particle swarm theory", Proceedings of the sixth International symposium on Micro Machine and Human Science, 1995 pp. 39-43

[25] Y. Shi and R. Eberhart, "A modified particle swarm optimizer," *IEEE World Congress on Computational Intelligence,* 1998 pp. 69-73.

[26] Y. Shi and R. C. Eberhart, "Empirical study of particle swarm optimization,"*Proceedings of the 1999 Congress on Evolutionary Computation,* vol. 3, p. -1950, 1999.

[27] R. C. Eberhart and Y. Shi, "Comparing inertia weights and constriction factors inparticle swarm optimization," *Proceedings of the 2000 Congress on Evolutionary Computation,* vol. 1, pp. 84-88, 2000.

[28] Haruna Musa, Sanusi Sani Adamu "PSO based DG sizing for improvement of voltage stability index in radial distribution systems" Proceedings of the IASTED International Conference Power and Energy Systems and Applications (PESA 2012) Pages 175-180

[29] Nabavi, S.M.H. Hajforoosh, S. ; Masoum, M.A.S., "Placement and sizing of distributed generation units for congestion management and improvement of voltage profile using particle swarm optimization" Innovative Smart Grid Technologies Asia (ISGT), 2011 IEEE PES , pages 1-6

[30] Amanifar, O. , "Optimal distributed generation placement and sizing for loss and thd reduction and voltage profile improvement in distribution systems using particle swarm optimization and sensitivity analysis Electrical Power Distribution Networks (EPDC), 2011 16th Conference on Publication Year: 2011 , Page(s):1-5

[31] Dias, B.H. ; Oliveira, L.W. ; Gomes, F.V. ; Silva, I.C. ; Oliveira, E.J., "Hybrid heuristic optimization approach for optimal Distributed Generation placement and sizing" Power and Energy Society General Meeting, 2012 IEEE Digital Object Identifier: 10.1109/PESGM.2012.6345653

[32] M. Gomez-Gonzalez, A. López, and F. Jurado, "Optimization of distributed generation systems using a new discrete PSO and OPF," *Elect. Power Syst. Res.*, vol. 84, no. 1, pp. 174–180, Mar. 2012.

[33] Wong, L.Y. ; Rahim, S.R.A. ; Sulaiman, M.H. ; Aliman, O. "Distributed generation installation using particle swarm optimization "Power Engineering and Optimization Conference (PEOCO), 2010 4th International , Page(s): 159 – 163

[34] Nguyen Cong Hien ; Mithulananthan, N. ; Bansal, R.C. Location and Sizing of Distributed Generation Units for Loadabilty Enhancement in Primary Feeder Systems Journal, IEEE Volume: 7 , Issue: 4 2013 , Page(s): 797 - 806 Digital Object Identifier: 10.1109/JSYST.2012.2234396

[35] Vladimiro Miranda, Nuno Fonseca, "EPSO Evolutionary self-adapting Particle Swarm optimization", internal report INESC Porto, July 2001

[36] Sedighizadeh, M. ; Fallahnejad, M. ; Alemi, M.R.; Omidvaran, M. ; Arzaghi-haris, D. "Optimal placement of Distributed Generation using combination of PSO and Clonal Algorithm" Power and Energy (PECon), 2010 IEEE International Conference 10.1109/PECON.2010.5697547, Page(s): 1 – 6

[37] Musa, H. ; Adamu, S.S., "Enhanced PSO based multi-objective distributed generation placement and sizing for power loss reduction and voltage stability index improvement" Energytech, 2013 IEEE Conference, pages 1-6

[38] Afzalan, M. ; Taghikhani, M.A. ; Sedighizadeh, M., "Optimal DG placement and sizing with PSO&HBMO algorithms in radial distribution networks" Electrical Power Distribution Networks (EPDC), 2012 Proceedings of 17th Conference on , 2012 , Page(s): 1 – 6.

[39] Ziari, I. ; Ledwich, G. ; Ghosh, A. ; Cornforth, D. ; Wishart, M. , "Optimal allocation and sizing of DGs in distribution networks Power and Energy Society General Meeting, 2010 IEEE Digital Object Identifier: 10.1109/PES.2010.5588114 2010 , Page(s): 1 – 8.

[40] A.Soroudi, M.Afrasiab; "Binary PSO-Based Dynamic Multi-Objective Model for Distributed Generation Planning under Uncertainty", IET Renewable Power Generation, 2012, Vol.6, No. 2, pp. 67 - 78.

[41] Wen Shan Tan, Hassan, M.Y., Rahman, H.A., Abdullah, M.P., Hussin, F. , "Multi-distributed generation planning using hybrid particle swarm optimisation- gravitational search algorithm including voltage rise issue" Generation, Transmission & Distribution, IET Volume: 7 , Issue: 9 , DOI10.1049/iet-gtd.2013.0050 , 2013 , Page(s): 929 – 942.

[42] Haruna Musa, Sanusi Sani Adamu "Optimal Allocation and Sizing of Distributed Generation for Power Loss Reduction using Modified PSO for Radial Distribution Systems" Journal of Energy Technologies and Policy Volume: 3 Issue: 3, 2013 Pages 1-8.

8

Optimum Design of Penstock for Hydro Projects

Singhal M. K., Arun Kumar

AHEC, Indian Institute of Technology, Roorkee, India

Email address:

mksalfah@iitr.ac.in (Singhal M. K.), akumafah@iitr.ac.in (Arun K.)

Abstract: Penstock, a closed conduit, is an important component of hydropower projects. Various methods are available for optimum design of penstock. These methods are either based on empirical relations or derived analytically by optimizing the friction loss in the penstock. These formulae produce different values of penstock diameter for same site. In this study, formulae available for penstock design have been compared to review their suitability. A new method has been developed for the optimum design of penstock based on minimizing the total head loss comprising of friction and other losses. By using new developed method, diameter and annual cost of penstocks for few Hydro Electric plants of varying capacity have been worked out and reduction in annual cost of penstocks have been found in comparison to penstock cost for these projects.

Keywords: Friction Losses, Total Head Losses, Annualized Penstock Cost, Optimum Diameter of Penstock

1. Introduction

Hydropower, a renewable and mature energy source utilizes water from higher to lower altitude to generate power. Hydro Power is one of the proven, predictable and cost effective sources of renewable energy. Hydropower system (Fig 1) comprises of hydro source, diversion/storage system, water conductor system (channel/tunnel/penstock), power house building, generating and control equipment. Penstock is a conduit or tunnel connecting a reservoir/forebay to hydro turbine housed in powerhouse building for power generation. It withstands the hydraulic pressure of water under static as well as dynamic condition. It contains the closing devices (gates /valves) at the starting (just after reservoir/forebay) and at the tail end just before turbine to control the flow in the penstock. The penstock material may be mild steel, glass reinforced plastic (GRP), reinforced cement concrete (RCC), wood stave, cast iron and high density polyethylene (HDPE) etc. However, in the most of the cases, mild steel has been used for penstock since long due to wider applicability and availability. The penstock cost contributes an appreciable percentage towards the total civil works cost of the hydroelectric project. By optimizing the penstock diameter, maximum energy generation can be obtained at minimum cost.

The aim of present study is to develop the new relation for the optimum design of penstock by considering all types of losses in the penstock.

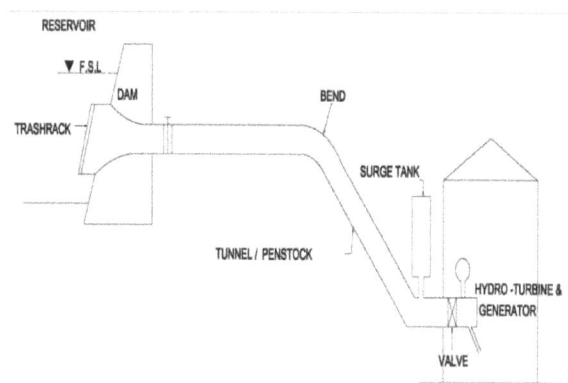

Fig. 1. Penstock and Hydropower system schematic layout.

2. Relations Available for Optimized Design of Penstock

Various researchers have proposed the methodology and relations for the optimum design of penstock. These relations are either empirical relations developed by analyzing and correlating statistical data of existing/installed projects designed as per past practice or derived analytically by minimizing annual penstock cost considering friction loss. The available analytical relations considered only friction loss whereas in addition to friction loss, there are other losses such as losses at specials (bends, transition piece, tri/bifurcation etc.), losses at valves and inlet which are not considered while optimizing the penstock design.

Annual penstock cost comprises of investment cost on penstock material, excavation for penstock, concreting, installation, operation, maintenance. depreciation and cost of energy loss caused by head loss.

The available empirical relations can be grouped in different categories based on parameters used to determine the optimum penstock diameter.

2.1. Empirical Relations Based on Penstock Discharge (Q)

Warnick et al. (1984) developed formula for optimum diameter of penstock pipe (D_e) for small hydro projects in terms of rated discharge (Q) as shown in eqn. (1)

$$D_e = 0.72 \, Q^{0.5} \qquad (1)$$

2.2. Empirical Relations Based on Project Installed Capacity (P) and Rated Head (H_r)

Bier (1945), Sarkaria (1979) and Moffat et al. (1990) have developed the formulae for estimation of economical diameter of penstock in terms of installed capacity (P) and rated head (Hr). Warnick et al. (1984) developed formula for optimum diameter of penstock pipe for large hydroelectric projects having rated head between 60 m to 315 m and power capacities ranging from 154 MW to 730 MW. These relations are shown in eqn. (2) to eqn. (5)

Bier (1945) relation

$$D_e = 0.176 \, (P / H_r)^{0.466} \qquad (2)$$

Sarkaria (1979) relation

$$D_e = 0.71 \, P^{0.43} / H_r^{0.65} \qquad (3)$$

Warnick et al. (1984) relation

$$D_e = \frac{0.72 P^{0.43}}{H_r^{0.63}} \qquad (4)$$

Moffat et al. (1990) relation

$$D_e = \frac{0.52 P^{0.43}}{H_r^{0.60}} \qquad (5)$$

2.3. Empirical Relations Based on Penstock Discharge (Q) and Rated Head (H_r)

Voetsch and Fresen (1938), USBR (1986), ASCE (1993) and Fahlbusch (1987) have developed the relations for economic penstock diameter in terms of rated discharge and rated head. These relations are shown as eqn. (6) to eqn. (9).

Voetsch and Fresen (1938) relation

This relation can be used for penstock having discharge more than 0.56 m^3/s . As per this method, coefficient (K) is computed by using eqn.(6)

$$K = \frac{k_s \, e \, f_l \, S \, e_j \, b}{a \, r \, (1 + n_s)} \cdot \qquad (6)$$

The loss factor (f_l) is obtained on the basis of plant load factor (p_f) from Graph A (16). K is used to select value of coefficient B and coefficient D' corresponding to penstock discharge and gross head from graph B and graph C respectively. Optimize diameter for penstock is calculated as $D_e = B \times D'$

USBR (1986) relation

$$D_e = \left(\frac{4Q}{0.125 * 3.14(2gH_r)^{0.5}}\right)^{0.5} \qquad (7)$$

After simplification, eqn. (7) may be written as shown in eqn. (8)

$$D_e = 1.517 \, Q^{0.5}/H_r^{0.25} \qquad (8)$$

Fahlbusch (1987) relation

$$D_e = 1.12 \, Q^{0.45} /H_r^{0.12} \qquad (9)$$

ASCE (1993) relation
The relation given by ASCE is shown in eqn. (10).This relation is in FPS system

$$D_e = 0.05 \left[\frac{S \, M \, h \, f \, e \, Q^3 P_{wf}}{W \, C \, H_r}\right] \qquad (10)$$

Present worth factor (P_{wf}) is computed from eqn. (11)

$$P_{wf} = ((1 + int)^{nr} - 1)/(int \, (1 + int)^{nr}) \qquad (11)$$

2.4. Analytical Relation for Optimize Design of Penstock

Analytical relation available for optimum diameter of penstock in Indian Standard (IS) 11625, is based on minimizing the total annual expenditure on penstock considering only friction loss. The relation is shown as eqn. (12).

$$D^{22/3} = \frac{2.36 \times 10^6 \, Q^3 \, n^2 \, e \, p_f \, C_p}{[1.39 C_e + 0.6 C_c + (121 \, H \, C_s (1+i))/(\sigma \, e_j)] \, p} \qquad (12)$$

3. Comparative Study of Various Relations Available for Optimum Diameter of Penstock

In this study, the comparative study of relations available for optimum diameter of penstock has been done by calculating penstock diameter for few Hydro Electric (HE) projects in India by these methods. The features of HE projects analyzed are shown in Table 1. For projects with higher gross head, the effect of head on computing penstock diameter becomes significant. Therefore, to compare the results of these relations, the projects are grouped in three categories based on gross head. The projects having gross head up to 100 m are grouped in Category A, 100 m to 200 m in category B and above 100 m in category C. The penstock diameter computed for these projects have been shown in Table 2. From the analysis of results shown in this table, it may be seen that the values of diameters obtained from these relations are different. The results have been compared in Fig. 2, Fig. 3 and Fig. 4 for projects in category A, category B and category C respectively. From these figures it may be seen that ASCE and Warnick relations generally provide the penstock diameters on higher side while Bier and USBR relations provide the penstock diameter on lower side.

Table 1. *Features of HE projects Analyzed in the Study.*

Sl. No.	Project	Capacity (kW)	Discharge (m³/s)	Penstock Length (m)	Gross Head (m)
Category A Projects					
1	Dugtu	25	0.17	360	31.25
2	Gaundar	100	0.38	105	49.16
3	Kuti	50	0.38	200	53.45
4	Kotijhala	200	0.48	120	80.60
5	Wachham	500	1.38	120	48.94
6	Debra	1500	1.84	138	95.35
7	Dhera	1500	2.12	170	85.20
8	Gaj	1500	4.88	121	38.44
9	Nyikgong	13000	23.75	156	78.00
10	Kamlang	24900	68.02	2260	44.92
Category B Projects					
11	Baram	1000	1.90	700	127.50
13	Divri	3200	3.06	191	118.42
14	Sarbari-ii	5400	3.65	365	191.37
12	Keyi	23000	20.50	1653	127.54
15	Thru	60000	37.18	8190	191.57
16	Phunchung	45000	37.46	4184	149.00
Category C Projects					
17	Jirah	4000	1.34	870	380.00
18	Ditchi	2500	1.45	453	204.35
19	Luni-II	5000	1.66	960	358.40
20	Luni-III	5000	1.70	1250	363.18
21	Pemashelpu	81000	34.07	350	289.00

Table 2. *Penstock Diameter obtained using Various available Methods.*

Sl. No.	Project	Penstock dia based on analytical method (IS) (m)	Penstock Diameter (m) based on empirical Methods							
			Moffat	Warnick	Sarkaria	Fahlbusch	USBR	Bier	ASCE	Voetsch & Fresen
Category A Projects										
1	Dugtu	0.38	0.27	0.29	0.31	0.33	0.27	0.16	0.53	-
2	Gaundar	0.53	0.37	0.44	0.41	0.45	0.35	0.25	0.70	-
3	Kuti	0.53	0.26	0.44	0.29	0.45	0.35	0.17	0.69	-
4	Kotijhala	0.58	0.37	0.50	0.40	0.48	0.35	0.27	0.73	-
5	Wachham	0.91	0.73	0.85	0.82	0.81	0.68	0.52	1.23	0.91
6	Debra	1.01	0.79	0.98	0.86	0.85	0.66	0.64	1.26	0.95
7	Dhera	1.07	0.84	1.05	0.92	0.92	0.73	0.67	1.37	1.03
8	Gaj	1.55	1.36	1.59	1.55	1.48	1.35	0.97	2.19	1.64
9	Nyikgong	2.82	2.24	3.51	2.47	2.76	2.49	1.91	3.90	2.86
10	Kamlang	4.64	4.34	5.94	4.92	4.79	4.94	3.48	6.62	4.92
Category B Projects										
11	Baram	1.00	0.56	0.99	0.60	0.84	0.63	0.46	1.23	0.88
13	Divri	1.22	0.96	1.26	1.03	1.05	0.81	0.82	1.52	1.14
14	Sarbari-ii	1.27	0.90	1.38	0.95	1.07	0.78	0.84	1.54	1.07
12	Keyi	2.68	2.18	3.26	2.34	2.45	2.06	2.01	3.41	2.50
15	Thru	3.34	2.64	4.39	2.79	3.06	2.54	2.66	4.15	2.90
16	Phunchung	3.41	2.67	4.41	2.84	3.16	2.69	2.58	4.32	3.05
Category C Projects										
17	Jirah	0.79	0.53	0.83	0.53	0.63	0.40	0.53	0.91	0.67
18	Ditchi	0.86	0.62	0.87	0.65	0.70	0.48	0.57	1.02	0.80
19	Luni-II	0.87	0.60	0.93	0.61	0.70	0.45	0.61	1.00	0.68
20	Luni-III	0.87	0.60	0.94	0.61	0.70	0.46	0.60	1.01	0.76
21	Pemashelpu	3.11	2.25	4.20	2.31	2.78	2.15	2.44	3.77	2.74

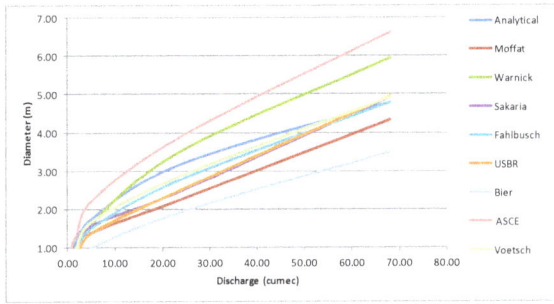

Fig. 2. *Relation between Penstock diameter and discharge for category A projects.*

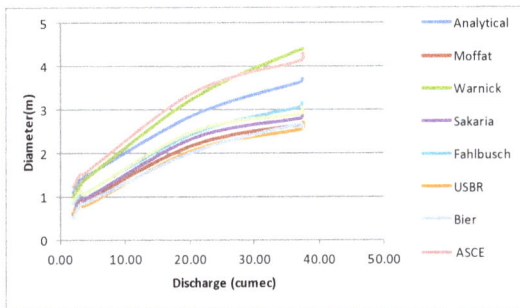

Fig. 3. *Relation between Penstock diameter and discharge for category B projects.*

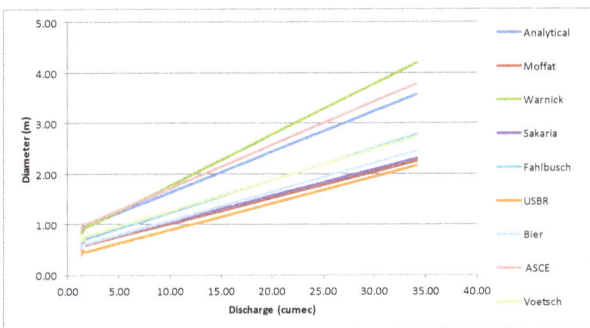

Fig. 4. *Relation between Penstock diameter and discharge for category C projects.*

4. Effect of Head Loss on the Computation of Optimum Diameter of Penstock

The optimum diameter of penstock is computed based on minimizing the total annualized cost of the penstock comprising of annualized expenditure on penstock and annual loss of revenue for energy loss caused by head losses in the pipe. For smaller diameter penstock, the expenditure on penstock will be less but, this will have higher head loss leading to higher energy and revenue loss. Similarly, the larger diameter penstock will have less head loss resulting in more energy generation and higher revenue but, will have more installation cost leading to higher annual cost of penstock. Therefore, head loss in penstock effects the determination of optimum diameter of penstock.

The head loss in penstock comprises of friction loss (h_f) due to resistance offered by penstock inner surface and other losses. The sources of other head losses (h_0) are specials comprising of intake provided at the entry of penstock with reservoir to allow smooth entry of discharge into penstock, control equipment (gate or valve) provided with penstock outside the reservoir/forebay to control the penstock discharge, bends provided at each location of change in penstock alignment, bifurcation (wye) piece provided at the end of penstock to connect it with its branches (one each for each unit), inlet valve provided at the end of each penstock branch to control the flow entering the turbine and transition piece (expansion or reducer) provided before or after inlet valve to connect it with penstock branch or with turbine.

The head loss also occurs at trash rack provided in front of penstock intake which checks the entry of floating and other material into the penstock. As the head loss at trash rack is not dependent on the flow velocity in the penstock and is not considered for the optimum design of penstock.

4.1. Determination of Friction Loss (h_f) in Penstock

The relation for determination of friction loss in penstock was first given by Chezy (1775). Subsequently, improved relations were given by various researchers. Some of these relations are for laminar flow having Reynold number (Re) less than 2000, some are for transition flow having Re in between 2000 and 4000 and some are for turbulent flow having Re more than 4000.

Poiseuille (1841) developed the relation for laminar flow. Hazen Williams (1902) and Blasius (1913) developed the relations for turbulent flow. The flow in penstock shall be turbulent flow as penstock with very small diameter like 0.1 m and flow velocity as 2.0 m/sec shall have Re as 2×10^5 which is in the range of turbulent flow. The relation developed by Hazen Williams is not much used for penstock as this relation can be effectively used for the flow velocity up to 1.0 m/sec which is much less than the flow velocity in penstock. The relation developed by Blasius is also not applicable for penstock flow as relation given by him is applicable for flow having Re up to 1×10^5 which is less than the Re of flow in penstock. The relations developed by Darcy-Weisbach (1850), Manning (1891) and Scobey (1930) are the most commonly used relations to determine the friction loss in penstock.

4.1.1. Friction Loss as Per Manning Relation

Manning relation computes the friction loss in pipe flow in terms of roughness coefficient, flow velocity, penstock length and hydraulic radius as shown in equation (13).

$$h_f = \frac{L\, v^2\, n^2}{R^{4/3}} \qquad (13)$$

In Manning equation, roughness coefficient is not dimensionless which lead to variation in Manning roughness coefficient for different penstock diameters under same flow conditions. Further, in Manning relation, flow characteristics are not accounted. The Manning roughness coefficient for steel pipe is taken between 0.008 to 0.012. By taking v = 4 Q/(πD^2) and R = D/4, eqn. (13) may be simplified as shown in eqn. (14)

$$h_f = \frac{10.29\, Q^2 n^2 L}{D^{16/3}} \qquad (14)$$

4.1.2. Friction Loss as Per Darcy Weisbach Relation

Darcy – Weisbach relation computes the friction loss in pipe flow in terms of friction coefficient and other parameters as shown in eqn. (15)

$$h_f = \frac{L f v^2}{2gD} \qquad (15)$$

By taking v = 4 Q/(πD^2) and g = 9.81, eqn. (15) may be simplified as shown in eqn. (16)

$$h_f = \frac{0.0826\ f\ Q^2\ L}{D^5} \qquad (16)$$

To determine friction coefficient in Darcy Weisbach equation, Nikuradse (1933) carried out experiments on uniformly roughened pipes and developed curves between friction coefficient and Re. Karman Prandtl (1932) developed the empirical relation for these curves. These curves and relation are having limited application as in practice, the roughness in pipes is not uniform but irregular and wavy.

Colebrook and White (1937) developed the relation between Re and f by carrying out experiments on commercially available pipes considering the effect of height of roughness in pipes. The equation given by them required various trials for solution as their equation was implicit. Moody's (1944) developed the curves corresponding to different values of relative roughness (k/D) for giving direct values of friction coefficient based on Colebrook White relation. Swami and Jain (1976) developed the explicit relation to determine the friction coefficient in terms of Re and relative roughness height (k/D) based on Colebrook and White relation as per eqn. 17.

$$f = 0.25 \log (k/3.7D + 5.74/Re^{0.9})^{-2} \qquad (17)$$

4.1.3. Friction loss as per Scobey Relation

Scobey relation computes the friction loss in terms of friction coefficient and other parameters as shown in eqn. (18). The value of this coefficient is taken in between 0.32-0.36 with average value as 0.34.

$$h_f = (1/1000)L\, K_s\, \frac{v^{1.9}}{D^{1.1}} \qquad (18)$$

The relation shown in eqn. (18) is in FPS system. This relation has been modified in MKS system and shown as eqn. (19)

$$h_f = 0.002586\, L\, K_s\, \frac{v^{1.9}}{D^{1.1}} \qquad (19)$$

4.2. Development of Relations Between Various Friction Coefficients

The relation given by Darcy Weisbach is mostly used for computation of friction loss in penstock. In Darcy Weisbach equation, friction coefficient is dimensionless and depends on Re representing the flow characteristics. Further, in case of Manning and Scobey equations, there is no guideline available to select the values of friction coefficients in these relations as per site requirement. In this study, new relations have been developed to select values of Manning and Scobey friction coefficients based on Darcy Weisbach coefficient

4.2.1. Development of Relation Between Manning and Darcy Weisbach Coefficients

To develop the relation between Manning and Darcy Weisbach friction coefficients, friction losses as per Manning relation (eqn. 14) and Darcy Weisbach relation (eqn. 16) have been compared as eqn. (20)

$$\frac{10.29Q^2 n^2 L}{D^{16/3}} = \frac{0.0826 L f Q^2}{D^5} \qquad (20)$$

Eqn. (20) may be simplified as eqn. (21)

$$n = 0.0896\sqrt{f\ D^{1/3}} \qquad (21)$$

4.2.2. Development of Relation between Scobey and Darcy Weisbach Coefficients

To develop relation between Scobey and Darcy Weisbach friction coefficients, friction losses as per Scobey relation (eqn. 19) and Darcy Weisbach relation (eqn. 16) have been compared as shown in eqn. 22

$$0.002586\, L\, K_s\, \frac{V^{1.9}}{D^{1.1}} = \frac{0.0826\, L\, f\, Q^2}{D^5} \qquad (22)$$

Eqn. (22) may be simplified as eqn. (23)

$$ks = 20.165\ f\ (Q/D)0.1 \qquad (23)$$

4.2.3. Comparison of Friction Coefficients Obtained from New Developed Relations

To compare the values of friction coefficients obtained from relations developed in eqn. (21) and eqn. (23), various options of penstock diameter have been considered. Minimum diameter of penstock has been taken as 0.1 m with increment in diameter as 0.1 m upto penstock diameter of 0.5 m then increment of 0.5 m upto penstock diameter of 8.0 m. The penstock discharges have been computed corresponding to all these diameters by taking the economic penstock flow velocity as 3.5 m/s.

Darcy Weisbach friction coefficient has been calculated as per Swami and Jain relation (eqn. 17). The average height of roughness in steel penstock has been taken for as 0.045 mm. Manning and Scobey friction coefficients have been computed as per eqn. (21) and eqn. (23) respectively. All these friction coefficients have been compared in Table 3 and it may be seen that as penstock diameter increases from 0.1 m to 8.0 m, Darcy Weisbach coefficient decreases from 0.0175 to 0.008 while Manning friction coefficient increases from 0.008075 to 0.011334. The values obtained for Manning coefficient are in conformity with the values given for this coefficient in para 4.4.1.2 of Indian Standard (IS) 11625 as 0.008 to 0.012. While carrying out various trials to compute the optimum diameter of penstock, average value of Manning friction coefficient is generally taken in these trials based on different penstock diameters. This average value of Manning coefficient may over estimate friction loss for smaller diameter pipes and underestimate for larger diameter pipes. However as per Gordon [1978], Manning equation may under estimate the friction loss for smaller diameter pipes. With regard to Scobey friction coefficient, it may be further seen from Table 3 that as penstock diameter increases, this coefficient varies (decreases or increases) at randomly. Further, the average value for this coefficient is generally taken as 0.32 which underestimates the friction loss in penstock as minimum value of Scobey coefficient is taken as 0.551261.

Table 3. Comparison of Friction Coefficients used in various formulae.

Sl. No	Penstock Diameter D (m)	Penstock Discharge Q (m³/s)	Relative Roughness (k/D)	Reynolds number (10^6)	Darcy Weisbach coefficient (f)	Manning coefficient (n)	Scobey coefficient (k_s)
1	0.1	0.03	0.0004500	0.35	0.01750	0.008075	0.778030
2	0.2	0.11	0.0002250	0.70	0.01500	0.008392	0.714748
3	0.3	0.25	0.0001500	1.05	0.01450	0.008828	0.719513
4	0.4	0.44	0.0001125	1.40	0.01350	0.008936	0.689443
5	0.5	0.69	0.0000900	1.75	0.01150	0.008560	0.600556
6	1.0	2.75	0.0000450	3.50	0.01000	0.008960	0.559704
7	1.5	6.18	0.0000300	5.25	0.00980	0.009490	0.571207
8	2.0	10.99	0.0000225	7.00	0.00950	0.009803	0.569882
9	2.5	17.17	0.0000180	8.75	0.00920	0.010012	0.564339
10	3.0	24.73	0.0000150	10.50	0.00900	0.010208	0.562229
11	3.5	33.66	0.0000129	12.25	0.00890	0.010416	0.564619
12	4.0	43.96	0.0000113	14.00	0.00880	0.010590	0.565779
13	4.5	55.64	0.0000100	15.75	0.00850	0.010614	0.552966
14	5.0	68.69	0.0000090	17.50	0.00845	0.010770	0.555536
15	5.5	83.11	0.0000082	19.25	0.00840	0.010910	0.557537
16	6.0	98.91	0.0000075	21.00	0.00835	0.011037	0.559062
17	6.5	116.08	0.0000069	22.75	0.00830	0.011151	0.560180
18	7.0	134.63	0.0000064	24.50	0.00820	0.011222	0.557548
19	7.5	154.55	0.0000060	26.25	0.00810	0.011282	0.554561
20	8.0	175.84	0.0000056	28.00	0.00800	0.011334	0.551261

4.3. Determination of Other Head Losses in Penstock (h_0)

The other head losses at specials comprises of intake loss (h_i), gate/valve loss (h_g), bend loss (h_b), bi/trifurcation loss (h_y), inlet valve loss (h_g) and transition piece loss (h_{tr}). All these losses can be represented as $k_a v^2/2g$ where v is the penstock flow velocity and k_a is the loss coefficient corresponding to that component.

Loss coefficient for penstock intake (k_i) depends upon shape of intake mouth. Loss coefficient for gate/valve (k_g) and control valve (k_v) depend upon the percentage reduction in flow area due to flap of closing device. Loss coefficient for bend (k_b) depends upon no. of bends, shape of bends, deflection angle and ratio of radius of bend with diameter of pipe. Loss coefficient at wye (k_y) is governed by angle of bi/trifurcation, ratio of cross-sectional area, type and shape of bi/trifurcation. The head loss in transition piece (h_{tr}) is represented as $k_{tr}(v_1^2-v_2^2)/2g$ where v_1 is penstock velocity or velocity in transition piece (whichever is higher) and v_2 is remaining velocity. By substituting v_1 and v_2 in terms of penstock velocity as $k_1 v^2/2g$ and $k_2 v^2/2g$ respectively, head loss due to transition piece may be taken as $k_{tre} v^2/2g$ where k_{tre} is effective head loss coefficient due to transition piece. The other head losses (h_0) at specials may be computed from eqn. 24.

$$h_o = h_i + h_g + h_b + h_y + h_v + h_{tr} \qquad (24)$$

Putting the values of head losses at specials in terms of $k_a v^2/2g$ in eqn. (24), h_o can be obtained from eqn. (25)

$$h_o = k_i \frac{v^2}{2g} + k_g \frac{v^2}{2g} + k_b \frac{v^2}{2g} + k_y \frac{v^2}{2g} + k_v \frac{v^2}{2g} + k_{tre} \frac{v^2}{2g} \qquad (25)$$

By taking $v^2/2g$ as common and writing sum of remaining terms as k_o, eqn. (25) may be written as shown in eqn. (26)

$$h_o = k_o v^2/2g \qquad (26)$$

4.4. Determination of Total Head Loss in Penstock (T_{hl})

The total head loss (T_{hl}) can be obtained as per eqn. 27.

$$T_{hl} = h_f + h_0 \qquad (27)$$

By putting value of h_f from Darcy Weisbach relation (eqn. 15) and h_o from eqn. (26) in eqn.(27), T_{hl} can be obtained from eqn. (28)

$$T_{hl} = \frac{fLv^2}{2gD} + ko\,\frac{v^2}{2g} \qquad (28)$$

By taking $v^2/2g$ common in eqn. (28), remaining terms shall be constant. Substituting sum of remaining terms as k_{tl}, T_{hl} can be obtained from eqn. (29)

$$T_{hl} = k_{t1}\, v^2/2g \qquad (29)$$

4.5. Development of Relation between Total Head Loss and Friction Loss

The total head loss comprises of friction loss and losses at specials and depends on diameter and length of penstock. For projects with high gradient, length of penstock required will be small. For projects having flatter gradient, penstock length is longer. As the length of penstock increases, total head loss in penstock also increases. The penstock can be categorized as longer and shorter penstock on the basis of the ratio (R_{plh}) of penstock length and gross head. For longer penstock, the main head loss is due to friction. For smaller penstock, the contribution of friction loss in total head loss is less. Therefore, ratio of total head loss and friction loss (R_{thf}) for longer penstock will be less and for shorter penstock, will be more. To study the variation in contribution of friction loss in total head loss corresponding to variation in length of penstock, the values of R_{plh} and R_{thf} for various hydro electric projects have been computed and shown in Table 4

Table 4. *Friction and Total Head Losses in Penstock of Hydroelectric Projects.*

Sl. No.	Project Name	Head Loss (m)			Ratio of penstock length/Gross head (R_{plh})	Ratio of Total head loss and Friction loss (R_{thf})
		Friction	Others (specials)	Total		
1	Pemashelpu	0.54	0.84	1.38	1.21	2.55
2	Debra	0.24	0.35	0.59	1.45	2.46
3	Kotijhala	0.25	0.37	0.62	1.49	2.45
4	Divri	0.33	0.47	0.79	1.62	2.41
5	Sarbari-ii	0.73	0.98	1.72	1.91	2.34
6	Dhera	0.29	0.39	0.68	1.99	2.32
7	Nyikgong	0.18	0.24	0.41	2.00	2.32
8	Gaundar	0.22	0.28	0.50	2.14	2.29
9	Ditchi	1.12	1.42	2.53	2.22	2.27
10	Jirah	2.88	3.62	6.50	2.29	2.26
11	Wachham	0.20	0.25	0.46	2.45	2.23
12	Luni-II	3.04	3.63	6.67	2.68	2.19
13	Gaj	0.17	0.19	0.35	3.15	2.13
14	Luni-III	4.11	4.48	8.58	3.44	2.09
15	Kuti	0.46	0.48	0.94	3.74	2.06
16	Baram	1.58	1.44	3.01	5.49	1.91
17	Dugtu	1.02	0.68	1.70	11.52	1.66
18	Keyi	2.81	1.76	4.56	12.96	1.63
19	Phunchung	7.52	3.03	10.54	28.08	1.40
20	Thru	17.08	5.04	22.12	42.75	1.30
21	Kamlang	2.98	0.76	3.75	50.30	1.26

From the analysis of results shown in Table 4, it may be seen that for shorter penstock (R_{plh} as 1.21) and longer penstock (R_{plh} as 50.30), ratio of total head loss with friction loss (R_{thf}) is 2.55 and 1.26 respectively. It means that contribution of other losses in total head loss for longer penstock may be as less as 26% of friction loss while for shorter penstock, this contribution may be as high as 155% of friction loss. The variation of R_{thf} and R_{plh} for various hydro electric projects has been shown in fig 5. The best fit curve has been drawn between values of R_{thf} and R_{plh} based on regression analysis. The equation of best fit curve is shown in eqn. (30).

$$R_{thf} = 2.644\,(R_{plh})^{-0.19} \qquad (30)$$

The correlation coefficient (R^2) for the above relation has been observed as 0.837 which is within permissible limit.

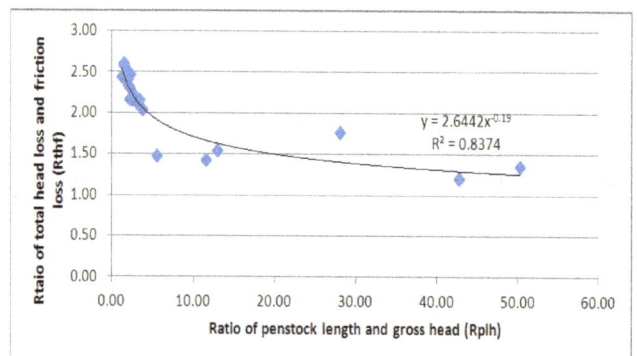

Fig. 5. *Relation of R_{thf} and R_{plh}.*

Therefore, relation shown in eqn. (30) may be used for

hydro electric projects. By placing R_{thf} as T_{hl}/h_f and R_{plh} as L/H in eqn. (30), the resulting eqn. is shown as eqn. (31)

$$T_{hl}/h_f = 2.644 (L/H)^{-0.19} \qquad (31)$$

For any project, the length of penstock and gross head are site specific parameters. These will remain same while carrying out various trials in order to determine optimum diameter of penstock. Therefore, term on right hand side of eqn. (31) shall be constant. By replacing right hand side in eqn. (31) as constant (k_t), the resulting eqn. is shown as eqn. (32).

$$T_{hl} = k_t h_f \qquad (32)$$

5. Development of New Relation for Optimum Design of Penstock Based on Total Head Loss

The available analytical relation for optimum design of penstock is based on minimizing the annualized cost of penstock system considering only friction loss. In this study, a new relation has been developed for optimum design of penstock considering losses at specials in addition to friction loss. The steps followed are as follows
- Computation of annualized cost of penstock
- Computation of annual loss of revenue due to energy loss caused by total head loss
- Development of relation for optimum diameter of penstock based on total head loss

5.1. Determination of Annualized Cost of Penstock (Ep)

Penstocks are laid on surface or buried/embedded in ground. The surface penstocks are laid above ground and supported on concrete blocks at regular interval. Buried penstocks are laid in tunnel or cut and fill section which requires excavation of trench/tunnel of slightly higher size than that of penstock so that penstock may be laid inside. After laying penstock in tunnel, concreting is done in extra portion. For surface penstock, the cost of excavation as well as concrete lining will be less. Total expenditure on penstock will comprise of cost of excavation for laying penstock, cost of concrete lining and cost of penstock.

5.1.1. Cost of Excavation for Laying Penstock (C_{ex})

For laying the penstock, trench/tunnel of higher diameter ($D_e + 0.33D_e$) is excavated. Cost of excavation for laying penstock is computed by multiplying volume of material excavated with unit rate of excavation as per eqn. (33).

$$C_{ex} = \frac{\pi}{4} (D_e + 0.33 D_e)^2 C_e L \qquad (33)$$

After simplifying, eqn. (33) may be written as per eqn. (34)

$$C_{ex} = 1.39 DD_e^2 C_e L \qquad (34)$$

5.1.2. Cost of Concrete Lining (C_{cl})

The cost of concrete lining for penstock may be computed by multiplying the quantity and unit rate of concrete as per eqn. (35). For computation of quantity, thickness of concrete lining may be taken as $0.165 D_e$.

$$c_{c1} = \pi (D_e + 0.165 D_e)0.165 D_e C_c L \qquad (35)$$

After simplifying, eqn. (35) may be written as per eqn. (36)

$$C_{cl} = 0.6 D_e^2 C_c L \qquad (36)$$

5.1.3. Cost of Penstock (C_{sp})

To compute cost of the penstock, the weight of penstock is required which can be computed in terms of penstock thickness, weight of penstock stiffeners as ratio (i) of penstock weight and density of penstock material (ρ) as per eqn. (37)

$$Wg = \pi D_e t (1 + i) \rho L \qquad (37)$$

The penstock thickness can be computed in terms of water pressure (p_r) developed in pipe due to gross head (H_{gross}), permissible stresses in pipe material, joint efficiency and other parameters as shown in eqn. (38)

$$t = \frac{p_r D_e}{2\sigma e_j} \qquad (38)$$

In case of sudden closure of the turbine, water pressure/head inside the penstock increases due to water hammer. Taking increase in head as ΔH, total design head inside the pipe (H) becomes ($H_{gross} + \Delta H$). By replacing p_r in eqn. (38) as 0.1H, t can be computed from eqn. (39)

$$t = \frac{0.1 H D_e}{2 \sigma e_j} \qquad (39)$$

Thickness computed from eqn. (39) should be more than the minimum handling thickness (t_{min}) of the penstock which is equal to ($D_e + 500)/400$ where D_e and t_{min} are in mm. ΔH can be computed in terms of velocity of pressure wave and other parameters as shown in eqn. (40)

$$\Delta H = v_a v/g \qquad (40)$$

velocity of pressure wave can be computed in terms of Bulk modulus of fluid and Young modulus of pipe material as per eqn. (41)

$$v_a = 4660/(1 + K_b D_e/E t) \qquad (41)$$

By placing the value of t from eqn. (39) in eqn. (37), Wg may be computed from the eqn. (42)

$$Wg = \pi \frac{0.1 H D_e^2}{2 \sigma e_j} (1+i)\rho L \qquad (42)$$

cost of penstock (C_{sp}) may be computed by multiplying penstock weight as per eqn. (42) with unit cost of pipe material and dividing by 9.81 to put the value of σ in MPa as shown in eqn. (43). The value of ρ has been taken as 7850 kg/m^3,

$$C_{sp} = \frac{121 H D_e^2 C_s (1+i)}{\sigma \ e_j} L \qquad (43)$$

5.1.4. Determination of Total Expenditure on Penstock (T_{exp})

Total expenditure on penstock is computed by adding cost of excavation, cost of concreting and cost of penstock as per eqn. (44).

$$T_{exp} = C_{ex} + C_{cl} + C_{sp} \qquad (44)$$

Putting the values of C_{ex}, C_{cl} and C_{sp} from eqn. (34), eqn. (36) and eqn. (43) respectively in eqn. (44), T_{exp} may be computed as per eqn. (45).

$$T_{exp} = 1.39 D_e^2 C_e L + 0.6 D_e^2 C_c + \frac{121 H D_e^2 C_s (1+i)}{\sigma \ e_j} L \qquad (45)$$

5.1.5. Annualized Cost of Penstock (E_p)

Annual expenditure on penstock may be taken as some percentage (p) of T_{exp}. Therefore, p is taken as the ratio of annual expenditure and total expenditure on penstock. E_p is computed by multiplying T_{exp} as per eqn. (45) and p as shown in eqn. (46)

$$E_p = D_e^2 \left[1.39 C_e + 0.6 C_c + \frac{121 H C_s (1+i)}{\sigma \ e_j} \right] p \ L \qquad (46)$$

5.2. Computation of Annual Loss of Revenue due to Energy loss caused by total Head Loss

To compute annual loss of revenue due to total head loss, the power loss (P_l) due to this head loss is required which is computed in terms of discharge, total head loss, and plant efficiency as shown in eqn. (47)

$$P_1 = 9.81 \ Q \ T_{hl} \ e \qquad (47)$$

The annual energy loss (E_l) can be computed by multiplying P_l as per eqn. (47) with total hours of operation in the year (h_r). h_r is computed by multiplying plant power factor (P_f) and total hours (8760) in a year. E_l is computed as per eqn. (48)

$$E_l = 9.81 \ Q \ T_{hl} \ e \ P_f \ 8760 \qquad (48)$$

The annual loss of revenue (E_t) due to energy loss can be computed by multiplying E_l as per eqn. (48) and cost of power (C_p) as shown in eqn. (49)

$$E_t = 9.81 \ Q \ T_{hl} \ e \ P_f \ 8760 \ C_p \qquad (49)$$

5.3. Development of Relation for Optimum Diameter of Penstock Based on Total Head Loss

In order to get the optimized penstock diameter (D_e) of penstock, the annualized penstock cost ($E_p + E_t$) is to be differentiated with respect to penstock diameter and equated to zero as shown in eqn. (50)

$$D_e = \frac{\delta (E_p + E_t)}{\delta D} = 0 \qquad (50)$$

E_t can be obtained by putting the value of T_{hl} from eqn. (32) in eqn. (49) as shown in eqn. (51)

$$E_t = 9.81 \ Q \ k_t \ h_f \ e \ P_f \ 8760 \ C_p \qquad (51)$$

value of h_f in eqn. (51) may be put up from eqn. (16) as shown in eqn. (52)

$$E_t = 9.81 Q \ k_t \frac{0.0826 f Q^2 \ L}{D^5} * e \ P_f \ 8760 \ C_p \qquad (52)$$

Placing the value of E_p and E_t in eqn. (50), from eqn. (46) and eqn. (52) respectively, the resulting eqn. is shown in eqn. (53)

$$\frac{\delta}{\delta D} \left(D_e^2 \left[1.39 C_e + 0.6 C_c + \frac{121 H C_s (1+i)}{\sigma \ e_j} \right] p L + 9.81 Q k_t \frac{0.0826 f Q^2}{D^5} L e P_f \ 8760 C_p \right) = 0 \qquad (53)$$

By differentiating eqn. (53), the relation obtained is shown in eqn. (54).

$$2 D_e \left[1.39 C_e + 0.6 C_c + \frac{121 H \ C_s (1+i)}{\sigma \ e_j} \right] p L - 5 \times 9.81 Q k_t \frac{0.0826 f Q^2}{D^6} L e P_f \ 8760 C_p = 0 \qquad (54)$$

After simplifying, eqn. (54) may be written as eqn. (55)

$$D^7 = \frac{0.0175 \times 10^6 \ Q^3 \ f \ e p_f \ Cp \ k_t}{[1.39 C_e + 0.6 C_c + 121 H \ Cs(1+i)/\sigma \ e_j] \ p} \qquad (55)$$

Putting the value of k_t in eqn. (55) from eqn. (31), the

relation obtained is shown in eqn. (56).

$$D^7 = \frac{0.04627 \times 10^6 \ Q^3 \ f \ e \ p_f \ Cp \ (L/H)^{-0.19}}{[1.39 C_e + 0.6 C_c + 121 H \ Cs(1+i)/\sigma \ e_j] \ p} \qquad (56)$$

The relation shown in eqn. (56) is the new relation for

optimum diameter of penstock based on total head loss

6. Applicability of Developed Relation for Hydro Electric Project

The relation developed in eqn. (56) has been used to determine the optimum penstock diameter of various HE Projects.

To compute the optimum penstock diameter, the values of parameters C_p, C_e, C_c and C_s have been taken as INR 5.5 per unit rate of electricity generation, INR 5150 per m^3, INR 8000 per m^3 and INR 100 per kg respectively. The values of parameters e, P_f σ, e_j and p have been taken as 0.85, 0.5 183.33 MPa, 1.0 and 0.16 respectively. The optimum diameter of penstock calculated for each project has been shown in Table 5. The saving in annual cost of each penstock may be seen in this table. The saving in cost is higher for penstock having lower ratio of penstock length with gross head. This saving decreases as the ratio of penstock length and gross head increases.

Table 5. Penstock Diameter and Annual Cost Saving As Per Developed Method for various HE projects.

Sl. No.	Project	Diameter as provided at site/DPR (m)	Diameter as per new developed relation (m)	% Increase in Diameter	Annual Cost of penstock (INR million)			
					as per site / DPR	as per new developed method	Net Saving	In % of cost of penstock
1	Pemashelpu	3.11	3.56	14.31	36.33	32.80	3.529	9.714
2	Debra	1.01	1.14	13.75	0.84	0.77	0.076	9.044
3	Kotijhala	0.58	0.66	13.67	0.23	0.21	0.020	8.939
4	Divri	1.22	1.38	13.41	1.85	1.69	0.160	8.639
5	Sarbari-ii	1.27	1.43	12.90	4.75	4.36	0.382	8.048
6	Dhera	1.07	1.20	12.77	1.09	1.00	0.086	7.892
7	Nyikgong	2.82	3.18	12.76	6.77	6.24	0.534	7.882
8	Gaundar	0.53	0.60	12.56	0.14	0.13	0.011	7.655
9	Ditchi	0.86	0.97	12.44	2.77	2.56	0.208	7.528
10	Jirah	0.79	0.89	12.35	6.55	6.07	0.486	7.418
11	Wachham	0.91	1.02	12.14	0.47	0.44	0.034	7.188
12	Luni-II	0.87	0.97	11.87	8.27	7.70	0.570	6.896
13	Gaj	1.55	1.73	11.38	1.28	1.20	0.082	6.377
14	Luni-III	0.87	0.97	11.11	10.82	10.16	0.660	6.098
15	Kuti	0.53	0.59	10.86	0.26	0.25	0.015	5.842
16	Baram	1.00	1.09	9.71	4.19	3.99	0.198	4.737
17	Dugtu	0.38	0.41	7.53	0.202	0.196	0.006	2.922
18	Keyi	2.68	2.87	7.18	66.92	65.13	1.789	2.674
19	Phunchung	3.41	3.58	4.96	278.61	274.94	3.672	1.318
20	Thru	3.34	3.47	3.77	577.00	572.48	4.513	0.782
21	Kamlang	4.64	4.79	3.31	178.57	177.47	1.094	0.613

7. Conclusion

The various relations available for optimum design of penstock have been compared and it was observed that these relations provide different values of optimum penstock diameter resulting in different cost. Some of these relations are based on minimizing annual cost of penstock considering friction loss only whereas in practice other losses in penstock also occurs and needed to be considered. In addition, the different relations available for friction loss also provides different values of losses in penstock. A new method has been developed to optimize the design of penstock for hydro power projects on the basis of minimizing annual project cost considering total head loss (friction and other losses). All these losses have been formulated using Darcy Weisbach formula. The newly developed relation has been used for 21 hydro power projects with capacity ranging from 25 kW to 60 MW to find out the optimum diameter and compared with results obtained as provided at site/DPR. By providing penstock diameter as per new method, though the penstock diameter increased in the range of 3.31 to 14.31%, it resulted in the net saving in annual cost of penstock. The saving has been obtained from 0.613 % to 9.714% of earlier penstock cost which justify the applicability of the new method for optimum design of penstock for hydro power projects.

Symbols Used

ρ = density of penstock material

n = Manning's roughness coefficient

σ = permissible stress in penstock

a = cost of pipe in $ per lb

b = cost of 1 kWh of energy in $

C = capital cost of penstock installed per unit weight

C_p = cost of 1 kWh of energy

C_{cl} = cost of concrete lining / cum

C_c = unit rate of concrete lining

C_e = cost of excavation / cum for laying penstock

C_{ex} = cost of excavation per unit length of penstock

C_s = cost of steel / kg

C_{sp} = cost of penstock

D = Diameter of penstock

D_e = economic diameter of penstock

e = turbine/generator efficiency

e_j = joint efficiency of penstock

E = Young modulus of elasticity of steel

E_l = annual loss of energy due to total head loss

E_p = annualised cost of penstock

E_t = annual loss of revenue

f = friction factor in Darcy Weisbach and ASCE relation

f_l = loss factor in Voetsch and Fresen relation

h = annual hours of operation

$h_{i..tr}$ = head loss at penstock components. i..tr denotes the respective component

h_o = other head losses in penstock

h_f = friction loss in penstock

H_r = rated head

HE = hydro electric

i = ratio of weight of stiffeners and weight of penstock

int = interest rate in percentage

k = average roughness in penstock

k_{hl} = coefficient of total head loss in penstock

k_{itr} = coefficient of head loss at penstock component. itr denotes the respective component

k_o = coefficient of other head losses in penstock

k_s = Scobey friction factor

K = coefficient in Voetsch and Fresen

K_b = Bulk modulus of elasticity of water

L = length of penstock

M = composite value of power

n_s = ratio of weight of stiffeners and weight of penstock

n_r = repayment period

p_r = water pressure inside the penstock

p_f = annual load factor/Plant load factor

p = ratio of annual charges to installation cost of penstock

P = Installed capacity of project

P_l = power loss due to total head loss

P_{wf} = present worth factor

Q = Penstock discharge

r = ratio of annual charges to installation cost of penstock

R = Hydraulic radius

R_{plh} = ratio of penstock length and gross head

R_{thf} = ratio of total loss and friction loss

S = allowable stress in psi

t = thickness of penstock

T_{exp} = total expenditure on penstock per unit length(INR)

T_{hl} = total head loss in penstock

v = flow velocity in penstock

v_a = velocity of pressure wave

W = specific weight of steel

W_g = weight of penstock per unit length

INR= Indian Rupees

References

[1] Colebrook, C.F. and White, C.M. (1937) "Experiments with Fluid Friction in Roughened Pipes", Proc. Roy. Soc. Series A, 161, 367.

[2] Colebrook, C.F. and White, C.M. (1937-38) "The Reduction of Carrying Capacity of Pipes with Age", J. Inst. C.E., 7, 99.

[3] Colebrook C.F. (1939) "Turbulent flow in pipes with particular reference to the transition region between the smooth and rough pipe laws", J Inst Civil Engineers, London, Vol. 11, pp. 133-156.

[4] Fahlbusch, F.(1982) "power tunnels and penstock the economics re-examined" International Water power and dam construction, Vol. 34 no 6, June 1982, pp 13-15.

[5] Fablbusch F (1987), "Determining Diameters for Power Tunnels ad Pressure Shafts" Water Power and Dam Construction February 1987.

[6] Indian Standard:11625-1986(2001); "Criteria for Hydraulic Design of Penstock". Bureau of Indian Standards, Delhi PP-16

[7] Manning, R. (1891) "On the Flow of Water in Open Channels and Pipes" Trans. Inst. C.E. of Ireland, 20, 161.

[8] Moody L.F. (1944) "Friction Factors for Pipe Flow" Transactions of the ASME, Vol. 66(8), pp. 671–684.

[9] Nikuradse, J. (1933) "Stroemungsgesetze in rauhen Rohren." Ver. Dtsch. Ing. Forsch., 361, 1933

[10] Gordon, J.L., (1978), "Design criteria for exposed hydro penstock" Canadian Journal of civil Engineering, 5, pp. 340 – 351.

[11] Sarkaria, G.S. (1957), "Penstocks sized Quickly" Engineering News Record, Aug.15, 1957, pp 78-79.

[12] Sarkaria G.S (1979), "Economic Penstock Diameter: A 20-year Review" Water Power and Dam Construction November 1979.

[13] ASCE Manuals and Reports on Engineering Practice No. 79, (1993), Steel Penstocks, ASCE, American Society of civil Engineers, New York, 1993

[14] Scobey F.C. (1930), " The flow of water in Riveted and analogous pipes", Vol 150, USDA, Washington, DC, 45

[15] Swamee P.K. and Jain A.K. (1976) "Explicit equations for pipe-flow problems" Journal of the Hydraulics Division (ASCE), Vol 102 (5), pp. 657–664.

[16] Voetsch Charles and Fresen M.H (1938),"Economic Diameter of Steel Penstock" ASCE Transaction Vol 103, paper no 1982.

[17] USBR, Engineering Monogram No. 3 (1986), Welded Steel Penstocks, United States Department of The interior Bureau of Reclamation, Washington, 1986.

9

Thermal Properties of $(Na_{0.6}K_{0.4})NO_3$ Thermal Storage System in the Solid-Solid Phase

Halima Ibrahim ElSaeedy[1], Maryam Ayidh Saad Al Shahrani[2], Karam Fathy Abd El-Rahman[1], Sayed Taha Mohamed Hassan[3]

[1]Department of Physics, Faculty of Science for Girls, King Khalid University, Abha, Saudi Arabia

[2]Department of Physics, Faculty of Science for Girls, Bisha University, Abha, Saudi Arabia

[3]Department of Physics, Faculty of Science, King Khalid University, Abha, Saudi Arabia

Email address:

ruba-222@hotmail.com (H. I. ElSaeedy), empress-321@hotmail.com (M. A. S. A. Shahrani), elrahman100@gmail.com (K. F. A. El-Rahman), profdrstahahassan@gmail.com (S. T. M. Hassan)

Abstract: The thermal behaviour for the DTA, DSC and TGA measurements have been carried on the solid phase transformation for the binary eutectic mixture of 60 wt% sodium nitrate ($NaNO_3$) and 40 wt% potassium nitrate (KNO_3). Thermal energy storage materials are important for the technology that is applied to reduce cost solar thermal power generation. The $NaNO_3$-KNO_3 system is a binary inorganic salt system and it is one of the most promising thermal storage materials. The methods are based on the principle that a change in the physical state of a material is accompanied by the liberation or absorption of heat. The various techniques of thermal analysis are designed for the determination of the enthalpy accompanying the changes in the physical properties of the material. The thermal measurements showed a reversible phase transition at ~114°C during heating process and at ~108°C during cooling process. It has been shown also the presence of thermal hysteresis during this transformation with a magnitude of the hysteresis temperature ~8°C. The thermogravimetery analysis (TGA) indicated that the eutectic system $(Na_{0.6}K_{0.4})NO_3$ is thermally stable up to the melting point at $\cong 225$°C. This means that the sample under study is structurally stable. DTA measurements were also carried out for the sample at different heating rates of (2, 5, 10, 15 and 20°C/min). Some thermal parameters such as the transition point, enthalpy and the activation energy for the transformation process were estimated at each heating rate. It has been also shown that these parameters are affected by the heating rate. The noticeable effect of heating rate on the thermal parameters means that the heating rate is a main factor to change the thermal interaction potential of the Na and K atoms around the nitrate group (NO_3^-) during the phase conversion for the eutectic $(Na_{0.6}K_{0.4})NO_3$ system.

Keywords: Phase Change Material, Thermal Storage, Calorimetry

1. Introduction

Thermal energy storage systems are a key technology for reduced cost solar thermal power generation [1, 2]. Also, they are one possibility for solar thermal power plants to compensate temporary divergences between the availability of sunlight and the demand for electricity. This high temperature application requires storage operation above 100°C. Therefore, thermal energy storage at high temperature (>120°C) is an efficient way for energy saving in industrial process. Several mixtures of alkali nitrates and nitrites have been used as a heat transfer medium because of its low cost and good compatibility with common structural materials [3]. Additionally, significant attention has been given to using salts as a phase change storage media (using encapsulation and otherwise) [4]. The $NaNO_3$-KNO_3 system is one of the most extensively investigated binary inorganic salt systems [3, 4]. Mixture of $NaNO_3$ and KNO_3 is promising from the viewpoints of cost and thermal stability and can be used for sensible heat storage. The application of this mixed salt to thermal storage, for load leveling, and for solar thermal electric power generation were postulated to be promising [7]. The effect of the thermal history has been investigated by DSC and X-ray diffraction and the enthalpy change has been measured for quenched and annealed samples of 50 mole% $NaNO_3$-50 mole% KNO_3 by a high-temperature calorimeter of the twin type [7, 8]. A binary mixture of 30 wt.% potassium nitrate (KNO_3) and 70 wt.% sodium nitrate ($NaNO_3$) has been studied by Martin et al [8]. The measurement systems include a differential scanning calorimeter, a melting point apparatus, a custom-built adiabatic calorimeter and a lab-scale storage unit. Recently, salt has been considered for use in trough and linear fresnel based solar collectors and has been shown to offer a reduction in levelized cost of energy as well as increased availability [10]. The phase diagram of $NaNO_3$-KNO_3 is determined by DSC and high-temperature x-ray diffractometry [11] Thermal and mechanical properties of three representative salts for use in thermal storage systems have been evaluated as a function of temperature, thermal conductivity, specific heat, and the apparent heat of fusion were obtained using a differential scanning calorimeter [9]. Using additives such as graphite to $NaNO_3$-KNO_3 mixture leads to insignificant thermal conductivity improvement in overall application and hence the thermo physical properties for energy storage will be improved [10]. The system KNO_3-$NaNO_3$ was also discussed in detail in terms of their thermo-physical properties in the liquid and solid phase [11]. Recent measurements of the electrical conductivity of the solidified KNO_3-$NaNO_3$ 50: 50 mole% composition and of the melting enthalpy have suggested that the solid consists of alternating areas of sodium nitrate-rich and potassium nitrate-rich composition [12]. Also, a detailed calorimetric study of the $NaNO_3$/KNO_3 phase behaviour as a function of cation composition was presented [13].

In spite of the interest, no or very few examples of commercial high temperature thermal energy storage were realized and the KNO_3-$NaNO_3$ system is still has the attention of researchers. Therefore, we studied in this work the thermal properties of 40 wt.% potassium nitrate (KNO_3) and 60 wt.% sodium nitrate ($NaNO_3$) by the measurements of DSC, DTA and TGA.

2. Experimental Details

Chemically pure $NaNO_3$ and KNO_3, Aualar from BDH, were used as constituent compounds for the mixture (NaK)NO_3 by the molecular ratio 60% $NaNO_3$ and 40% KNO_3. The binary system was formed by the pre-annealing of the pure nitrates up to 400°C in a muffle furnace for three hours. The mixture was then allowed to be equilibrated in air at 400°C before casting into a stainless steel mould to quench and solidify to form the eutectic system ($Na_{0.6}K_{0.4}$)NO_3. Differential Scanning Calorimetry (DSC) and Differential Thermal Analysis (DTA) measurements were performed using a DSC-DTA apparatus (model Shimadzu DSC-50).

However Thermogravimetry Analysis (TGA) was performed using TGA apparatus (model Shimadzu TGA-50). It was found that this eutectic system has a melting point of $\cong 225$°C which is less than melting points of both $NaNO_3$ and KNO_3. The thermal measurements for the eutectic ($Na_{0.6}K_{0.4}$)NO_3 system were carried out by using (DSC) during heating and cooling at rate of 10°C/min, (TGA) and Differential Thermal Analysis (DTA) at heating rates of 2, 5, 10, 15 and 20°C/min.

3. Results and Discussion

3.1. Differential Scanning Calorimetry (DSC)

Differential Scanning calorimetry (DSC) technique is used in a wide range for the thermal investigation. Usually, a material such as (NaK)NO_3 will show definite and characteristic effects on heating which relate to its nature, composition and history. These observations are informative about its properties. The program may involve heating or cooling at a fixed rate of temperature change, or holding the temperature constant, or any sequence of these. Differential techniques involve the measurements of a difference in the property between the sample and a reference material. Thermal analysis is now generally recognized as one of the basic analytical tools for characterizing the compounds. The nature of phase transitions even in complex compounds and their thermal stability is often studied by DTA and DSC. Endothermic or exothermic band may be due to chemical or physical or physical reaction. DTA and DSC are completely to each other. Both of them provide a rapid method for studying the thermal kinetics of the material under test.

DSC which is performed in the present work for the eutectic ($Na_{0.6}K_{0.4}$)NO_3 samples was recorded during heating and cooling process. Various kinetic parameters controlling such phase transitions and thermal information are to be determined by such thermal analyses.

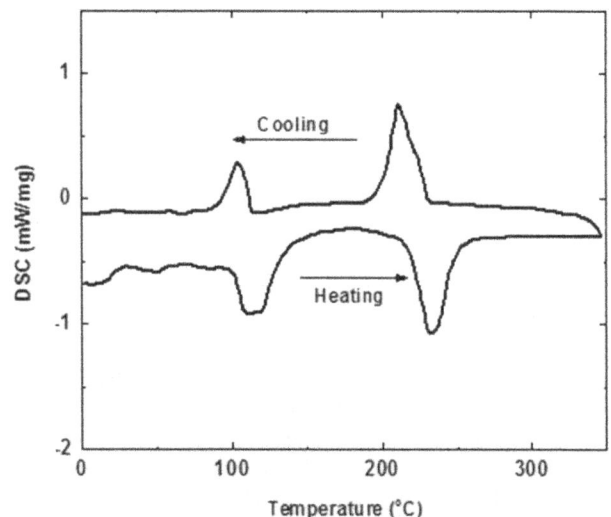

Figure 1. *DSC thermogram record during heating and cooling of the eutectic system of ($Na_{0.6}K_{0.4}$)NO_3 sample.*

The differential scanning thermal analysis curve shown in figure 1 was recording during heating and cooling process at a rate of 10°C /min. It is clear that it supports the presence of thermal hysteresis phenomena. This figure indicates that the endothermic peak of transformation for eutectic $(Na_{0.6}K_{0.4})NO_3$ system takes place at ~114°C during heating. However the reversibility of the exothermic peak of this transformation appeared at ~106°C. The difference in the peak height, in the forward and reverse direction i.e. during heating and cooling runs, is due to the difference in the heat of transformation (ΔH) during heating and the heat evolved during cooling. However, it is clear that there is a pronounced thermal hysteresis during the reversible transformation of the eutectic $(Na_{0.6}K_{0.4})NO_3$ system [14].

3.2. Thermogravimetry Analysis (TGA)

The thermogravimetry analysis (TGA) is basically quantitative in nature where the mass-changecan be accurately determined. This method is a useful technique for studying the ability of a substance to maintain its mass under a variety of conditions [17]. Figure 2 a, b show a high accurate thermal gravimetric analysis measurement for $(Na_{0.6}K_{0.4})NO_3$ sample during its phase transformation. It is note the analysis of this curve and its derivative that there is no nearly any mass loss in the temperature range (30 – 20°C) i.e. during the transformation process for this sample. This means that the $(Na_{0.6}K_{0.4})NO_3$ sample is structurally stable contributing to ant mass loss during the transformation process. It demonstrated also this transformation to have a thermal stability up to the melting point [18].

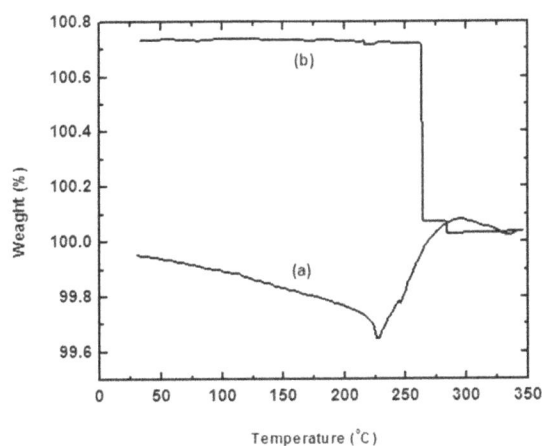

Figure 2. TGA analysis (TGA and derivative curves) for thermal stability of the phase transition for $(Na_{0.6}K_{0.4})NO_3$ sample. (a) TGA curve, (b) Derivative curve.

3.3. Differential Thermal Analysis (DTA)

Differential Thermal Analysis (DTA) was also successfully used, in the present work for the eutectic $(Na_{0.6}K_{0.4})NO_3$ samples, to characterise the kinetic parameters controlling such phase transitions. Figure 3 shows the DTA thermograms for $(Na_{0.6}K_{0.4})NO_3$ samples at different heating rates (2, 5, 10, 15, and 20°C /min) during heating run up to 250°C. An endothermic peak at approximately 114°C is shown in figure

for the measurements that recorded at 2°C/min. The calculated thermal energy of transformation was found to be 26.32 J/g. Further, heating of $(Na_{0.6}K_{0.4})NO_3$ sample up to 220°C gives no change indicating to the sample stability against the thermal agitation up to 220°C. As the temperature increases, an endothermic peak near to ~225°C which is due to melting process of the sample. The DTA thermograms recorded at 5°C/min represented also an endothermic peak at ~114°C and the thermal energy of transformation was found to be 18.87 J/g. The endothermic peak that appeared with The DTA thermograms recorded at 10°C/min was found to be at ~112°C and the thermal energy of transformation was found to be 41.3 J/g. However the endothermic peak that appeared with the DTA thermograms recorded at 15°C/min was found to be at ~117°C and the thermal energy of transformation was found to be 11.9 J/g. For the DTA thermograms recorded at 15°C /min, the endothermic peak that appeared at ~118°C and the thermal energy of transformation was found to be 9.52 J/g.

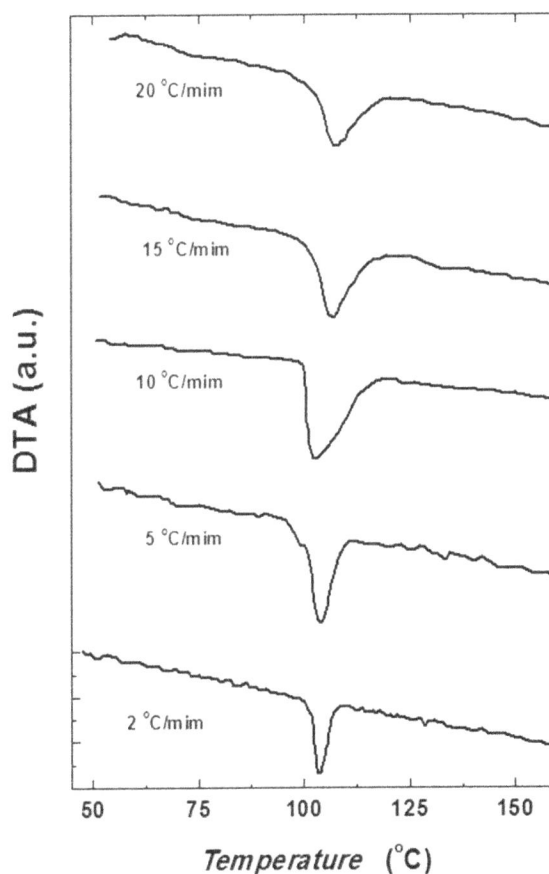

Figure 3. The variation of DTA thermogram with different heating rates (2, 5, 10, 15 and 20°C/min).

The effect of the heating rate on the band shape of the DTA curves for the phase-phase transformation process of $(Na_{0.6}K_{0.4})NO_3$ sample is clearly seen in figure 3. It could be observed from these curves that the increasing of the heating rate of the transformation process leading to increase of the band width and the maximum differential temperature ΔT_{max} up to heating rate of 10°C/min. After this rate i.e. at rates of 15°C/min and 20°C/min, the area under the DTA curves does not nearly change. While the phase transition point is varied according to the heating rate on the sample during the transformation process. It is also clear from the superimposed thermograms that represented in figure 3 that the peak symmetry is sensitive to the change of the heating rate. It is observed that the transformation peak at heating rate of 10°C/min has a large symmetry differ from the peaks obtained with the other heating rates. This can be attributed to the increase of the different dynamical situations of Na^+ and K^+ ions around the (NO_3^-) group. This is because the (NO_3^-) group has a large relaxation time, high enthalpy and approximately low transition point [19].

Figure 4 shows the relation between the rate of heating and peak position of transition point. It is shown that the peak position is changed with increasing the heating rate. The location of the peak can be related to the activation energy provided to the phase transformation in the eutectic $(Na_{0.6}K_{0.4})NO_3$ system at different heating rates. The activation energy at each heating rate was calculated and listed in table 1. The variation of the activation energy with the heating rate for the $(Na_{0.6}K_{0.4})NO_3$ sample seems to be consistent with the expectation that the temperature dependence of the activation energy is mainly caused by the lattice thermal expansion. The temperature dependence is also caused by the formation of polarizability arising from the cationic Na^+ and K^+ ions sphere around the (NO_3^-) group [19, 21].

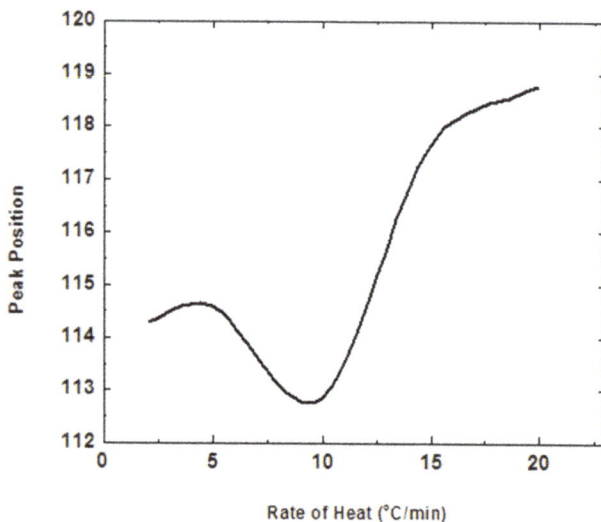

Figure 4. Variation of the peak position (transition point) of $(Na_{0.6}K_{0.4})NO_3$ sample with heating rate.

Figure 5 shows the relation between the enthalpy, ΔH, and the heating rate. It is clear that the enthalpy increasing with increasing the heating rate reaching a maximum of value equals to 41.3 J/g at heating arte of 10°C/min. Then the enthalpy decreases with increasing the heating rate.

The effect of heating rate on the thermal characteristic behaviour can be clearly observed. This indicates to the presence of a certain dependence on the heating rate for the thermal phase-phase transformation of $(NaK)NO_3$ sample which was detected by the DTA measurements. This means that the effect of heating rate is mainly associated with changes of the thermal interaction potential of Na^+ and K^+ ions around the (NO_3^-) group during the phase transformation. This

leads to a change in both the transition point and the enthalpy for the $(NaK)NO_3$ sample [20].

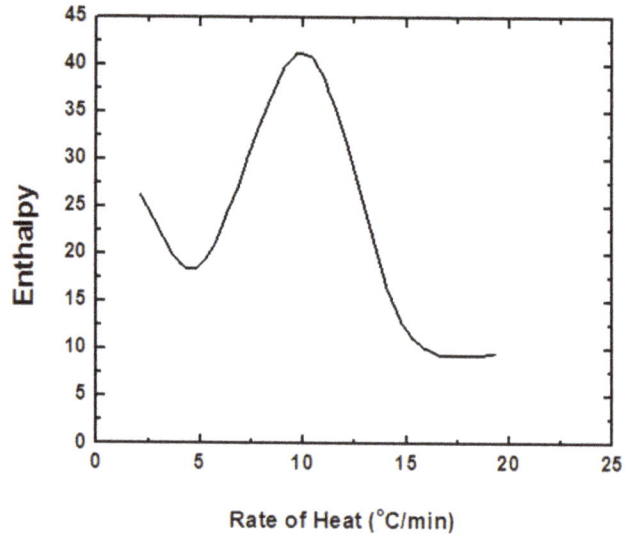

Figure 5. Variation of the enthalpy (ΔH) of $(Na_{0.6}K_{0.4})NO_3$ sample with heating rate.

Table 1. The activation energy at different heating rate of the $(NaK)NO_3$ sample.

Heating Rate (°C/min)	E x 10^{23} eV/mole
2	49.109
5	13.261
10	14.183
15	23.970
20	8.456

4. Conclusion

A mixture $(Na_{0.6}K_{0.4})NO_3$ eutectic system of molecular ratio 60% $NaNO_3$ and 40% KNO_3 was prepared from chemically pure $NaNO_3$ and KNO_3. The various techniques of thermal analysis such as Differential Thermal Analysis (DTA), Differential Scanning calorimetry (DSC) and Thermogravimetry Analysis (TGA) were applied to study the thermal properties of the prepared mixture. The DSC measurements showed a thermal hysteresis during this transformation with a magnitude of the hysteresis temperature ~8°C. Where the measurements showed a reversible phase transition at ~114°C during heating process and at ~108°C during cooling process for the solid phase transformation for the eutectic $(NaK)NO_3$ system. The thermogravimetery analysis (TGA) indicated and demonstrated the phase transition inside this eutectic system having thermal stability up to the melting point at $\cong$225°C indicating to the structural stability of the sample under study. It has been also shown from the DTA measurements that the thermal parameter such as the transition point, enthalpy and the activation energy for the transformation process are affected by the heating rate of (2, 5, 10, 15 and 20°C/min). The noticeable effect of heating rate on the thermal parameters means that the heating rate is a main factor to change the thermal interaction potential of the Na and K atoms around the nitrate group (NO_3^-) during the phase conversion for the eutectic $(Na_{0.6}K_{0.4})NO_3$ system.

References

[1] A. Kere, V. Goetz, X. Py, R. Olives, N. Sadiki, E. Mercier, Energy Procedia, 49(2014) 830-839

[2] G. Zanganeh, R. Kanna, C. Walser, A. Pedretti, A. Haselbacher, A. Steinfeld, Solar Energy, 114 (2015) 77-90

[3] Berul and A. G. Bergman, Izv. Sekt. Piz. Khim. Anal., Inst. Obsbch. Neorg. Khim., Akad. Nauk SSSR, 25 (1954) 233

[4] D. Y. Goswami, C. K. Jotshi, M. Olszewski, 1990. Analysis of thermal energy storage in cylindrical PCM capsules embedded in a metal matrix. In: 25th Intersociety Energy Conversion Engineering Conference. IECEC Reno, NV, USA (August 12–17)

[5] N. K. Voskresenskaya: "Handbook of Solid-Liquid Equilibria in Systems of Anhydrous Inorganic Salts," Kete Press, Jerusalem, 1970, pp. 431-37

[6] Y. J. Zhao, R. Z. Wang, L. W. Wang, N. Yu, Energy, 70(2014) 272-277

[7] M. D. Silverman and J. R. Engel, US_ Dep. Commer. ORNL/TM-5682, 1977

[8] M. Kamimoto, ThermochimicaActa, 49 (1981) 319-331

[9] C. Martin, T. Bauer, H. Müller-Steinhagen, Applied Thermal Engineering 56 (2013) 159-166

[10] D. Kearney, U. Herrmann, P. Nava, B. Kelly, R. Mahoney, J. Pacheco, R. Cable, N. Potrovitza, D. Blake, H. Price, J. Sol. Energy Eng. 125 (2) (2003) 170–176

[11] X. Zhang, J. Tian, K. Xu, and Y. Gao, Journal of Phase Equilibria, 24, 5 (2003) 441-446

[12] B. D. Iverson, S. T. Broome, A. M. Kruizenga, J. G. Cordaro, Solar Energy 86 (2012) 2897–2911

[13] Z. Acem, J. Lopez, E. Palome Del Barrio, Applied Thermal Engineering 30 (2010) 1580-1585

[14] T. Bauer, D. Laing and R. Tamme, Advances in Science and Technology, 74 (2010) 272-277

[15] E. I. Eweka, D. H. Kerridge, Phys. Lett. A 174 (1993) 441–442

[16] W. Pinga, P. Harrowell, N. Byrne, C. A. Angell, ThermochimicaActa 486 (2009) 27–31

[17] W. Hemminger, Int. J. Thermophys., 10 (1998) 765

[18] R. Berg and D. H. Kerridge, Dalton Trans. (2004) 2224-2229

[19] S. Taha, F. El-Kabbany, Y. Bader and M. Tosson, Annalen der Physik, 7, 46, 5 (1989) 355-366

[20] S. Taha and M. Tosson, ThermoChemicaActa, 236 (1994) 217-226

[21] S. Mahadevan, A. Giridhar and A. Singh, J. Phys. Soci. Jap., 35, 1 (1973)

10

Determination of the Penetration Level of ASVT Sub-stations on 132kv Line Without Voltage Profile Violation

Kitheka Joel Mwithui[1], Michael Juma Saulo[2], David Murage[3]

[1]Jomo Kenyatta University of Agriculture and Technology, Nairobi, Kenya

[2]Department of Electrical and Electronic Engineering/Faculty of Engineering and Technology, Technical University of Mombasa, Mombasa, Kenya

[3]Department of Electrical and Electronic Engineering, Jomo Kenyatta University of Agriculture and Technology, Nairobi, Kenya

Email address:

kithekajoelmwithui@tum.ac.ke (K. J. Mwithui), michaelsaulo@tum.ac.ke (M. J. Saulo), dkmurage25@yahoo.com (D. Murage)

Abstract: In developing countries, there are many high voltage transmission lines which transverse villages not supplied with electricity to supply main towns and industrial areas. The conventional substations are too expensive and the power distributor can only set them up if return on investment is assured. Non-conventional (ASVT) sub-stations have been tried and found to be technically successful in stepping down 132kv to low voltages like 240volts in one step to supply single phase loads. Though this technology is cheap and technically fit to be applied in areas of low demand were setting up conventional sub-station will be uneconomical, the technology is not fast spreading in Sub-Sahara Africa (SSA) where there are well established transmission line but poor distribution network. More so the technology remains as a pilot project in countries like Congo were they were first tried. This research aimed at investigating whether violation of voltage profile of the transmission line could have led to low spread of ASVT sub-station technology in Sub-Sahara Africa. The investigation of the maximum number of ASVT substations which could be terminated on 132kv line to supply these villages with electricity without voltage profile violation was carried out. In this research, transmission line and ASVT substation models were implemented using SIMULINK software in MATLAB environment. Surge impedance curves were also used to identify the point of voltage instability or voltage collapse in the system.

Keywords: Auxiliary Service Voltage Transformer (ASVT), Voltage Profile (VP), Transmission Line (TL), Penetration Level (PL)

1. Introduction

In most Sub-Sahara Africa rural areas, the concentration of electricity users is low and the cost of deploying a conventional sub-station is very high. As a result power utility cannot be able to generate an adequate return on investment necessary to bring a conventional distribution sub-station on line [1, 2], on the other hand there are large numbers of rural communities in these areas living around or in close proximity to high transmission lines but are not supplied with electricity. The main obstacle is that these transmission lines have very high voltages that cannot be directly and cheaply be used for electrification [3, 4, 10].

To address the prohibitive costs incurred with the use of conventional sub-stations, non conventional substation namely; Auxiliary Service Voltage transformer (ASVT) sub-station is explored in this journal.

The auxiliary service voltage transformer also known as station service voltage transformer (SSVT) combines the characteristics of instrument transformer with power distribution capability. In this transformer, the high voltage side is connected directly to the overhead transmission line of either 220kV or 132kV, while the secondary side may be of typical voltage ratings of 240V, 480V, 600V or any other voltage level supplies designed on order. One step down principle is applied to achieve the low voltages just like in instrument transformers [5].

The Auxiliary service voltage transformer can either be used with its low voltage output to directly supply needed power near transmission lines or simply step up the ASVT low voltage output through distribution transformer for a local distribution network.

In developing countries where transmission line infrastructure is already in place but a wide spread distribution infrastructure is lacking, the non conventional distribution sub-station technologies can be used to greatly reduce the electrification costs for small villages [6, 9].

2. The Auxiliary Service Voltage Transformer

The ASVT, sometimes known as a station service voltage transformer (SSVT) is insulated in sulfur hexafluoride (SF6) gas and combines the characteristics of instrument transformer with power distribution capability [3, 5, 7]. All the dielectric characteristic of the conventional instrument transformer are applicable to ASVT even though these are hybrid apparatus which are between an instrument transformer and a distribution transformer. These transformers fulfill the standards for both types, i.e. IEEEC 57.13.1993 and IEEE C57.12.00 [8]. This inductive transformer has a very high thermal power in comparison with conventional instrument transformer, in general from 20 up to 60 times more than the design of new generation,

without reaching the capacity of a power transformer. [8]

The ASVT allows directly connection from high voltage line and transforms voltage from 230kV to 600V or even smaller in one step, with a thermal power of 50kVA up to 330kVA per phase [1, 3, 6].

ASVTs were originally designed to suit supply for auxiliary services within the substation such as lighting loads, motor loads and instrument purposes [4, 8]. In developing countries where transmission lines infrastructure is already in place but a wide spread distribution infrastructure is lacking, the non conventional ASVT sub-station technologies can be used as a compact transformer to greatly reduce the electrification cost for rural electrification. The ASVT can be used to supply loads directly with its low voltage or simply step up the ASVT low voltage output through distribution transformer for local distribution network.

Tapping the high voltage transmission line and connecting an ASVT with a small foot print sub-station will provide affordable, readily available electricity to many rural dwellers in close proximity to high voltage lines and presently without power [4, 6].

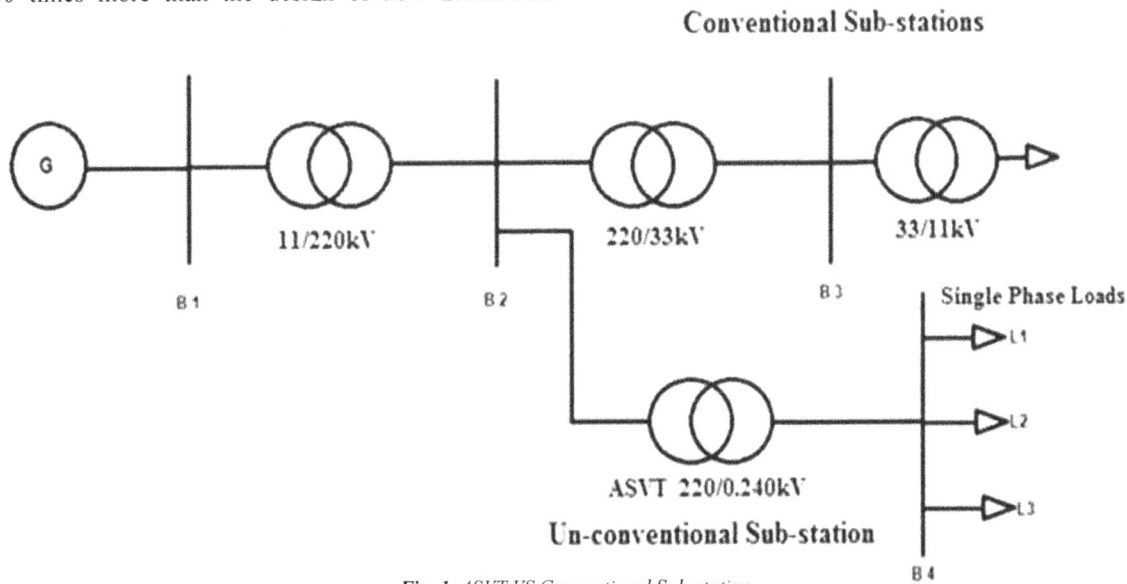

Fig. 1. *ASVT VS Conventional Sub-station.*

3. Methodology

3.1. Penetration Level of ASVT Substations

This research was aimed at determining the maximum number of substations that can be terminated on a 440KM, 132kv transmission line to supply power to the villages without violating the voltage profile of the transmission line.

To carry out the simulation, SIMULINK software in a MATLAB environment was used. The transmission line parameters used were calculated as shown below.

The ASVT substations were terminated on the transmission line from the generation point till the voltage profile of the network was violated. The maximum numbers of ASVT substations terminated were noted.

3.2. Transmission Line Parameters

(a). Inductance of a transmission line,

$$L = 4 \times 10^{-7} \ln \frac{D}{r'}$$

Where

$$r' = 0.7788 \, r$$

D = distance between conductors (1 metres)
R = radius of each conductor (0.03 metres)

$$L = 4 \times 10^{-7} \ln \frac{D}{0.7788 \, (0.03)} = 1.5 \times 10^{-6} \text{ H/M}$$

L = 1.5mH/KM
(b). Resistance of the transmission line

$$R = \frac{\varrho L}{A}, \; A = \pi r^2 = \pi (0.03)^2 = 2.827 \times 10^{-3} \; m^2$$

$$R = \frac{2.83 \times 10^{-8}}{2.83 \times 10^{-3}} \times 1000 \; m = 0.01\Omega/KM$$

$$C = \frac{2 \pi \, 8.85 \times 10^{-12}}{\ln \frac{1}{0.03}}$$

$$C = 1.586 \times 10^{-8} \; F/KM$$

(c). Capacitance of transmission line

$$C = \frac{2\pi}{\ln \frac{D}{r}}$$

3.3. Termination of Reactive Components on a Transmission Line

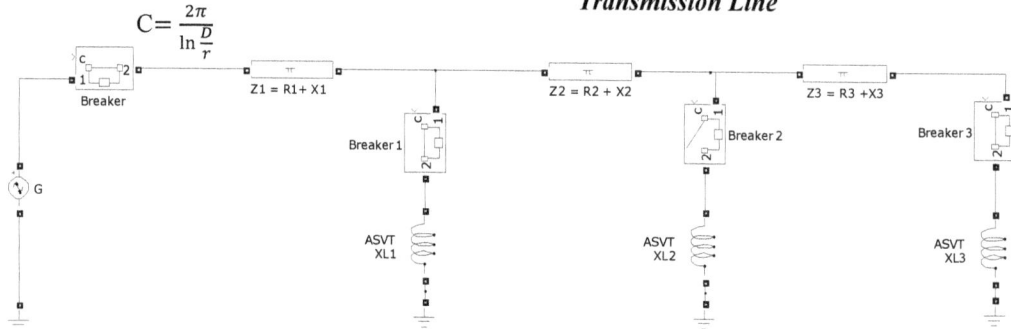

Fig. 2. *ASVT termination on a transmission line.*

Considering breaker 3 alone to be closed, with 1 and 2 open, the impedance of the transmission line shall appear as follows: [11]

$$Z_l = R_l + X_l \tag{1}$$

Where Z_l is the total impedance of the transmission line

$$R_l = R_1 + R_2 + R_3 \tag{2}$$

$$X_l = X_1 + X_2 + X_3 + Xl_3 \tag{3}$$

Considering breaker 3 and breaker 1 to be closed;
The output reactance $(x_o) = (X_2 + X_3 + Xl_3)//(Xl_1)$
If $X_1 = X_2 = X_3$ and $Xl_1 = Xl_2 = Xl_3$ then expression (3) can be reduced to:

$$(2X_1 + Xl_1)//(Xl_1) \cong Xl_1 \tag{4}$$

While the input reactance

$$(x_i) = X_1 \tag{5}$$

In a transmission conductor, the resistance of the line is supposed to be very small to reduce power losses i.e. $p = I^2 R$ watts. Thus resistance can be considered negligible. This means in this case equivalent reactance of the line will be given by:

$$X_l = X_1 + Xl_1 \tag{6}$$

Comparing equation (3) and (6), it's clear that the reactance of the line has been reduced by $2X_1 = (X_2 + X_3)$ by terminating a reactive machine (ASVT) on the line. This leads to change of capacitance and reactance of the transmission network as well as the voltage profile of the line.

If breaker 2 is closed implying termination of another ASVT on the line, the reactance of the line will further change implying a further change on the capacitance and inductance of the line and voltage profile of the entire system.

The surge impedance loading (SIL) of the line can be determined by an equation involving capacitance and inductance, as shown in equation (7).

$$SIL = \frac{(VL)^2}{\sqrt{L/C}} \tag{7}$$

When the line is loaded below SIL value, the line will be providing reactive power and when it is loaded above SIL, it absorbs reactive power. This requires additional sources of reactive power to be supplemented.

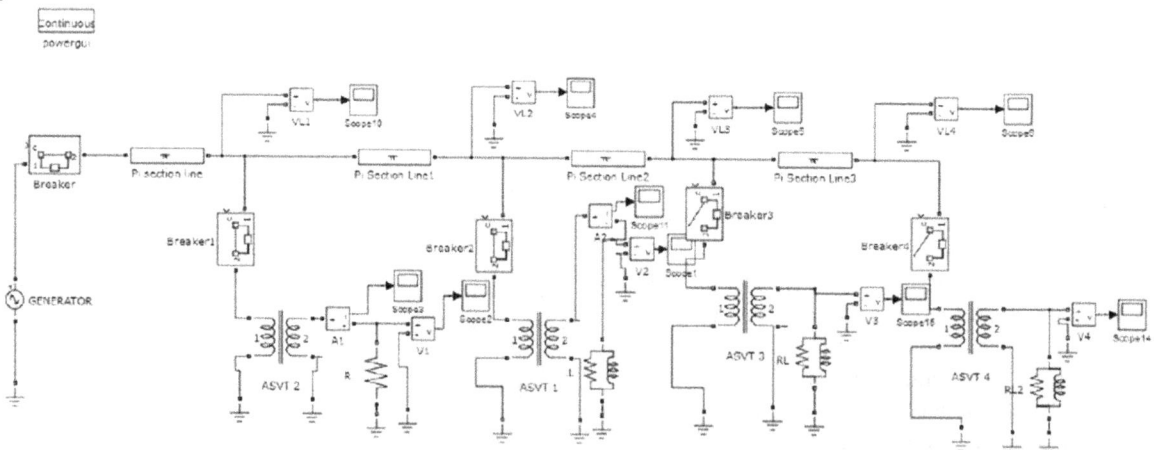

Fig. 3. *ASVT substations terminated on a transmission network.*

3.4. ASVT Substations Terminated on 132kv Transmission Line

This research was carried out to investigate the maximum number of ASVT sub-stations that can be terminated on a 440KM, 132kv transmission line to supply villages living within a radius of 500metres without violating the voltage profile of the line. The transmission line model shown in Fig 3 was constructed in SIMULINK software using MATLAB environment. The ASVT sub-stations were terminated to the transmission line network via a circuit breaker. To monitor the changes of the voltage profile of the network on terminating sub-stations, ammeter, voltmeters and oscilloscopes were used.

3.5. Voltage Stability Limit

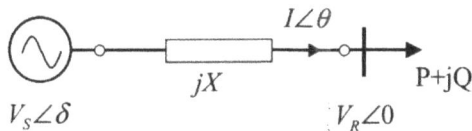

Fig. 4. Single line diagram for voltage stability limit.

Where; [16]

$$P = \frac{V_R V_S}{X} \sin\theta$$

$$Q = \frac{V_R V_S}{X} \sin\theta - \frac{(V_R)^2}{X}$$

At the receiving end,

$$V_R I^* = P + jQ$$

Power delivered to the load as a function of the receiving end voltage when Q = 0 is given by $p = \sqrt{\frac{(v_S) - (v_R)}{X}} V_R$

The maximum power is attained when $\frac{d_p}{d_{VR}} = 0$

$P_{max} = \frac{V_S^2}{2x}$, this is the voltage transmission limit of the power transmission line. Where; $V_{nose} = \frac{V_S}{\sqrt{2}}$

The transmission capacity of a particular line is limited by its thermal capacity. However, in case of long high voltage AC lines efficiency of transmission capacity is below its thermal limit and restricted by angular and voltage stability limits which restricts line load ability up to its Surge Impedance Loading (SIL) level. [15]

3.6. Results and Discussion

The research was carried out by considering a transmission line of 440KM and the calculated transmission parameters used were L= 1.5mH/KM, C= 158.6µF/KM, R= 0.01Ω/KM.

The number of ASVT terminated on the high voltage transmission line was varied to monitor the number of ASVT sub-stations that can be terminated on a transmission line without violating the voltage profile. Table 1 displays the steady state voltage of the transmission line. Voltage wave forms were also displayed and used to monitor the response of voltage levels with variation of the number of ASVTs terminated.

Table 1. Voltages levels vs pi-section distance with 3 ASVT, 7ASVT, 9ASVT & 10ASVT sub-stations terminated.

Length of HV line from generation station(KM)	Voltmeter reading (V), 3ASVT terminated	Voltmeter reading (V), 7ASVT terminated	Voltmeter reading (V), 9ASVT terminated	Voltmeter reading (V), 10ASVT terminated
62.857	133363.67	133353.73	133065.53	121237.56
125.714	134522.34	134502.14	134007.87	110609.76
188.571	135474.26	135444.50	134826.18	100191.11
251.428	136217.95	136178.36	135519.70	90090.07
314.285	136749.90	136702.91	136087.78	80461
377.142	137069.00	137017.00	136845.64	71530
391.12	-	-	137034.71	57191.69
440	-	-	137096.93	52800

The transmission line voltage profile was to be maintained at 6% as per the Kenya power and lighting company recommendations. This means that the voltage levels were to be in the range 132000 ±7920 Volts.

From table 1, the ASVT sub-stations were terminated at a pi-section of 62.857KM from each other and the ASVT sub-stations terminated in the order of 3, 7, 9 and 10. The above results shows that the voltage profile of the 440KM, 132kv transmission line was maintained when 3ASVT sub-stations, 7ASVT sub-stations and 9ASVT sub-stations were terminated. On terminating the tenth ASVT sub-station, the voltage levels of the transmission line appears to have decreased drastically. This means a maximum of nine ASVT sub-stations can be used to supply villages living at a radius of 500metres with electricity without violating the voltage

profile of a 440KM, 132kv transmission line.

The data captured in table 1was plotted on a graph for further analysis and the following results are as shown in Fig 5.

From Fig 5, it is clear that the voltage profile of the 440KM, 132kv network was maintained when up to nine ASVT sub-stations were terminated. The transmission line voltage levels changed drastically on terminating the tenth ASVT sub-station as shown by the blue line on the graph. This means the transmission line voltage profile was violated on terminating the tenth ASVT sub-station.

To further analyse the situation, the voltage waveforms of the transmission line and ASVT sub-stations were displayed before and after the voltage violation.

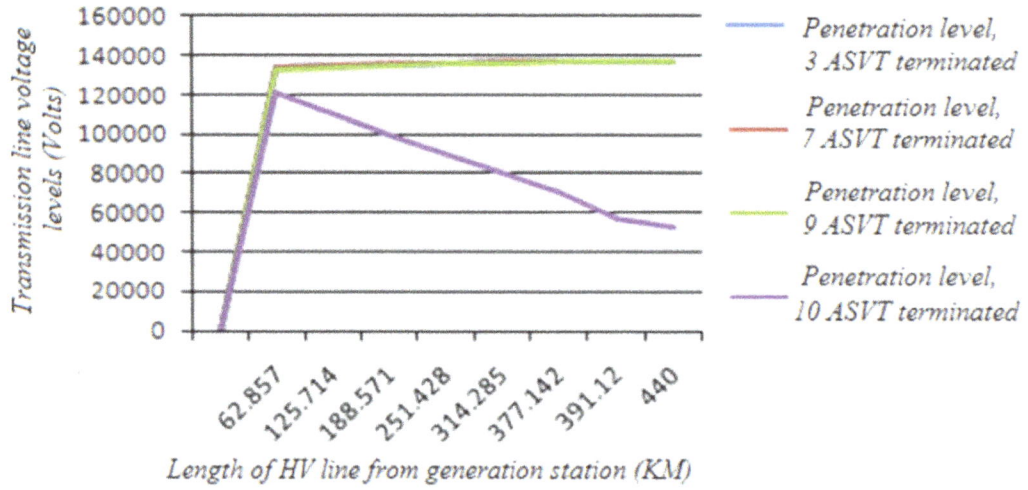

Fig. 5. *Penetration level graph.*

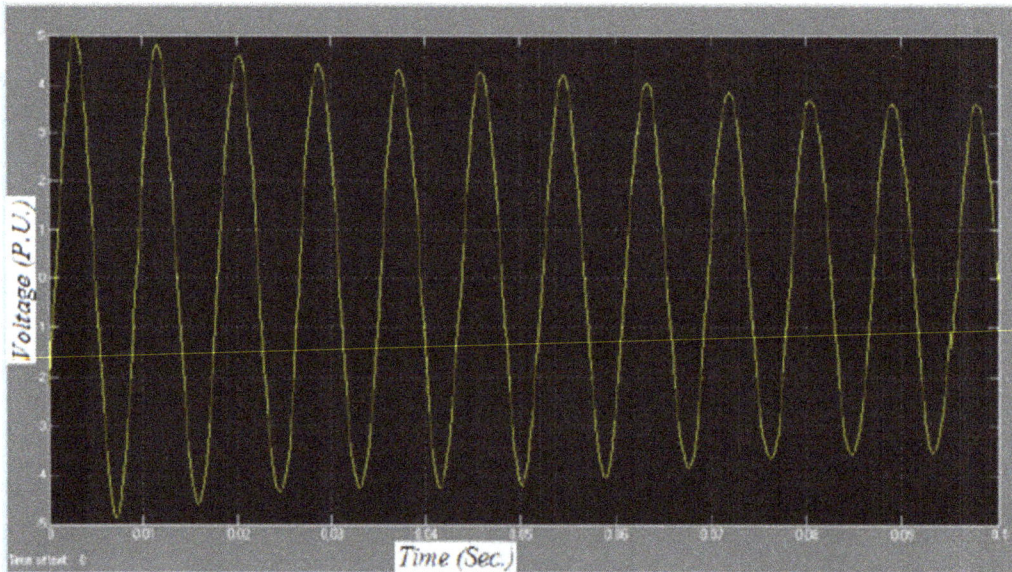

Fig. 6. *Transmission line voltage waveform when maximum of nine ASVT sub-stations were terminated.*

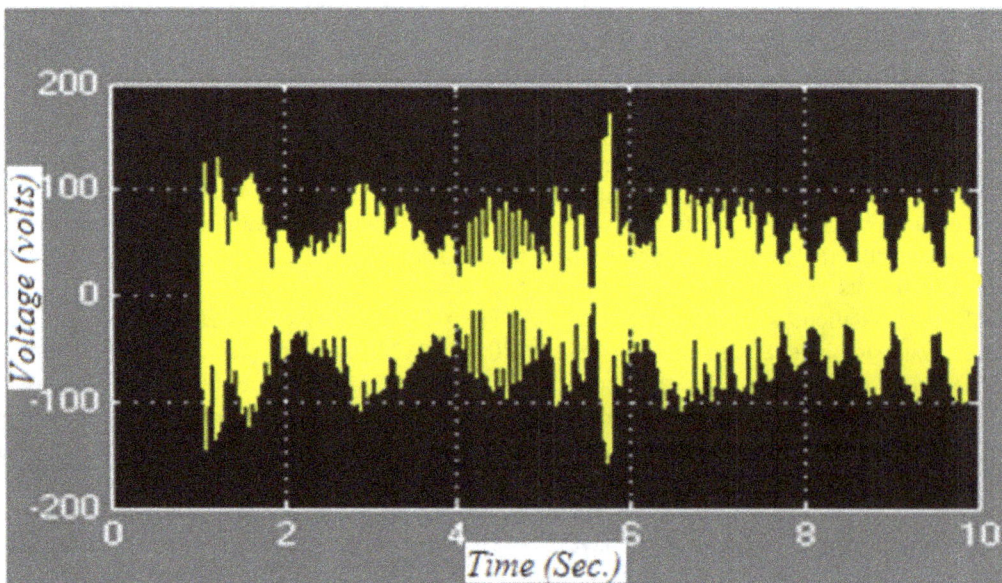

Fig. 7. *Voltage waveform when ten ASVT sub-stations are terminated.*

When a maximum of ten ASVT sub-stations were terminated on the 440KM, 132kv transmission line network the line voltage waveform appeared as shown in Fig 7.

Fig 7: transmission line voltage waveform after ten ASVT sub-stations were terminated.

The change of the transmission line waveform when the tenth ASVT substation was terminated on the 440KM, 132kv transmission line was also noted.

This implied that the surge impedance loading of the transmission line had changed and could not contain an additional ASVT sub-station in the system. This led to voltage instability or voltage collapse.

Further investigations were carried out to display the output voltage waveforms of the ASVT sub-stations and results were as captured in fig 8 and 9.

From the study of Fig 8 and 9, it was observed that when the voltage profile of the 132kv transmission line was violated, the ASVT sub-station transmission line was also distorted. This implies that for the villages located 500metres from the transmission line to be supplied with electricty then the ASVT sub–stations terminated to the transmission line should not be more than nine.

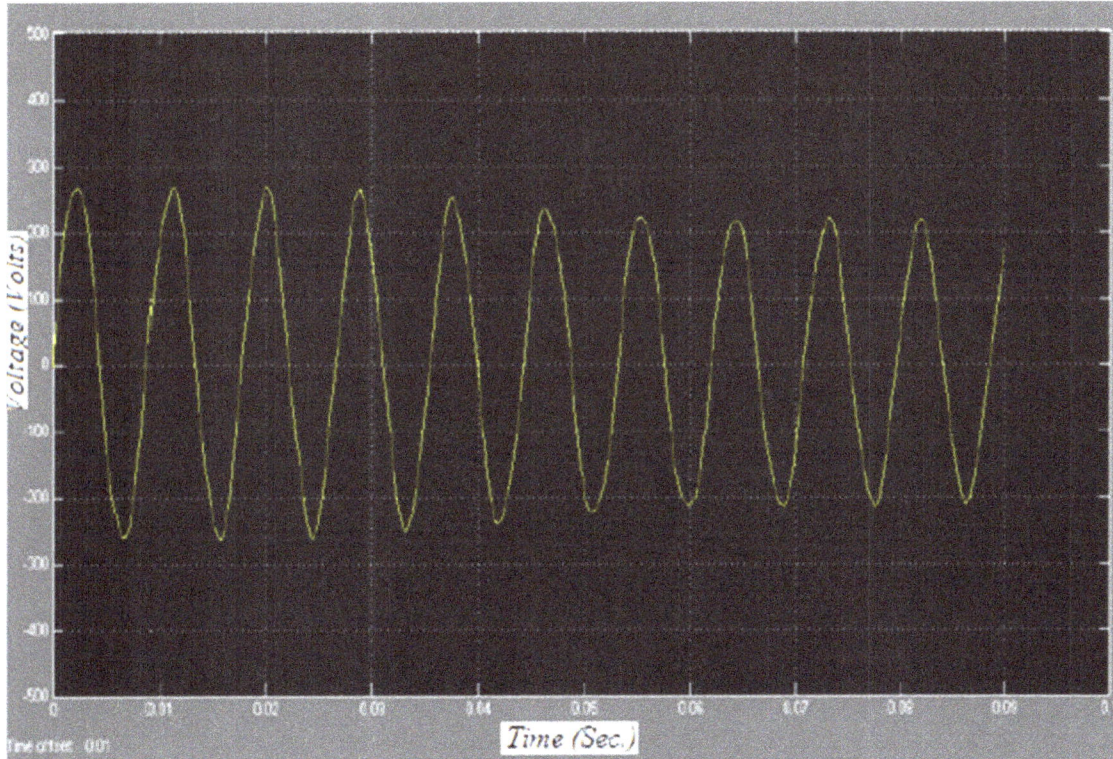

Fig. 8. *ASVT sub-station output before violation of the line voltage profile.*

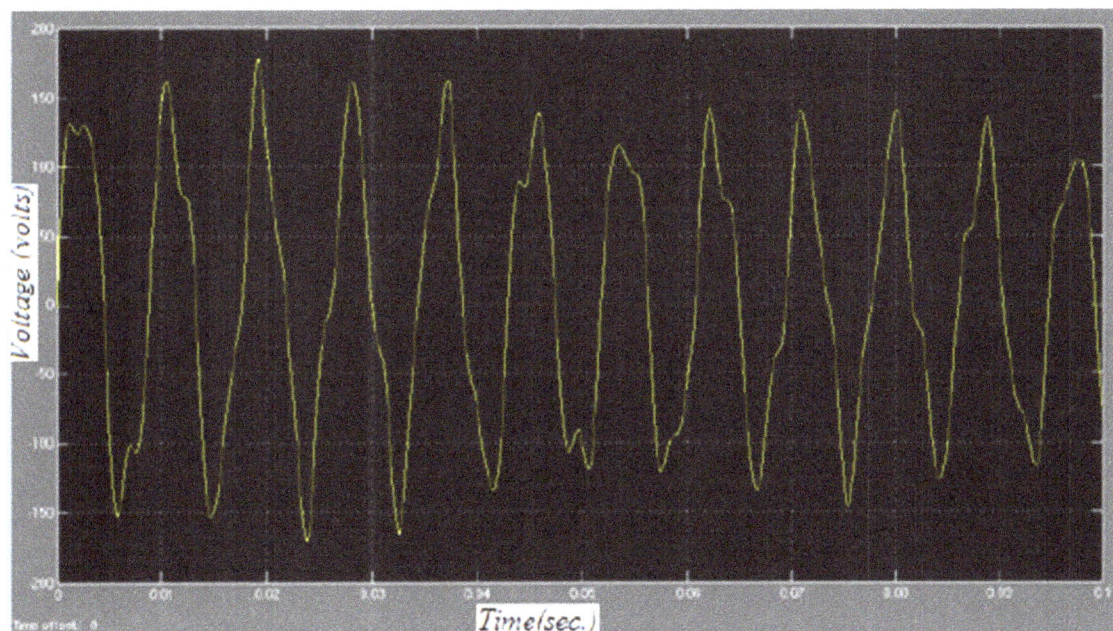

Fig. 9. *ASVT sub-station output after violation of the line voltage profile.*

3.7. *Voltage Stability*

The 440KM, 132kv transimission voltage was investigated using surge impedance loading curve. The SIL curve was used to investigate the effect of violation of transmission line voltage profile to its voltage stability and the results are as depicted on fig 10.

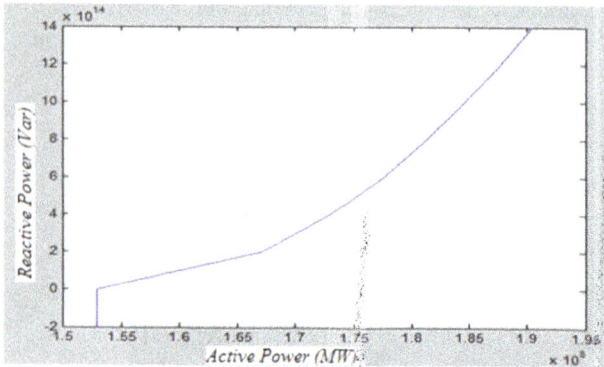

Fig. 10. *SIL curve.*

Fig 10 shows a surge impedance loading curve when the voltage profile of the network is maintained and when the profile is violated.

From 1.53×10^8 to 1.7×10^8 MW, the voltage stability of the network appears to be sound, beyond this point there is an exponential growth of the curve depicted system voltage instability which further led to a voltage collapse.

This resulted from the fact that further termination of the ASVT sub-station led to changes of the transmission line inductance and capacitance which brought forth changes in line impedance. These changes in line impedance interfered with the surge impedance loading of the transmission line resulting to voltage instability or voltage collapse.

4. Conclusion

A maximum of nine auxiliary service voltage transformer sub-stations can be used to supply villages living at close proximity to 440KM, 132kv transmission line without violating the voltage profile of the system.

References

[1]   Saulo M. J, Gaunt C. T, and Mbogho M. S (2012): Comparative assessment of capacitor coupling substation and Auxiliary Service Voltage Transformer for Rural Electrification, 2nd Annual Kabarak University, Nakuru, Kenya.

[2]   Dagbjartsson G., Gaunt C. T, Zomers A. N: Rural Electrification, A scoping report.

[3]   Gomez, R. G, Solano, A. S, Acosta, E. A (2010): Rural Electrification project development, using Auxiliary Transformers, location of Tubares, Chihuahua, Mexico.

[4]   Saulo M. J, Gaunt C. T (2014) "implication of using Auxiliary Service Voltage Transformer substation for Rural Electrification." *International journal of energy and power engineering*. [on – line] 4(2-1) pp 1-11. Capetown, South Africa

[5]   Arteche Instrument Transformer manual (2010): ASVT – 245 and ASVT -145 manual and technical brochures

[6]   Anderson, G. O, Yanev K. (2010): Non convectional substation and distribution system for rural Electrification. 3rd IASTED Africa PES 2010. Gaborone, Botswana.

[7]   Omboua A. (2006) Application report "the high voltage line becomes a power distributor: A successful test in Congo – Brazzaville" Congo.

[8]   Omar C., Gomez R. Solano A. Acosta E. (2010) Eradicating energy poverty "Rural Electrification in Chuahuahua, Mexico at one third of the cost versus a conventional substation" Mexico.

[9]   Michael J. S, Mbogho M. S (2014): Implication of capacitor coupling substation on Rural Electrification planning in Kenya. Proceedings of 3rd international Kenya society of Electrical and Electronics Engineers conference, KSEEE 2014. Mombasa, Kenya.

[10]   Haanyika C. M, (2008): *Rural electrification in Zambia*: A policy and institutional analysis: Energy policy 36 (2008) pp 1044 – 1058.

[11]   J. Kitheka, M. Saulo, D. Murage, The penetration level of auxiliary service voltage transformer substations on a power network for rural electrification.," in Kabarak University 5th Annual International Conference., July 2015.

[12]   Jacobson D. A. N (2000) example of ferroresonance in high voltage power system proceedings of the 1999 international conference on power system transients.

[13]   Namsil K, Lorenz P. (2008): Appropriate Distributed Generation Technology for Electrifying the village. Hague, Nertherlands.

[14]   MNES (2002 -2003): Ministry of nonconventional energy Sources, INDIA.

[15]   M. Globler, "Determination of transmission line parameters from time stamped data," masters thesis, university of Pretoria, July 2007.

[16]   M. Saulo, C. Gaunt, penetration level of unconventional rural electrification technologies on power networks. PhD thesis, University of Capetown, May 2014.

11

Municipal Solid Waste as Sustainable Energy Source for Brazil

Fátima Aparecida de Morais Lino[*], Kamal Abdel Radi Ismail

Energy Department, Faculty of Mechanical Engineering, State University of Campinas, Barão Geraldo, Campinas, Brazil

Email address:

fatimalino@fem.unicamp.br (F. A. M. Lino), kamal@fem.unicamp.br (K. A. R. Ismail)

Abstract: Depositing municipal solid waste (MSW) in dumps has provoked serious impacts in Brazil in the last decades because of the gas and liquid effluents which contaminate the soil and underground water resources in addition to emitting greenhouse gases (GHG) to the atmosphere. To mitigate these impacts, this paper presents proposals for treatment of solid waste by recycling, incineration and biodigestion with the objective of showing the decision makers that solid waste is not a problem and can be a solution as a source of renewable energy. The results of the study show that the MSW proposed treatments represent a forward march for sustainability and environment preservation. The biological treatment option can produce about 221.7 GWh/month or energy enough for 1.26 million homes. The incineration treatment option can produce energy of about 2902.6 GWh/month. The generated ash of about 10% can be used for manufacturing bricks, biofuels and other products. In addition, in all processes CO_2 emissions are significantly reduced.

Keywords: Municipal Solid Waste, Sustainability, Energy, Recycling, Incineration, Biodigestion

1. Introduction

Collection and treatment of municipal solid waste (MSW) is one of the essential sanitary services for the society. Due to the world demographic explosion during the last 70 years, this service was realized partially in many countries [1]. In Brazil, which is the fifth most populated country in the world, the situation is not different. Collecting and treating the solid waste generated by about 202 million inhabitants [2] is a real challenge for public administration. The country population increases at the rate of about 1.17% and the urban population amounts to 84.9% [3].

The amount of 259,547 t/day of MSW collected in 2008, was generated by 98% of the population and was deposited in 2906 open dumps, 1310 covered dumps and 1254 sanitary landfills [4]. The first two methods of disposal of MSW are inadequate because of soil and underground water contamination, and emission of GHG (greenhouse gas) to the environment. Some countries in Europe and Japan treat MSW including energy production and other applications to reduce its offensive impacts and their dependence on landfills [5, 6, 7, 1].

MSW is composed of organic degradable matter such as food leftovers, paper, cardboard, and pruning of plants; organic matter non degradable such as plastics and inorganic matter such as glass and metals which takes hundreds of years to decompose.

In general, solid waste has energy and economic potential. It can be used as a renewable source of energy when incinerated producing heat or biodegraded producing biogas. Also part of it can be recycled resulting in financial gain, energy and raw material savings, in addition to the reduction of CO_2 emissions. Additional benefits can be obtained such as reducing public expenditure for treating the solid waste, creating jobs, enhancing the social inclusion of poor families and increasing the life useful of landfills [8, 9, 4, 10, 11]. Actually Brazil recycles less than 2% of the collected MSW and hence most of the inherent potential of MSW is wasted.

As an attempt to improve the actual situation of waste treatment in the country and reduce its environment and public health negative impacts, the Federal Government launched in 2010 the Solid Waste National Policy (PNRS) law number 12.305. Among the actions established by this law are the increase of the recycled mass to 20% until 2015 and the extinction of dumps in august 2014, what hasn't happened yet [12].

2. Literature Review: Treatment of MSW

Developed countries as Netherland, Germany, Sweden, Switzerland and Japan have given high priorities in their public policies to incineration and biodigestion of MSW with or without energy recovery and incentives to recycling with the result that a big part of generated MSW is deviated from landfills. Data from Unstat [5] and OECD [6] show that deposition of MSW in landfills in 2009 in Germany was 0.4% and in Netherland was 0.7%. Switzerland treats nearly100% of MSW generated by the population.

2.1. Recycling

Recycling is a way to reuse the solid waste and benefit from its inherent financial and energy potential. This involves a number of processes such as separation of the recyclables in residences, selective collection, sorting by type of material, pressing, packaging and selling to recycling industry [8, 9, 10, 11].

Many reports on recycling programs and experiences of different countries are available in the literature as Read [13] and Defra [14] in the UK; Themelis et al. [15] and EPA [16] in the USA; Okuda et al. [17] and Ministry of the Environment [18] in Japan; Pariatamby and Tanaka [19] in China; Lino and Ismail [11]; Lino et al. [9] and Lino [8] in Brazil.

In Brazil, the selective collection started in 1980, and after 30 years, only 994 of the 5,564 Brazilian municipalities implemented recycling programs with the result that MSW selectively collected is about 1.2% [4]. The selective collection is implemented in capitals and big urban centers, mainly in the South and Southeast regions where this service is essentially realized by the public sector, organized cooperatives and associations [8].

2.2. Landfilling

The deposition of organic waste in landfill generates biogas which is essentially a mix of carbon dioxide (CO_2) and methane (CH_4). The use of biogas for electricity generation is a common practice in many countries. Germany had, for example, in 2008, about 4,100 biogas plants supplying approximately 1,435 megawatts of electric energy [20]. In the United States in 2011, landfill methane capture installations produced 14.3 TWh of electricity, enough to provide power to more than 1 million homes [21]. The number of installations for the commercial utilization of biogas increased in developing countries especially in China, India and Nepal where the biogas from waste treatment has been used as energy source [22].

In Brazil, experimental studies are realized to show the potential of energy generation from landfill biogas as the pilot plant from landfill in Muribeca, Recife (PE), producing about 80 m^3/h of biogas, but with potential about 100 kW [23]. In the state of São Paulo, three thermoelectric power plants are installed in landfills sites producing about 75 MW while other installations in different localities are producing about 109 MW [24]. The rate of energy production from MSW varies between 0.66 to 1.45 MWh/t [24].

MSW landfilling is a common practice in many countries. One of the major problems associated with it is the fugitive biogas which escapes through cracks and voids in the covering layer and sidewalls and released freely to the atmosphere. Monni et al. [25] considered that the rate of fugitive biogas flow varies between 25 to 50% depending on the construction conditions of the landfill. An alternative way to treat solid waste, improve the public cleaning service, and at the same time produce energy is through incineration, considering that the flue gases will be monitored and carefully treated before its release to the ambient.

2.3. Incineration of MSW

The first incineration plant for treating MSW was developed and operated in Manchester (UK) in 1876, and since then incineration is considered as an effective tool to treat dangerous and infectious material from hospitals and similar establishments. This process reduces the original volume of the material to about 10% and can be a viable option for municipalities where there are no suitable and cheap areas for constructing well engineered landfills (sanitary landfills).

Incineration is extensively used in Japan [26], Europe [27], and Korea [28]. The utilization of thermal energy from incineration for heating and generation of electricity is a viable option accepted by the public opinion due to the evolution of pollution control equipments, instrumentation, and the development of highly efficient systems for treating effluents [29]. In Brazil, incineration is only used for treatment hospitals wastes [30]. In modern incineration units one ton of MSW can produce up to 600 kWh [31]. This amount of energy is sufficient for four average Brazilian residences [32].

Considering the huge amounts of MSW generated daily, the energy, financial losses and the negative ambient and public health impacts, this paper presents assessment of proposals for MSW treatments based upon recycling, biodigestion and incineration showing the energy, economic and ambient potential of MSW.

3. Materials and Methods

3.1. Materials

From the literature review, it is clear that recycling part of MSW together with landfilling and incineration including energy recovery are possible treatment routes to ensure environmental sustainability, additional energy and financial gains.

In this way, this paper presents an assessment of two possible routes for the treatment of MSW: biogestion and incineration and both with 10% of reuse of recyclables generated in the country, as show Figs. 1 and 2. Recognizing that there are different consuming habits in Brazil, an average MSW composition based upon [33] is used since there is no other official data available after the above date.

The data used is only for MSW generated in urban areas. Since the main objective is assessing the relative merits and drawbacks of each proposal, the sensitivity of the results due to the uncertainty of the data will affect equally the results from both routes. Were also used global values obtained from reports for the production and composition of biogas from landfills, emission rates and fuel consumption aid in incineration. Again, the uncertainty in these global values may affect the results.

The parameters and data used in the calculations are presented in Table 1 while Table 2 presents the composition of solid waste in Brazil.

Table 1. Data used in the calculations.

Description	Reference value	Adopted value	Reference
Biogas production from landfill (L/kg)	35 - 45	40	[34]
Specific mass of CO2 (kg/m^3)		1.83	[35]
Emission of MSW incinerated (tCO$_2$ /TJ)	10 - 40	25	[25]
LCV of CH4 (MJ/m^3)[1]		33.95	[35]
Avoided emissions in recycling (CO$_2$/t)		1.971	[11]
Avoided energy in recycling (GJ/t)		31.629	[11]
LCV of MSW incinerated (kJ/kg)	5250 – 10,264	6,130	[31]
Auxiliary fuel for incineration LPG (kg/t)[2]		8.0	[36]
Efficiency of recovered biogas (%)	50 - 75	75	[25]
Emissions due to combustion of LPG (kg CO$_2$/kg)		3.019	[36]
LCV of commercial LPG (MJ/kg)	40.05 - 46.05	40.05	[35]

(1) Lower calorific value; (2) Liquefied Petroleum Gas.
Source: Prepared by the authors.

Table 2. Typical composition of solid waste in Brazil.

Material	Organic Matter (%)	Paper/cardboard (%)	Plastics (%)	Glass (%)	Metal (%)	Others (%)
	52.5	24.5	2.9	1.6	2.3	16.2

Source: [33]

3.2. Methods

This Section Presents the Simplified Diagrams of the Proposals, Explanations and Equations used in the Calculations

3.2.1. Landfilling of MSW with Biogas Utilization

Fig.1 shows a simplified representation for the biological treatment of MSW. The amount of recyclables separated and collected selectively for reuse is 10%. Organic matter and the rest of uncollected recyclables are transported to landfills equipped for biogas collection and utilization for heat and electricity production. The biodigestion of the organic matter in MSW can reduce its volume by about 20 to 25% [36].

From the gravimetric analysis of MSW, (Table 2) it is possible to determine the amount of recyclables from equation 1

Quantity of recyclables = Recyclables fraction x Collected MSW   (1)

The financial gain from commercializing the recyclables is obtained from equation 2.

Financial gain = Price of recyclables U$ / ton x quantity of recyclables   (2)

Recycling eliminates the necessity of energy to process raw material e consequently the associated emissions. Lino and Ismail [11], by using the data from [37, 38, 39] together with the recyclables composition, calculated the energy savings per ton of recyclable mix (Table 1). The same procedure is used to calculate the amount of avoided CO_2 due the reuse of the recyclables (Table 1).

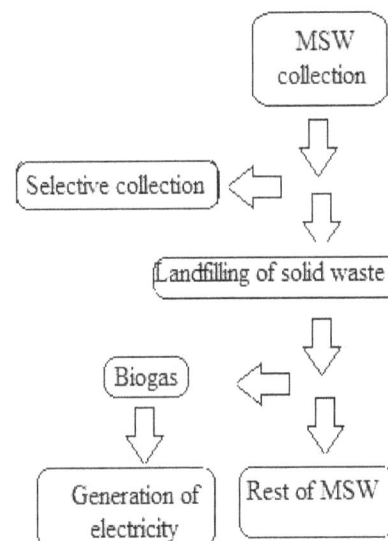

Fig. 1. Recycling and biological treatment of MSW.

The avoided energy and emissions due to recycling can be calculated form equations 3 and 4 [11],

Avoided energy = Avoided energy factor x   Recyclable mass   (3)

Avoided emissions = Avoided emissions factor x Recyclable mass   (4)

The rest of MSW is transported to landfill. The rate of biogas production depends on MSW composition, ambient conditions, humidity and the average pH value. The average quantity of biogas production [34] can be calculated from equation 5 as

Quantity of generated biogas = Rate of biogas production x biodegradable mass in MSW   (5)

The generated biogas is collected, cleaned and the forwarded for utilization or for energy generation. Not all the biogas generated is collected, some of it escapes and the recovery efficiency may vary and, in the present work the value of 75% was used [25]. The collected gas can be calculated from equation 6

Collected biogas = Recuperation efficiency x volume of generated biogas   (6)

The generated biogas is principally composed of (CH_4) and (CO_2) and small quantities of other gases. In the present study a composition of biogas of 45% CH_4 and 55% CO_2 is adopted. The energy contained in the collected biogas can be calculated by using the lower calorific value (LCV) of the methane [35] or by using an average value for the biogas LCV. Equation 7 can be used to calculate the energy content of the collected biogas.

Energy content of the collected biogas = mass of collected biogas x LCV of the biogas   (7)

3.2.2. Calculation of Emissions
The combustion of CH_4 produces the same quantity of CO_2 according to equation 8

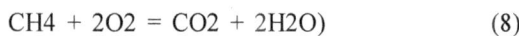

CH4 + 2O2 = CO2 + 2H2O)   (8)

Hence the quantity of CO_2 generated due to the combustion of collected biogas is equal to quantity of collected biogas or

CO2 generated from the combustion of collected biogas = quantity of collected biogas   (9)

The calculations of the fugitive biogas [25] and the equivalent emissions (CO_2e) can be calculated from equations 10 and 11

Quantity of fugitive biogas = (1-η) x Quantity of generated biogas   (10)

Equivalent CO2e   of CH4 = GWP x CH4 quantity   (11)

where η is the biogas recovery efficiency and equals 75%, and GWP = 25 is the GWP of methane.

3.2.3. Incineration of MSW
Fig. 2 shows a simplified flow chart of the thermal process for treating MSW. Subtracting 10% of the recyclables, the rest of MSW is directed to the mass incineration plant where it is burnt with the help of auxiliary fuel, considered here as LPG. Energy generated from the combustion of MSW can be used to generate steam and electricity. The heat content of the hot flue gases can be recovered and used for heating admission air, feed water for the boilers and other applications. After the cleaning processes, the gases are discharged into the ambient. The remains of the incineration process can be recycled or used in road paving and civil construction etc.

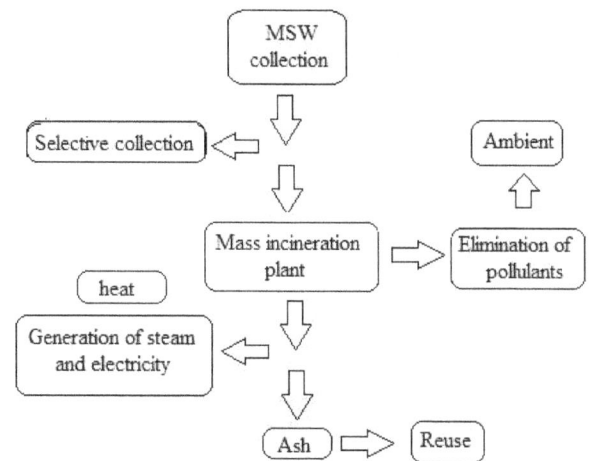

Fig. 2. Recycling and incineration of MSW.

As in the previous case, financial, ambient and energy gains by recycling can be calculates from equations 1 to 4. The mass of MSW sent for incineration is the mass of collected MSW minus the mass of commercialized recyclables (10%). The heat released during incineration depends upon the heat content of MSW and can be calculated from equation 12

Heat released during incineration = Mass of MSW incinerated x Heat content of MSW   (12)

To start incineration and maintain the temperature level in the furnace, LPG is used in the present work. Equation 13 can be used to determine the amount of heat released by the auxiliary fuel.

Energy released by the auxiliary fuel = mass of the auxiliary fuel x LCV of the auxiliary fuel   (13)

The net heat released from the incineration process is the difference between the heat released by incineration of MSW and the heat released due to the combustion of the auxiliary fuel as in equation 14

Net heat released in the incineration process = Heat of combustion of MSW – Heat the auxiliary fuel   (14)

This energy will be converted to electricity with an average conversion efficiency of about 30%.

The amount of CO_2 generated due to the combustion of both MSW and the auxiliary fuel can be calculated by using data (Table 1) and equations 15 and 16.

$$\text{Quantity of CO2 generated due to the combustion of LPG} = \text{Emission factor x mass of LPG} \quad (15)$$

$$\text{Quantity of CO2 generated due to the combustion of MSW} = \text{Emission factor x mass of MSW} \quad (16)$$

4. Results

In this section, the actual situation and the results from the proposals are calculated, presented and analyzed.

4.1. Evaluation of MSW Actual Treatment

The collected MSW [4] is 259,547 t/ day and from Table 2 the organic matter in MSW is 52.5% and paper and cardboard is 24.5%. The total amount of biodigestable matter is 77% or 199,851.2 t/day. The selective collection of recyclables [4] is 3,122 t/day and the quantity of paper and cardboard in the selective collection is 24.5% or 765 t/day. Hence the amount of biodegradable matter for landfilling is 199,851.2 − 765 = 199,086.2 t/day.

The rate of biogas production (Table1), varies according to its composition, ambient conditions, and humidity. Due to the actual conditions of landfills a biogas production rate average value of 0.030 m^3/kg is used, the amount of generated biogas is 5.972586 x 10^6 m^3/day = 2.1799939 x 10^9 m^3/year. Considering the composition of biogas as 45% CH_4 and 55% CO_2, it is possible to calculate the total amount of CO_2e as 25.72 x 10^9 m^3/year or 47.1 Mt CO_2/year.

The amount of recyclables collected [4] is 3122 t/day and the selling price of a ton of recyclables mix is R$ 450 (US$ 203,71). Hence the financial gain is 1,404,900 R$/day (US$ 635,982.62) or R$ 42,147,000 /month (US$ 19,079,479).

The calculations show that the energy savings amounts to 98745.7GJ/day = 10011.72 GWh/year while the avoided emissions comes to 6153.5 tCO_2/day = 2.24 MtCO_2/ year. Summary of the calculations is presented in Table 3.

4.2. Proposals for Treatment of MSW

The models used for the evaluation of the proposed treatment routes are presented with the respective equations in section 3.2. Following the procedures outlined in sections 3.2.1, 3.2.2 and 3.2.3 it is possible to obtain the results shown in Table 3 for the proposed treatment routes

5. Discussion

5.1. Actual Situation of MSW Treatment in Brazil

Analysis of the results shows that the actual situation of

MSW treatment is critical and challenging considering the serious risks to public health and to ambient in addition to the huge economic and energy losses. The MSW actual treatment schemes are not adequate or sufficient to cope with the needs of the population. The national panorama demonstrates that MSW generated by 200 million inhabitants is essentially dumped provoking disastrous consequences to ambient and the population.

Recycling can be considered as an additional instrument for creating jobs, income and conservation and better use of natural resources. Irrespective of the extensive official efforts to encourage the recycling industry these activities are still incipient, about 1.2%.

5.2. Proposal of MSW Treatment in Brazil

Benefits due to recycling include economy of energy, water and raw materials, reduction of emissions, reduction of MSW destined to landfills and can be considered as a source of jobs and income for poor families [40]. In the present study an initial target of recycling 10% is proposed, (observe that the national solid waste policy adopts a target of 20%). The proposed target value is considered acceptable since it can be achieved within reasonable period of time by adopting adequate measures such as priorities in public policies and implementation of public awareness programs in a way similar to what was done in UK [13].

Recycling 10% of the available potential of recyclable, according to Table 3, corresponds to about 8000 t / day and when commercialized can render a sum of US$ 50 million per month which corresponds to 151,480 minimum national salary (R$ 724 or US$ 327.75). Part of these funds can be used to promote selective collection and upgrade infrastructure for MSW treatment. The avoided energy and emissions come to about 256,948.3 GJ/day and 5.84 MtCO_2 /year, respectively.

Using the same methodology of calculation, a comparison between recycling in 2008 and this proposal, that is 1.2 and 10%, respectively, shows that the energy, financial gains and the avoided emissions obtained from the sale of recyclables are found to be 2.6 times that of 2008.

5.2.1. Landfilling of MSW

In the proposal of landfilling nearly199 thousand tons of MSW, according to Table 3, can generate about 7 million cubic meters of biogas, which after subtracting the fugitive biogas, can be converted to about 221 million kWh per month sufficient for the consumption of 1.26 million residences. Considering the electricity tariff of R$ 0.389 kWh or US$ 0.176 per kWh in the municipality of Campinas (SP) in 2014, the cost of energy generated yearly is nearly equal to the value of three mass incineration plants.

The remaining heat contained in the combustion products can be used for preheating combustion air, feed water for boilers and other applications.

Table 3. Summary of the results of the proposed treatment routes for MSW.

Description	Situation in 2008	Proposed MSW treatment	
Recycling			
Collected MSW (t/day)	259,547	259,547	
Available recyclables (t/day)	81,238.2	81,238.2	
10% of available recyclables (t/day)	-	8,123.82	
Recyclable collected in 2008 (t/day)	3,122	-	
Monthly gain from the recyclables (R$ or US$)*	42,147,000 (19079,675)	109,671,585 or (49647,617)	
Avoided energy due to recycling (GJ/day)	98,745.7	256,948.3	
Emissions avoided by recycling (MtCO2 /year)	2.24	5.84	
Treatments for MSW		Landfilling with biogas recovering	Incineration
Organic matter for landfilling (t/day)	199,086.2	193,492.3	
Waste for incineration (t/day)			199.851,2
Biogas generated (m3/dia)	5.9726 x 106	7.73969 x 106	
Recovered biogas (m3/day)		5.80477 x 106	
Energy generated by the biogas (J/day)		88.68 x 1012	
Energy due to incineration (J /day)			1225.0879 x 012
Energy due to incineration (J/ month)			36752.64 x 1012
Net energy due to incineration (Je /month)			10449.4999 x 1012
Energy due to biogas (Je /month)		798.097 x 1012	
Energy due to biogas (GWh /month)		221.694	
Energy due to incineration (GWh/month)			2902.639
Number of homes attended by generated energy**		1,258,430	16,476,663
CO2 emissions (Mt CO2 /year)	47.0748	3.877	12.941
Fugitive biogas (m3/day)		1.93492 x 106	
Equivalent emissions due to fugitive biogas (m3CO2e/day)		22.8321 x 106	
Mass of fugitive biogas (MtCO2e /year)		15.251	

* Conversion rate to Dollar = US$ 2.209; **National average residential electric energy consumption = 0.6342 G Jel /month.
Source: Elaborated by authors.

In this proposal, emissions due to the combustion of biogas are about 3.88 Mt CO_2 /year in addition to the quantity of 15.25 $MtCO_2e$ /year due to fugitive biogas. This sum is 2.5 times lower than the 2008 emissions data.

5.2.2. Incineration

Incineration of 200,000 t / day of MSW can produce a net thermal energy of 37,000 TJ / month or 10,000 TJ_{el} / month. This amount of energy is sufficient for the consumption of 16.5 million residences each consuming an average of 0.6342 GJ_{el}/ month. Emissions due to incineration of MSW are of the order of 12.94 Mt CO_2/ year compared to nearly 19 $MtCO_2e$/ year in the case of landfilling.

From these results it is possible to conclude that the energy and environment benefits from incineration are much more than those of landfilling. The thermal treatment leaves about 10% of ash which can be reused. This can be a solution for many cities as Campinas (SP) where there is no available land to construct landfills. It is important to mention that incineration is a viable option but must have adequate installations equipped with equipments for monitoring, control and treatment of effluents to ensure safe and adequate operation [40].

6. Conclusions

The results show that either of the proposals can produce favorable impacts such as reduction of emissions, reduction of contamination of soil and water resources, more energy generation, saving raw materials and water resources.

The proposal of landfilling MSW with biogas capture can generate energy enough for about 1.8% of the total 65 million of Brazilian residences and emit to the atmosphere about 15.25 $MtCO_2e$ /year.

One inconvenient aspect is the fugitive gas (biogas) which escapes at the site and released freely to ambient aggravating the greenhouse effects. There is always a risk of leachate leakage which could contaminate the soil and underground water sources. Even after closing the landfill site the remaining biodegradable matter continue producing biogas at smaller rates which needs to be continually monitored for many years to avoid risks of explosion.

Incineration, on the other hand reduces MSW mass to about 10% of ash which can be reused. The capital costs for an incinerator will depend on the quality of waste to be processed and the technology employed. Costs will not only comprise those associated with the purchase of the incinerator plant, but also costs for land procurement and preparation prior to building and also indirect costs, such as planning, permitting, contractual support and technical and financial services over the development cycle. Facilities in operation after 2000 report a cost of £82 per tonne (£44-£101 range) [41].

Based on this assessment, the authors consider that incineration is the a most viable system for treating MSW in Brazil, because it reduces the mass of solid waste dumped in soil, avoids problems as contamination of soil, air, and underground water and finally avoids risks to public health. In addition, one should forget that landfill needs of continuous monitoring for many years after its deactivation.

Acknowledgements

The authors wish to thank the CNPQ for the Doctorate scholarship to the first author and the PQ Research Grant to the second author.

References

[1] World Bank, Urban Development Serie. What a waste: A global review of solid waste management, Hoornweg, D. and Bhada-Tata (authors), World Bank, 15p. USA, 2012.

[2] IBGE - Instituto Brasileiro de Geografia e Estatística, Projeção população brasileira: abril de 2014. Available at: http://www.ibge.gov.br/home/

[3] IBGE - Instituto Brasileiro de Geografia e Estatística, Banco de dados Brasil: população, 2012. Available at: http://www.ibge.gov.br/paisesat/main_frameset.php.

[4] IBGE - Instituto Brasileiro de Geografia e Estatística, Pesquisa Nacional de Saneamento Básico 2008. BR, 2010, Available at: http://www.ibge.gov.br/home/estatistica/populacao/condicaod evida/pnsb2008/PNSB_2008.pdf.

[5] UNSTAT -United Nation Statistic Division, Environmental indicators, Waste: Municipal waste treatment 2011. Available at: https://unstats.un.org/unsd/environment/wastetreatment.htm

[6] OECD - Organization for Economic Co-operation and Development Environmental Outlook to 2050, The consequences of inaction 2012, OECD Publishing. Available at: http://dx.doi.org/10.1787/9789264122246-en.

[7] OECD - Organization for Economic Co-operation and Development, Total amount generated of municipal waste, in Factbook Country Statistical profiles 2013 edition/Environmental, OECD. StatExtracts. Available at: http://stats.oecd.org/Index.aspx?DatasetCode=CSP2013.

[8] FAM Lino, Consumo de energia no transporte da coleta seletiva de resíduo sólido domiciliar no município de Campinas (SP), Master Thesis, Universidade Estadual de Campinas: Unicamp. Brasil, 2009.

[9] FAM Lino, WA Bizzo, EP Silva, KAR Ismail, Energy impact waste recyclable in a Brazilian Metropolitan, Resources, Conservation and Recycling, vol. 54, pp. 916-922, 2010.

[10] FAM Lino, KAR Ismail, Energy and environmental potential of solid waste in Brazil. Energy Policy, vol. 39, pp. 3496-3502, 2011.

[11] FAM Lino, KAR Ismail, Analysis of the potential solid waste in Brazil. Environmental Development, Vol.4, 2012 pp. 105-113, 2012.

[12] Brasil/ MMA - Ministério do Meio Ambiente, Política Nacional de Resíduo Sólido, 2014, Available at: http://www.mma.gov.br/cidades-sustentaveis/residuos-solidos/ politica-nacional-de-residuos-solidos

[13] AD Read, A weekly doorstep recycling collection, I had no idea we could, Resources, Conservation and Recycling, vol. 26, pp. 217-249, 1999.

[14] DEFRA - Department for Environment Food and Rural Affairs, National Statistic. Statistics on waste managed by local authorities in England in 2012/13. 2013a, Available at: https://www.gov.uk/government/uploads/system/uploads/attac hment_data/file/255610/Statistics_Notice1.pdf.

[15] NJ Themelis, CE Todd, 'Recycling in a megacity', Journal of the Air & Waste Management Association, vol. 54, pp. 389-395, 2004.

[16] EPA- Environmental Protection Agency, Municipal Solid Waste Generation, Recycling, and Disposal in the United States: Facts and Figures for 2012, 2014. US EPA. Available at: http://www.epa.gov/waste/nonhaz/municipal/pubs/2012_msw _fs.pdf

[17] I Okuda, VE Thomson, 'Regionalization of municipal solid waste management in Japan: Balancing the proximity principle with economic efficiency, Environment Management, vol. 40, pp. 12–19, 2007.

[18] Ministry of the Environment, Solid Waste Management and Recycling Technology of Japan: Toward a Sustainable Society, 2012. Available from: http://www.env.go.jp/recycle/circul/venous_industry/en/broch ure.pdf.

[19] A Pariatamby, M Tanaka, Municipal Solid Waste Management in Asia and the Pacific Islands: Challenges and Strategic Solutions, Springer Singapore Heidelberg New York Dordrecht London, 2014.

[20] P Taglia, Biogas: Rethinking the Midwest's Potential. Clean Wisconsin, 2010. Available at: http://issuu.com/cleanwi/docs/biogas.

[21] REN21 - Renewable Energy Policy, Network for the 21st Century, Global Status Report, Paris: 2012. Available at: http://www.martinot.info/REN21_GSR2012.pdf.

[22] REN21 - Renewable Energy Policy: Network for the 21st Century, Global Status Report, Paris: 2007. Available at: http://www.ren21.net/Portals/0/documents/activities/gsr/RE20 07_Global_Status_Report.pdf

[23] FJ Maciel, JF Jucá, A Codeceira Neto, PB Carvalho Neto, Recuperação de biogás em aterros de resíduos sólidos urbanos – Projeto piloto da Muribeca, V Congresso de Inovação Tecnológica em Energia Elétrica (V CITENEL), Belém-PA: 2009.

[24] MME/Aneel, Atlas de Energia Elétrica do Brasil- Biogás, Ministério de Minas e Energia/Agência Nacional de Energia Elétrica, 3ª ed. Brasil, 2008. Available from: http://www.aneel.gov.br/arquivos/PDF/atlas3ed.

[25] S Monni, R Pipatti, A Lehtilä, I Savolainen, S Syri, Global climate change mitigation scenarios for solid waste management, VTT publications 603, ESPOO, 2006.

[26] H Cheng, Y Hu, Municipal solid waste (MSW) as a renewable source of energy: current and future pratices in China, Bioresource Technology, vol. 101, pp. 3816-3824, 2010.

[27] CEWEP- Confederation of European Waste-to-Energy Plants, A decade of Waste-to-Energy in Europe (2001-2010/11); 2013, Available at: http://www.cewep.eu/information/publicationsandstudies/state ments/ceweppublications/m_1174

[28] JM Park, SB Lee, MJ Kim, OS Kwon, DI Jung, Behavior of PAHS from sewage sludge incinerators in Korea, Waste Management, vol. 29, pp. 690-695, 2009.

[29] FEA - Federal Environment Agency. Biogas production in Germany: Umweet Bunder Amt. by Graaf, D. and Fenler, R., 2010, Available at: http://www.spin-project.eu/downloads/QBlackground_paper_ biogas_germany_en.pdf.

[30] IBGE - Instituto Brasileiro de Geografia e Estatística. Atlas de saneamento, BR: Rio de Janeiro, 2011.

[31] WR Niessen, Combustion and Incineration Processes, 3rd Edition: Marcel Dekker, Inc. New York, 2002.

[32] Brasil/MME- Ministério de Minas e Energia, Anuário Estatístico de energia elétrica 2013, BR: Rio de Janeiro, Epe 2013, Available at: http://www.epe.gov.br/AnuarioEstatisticodeEnergiaEletrica/20 130909_1.pdf

[33] IPT/Cempre, Lixo municipal: manual de gerenciamento integrado, Instituto de Pesquisa Tecnológica & Compromisso Empresarial para Reciclagem, BR: São Paulo, 2000.

[34] A Karagiannidis, Waste to Energy: Opportunities and challenges for developing and transition economies, editor Avraam Karagiannidis, London: Springer-Verlag; 2012.

[35] JW Rose, JR Cooper, Technical data on fuels, The British National Committee: 7th edition, London, 1977.

[36] Brasil/MCT- Ministério da Ciência, Tecnologia e Inovação, Projeto Usina Verde – Incineração de resíduos sólidos urbanos, com carga de composição similar ao RDF, evitando emissão de metano e promovendo geração de eletricidade para autoconsumo, BR: Brasília, MCT, 2005. Available at: http://www.mct.gov.br/upd_blob/0018/18123.pdf

[37] F McDougall, P White, M Franke P, Hindle, Integrad solid waste management: a life cycle inventory. Blackwell Science published: 2ª ed., USA, 2001.

[38] MP Hekkert., LAJ Joosten, E Worrell, Reduction of CO2 emissions by improved management of material and product use: the case of primary packaging, Resources, Conservation and Recycling, vol. 29, pp. 33-64, 2000a.

[39] MP Hekkert, LAJ Joosten, E Worrell, Reduction of CO2 emissions by improved management of material and product use: the case of transport packaging, Resources, Conservation and Recycling, vol. 30, pp. 1-27, 2000b.

[40] FAM Lino, Proposta de aproveitamento do potencial energético do resíduo sólido urbano e do esgoto doméstico com minimização dos impactos ambientais, Doctorate Thesis, State University of Campinas, BR, 2014.

[41] DEFRA - Department for Environment Food and Rural Affairs. Incineration of municipal solid waste, 2013b. Available at: http://www.defra.gov.uk/publications/

12

Modelling and Analysis of Thermoelectric Generation of Materials Using Matlab/Simulink

K. P. V. B. Kobbekaduwa, N. D. Subasinghe*

National Institute of Fundamental Studies, Hanthana Road, Kandy, Sri Lanka

Email address:

deepal@ifs.ac.lk (N. D. Subasinghe)
*Corresponding author

Abstract: This paper presents several models and implementations on measuring the thermoelectric behaviour of an unknown material using Matlab/Simulink. The proposed models are designed using Simulink block libraries and can be linked to data obtained from an actual experimental setup. This model is unique, as it also contains an implementation that can be used as a laboratory experiment to estimate the thermal conductivity of the unknown material thus, making it easy to use for simulation, analysis and efficiency optimization of novel thermoelectric material. The model was tested on a natural graphite sample with a maximum output voltage of 0.74mV at a temperature difference of 25.3K. Thus, according to the collected data, an experimental mean value of 68W/m.K was observed for the thermal conductivity while the Seebeck coefficient had a mean value of -3.1μV/K. Hence, it is apparent that this model would be ideal for thermoelectric experimentation in a laboratory based environment especially as a user interface for students.

Keywords: Seebeck Effect, Thermoelectric Power, Thermal Conductivity, Electrical Conductivity, Simulink Modelling

1. Introduction

Thermoelectric effect is a simple phenomenon based on the thermal and electrical characteristics of a material. Discovered in the early 1800s it was thought to be an interesting form of energy conversion that relies on the physical characteristics of materials. This phenomenon is observed when two different types of material are combined and a temperature gradient is applied between the joint and open ends. When the fore mentioned conditions are met, a thermal current flows through the combined thermocouple, hence a voltage is generated between them. The thermoelectric effect, generally known as the Seebeck effect, gives rise to this inherent EMF or voltage due to the material property known as the Seebeck coefficient. These individual thermocouples can be combined in series to increase the output voltage to create a single thermoelectric module commonly known as a *Peltier* module. These can be used as voltage generators known as Thermoelectric Generators (TEG) or as coolers known as Thermoelectric Coolers (TEC). The main advantage of thermoelectric power is, it is a solid state energy conversion that does not have mechanical or liquid based moving parts. Hence, modules can be designed to be compact, stable as well as being reliable and noiseless. The main drawback is the efficiency of these materials thus, their uses have been confined to relatively smaller applications. However, when it comes to energy scavenging, thermoelectric generators always improve the overall energy efficiency of an existing system. For example, certain car manufactures could increase the fuel efficiency by over 5% simply generating electricity from the exhaust heat [1]. With the advent of semiconductor materials as well as improvements in synthesis, materials with larger efficiency values have been discovered in the recent past, hence the renewed interest in this form of energy conversion. This has resulted in a plethora of new applications in numerous fields such as the automotive industry, space exploration and wearable nano-based technology.

Modelling of TE device or the behaviour of individual materials is an important prerequisite for the design and control verification of the final output device. Hence, the model has to be integrated seamlessly in to the overall system model that may contain other electrical, thermodynamic, or even mechanical components [2]. Numerous research work on modelling a thermoelectric material or module has been done using software such as SPICE [3, 4] as well as Dynamic and static modelling of TE modules using MATLAB/Simulink [5, 6]. The modelling of thermal and power generation behaviour of these TEGs have been extensively studied [7] and the output values have also been modelled and estimated using novel techniques such as Artificial Neural Networking [8]. In most of these prior work the main focus has been on modelling TEG modules where as in this research we focus on modelling the behaviour of a particular material.

2. Principle of Thermoelectric Generation

The physical process of the Seebeck effect can be characterized in 5 distinct steps,

- Temperature difference generates a difference in Fermi level
- Bandgap distance changes with temperature
- Diffusion coefficient is a function of temperature
- Charge carriers move from the heated side to cold side - thermodiffusion
- Electric field will be generated due to the transport of charge carriers

In this paper we specifically look at modelling the TEG of materials. The following equations govern the thermoelectric behaviour of any material [9, 10]. The Seebeck coefficient S is defined as:

$$S = \frac{dV}{dT} \qquad (1)$$

where V is the Seebeck voltage or electromotive force (EMF) and T is the temperature. Apart from the Seebeck effect there are 3 other forms of energy conversion taking place in a TEG material these are, thermal conduction described by

$$Q_{th} = -\kappa_{TH}\Delta T \qquad (2)$$

where κ_{TH} is the thermal conductivity of the material and ΔT = T_H (hot side temperature) $-T_C$ (cold side temperature). Joule heating, which is the heat dissipation due to the internal resistance of the material given by,

$$Q_J = I^2 R \qquad (3)$$

where R is the electrical resistance and I is the current. Peltier cooling/heating effect, which is a phenomenon of heat absorption/dissipation by a junction between two dissimilar materials when electrical current flows through the junction is given by,

$$Q_{E/A} = SIT_{H/C} \qquad (4)$$

Apart from the above the additional Thompson effect, which is described by the Thompson coefficient τ = dS/dT is

small enough to be neglected. Thus, heat flow at the hot and cold end respectively can be expressed as,

$$Q_H = K_{TH}\Delta T + SIT_H - \frac{1}{2}I^2 R \qquad (5)$$

$$Q_C = \kappa_{TH}\Delta T + SIT_C + \frac{1}{2}I^2 R \qquad (6)$$

Thus, the net power is given by $P = Q_H - Q_C = [S\Delta T - IR]I$. Hence, the output voltage is,

$$V = S\Delta T - IR \qquad (7)$$

Apart from the output values the usefulness of a thermoelectric material is dependent on the power factor of the said material this is calculated using the Seebeck coefficient and the electrical conductivity. Thus power factor P_f is given by,

$$P_f = \sigma S^2 \qquad (8)$$

The efficiency of a thermoelectric material is described using a dimension less figure of merit Z. A good TEM must combine a large Seebeck coefficient S with high electrical conductivity σ and low thermal conductivity κ_{TH}. Hence FOM is given by,

$$Z = \frac{\sigma S^2}{\kappa_{TH}} \qquad (9)$$

The efficiency of these materials is described according to the output electric power compared to the applied heat energy Q_H. Thus,

$$\phi = \frac{I^2 R_L}{Q_H} \qquad (10)$$

where R_L is the load resistance. Using the FOM value the maximum efficiency of the TE device is written as,

$$\phi_{max} = \frac{\Delta T}{T_H}\left(\frac{\sqrt{1 - ZT_{avg}} - 1}{\sqrt{1 + ZT_{avg}} + \frac{T_C}{T_H}}\right) \qquad (11)$$

where $T_{avg} = \frac{T_H + T_C}{2}$

3. Measurement of Thermal Conductivity

There are several methods to measure thermal conductivity. The measurement of heat flow is done directly which is known as absolute method and indirectly known as comparative method. Apart from measuring heat flow there are several other methods to measure thermal conductivity at sub ambient temperatures and higher. The most commonly used method is the axial flow type where heat flow is considered as axial and conductivity calculated accordingly.

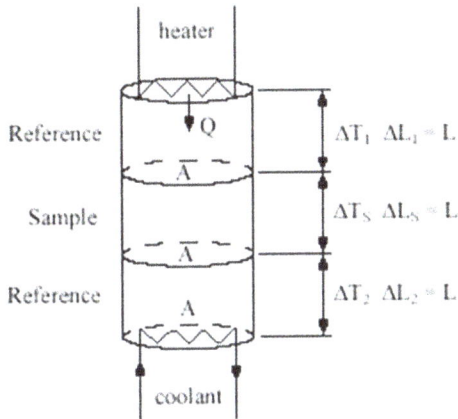

Figure 1. Comparative cut bar method for measuring thermal conductivity [11].

In this paper we consider the comparative cut bar method which is widely used for determining axial thermal conductivity.

In this, the principle of the measurement lies with passing the heat flux through a known sample and an unknown sample and comparing the respective thermal gradients, which will be inversely proportional to their thermal conductivities [11, 12]. As shown in figure 1 this method involves the measurement of 4 separate temperature values T_1, T_2, T_3, T_4

where, $\Delta T_1 = T_1 - T$, $\Delta T_S = T_2 - T_3$ and $\Delta T_2 = T_3 - T_4$. Thus, the heat flux can be calculated as,

$$\frac{Q}{A} = \kappa_S \frac{\Delta T_S}{L} = \kappa_{ref} \frac{\Delta T_1 + \Delta T_2}{2L} \qquad (12)$$

where κ_S and κ_{ref} are thermal conductivities of the sample and reference material respectively. Thus, κ_S can be calculated as,

$$\kappa_S = \kappa_{ref} \frac{\Delta T_1 + \Delta T_2}{2\Delta T_S} \qquad (13)$$

(a)

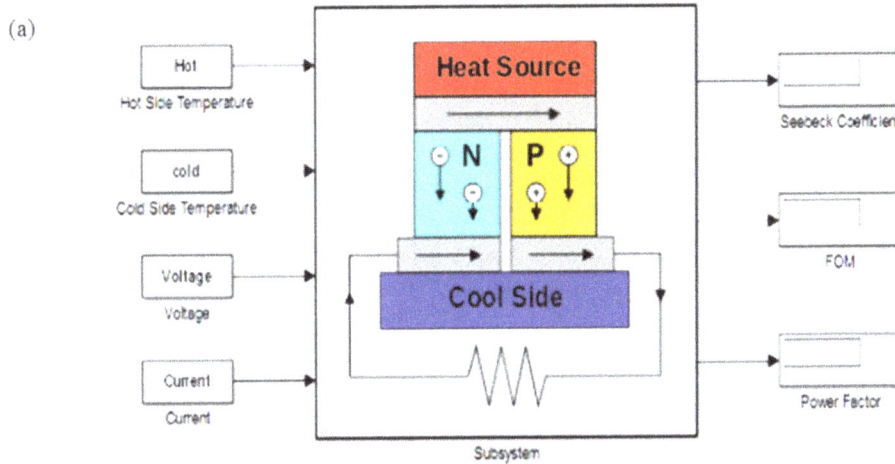

(b)

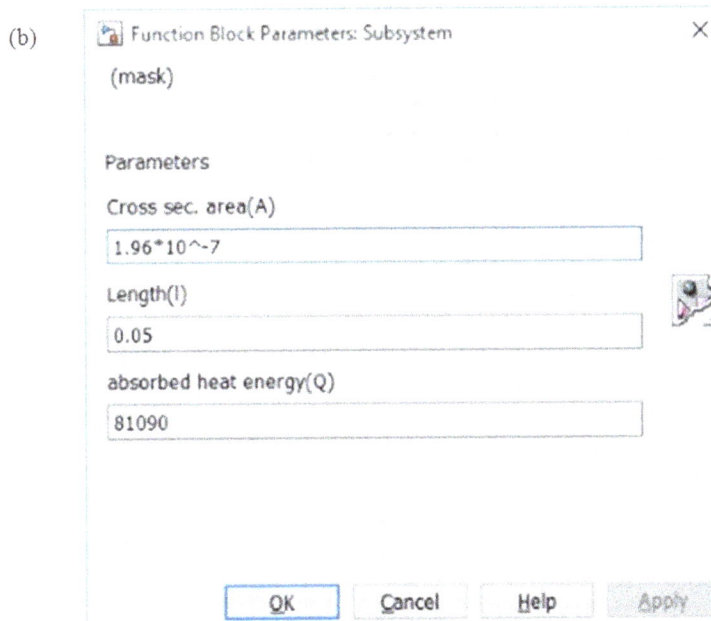

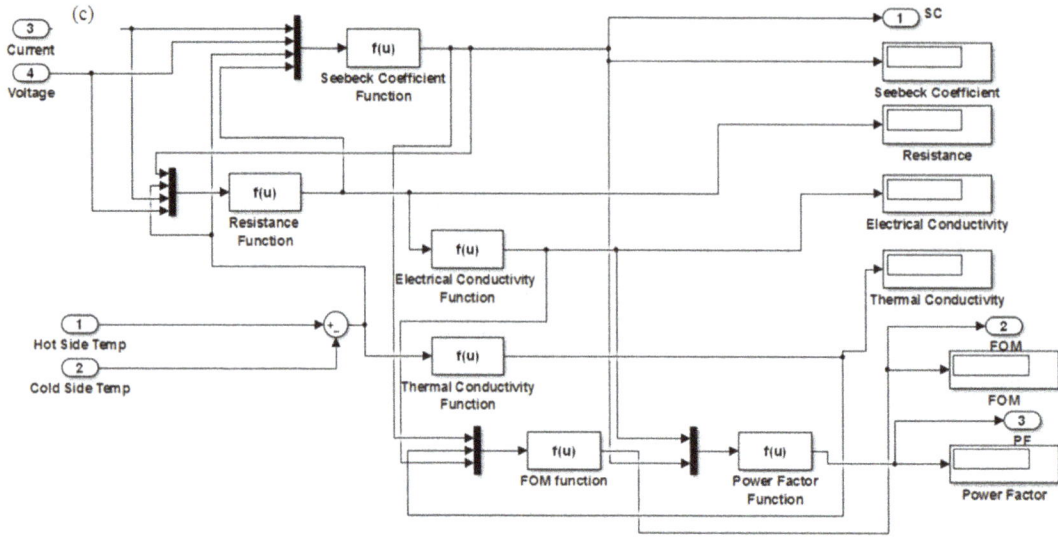

Figure 2. (a) *Mask implementation,* (b) *Initial conditions and* (c) *Subsystem of the proposed TE model*

4. Model Building and Implementation

The required data about the material should be obtained experimentally and the model is a guideline to implement and find the pertinent values related to thermoelectricity i.e. Seebeck coefficient, Power Factor and the figure of merit. MATLAB/Simulink software is used to construct the model. Previous work [9] includes a model based on the specifications related to Peltier modules, hence this paper looks at the feasibility of using an extended model to calculate the important factors related to thermoelectric behaviours of a novel material. A secondary model is used for calculating the thermal conductivity of the material based on the comparative cut bar method mentioned previously.

Figure 2 shows that in the initial theoretical model resistance is also calculated from the current, voltage and temperature data as well as the user inputs. In this case the thermal conductivity value also needs to be calculated using a given absorbed heat energy value. As the accuracy of this value is questionable (due to various heat loss factors) it is better to use a tested method as mentioned above to calculate the thermal conductivity. Thus, in this model we have used the aforementioned comparative cut bar method to find an experimental value for thermal conductivity. The standalone Simulink implementation for this method is as shown in figure 3 below.

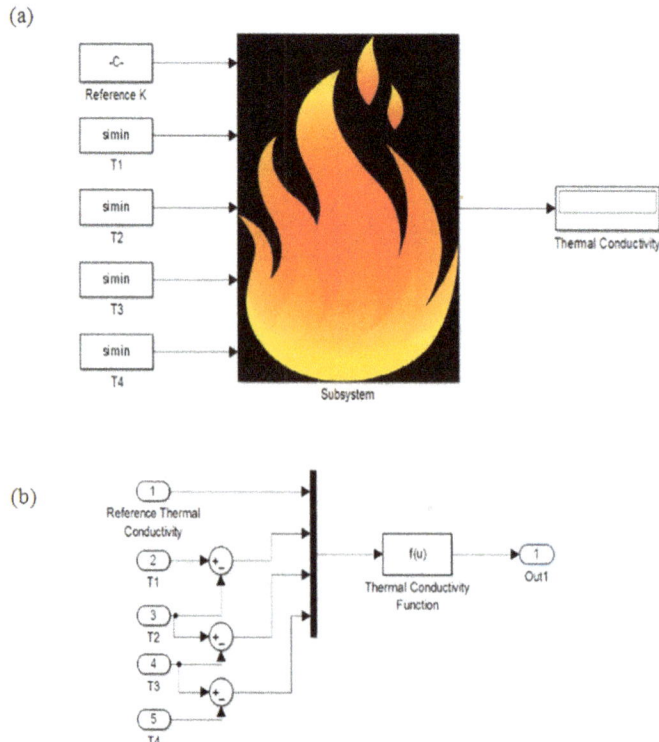

Figure 2. (a) *Masked implementation and* (b) *Subsystem of thermal conductivity model*

In this model we need 4 separate temperature measurements along with the thermal conductivity of the reference material used in the experiment. Thus, the front view and the parameter block used to enter the above values is shown in figure 4 (a) and (b) respectively. We can combine both these models to obtain the final implementation and the complete subsystem as seen in figure 5. Thus, the reference material thermal conductivity, length and cross sectional area of the sample are given as initial conditions and entered in the dialog box. In this revised model resistance can be added as a set of measured data instead of a theoretical calculation.

The thermal conductivity implementation is added to the subsystem of the final model. The data is read off the MATLAB workspace and by adjusting the 'simulation stop time' we can observe the resulting values at each of the measured data points. To obtain the Seebeck coefficient, FOM and Power factor of a given data set the inputs should be obtained from the MATLAB workspace hence we can observe variations in the final output values. The fore mentioned outputs are obtained according to the equations 1 through 13.

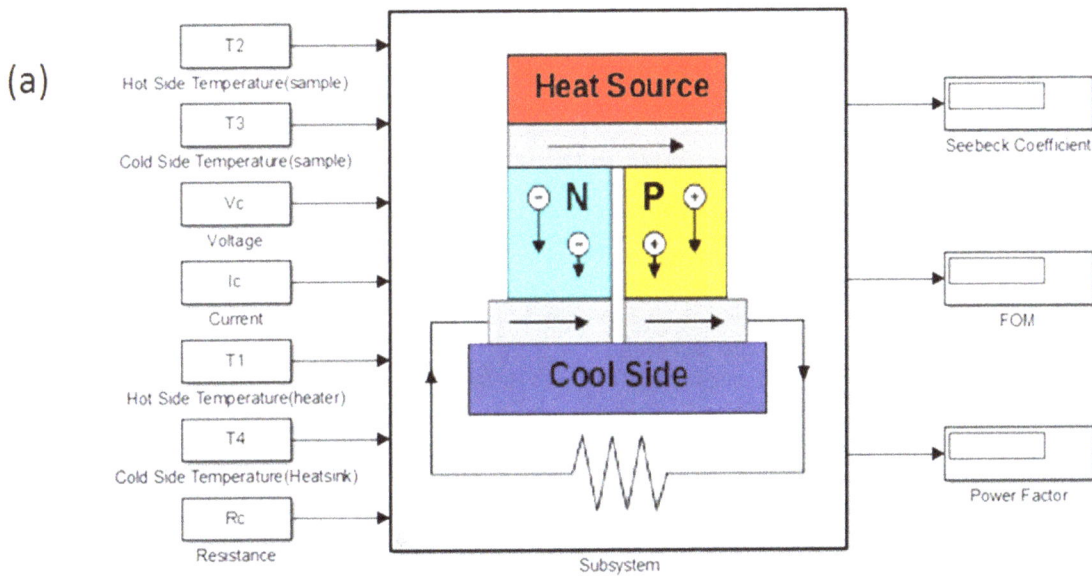

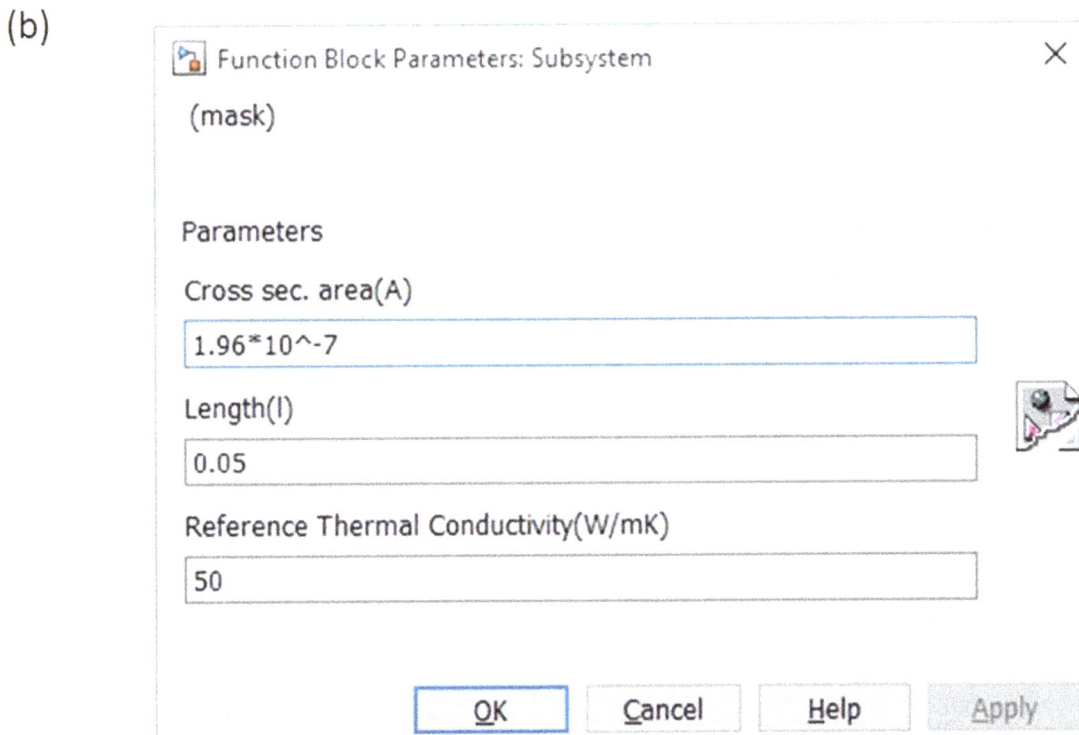

Figure 3. The (a) masked implementation, (b) Parameter block for the thermoelectric model.

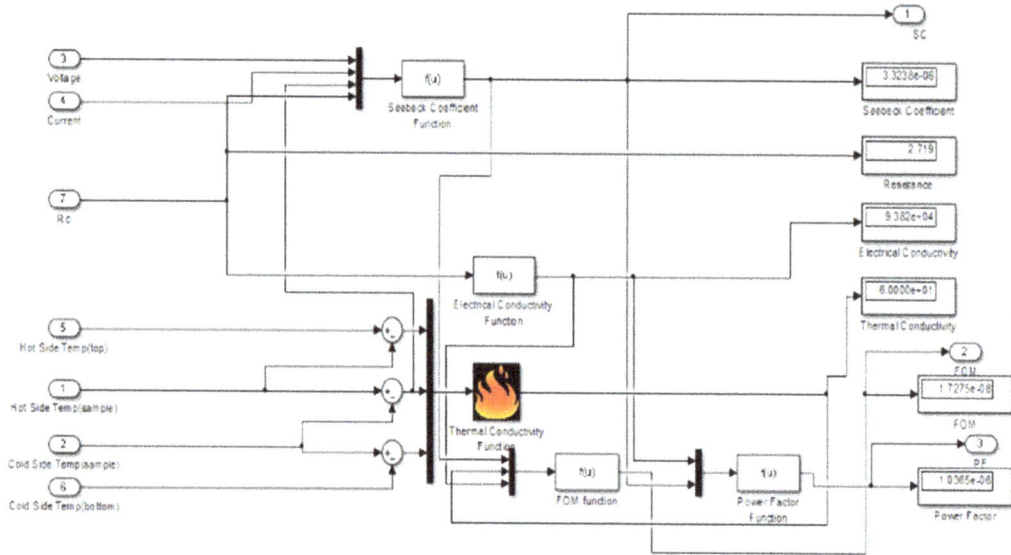

Figure 5. The subsystem for the thermoelectric model combined with the model for thermal conductivity.

5. Simulation

The model is used to simulate and find the Seebeck coefficient of a graphite sample which has dimensions of 4mm thickness and 1.17cm diameter (A = 1.075cm^2). This sample is created using a powdered graphite sample ground using a ball mill and then compressed at 1.5 ton/m^2 pressure. The reference samples are soldering lead melted and solidified to form a sample which has similar dimensions as above. The setup is insulated to reduce thermal leakage using wood and cotton. The experimental setup is as shown below in figure 6.

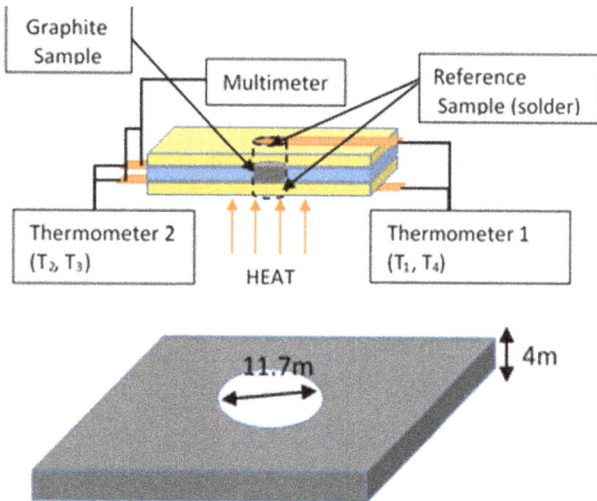

Figure 6. Experimental setup to measure temperature values and output voltage and current.

The thermal conductivity value of solder lead (Sn 63% Pb 37%) used (κ_{reff}) is 50 W/mK [13]. Using the above set up voltage, current and resistance values are obtained along with 4 temperature values. Figure 7 shows variations of the fore mentioned values with varying temperature difference for the graphite sample while the VI curve is shown in figure 8.

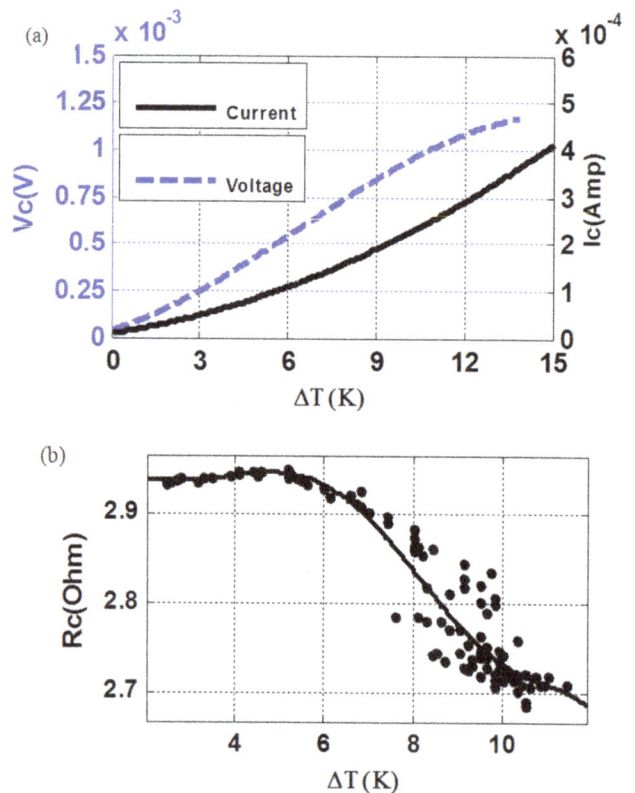

Figure 7. (a) Output voltage, Current (b) Resistance vs. Temperature difference.

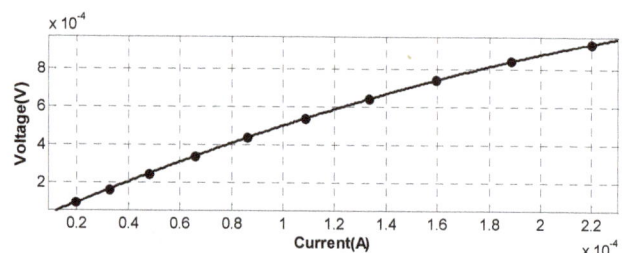

Figure 8. VI curve related to the graphite sample.

In this example we have measured and collected a data set which has 100 points. By reading the obtained values from the MATLAB workspace we can estimate values for thermoelectric generation of the sample. Figure 9 (a) and (b) shows the thermal conductivity and Seebeck coefficient values obtained at several different temperature differences for the natural graphite sample

From the graphs in figure 9 we can observe that the thermal conductivity varies with temperature difference with a maximum value of 95W/m.K and a mean value of 68W/m.K. These values are in line with previously observed thermal conductivity values of graphite ranging from 25-470W/m.K [14, 15].

The Seebeck coefficient varies between 0 and -10µV/K with a mean value of -3.1µV/K. Values for the Seebeck coefficient of graphite have been previously reported for flexible graphite [16], Carbon nanotubes [17] and Graphite Intercalation compounds and composites [18, 19] and in most cases they vary from positive to negative due to the presence of other material combined with graphite. In this case the above results tend to agree with flexible graphite as the Seebeck coefficient reported was -2.6µV/K at 300K. These variations in output results can attributed to the fact that like flexible graphite, the used natural graphite sample contains impurities which will change the overall behaviour of the sample.

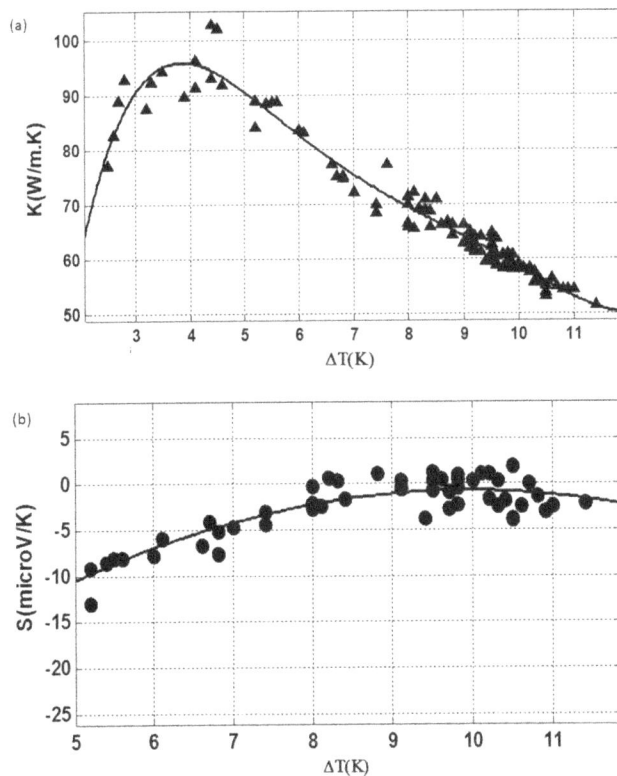

Figure 9. Calculated variation for (a) thermal conductivity and (b) Seebeck coefficient vs. temperature difference for the graphite sample.

Apart from the Seebeck coefficient we can also look at the figure of merit (FOM) and the overall efficiency of a material using this model. In this case we have calculated the FOM

and the efficiency of the natural graphite sample and the variations with temperature difference are shown in figure 10 (a) and (b).

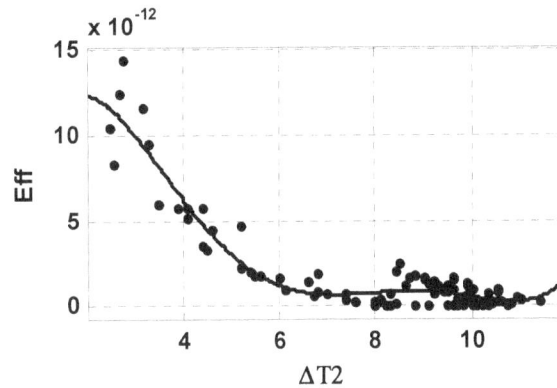

Figure 10. Calculated variation for (a) Figure of merit (Z) and (b) Efficiency vs. temperature difference for the graphite sample.

As the two graphs above depict graphite in its raw natural form is not efficient or does not provide adequate figure of merit to use as an energy efficient thermoelectric material. Though this may be used in low tech devices which does not require high efficiency or high output values to operate. As it is cheap and easily found we can also manufacture other forms of graphite such as GICs, nanotubes and graphene all of which have exhibited larger FOM and efficiency vales in previous work.

6. Discussion and Conclusion

The discussed model for calculating the thermoelectric behaviour of a material has been simulated for a laboratory based environment. External errors may exist especially when calculating thermal conductivity of a material using the comparative cut bar method mainly due to thermal insulation issues. The model is also dependent on external data specifically from the MATLAB workspace. Hence, it is also possible to input a reference dataset based on previous work to verify certain output values.

As the resistance values are also obtained thorough a measured dataset we can also extend the model to calculate physical parameters such as the temperature coefficient of resistance. Another advantage of this model is that we can adjust it for a different method for calculating thermal conductivity thus we can use it as a laboratory interface especially for students who are calculating these values. As we are using experimentally obtained datasets the accuracy of the output values is higher and these values can be exported to MATLAB to be further analysed. The model is also user friendly with a dialog box which allows the user to change parameters of the sample as well as include data from a reference material. Future work includes the further verification of this model for more extensive experimental work as well as extension of this model to predict output values when various cooling techniques are used to increase the temperature difference.

Acknowledgements

The authors wish to thank the director and members of the academic and non-academic staff at the National Institute of Fundamental Studies for their gracious support in completing this research work.

References

[1]   B. Orr, A. Akbarzadeh, M. Mochizuki and R. Singh. "A review of car waste heat recovery systems utilising thermoelectric generators and heat pipes," in Applied Thermal Engineering, 2016, p. 13.

[2]   F. Felgner, L. Exel, M. Nesarajah, and G. Frey." Component-Oriented Modeling of Thermoelectric Devices for Energy System Design" in IEEE Transactions on Industrial Electronics, vol. 61 (3), 2014, p. 1301-1310

[3]   Y. Moumouni and R. J. Baker. "Improved SPICE Modeling and Analysis of a Thermoelectric Module," in International Midwest Symposium on Circuits and Systems (MWSCAS), Fort Collins, CO, IEEE, 2015.

[4]   C. Li, et al, "Thermoelectric Cooling for Power Electronics Circuits: Modeling and Active Temperature Control," IEEE Transactions on Industry Applications, vol. 50 6), 2014, p. 3995 - 4005.

[5]   A. Kane, V. Verma, and B. Singh. "Temperature Dependent Analysis of Thermoelectric Module using Matlab/SIMULINK," in IEEE International Conference on Power and Energy (PECon), Kota Kinabalu Sabha, Malaysia, 2012.

[6]   A. M. Yusop, et al., "Dynamic Modeling and Simulation of a Thermoelectric-Solar Hybrid Energy System Using an Inverse Dynamic Analysis Input Shaper" in Modelling and Simulation in Engineering, 2014: p. 13.

[7]   A. M. Yusop, R. Mohamed and A. Ayob "Model Building of Thermoelectric Generator Exposed to Dynamic Transient Sources" in IOP Conf. Series: Materials Science and Engineering, vol. 53, 2013.

[8]   B. Ciylan. "Determination of Output Parameters of a Thermoelectric Module using Artificial Neural Networks," in Elektronika ir Elektrotechnika, vol. 116 (10), 2011, p. 63-66.

[9]   H. L. Tsai and J.-M. Lin, "Model Building and Simulation of Thermoelectric Module Using Matlab/Simulink," Journal Of Electronic Materials, vol. 39 (9), 2010, p. 2105-2111.

[10]   Y. Apertet and C. Goupil, "On the fundamental aspect of the first Kelvin's relation in thermoelectricity," in International Journal of Thermal Sciences, vol. 104, 2016, p. 225-227.

[11]   Editors, "Thermal Conductivity - Different Methods for Measurement of Thermal Conductivity," 2011, 11th June 2013 [cited 8th March 2016]; Available from: http://www.azom.com/article.aspx?ArticleID=5615.

[12]   TA Instruments, "Principal Methods of Thermal Conductivity Measurement," 2012, p. 5.

[13]   J. Wilson, "Thermal conductivity of solders," in Electronics Cooling, vol. 12 (3), 2006, p. 4-5.

[14]   S. Desai and J. Njuguna, "Thermal properties of natural graphite flake composites," in International Review of Mechanical Engineering, 4th ed., vol. 6, 2012, p. 923-926.

[15]   M. Smalc, et al. "Thermal performance of natural graphite heat spreaders," in ASME 2005 Pacific Rim Technical Conference and Exhibition on Integration and Packaging of MEMS, NEMS, and Electronic Systems collocated with the ASME 2005 Heat Transfer Summer Conference, San Francisco, CA, USA, American Society of Mechanical Engineers, 2005.

[16]   Y. M. Hoi and D. D. L. Chung, "Flexible graphite as a compliant thermoelectric material," in Carbon, vol. 40 (7), 2001, p. 1134-1136.

[17]   M. Penza, et al., "Thermoelectric Properties of Carbon Nanotubes Layers," in Sensors and Microsystems, vol. 91, 2011, Springer, p. 73-79.

[18]   R. Matsumoto, Y. Okabe, and N. Akuzawa, "Thermoelectric Properties and Performance of n-Type and p-Type Graphite Intercalation Compounds," in Journal of Electronic Materials, vol. 44 (1), 2015, p. 399-406.

[19]   R. Javadi, P. H. Choi, H. S. Park, and B. D. Choi, "Preparation and Characterization of P-Type and N-Type Doped Expanded Graphite Polymer Composites for Thermoelectric Applications," Journal of Nanoscience and Nanotechnology, 11th ed., vol. 15, November 2015, p. 9116-9119.

13

Numerical Analysis of Thermosyphon Solar Water Heaters

Samuel Sami, Edwin Marin, Jorge Rivera

Research Center for Renewable Energy, Catholic University of Cuenca, Cuenca, Ecuador

Email address:

dr.ssami@transpacenergy.com (S. Sami)

Abstract: This paper presents the modeling and simulation as well as validation of a natural circulation closed thermosyphon glass tube solar collector water heater. Energy conservation equations for the heat transfer fluid flow and the storage tank were written in finite-difference form, integrated and solved to yield the characteristics of the thermosyphon system at different solar insolations and water mass flow rate conditions as well as water temperatures. Comparison between experimental data and numerical prediction of the proposed showed that the model predicted fairly the evacuation of storage tank temperature at various initial temperature of the water at the storage tank.

Keywords: Thermal Solar Collector, Thermosyphon, Water Heater, Modeling, Analysis Validation

1. Introduction

Water heating typically represents a significant percentage of energy consumption in domestic, industrial applications. Carbon baseline fuels water heating produce emissions of greenhouse gases and other pollutants. Due to high cost primary energy resources and their associated with serious environmental issues, solar energy is an alternative viable source of energy for water heating. Solar water heating can be characterized as active or passive [1-5]. An active system is based on an electric pump to circulate the working fluid through the solar collector. In passive solar water heating, heat-transfer fluid uses thermosyphoning phenomenon to circulate the water by the buoyancy forces and is replaced by colder water from the bottom of the tank [5]. This continuous circulation continues until heats up water in the storage tank (Figure 1).

Abgo reported on the performance profile of a thermosyphon solar water heater [6]. The performance evaluation was based on the mathematical models that describe the test system and some measured experimental data. The effect of some of the design and operating parameters that have been shown to affect the system's performance was investigated. Numerical simulation of steady state natural convection heat transfer in a 3-dimensional single-ended tube subjected to a nanofluid has been presented by Shahi et al. [7]. It was a simplified model

for single-ended evacuated solar tube of water glass evacuated solar water tube heater. It was assumed in the model that the water sealed tube is adiabatic and also the tube opening is subject to copper-water nanofluid. Governing equations were based upon cylindrical coordinate system.

Another numerical analysis of a modified evacuated tubes solar collector was presented by Sato, et al. [8]. This paper proposed a study of solar water heating with evacuated tubes, their operation, characteristics and operating parameters. Furthermore, another analytical study was reported by Hammadi [9] on the study of solar water heating system with natural circulation in Basrah. The results show that the performance of the solar water heater depends on parameters such as tilt angle and orientation of the collector, wind velocity, area of the collector, latitude, and solar time.

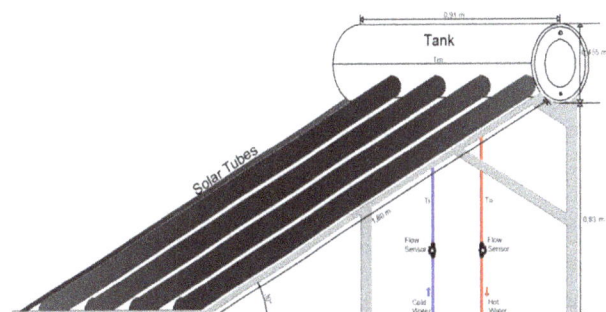

Figure 1. Thermosyphon solar water heating system.

2. Mathematical Model

A schematic of the system under study is depicted in Figure 1. The system consists of a thermal solar panel, with solar glass tubes as water tube heater and control valves. Specifications of the solar glass tube collector are given in Table.1. As in passive solar water heating system, the water as heat-transfer fluid uses thermosyphoning phenomenon to circulate between the solar tubes and tank. With the buoyancy forces colder water is replaced from the bottom of the tank. This continuous circulation continues until heats up water in the storage tank (Figure 1). The storage was filled with 150 liters of water and the working fluid tubes were filled with working fluid. The working fluid flows inside the tubes in the collectors and by thermosyphoning back to the solar water storage tank. In the solar collector the solar radiation is absorbed by the working fluid. Due to the absorption of solar radiation, working fluid temperature increases. Thus, because of the difference of working fluid densities, the heated working fluid moves upward to the storage tank. The heat from the working fluid is transferred to the cold water and increases its temperature in storage tank. The following parameters were recorded at 1 minute intervals for the purpose of validation of the proposed model; cold water temperature working fluid temperature, solar collector temperature, storage tank temperature, solar insolation (W/m^2), ambient temperature, wind velocity (m/sec.), hot water temperature and relative humidity (%).

The conservation equations and heat transfer equations were written for each water solar glass tube collector, storage tank and heat transfer fluid as follows;

Energy conservation and heat transfer equations:
The mass flow rate circulating through the solar glass tube collector can be calculated using the following formula [9];

$$128 \frac{v L_c m}{\pi N d_c} = \rho_0 \, g \, \beta' \, (T_o - T_i) \left[\frac{L_c \sin \varphi}{2} + H \right] \quad (1)$$

Equation (1) can be rewritten as follows;

$$m = \frac{\rho_0 \, g \, \beta' \, (T_o - T_i) \left[\frac{L_c \sin \varphi}{2} + H \right] v L_c m}{128 \, \pi N d_c} \quad (2)$$

and;

$$T_o - T_i = \frac{I \alpha A_c - h \, A_c (T_m - T_\infty)}{m c_p} \quad (3)$$

Where T$_m$ is;

$$T_m = \frac{T_o + T_i}{2} \quad (4)$$

The heat transfer coefficient of wind is given by [9, 10];

$$h = 2.8 + 3 \, u_{wind} \quad (5)$$

In addition, the following parameters; density, specific heat, coefficient of water expansion, as well as volume of storage tank are defined as follows respectively;

$$\rho = 1001 - 0.08832T - 0.003427T^2 \quad (6)$$

$$C_p = 4226 - 3.244T + 0.00575T^2 - 0.0003656T \quad (7)$$

$$\beta' = (0.3 + 0.116T - 0.0004T^2)10^{-4} \quad (8)$$

$$v = \left(\frac{1}{0.5155 + 0.0192T} - 0.12 \right) 10^{-6} \quad (9)$$

Meanwhile, the energy equation for the thermosyphon solar water heater during daily solar radiation can be written as follows [6];

$$W \frac{dTm}{dt} + U(T_m - T_o) + BM(t)C_p (T_m - T_o) = I_o(\tau \alpha)_c \, F' \, A_c \quad (10)$$

The rate of water evacuated from the storage tank is given by;

$$\frac{dTm}{dt} = \frac{I_o(\tau \alpha)_c \, F' \, A_c - [U(T_m - T_o) - Bm(t)C_p (T_m - T_o)]}{W} \quad (11)$$

Where $\tau \alpha\_$ is effective transmittance-absorptance and F^1 represents collector efficiency factor, and U is collector overall heat loss coefficient.

W, and *Tm* are the total heat capacity of the system, and the mean system temperature respectively. *B* is a constant which can be taken as 1 if water is drawn from the middle of the tank and 2 if it drawn from the top of the tank [6].

The solar collector energy conversion efficiency can be calculated as follows;

$$\eta_c = Q_u / A_c \, I_o \quad (12)$$

Where Q_u is energy absorbed by the collector.
The energy absorbed by the collector can be obtained as follows [6];

$$Q_u = A_c \, F^1 [\, I_o \, (\tau \alpha) - U_1 (\, T_m - T_a)] \quad (13)$$

Where,
F^1: is the collector efficiency factor
U$_1$: is the collector overall heat loss coefficient (W/m^{2}K)
Solving the aforementioned mentioned equations yield the water mass flow rate evacuated from the storage tank, energy absorbed by solar collector, energy conversion efficiency as well the storage tank temperature during the water evacuation from the tank.

3. Numerical Procedure

The energy conversion and heat transfer mechanisms taking place during the thermosyphoning process in a thermal solar panel glass tubes have been outlined in equations (1) through (13). The aforementioned equations have been written in finite-difference form, solved as per the logical flow diagram shown in Figure 2 as a function of time. First the input independent parameters are defined and other dependent parameters were calculated and integrated to yield evacuated water mass flow rate as per equation (2) and the temperature gradient in the storage tank as per equation (11). Iterations were performed until a solution is reached with acceptable iteration error. The numerical procedure starts with using the solar radiation to calculate the mass flow water circulating in the solar panel due buoyancy forces. This follows by predicting the water temperature profile as a function of time.

Table 1. Specifications of the water glass solar tubes collector.

SUNSHORE VACUUM TUBES SPECIFICATION	
Model	58*1800
Material	Borosilicate glass
Outer tube dia	∅ 58 mm
Inner tube dia	∅ 47 mm
Tubes length	1800 mm
Tubes number	10
Thickness	1.6 mm
Vacuum inner pressure	≤ 5.0 x 10-3 Pa
Absorptivity	≥ 0.92 (AM 1.5)
Emissivity	≤ 0.08 (80°C ± 5°C)
Absorptivity coating	Al/Copper/Stainless or Al/N/AL
Thermal expansion	3.3
Stagnation temperature	≥ 270°C
Heat loss	22 W/(m3, K)
Maximum strength	30mm iron ball dropped directly against the tubes from 450mm high attitude, the tube without any damages
Anti-freezing	- 30°C
Resist Hailstone	25 mm diameter hailstone
Life time	15 years

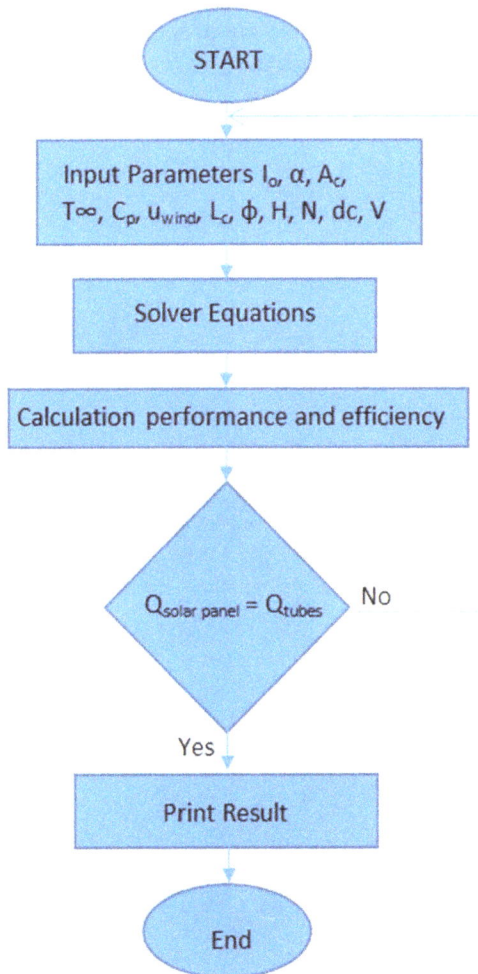

Figure 2. Logical flow diagram.

4. Results and Discussion

The thermosyphon solar water heater shown in Figure. 1, has specifications outlined in Table. 1. The storage tank capacity of 150 liters was connected to plastic pipes connections for the circulation of the water from and to the collector. In this section, we first present numerical simulation results of the solar water heater system presented in reference [6] under various conditions and secondly validate our proposed model with experimental data.

The aforementioned system of equations (1) through (13) in finite-difference formulation has been numerically solved and samples of the predicted results are plotted in Figures 3 through 15 under different inlet conditions such as insolation, heat transfer fluid flow rates and heat transfer fluid temperatures as well as solar panel inclination angles. In particular, Figure. 3 presents the daily time variation of solar insolation (W/m^2) measured at the site and employed in the validation of the proposed model. It is quite clear that the intensity of radiations depends upon the hour of the day and the month of the year. The average values of radiation insolation were used in the present study.

In general, it is quite clear from these figures (4) through (11) that temperature of water in the storage tank decreases during the evacuation process and the rate of temperature drop has functional dependence on the various parameters outlined in equations (1) through (13). In the following sections, the simulation results of the storage tank temperature will be presented and analyzed under different conditions.

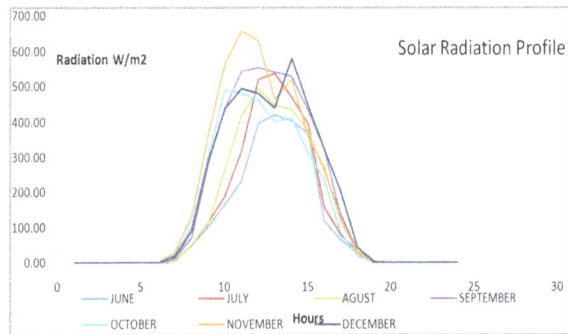

Figure 3. Time variation of solar insolation during during 2015.

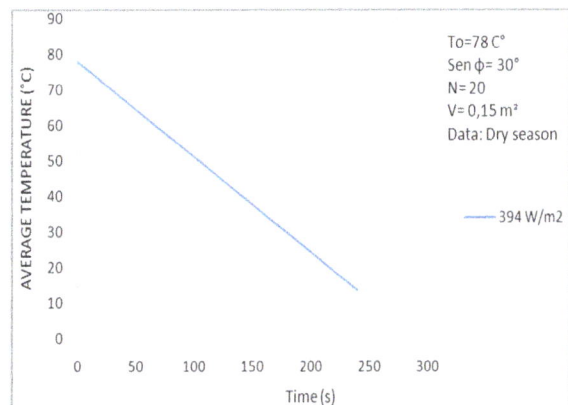

Figure 4. Time variation of initial water temperature in storage tank.

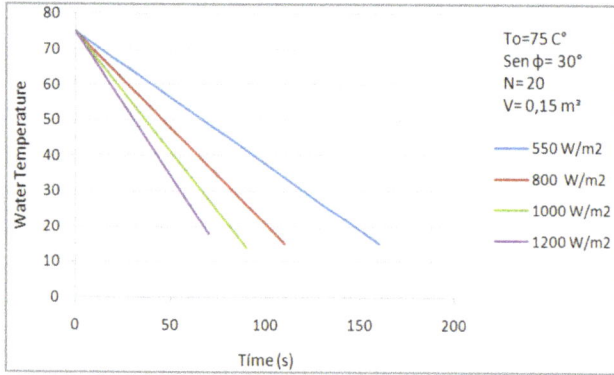

Figure 5. Time variation of initial water temperature in storage tank at various solar radiations.

The simulated results presented in Figures. 5 through. 11 show that the water temperature in the storage tank is principally a function of solar radiation and the ambient air conditions. However, a typical analysis of the simulated results presented in Figures 7 through 11 also shows that the water temperature in the storage tank depends upon other parameters such as collector orientation, number of pipes in the solar collector and tank volume, the rate of evacuation of the water from the tank and the makeup water.

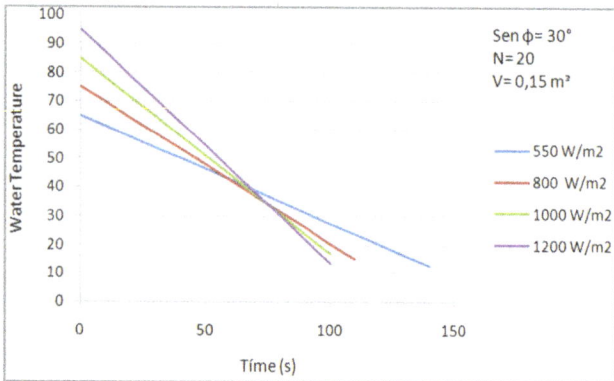

Figure 6. Time variation of different initial water temperature in storage tank at different solar radiations.

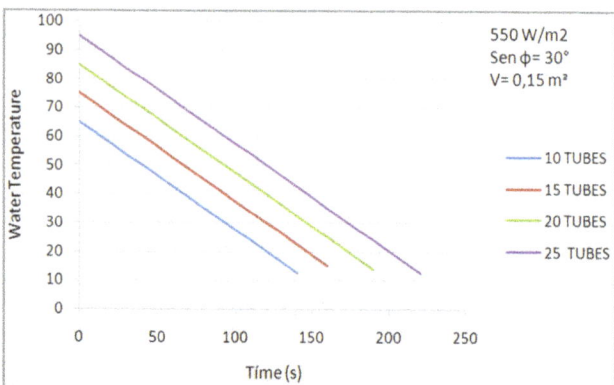

Figure 7. Time variation of different initial water temperature in storage tank with various heat tubes in solar collector.

In particular, Figure. 5 shows that impact of solar intensity on the rate of decrease temperature inside the storage tank. It can be also observed from this figure that the lower the

radiation insolation is the slower rate of temperature reduction.

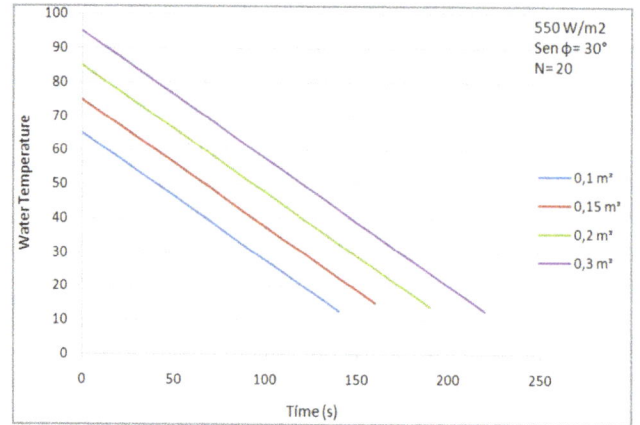

Figure 8. Time variation of initial water temperatures in different storage tank volumes.

The simulation results presented in Figure. 6 illustrate the impact of initial storage tank temperature on the rate of decrease of temperature in the storage tank under different solar insolation at constant water evacuation flow rate from the tank. It is quite clear from this figure that longer period for water evacuation occurs at lower solar insolation and lower initial storage tank temperature.

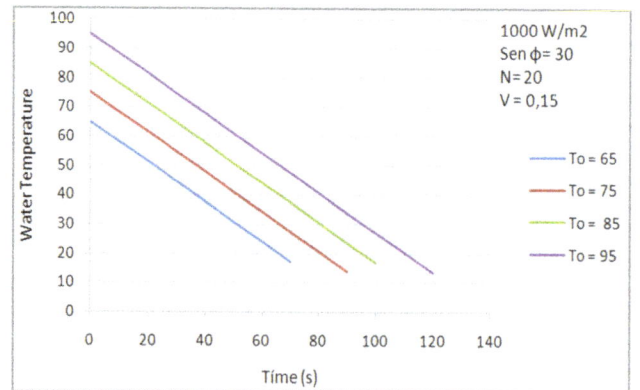

Figure 9. Time variation of initial water temperature in storage tank at different outlet water temperatures.

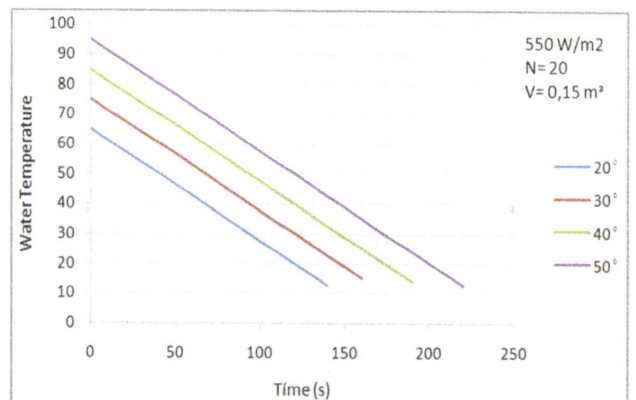

Figure 10. Time variation of initial water temperature in storage tank at different solar tubes inclination angles.

However, Figure. 7 showed that increasing the number of solar tubes will prolong the water evacuation time of the storage water tank at constant solar insolation. It is also evident from the results presented in this figure that the higher the number of solar tubes the higher the water storage tank temperature.

Figure 8 has been constructed to examine the impact of the storage tank volume on the time variation of initial water temperature and the water evacuation time from the storage tank. It is obvious that the bigger the storage tank volume the longer period the evacuation takes.

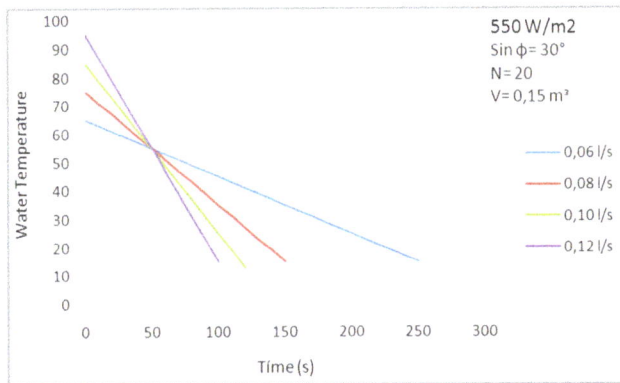

Figure 11. Time variation of initial water temperature in storage tank at various evacuation flow rates.

It was also observed from the simulation results that the initial maximum storage tank temperature exhibits significant influence on the evacuation time, therefore, this figure is significant in presenting the impact of the initial temperature in the storage tank. The figure also shows that the higher the initial temperature the longer the evacuation period.

On the other hand, Figure 10 presents the effect of changing the inclination angle of the solar tubes collector. It appears from this figure that the bigger the angle the longer is the evacuation period. Furthermore, Figure. 11 has been presented to illustrate the impact of water evacuation flow rates at constant solar insolation on the evacuation period. As shown in this figure, the higher evacuation flow rate is the quicker the evacuation time. It is worthwhile mentioning that the evacuation flow rate depends upon the solar water heater application. Normally, this type of solar water heater is equipped with a control system panel and temperature setting controller and a temperature sensor inside the tank as well as an electric heater to ensure that the desired temperature is always supplied. This electrical heater can be used when solar intensity is very low or at extreme cold water conditions to ensure the desired temperature is supplied.

In order to validate the prediction of the proposed model presented in the aforementioned equations (1) through (13), Figures 12 through 14 have been constructed to simulate the data presented in these figures. The predicted results were based upon the following experimental set up for the water glass solar tubes [(Table. 1) and Figure. 1]; inclination angle of 30° (Sin φ= 30°), number of solar tubes = 10, and V= 0.15 m³. The flow of water leaving the tank was kept constant

during the simulations presented in these figures. The temperatures were measured with sensors Tip 3-pin transistor DS18B20 with precision ±0.5°C, and analog multi meter EM5510 with resolution 0.1°C and resolution accuracy ± 1.5% ± 3°C.

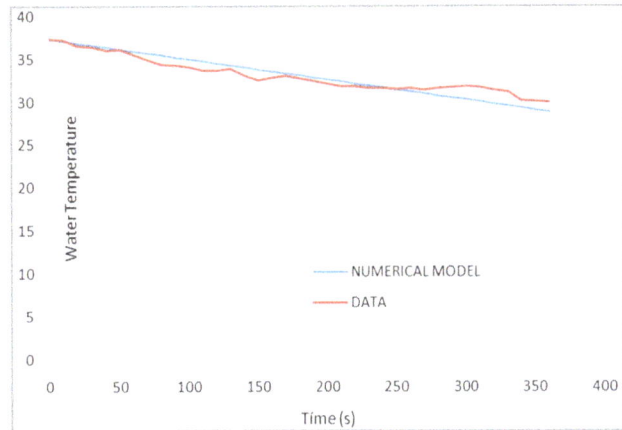

Figure 12. Comparison between model prediction and experimental data.

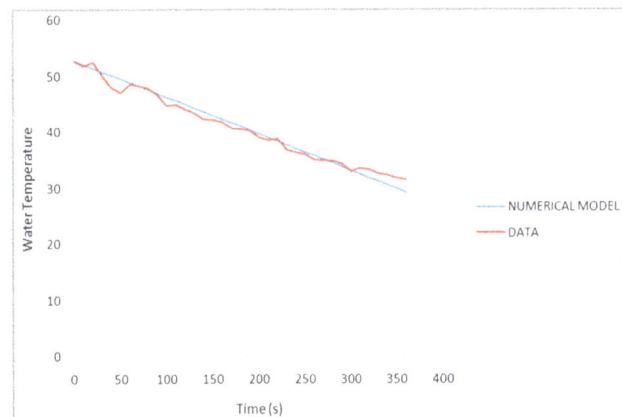

Figure 13. Comparison between model prediction and experimental data.

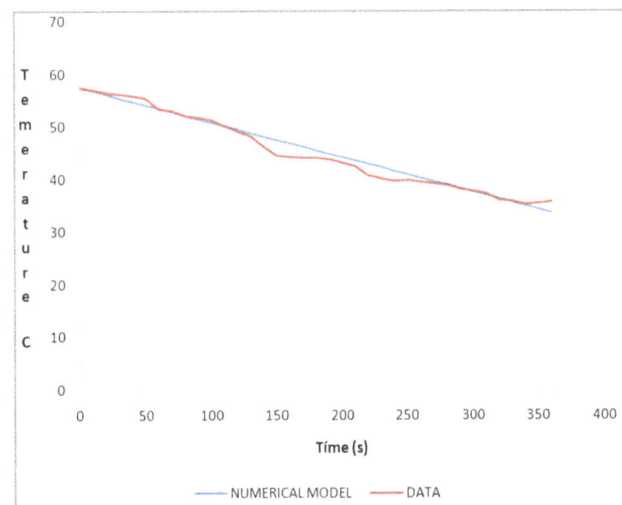

Figure 14. Comparison between model prediction and experimental data.

The water flowrates were measured with flow meters type Hall effect model number YFG1 with precision +/- 3% of the flow rate in the range of 1-10 l/min. The solar intensity was measured at the site using a HOBO ware software package for HOBO data loggers.

It is quite clear from these figures that the water temperature in the storage tank decreases as water is evacuated from the tank and makeup city water is supplied to the storage tank. It is also evident from the comparison presented in these figures that the proposed model predicted fairly the evacuation storage tank time variation temperature at various initial temperatures of the water at the storage tank and solar radiations as measured by the Hobo data acquisition system. It can also be pointed out that the results presented in these figures show minor discrepancies between the predicted results and data which are attributed to heat losses. The higher the storage tank temperature the higher the heat losses.

The energy conversion efficiencies of the solar collector were calculated using equations (12) and (13) and presented in figure. 15, for various solar radiations and initial water temperatures in the storage tank. The results illustrated, in this figure, clearly show that higher solar radiation enhances the solar collector energy conversion efficiency. Also as shown, lower water temperature in the storage tank is associated with higher efficiency. This is attributed to less heat losses from the storage tank during the water heating process.

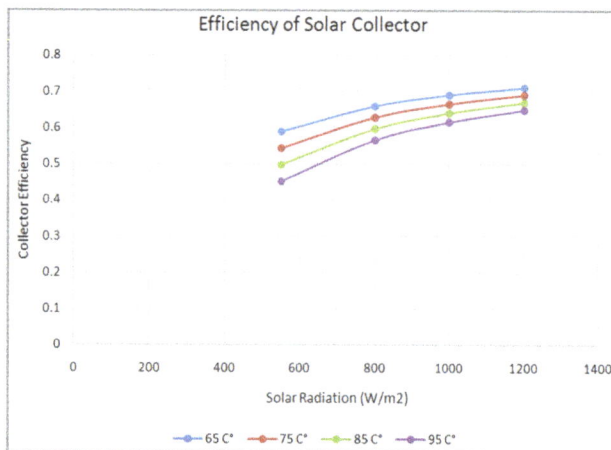

Figure 15. Energy conversion efficiency of solar collector.

5. Conclusions

During the course of this study, modeling and simulation as well as validation of a natural circulation closed thermosyphon glass solar tube collector water heater have been presented and discussed. Energy conservation equations were written for heat transfer fluid flow from the storage, in finite-difference form, integrated and solved to yield the behavior of the thermosyphon system at different solar insolations and water mass flow rate conditions as well as water storage tank temperatures.

The results presented in this study showed that higher solar radiation enhances the solar collector energy conversion efficiency. It is quite evident from the comparison presented

in hereby that the proposed model predicted fairly the evacuation storage tank time variation temperature at various initial conditions of the water at the storage tank and solar radiations.

Acknowledgement

The research work presented in this paper was made possible through the support of the Catholic University of Cuenca.

Nomenclature

A	Collector area (m^2)
d	Diameter of glass solar tube
g	Acceleration due to gravity (m/s^2)
H	Height of tank (m)
I	Solar intensity (W/m^2)
L	Length of the collector (m)
m	Water mass flow rate (kg/s)
N	Number of tubes in the collector (–)
u	Velocity (m/s)
T	Temperature (°C)
t	Time (hr)

Greek symbols

a	Absorbitivity (-)
τα	Effective transmittance-absorptance
ρ	Water density (kg /m^3)
β′	Coefficient of expansion of water (k-1)
υ	Mean kinematics viscosity of water (m^2/s)

Indices

c	Collector
s	Storage tank
o or i	Outlet and Inlet of the collector, respectively

References

[1] Mahendra S. Seveda, "Performance analysis of solar water heater in NEH region of India", International Journal of Renewable and Sustainable Energy, 2013; 2 (3): 93-98.

[2] O. B. Bukola, "Flow design and collector performance of a natural circulation solar water heater" Journal of Engineering and Applied Science, Vol. 1, Issue: 1, pp. 7-13, 2006.

[3] B. Sitzmann, "Solar Water Heater with Thermosyphon Circulation", Appropriate Technology, Vol. 31, Issue: 1, pp. 66-70, 2004.

[4] D. J. Close, "The Performance of Solar Water Heaters with Natural Circulation", Solar Energy, Vol. 6, Issue: 1, pp. 33-40, 1962.

[5] J. Huang, S. Pu, W. Gao, and Y. Que, "Experimental investigation on thermal performance of thermosyphon flat-plate solar water heater with a mantle heat exchanger", Energy, Vol. 35, pp. 3563-3568, 2010.

[6] S. Abgo "Analysis of the performance profile of the NCERD thermosyphon solar water heater", Journal of Energy in Southern Africa, Vol 22 No 2, May 2011.

[7] M. Shahi, A. Mahamoudi, and F. Talebi, "Numerical simulation of steady natural convection heat transfer in a 3-dimensional single-ended tube subjected to nanofluid", International Communications in Heat and mass Transfer, 37, pp. 1535-1545, 2010.

[8] A. I. Sato, V. L. Scalon and A. Padilha, "Numerical analysis of a modified evacuated tubes solar collector" International Conference on Renewable Energies and Power Quality (ICREPQ'12), Santiago de Compostela (Spain), 28th to 30th March, 2012.

[9] S. Hammadi, "Sudy of solar water heating system with natural circulation in Basrah" Al-Qadisi Journal for Engineering Sciences Vol. 2, No. 3, 2009.

[10] T. T. Chow, W. He, and J. Ji, "Hybrid photovoltaic thermosyphon water heating system for residential application" J. Solar Energy, Vol., 80, pp. 298-306, 2006.

14

A Novel Technique to Design Flat Fresnel Lens with Uniform Irradiance Distribution

Pham Thanh Tuan, Vu Ngoc Hai, Seoyong Shin[*]

Department of Information and Communication, Myongji University, Yongin City, Republic of Korea

Email address:
sshin@mju.ac.kr (S. Shin), pttuan1412@gmail.com (P. T. Tuan)
[*]Corresponding author

Abstract: In this paper, we propose a novel technique to design flat Fresnel lens to achieve big concentration ratio, high uniform irradiance distribution while keeping F-number small using both refraction and total internal reflection (TIR) phenomena. Also, this method can be used to design concentrated photovoltaic (CPV) system without secondary optics lens (SOE). In this technique, Fresnel lens is constructed by many prisms, which are built by ideal Cartesian oval. The design process of Fresnel lens was performed using Matlab program. Ray tracing technique has been used to optimize the structure of lens using LightTools™ software. The simulation results have been shown that Fresnel lens has a good optical property such asconcentration ratio is 2500x, F-number =0.4, and high uniform irradiance distribution.

Keywords: Fresnel Lens, CPV System, Total Internal Reflection, Second Optics Lens

1. Introduction

Nowadays, the advantages of photovoltaic technology help improve efficiency of solar cell significantly. The conversion efficiency of solar cell with GaInP/GaInAs/Ge multi-junction structure can be reached to 42% [1, 2]. However, the high cost of multi-junction cells leads to high price of solar systems. An effective way to reduce cost is to cut down the amount of required cell area while still increasing the cell efficiency by using concentrated photovoltaic system (CPV) [2, 3].

Using CPV systems is a good promising option to reduce cost and increase application ability in reality of photovoltaic systems. However, almost CPV systems still have some challenges such as non-uniform irradiance distribution, creating hot spot point over cell surface. Non-uniform irradiance and hot spot point degrade the reliability, the conversion efficiency, and life time of solar cell [1]. To solve these problems, CPV has been constructed with two lenses [4, 5]. The first lens is Fresnel lens [6, 7, 8] which concentrates sunlight to a point onto second lens (SOE). The second lens which is close to cell surface has been used to re-distribute sunlight uniformly over solar cell surface, which

makes the design process become more complex [9]. To overcome this problem, there have been some efforts to design CPV system without second lens [1], [10]. However, these methods still have some limits in terms of increasing concentration ratio.

The uniformity and concentration ratio are two crucial parameters in designing Fresnel lens for CPV system in two reasons. First, the high uniformity of concentrated beam can increase conversion efficiency in multi-junction solar cell [11]. Second, the cost of solar system can be reduced if CPV with high concentration ratio is used. In this paper, we propose a novel idea to design CPV system with uniform irradiance distribution using flat Fresnel lens without SOE. We also propound a new technique to increase concentration ratio of CPV while keeping F-number small. The Fresnel lens shape can be built in 3D by using Solid Works and LightTools™ software. The optical performances of the designed Fresnel lens are simulated by using the non-sequential ray tracing software, LightTools™. An optimized structure is designed by modulating the geometric parameters of the Fresnel lens. The concentration ratio of 2500x, F-number of 0.4, and high uniformity were achieved in the optimized design of Fresnel lens.

2. Basic Ideas

The optics systems applied to CPV require large area, light weight, mass production, and low image quality to reduce cost of photovoltaic systems. So, Fresnel lens is a best choice and it has been used widely in CPV. Typical structure of Fresnel lens consists of series of concentric grooves which act as individual refracting surfaces, bending parallel light rays to a common focal point. Figure 1(a) shows a traditional Fresnel lens. By the nature of optical property of conventional Fresnel lens, irregular illumination on solar cell is non-uniform, which produces a local hot point on cell surface. At that hot point, the temperature is so high that it affected Ohmic contact, which degraded conversion efficiency and life time of solar cell significantly. The decrease of conversion efficiency due to non-uniform irradiance can be compensated partly by using unique structure solar cell. That is achieved by connecting solar cell units in serially to match the voltage level at each sub region. Then these sub regions with equal voltage level are connected in parallel to collect the electric current from the cell assembly [10]. However, that is not really a fundamental solution to solve these challenges of solar cell system.

To overcome disadvantages of conventional Fresnel lens we propose a novel technique to design Fresnel lens. The basic idea of the Fresnel lens for the 2-D CPV system is depicted in Figure 1(b). Each groove of Fresnel lens focus bundle of parallel rays to focal point P_{ng}, then the bundle of light has been distributed over receiver with uniform irradiance.

In our design, the process consists of four steps described as follows. That is shown in Figure 1(b).

- Step 1: The light which comes to extreme left position of groove P_s will be refracted and directed to the extreme right position of receiver.
- Step 2: The light which comes to extreme right position of groove P_n will be refracted and directed to the extreme left position of receiver.
- Step 3: The light at P_n and P_s position will be focused at P_{ng} which is in-between lens and receiver. The rest of bundle rays coming to groove between P_n and P_s will be refracted and directed to P_{ng}, then distributed uniformly over receiver.
- Step 4: The same procedure is repeated for every groove. One discontinuity in the normal of exit surface at P_s and P_n is necessary. That helps lens shape becomes continuous.

Every bundle of rays coming to groove will be distributed over cell surface with uniform irradiance; therefore, light coming to Fresnel lens will be distributed uniformly over cell surface.

The advantage of this design is that we can distribute light uniformly over the receiver with any size. However, there is a challenge of finding out P_{ng} position accurately. If the position of P_{ng} is not accurate then distribution irradiance will not be uniform over the receiver. Some different positions of P_{ng} and irradiance distributions are shown respectively in Figure 2.

Fortunately, we can overcome the challenge of finding P_{ng} position by using computer programming. Nowadays, computer technology has been developed powerfully. Computer and computation program can solve millions of calculations per second. Thus, finding the position of P_{ng} accurately is not impossible. More accurate position of P_{ng} requires more time. The schematic of algorithm of designing the lens and finding P_{ng} position is shown in figure 3.

Every groove of lens needs two surfaces to distribute light uniformly over receiver. First is entry surface which is perpendicular with bundle of parallel rays and second is exit surface which is a Cartesian oval surface. The light coming to entry surface will exit at exit surface and become focused at P_{ng} position. Every light must have the same optical path length (OPL) satisfying following equation.

$$a_1 = n \times a_2 + a_2' = n \times a_3 + a_3' = ...$$
$$... = n \times a_m + a_m' = n \times a_{end} + a_{end}' = OPL \quad (1)$$

Where n is refractive index. a and a' are path lengths of a ray in medium (Figure 4). From equation (1) we can calculate the small pieces of exit surface. These small pieces correspond to Cartesian ovals (see Figure 5 (a)), and in this way, we can build whole flat Fresnel lens.

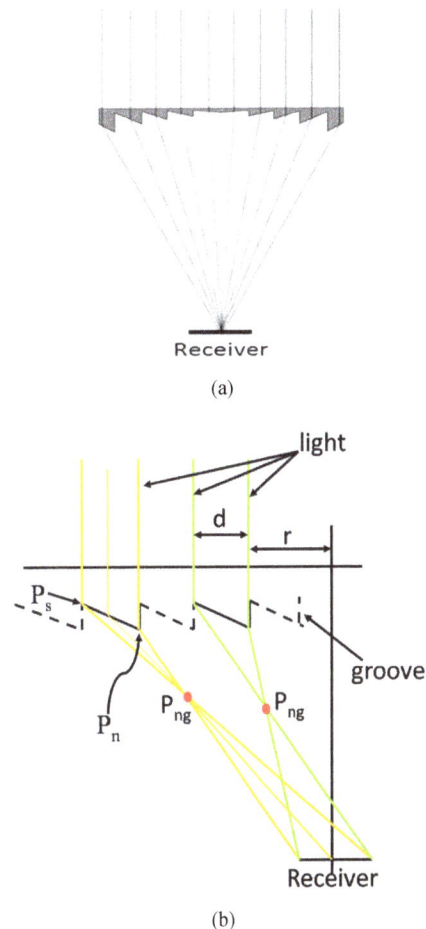

(a)

(b)

Figure 1. The difference between a) Conventional Fresnel lens [12] and b) Newly designed Fresnel lens.

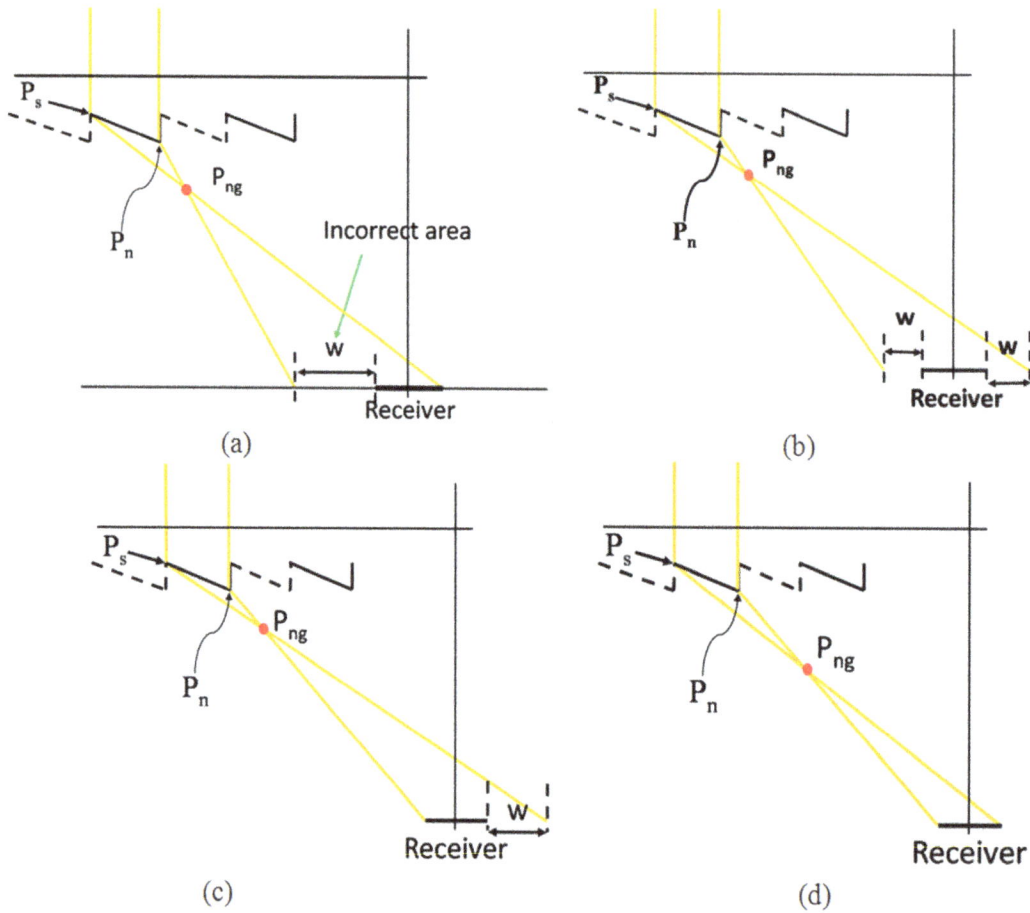

Figure 2. Some different positions of P_{ng} and distribution irradiance over receiver: a) irradiance distribution shifted to left of receiver, b) irradiance distribution shifted to both sides of receiver, c) irradiance distribution shifted to right of receiver, d) irradiance distribution is exactly over receiver.

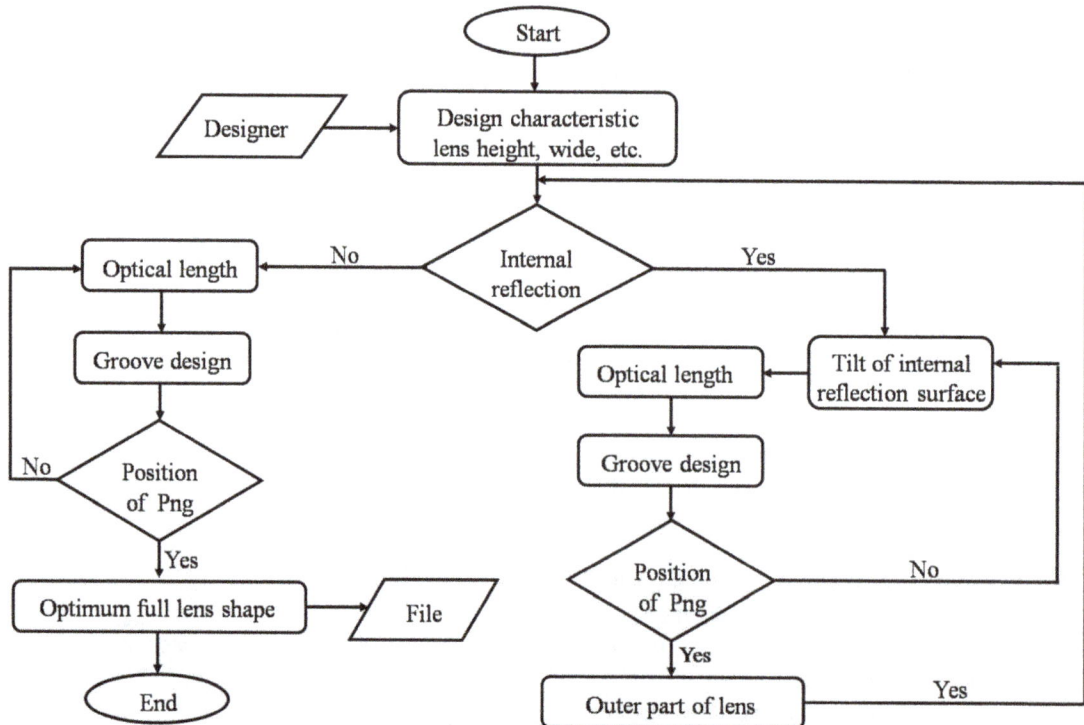

Figure 3. Flow chart of optimum lens calculation.

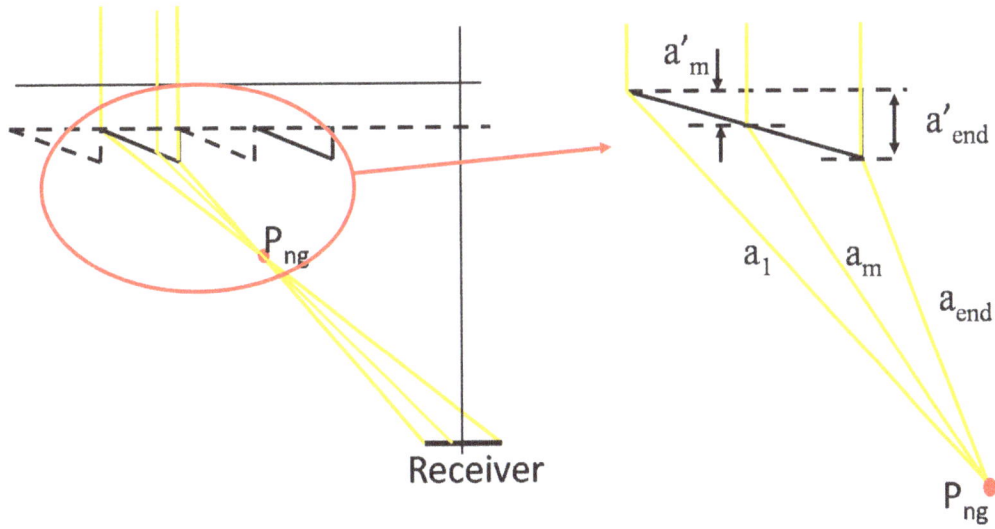

Figure 4. Optical path length of ray in groove.

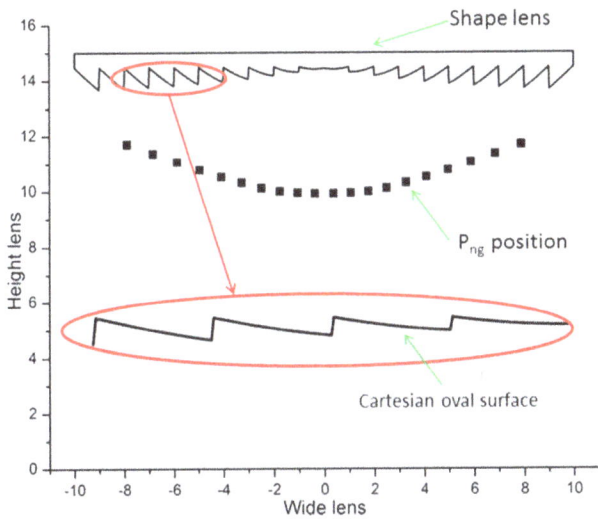

(a)

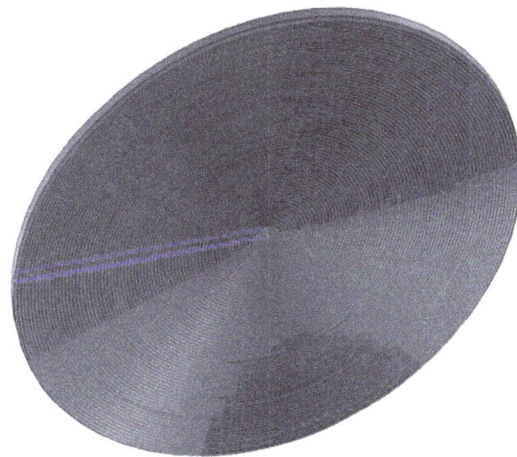

(b)

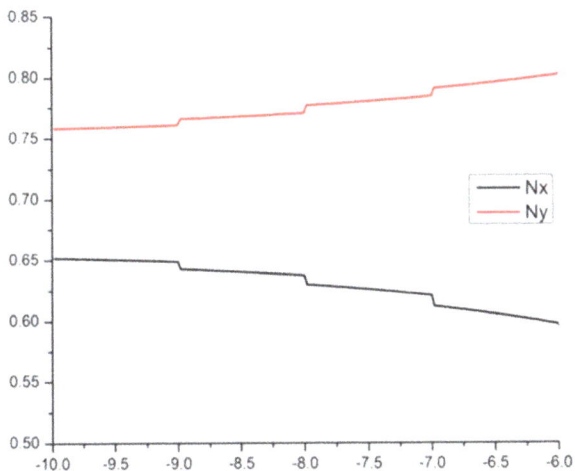

(c)

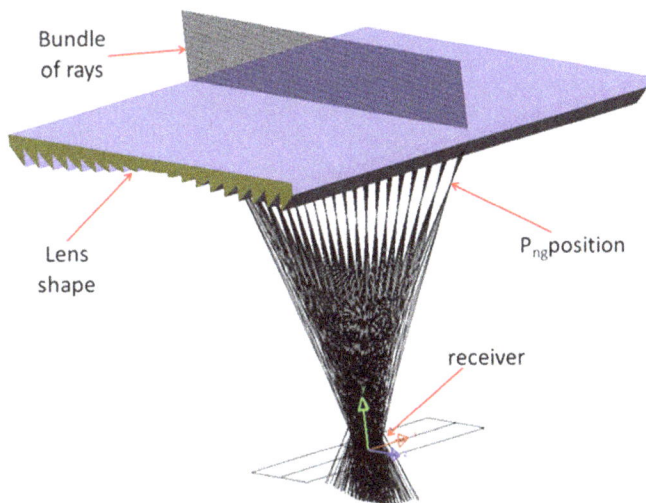

(d)

Figure 5. Shape and feature of flat Fresnel lens. a) Lens shape, P_{ng} position, and exit surface as Cartesian oval in 2-D. b) Circular Fresnel lens in 3-D. c) The two components (N_x, N_y) of the normal vector of the exit surface. d) Square Fresnel lens shape, P_{ng} position in 3-D.

After design process, flat Fresnel lens has a shape as shown in Figure 5 (a). The flat Fresnel lens has been designed with large groove to show the profile clearly. Figure 5 (a) also shows exit surface shape of Cartesian oval and P_{ng} positions. In addition, the normal vector of Cartesian changed through exit surface and it can be calculated by two component N_x (follow x direction) and N_y (follow y direction). Figure 5 (c) shown the volume of Cartesian normal vector with two components N_x and N_y. In this design method, the discontinuity of normal vector at each groove edge is needed for the shape of lens continuity. Moreover, the lens with square or circle shape in 3-D can be built by using Solidworks and LightTools[TM] software and was shown in Figure 5 (b) and (d).

3. Increasing Dimension of Lens, Loss Problem, and a Solution

Lens designers are intending to increase concentration ratio as big as possible while keeping F-number small. In this part, we introduce a new technique to increase concentration ratio and decrease the optics loss.

In our design, the exit surfaces of lens are Cartesian ovals. Dimension of flat Fresnel lens will be increased until total internal reflection (TIR) appears. That is shown in Figure 6. If the TIR happens, the bundle of parallel rays remains inside

the lens and will not reach to the receiver.

TIR phenomenon is really a challenge to increase dimension of Fresnel lens. Fortunately, we can solve this problem by changing the shape of groove in Fresnel lens. Some techniques have been used to avoid TIR inside the lens [4], [13]. In [4], exit surface consists of two tilt surfaces where total internal reflection happens. This design, the loss appears at extreme right position of exit surface which shown in Figure 7 (a). In [13], whereas, the TIR repeats two times at output surface, however, this leads to the calculation become more complex. To decrease loss and increase dimension of Fresnel lens, we propose a new technique in the design of lens as shown in Figure 7(b).

In this design, the groove of Fresnel lens consists of flat surface S1 and Cartesian oval surface S2. The bundle of parallel rays will be reflected (TIR) at S1 surface, and become refracted at S2 surface, then the rays will be focused on P_{ng} position and distributed over the receiver (see Figure 7 (b)).

Figure 8 shows some grooves designed by using TIR phenomenon. If the tilting angle of S1 surface is $\alpha = 45^0$, the TIR ray will be refracted 90 degrees. In this design, S2 surface is too close to S1 surface of next groove (Figure 8). Therefore, the bundle of rays exit from S2 surface will be refracted again at S1 surface of next groove, which makes the rays not reach to the receiver.

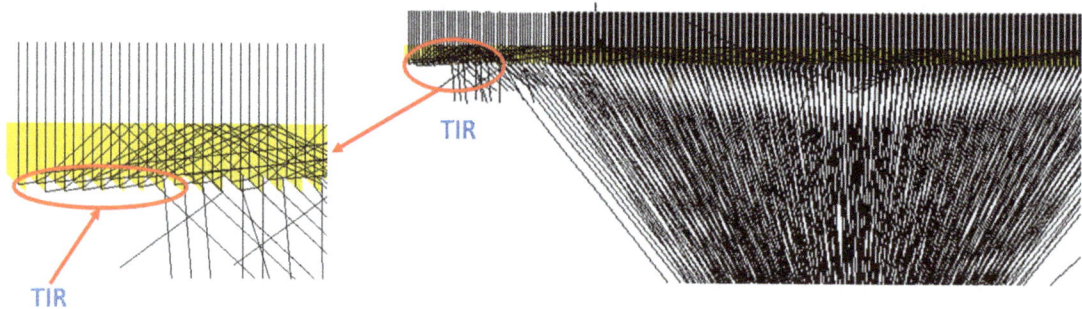

Figure 6. *Total internal reflection appears at positions indicated.*

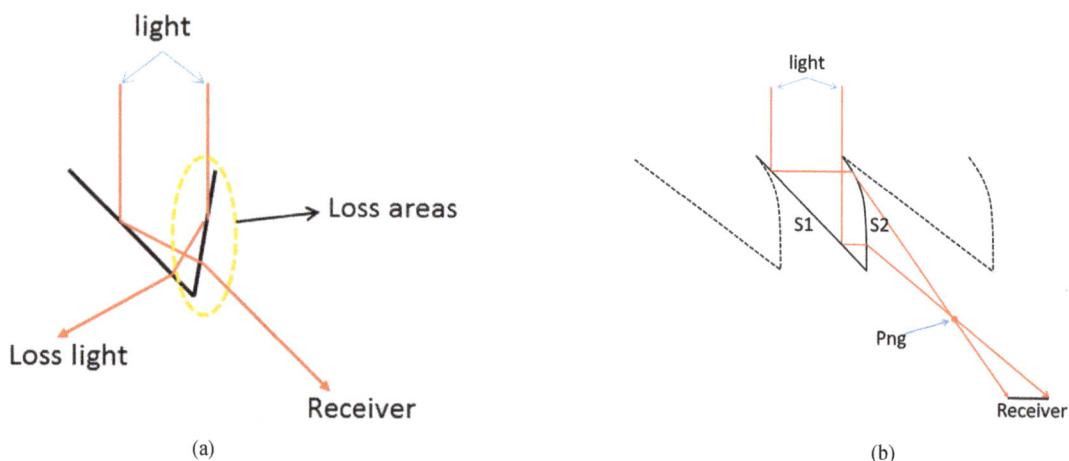

(a)   (b)

Figure 7. *Schematic of groove design using TIR a) Conventional groove [4] and b) our groove design (the picture is not scale).*

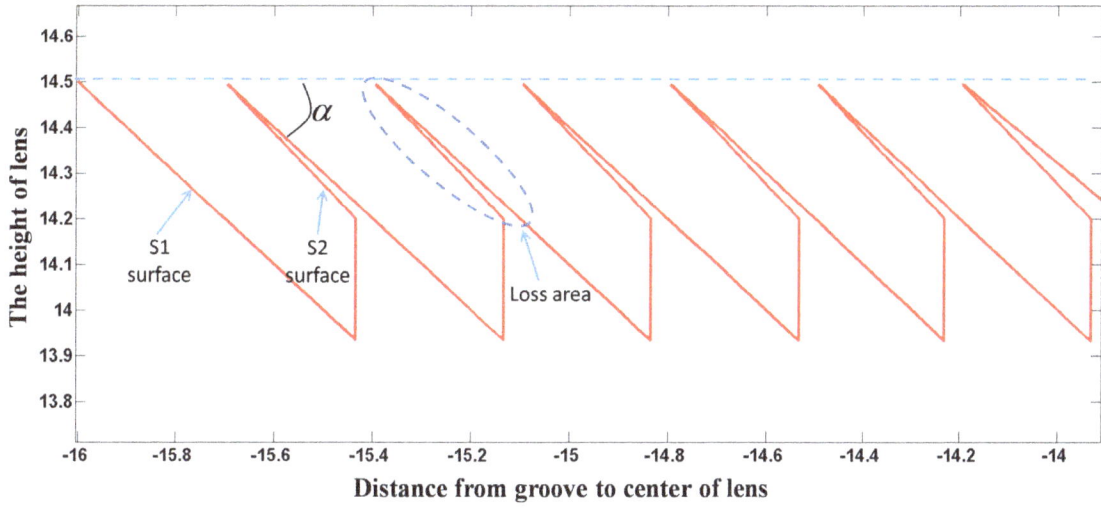

Figure 8. *The grooves of flat Fresnel lens have been designed with the tilt of S1 surface is α = 45⁰.*

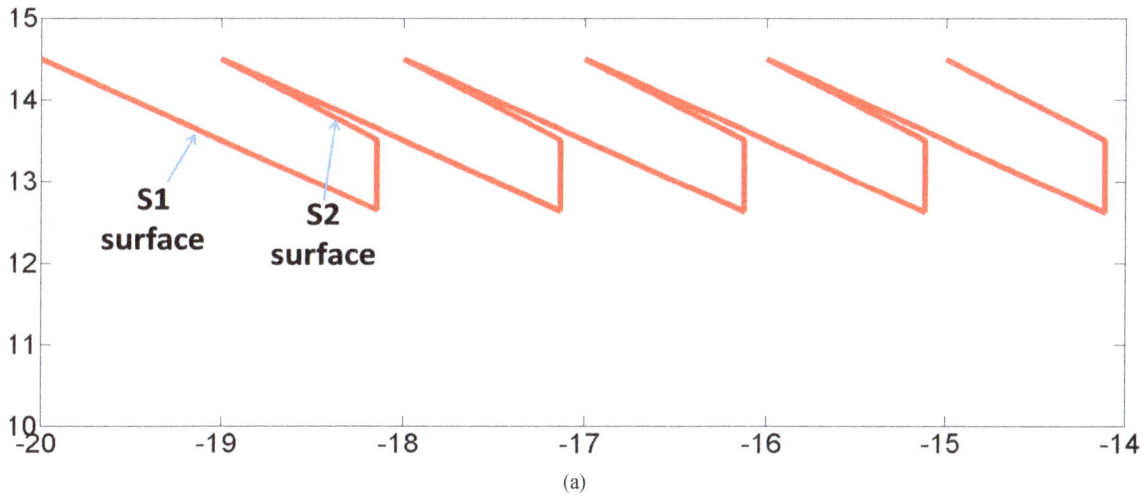

(a)

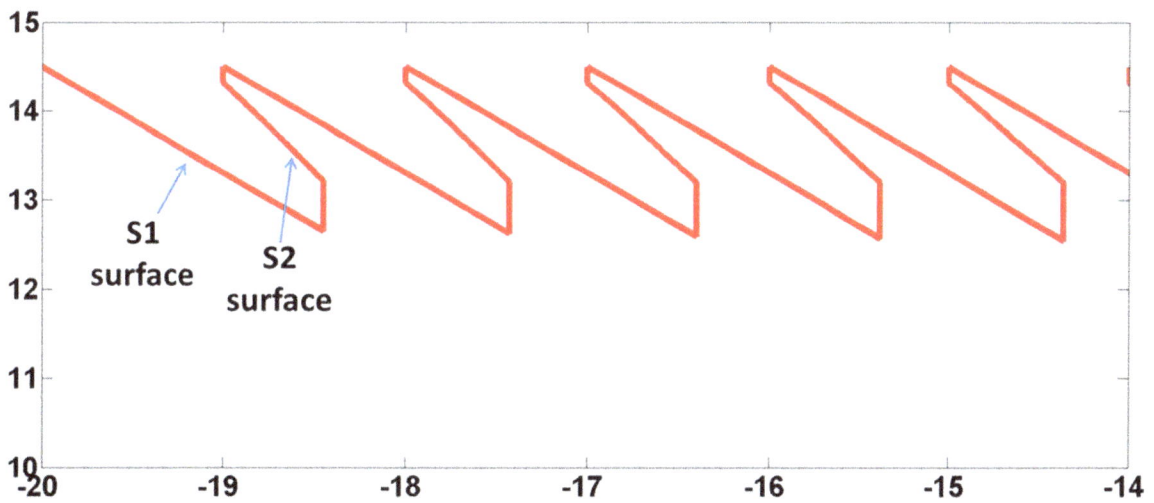

(b)

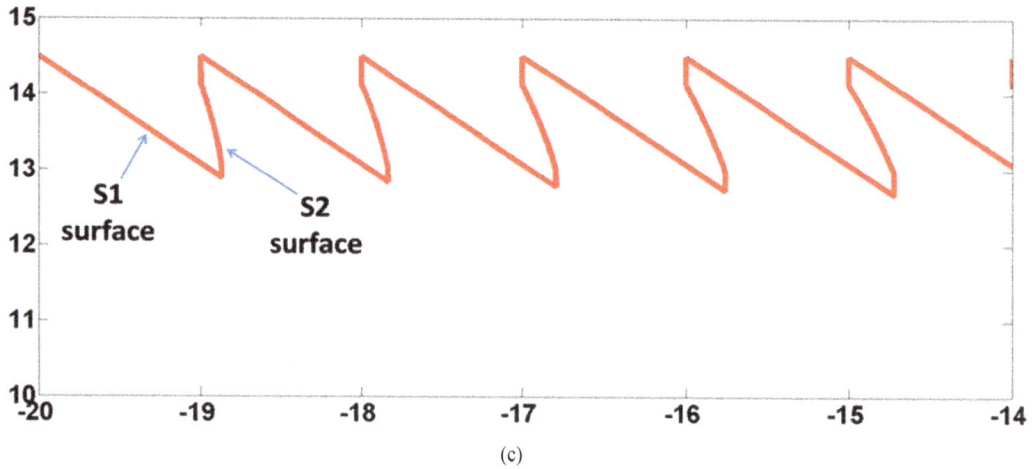

(c)

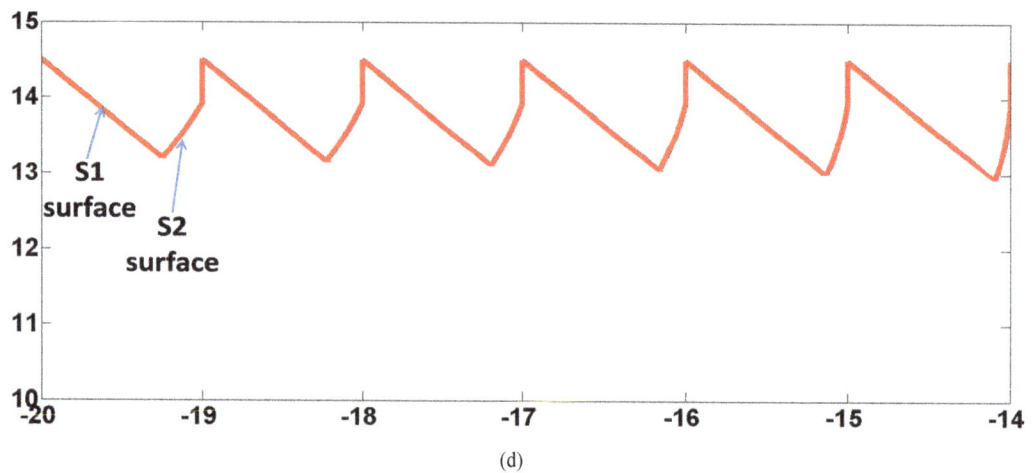

(d)

Figure 9. *Groove shape (exit surface) of Fresnel lens with different tilt angles of S1 surface a) α = 45°b) α = 50° c) α = 55° d) α = 60°.*

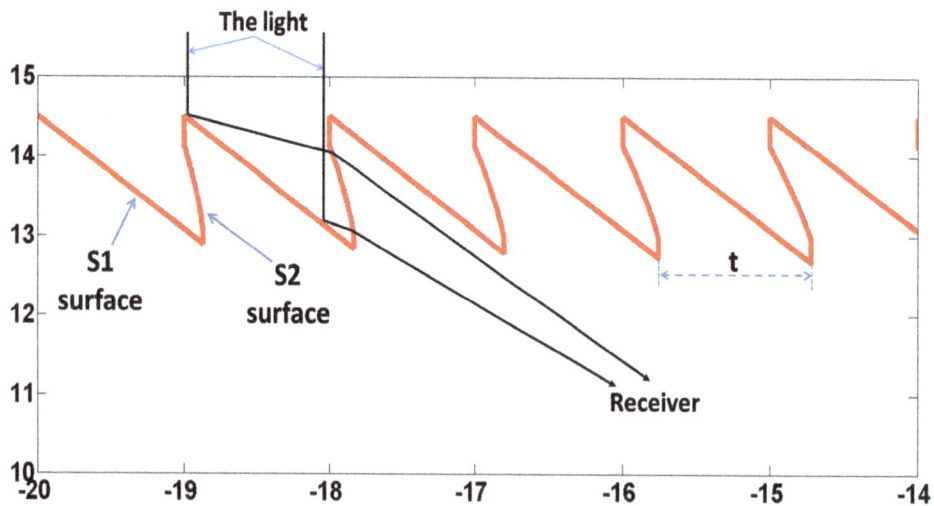

Figure 10. *The way of light inside Fresnel lens with tilt angles of S1 surface α = 55°.*

Consequently, the tilt of S1 surface has been investigated to decrease optics loss at loss area (see Figure 8). Figure 9 shows the tilt of S1 surface at different angles (α = 45°, 50°, 55°, 60°). Fresnel lens with tilt angle S1 α = 55° is the least loss comparing with the rest because of two reasons. First reason: distance t (see figure 10) is big enough for the rays to exit from S2 surface and come to receiver without being refracted at S1 surface of next groove. Second reason: S2 surface is tilted forward to the right direction, as a result, there is no rays to be internally reflected at S2 surface.

However, there still exits loss at the tip of groove. The lights coming from the upper position of S2 surface of the groove were interrupted by the surface (S1) of next groove (see Figure 11(a)). We can decrease optics loss more by changing the sharp of groove. The tip of groove has been cut (see Figures 11(b) and (d)). That helps the light come to the receiver easier.

In brief, we introduced a new technique to increase dimension of Fresnel lens and decrease optics loss in the design process by using TIR.

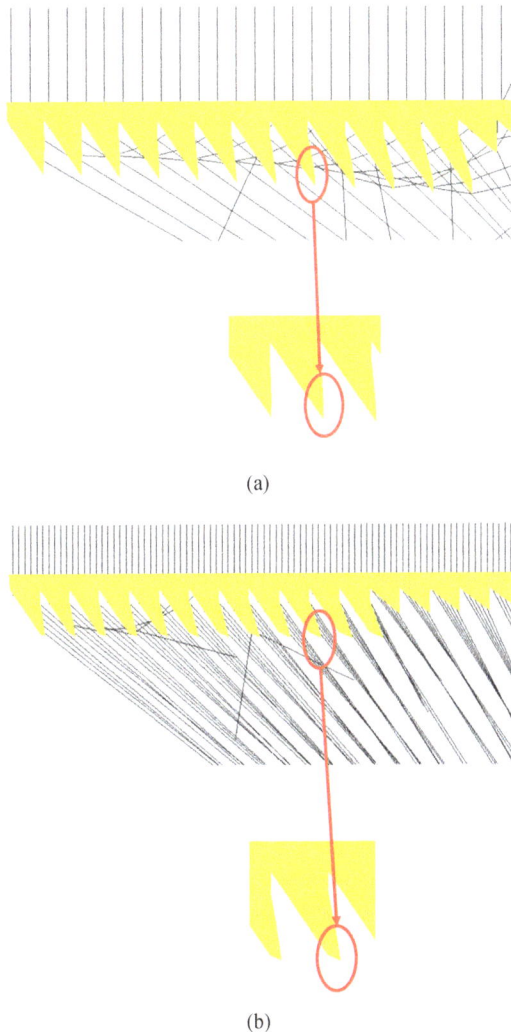

4. Performance and Discussion

In our design, the flat Fresnel lens has diameter L=50cm. It concentrates and distributes light over receiver with diameter l=1cm, 2cm, and 3cm, alternately. With the definition of the geometrical concentration $C=(L/l)^2$, the concentration ratios becomes 2500x, 1100x, and 625x sun, alternately. The irradiance distributions at the cell lens are estimated to be high uniform. Lens has height 20cm and F-number 0.4 (height / diameter of lens).

Simulated process was performed by LightTools™ software. Simulation parameters are following: power of light source is 2000Watt, the number of rays viewed in ray tracing program is 400. In addition, the ray tracing technique has been used to characterize lens' performance.

Figure 12 shows the race tracing results of flat Fresnel lens. Figure 12 (a) is the result without TIR based design of groove.In this structure, the diameter of effective area of Fresnel lens where the light becomes concentrated at the receiver is just only 28cm. Figure 12 (b) shows the result with TIR based design of groove. The diameter of lens becomes increased to 50cm, which means the whole lens contributes to the concentration of light perfectly.

Figure 13 shows the way of rays exits from flat Fresnel lens. At outer areas, the light exits the lens at right side of groove (S2 surface) while the light exits the lens at left side of groove at inner areas (S1 surface). This is because we used refracting phenomenon to design inner groove, whereas we used total internal reflecting phenomenon to design outer groove of lens (told in third part). The interface between inner and outer area can be estimated by solving mathematics. Refracted angle will be increased if radius of lens increases. When it reaches 90°, the total internal reflection appears then the position of interface will be determined. The inner area has been limited by TIR, whereas the size of outer doesn't have a limit clearly. The limit of outer area depends on designer and efficiency of optical loss. The simulation process help designer decide how big lens should be.

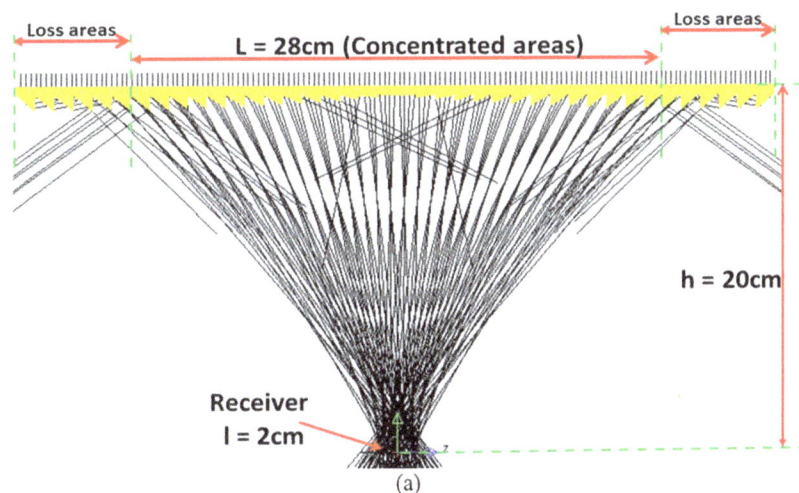

(a)

(b)

Figure 11. The shape of groove a) the sharp groove and ray tracing, b) the cutting groove and ray tracing.

(a)

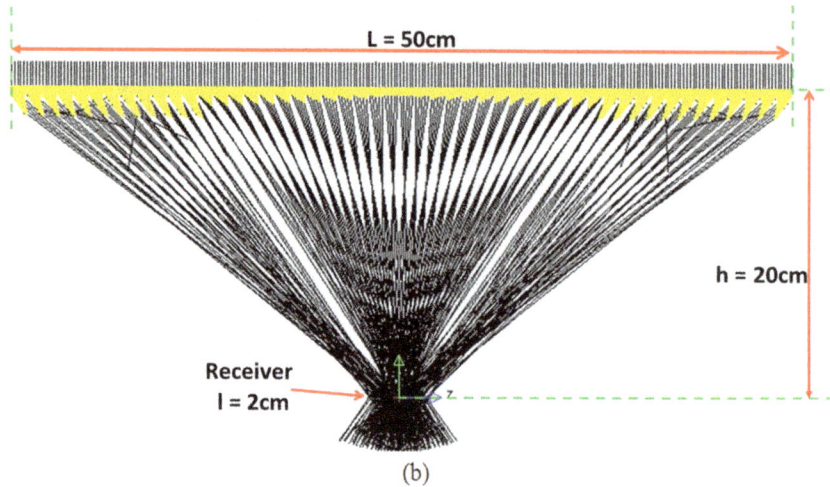

Figure 12. *The flat Fresnel lens a) not using TIR and b) using TIR to design groove.*

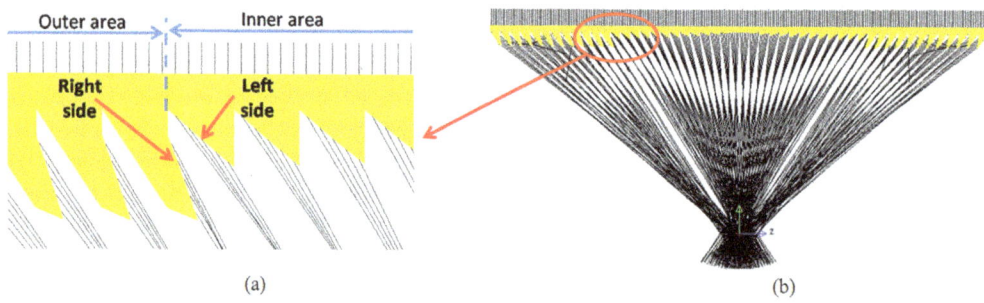

Figure 13. *The ray tracing of a) outer and inner areas of flat Fresnel lens and b) whole lens.*

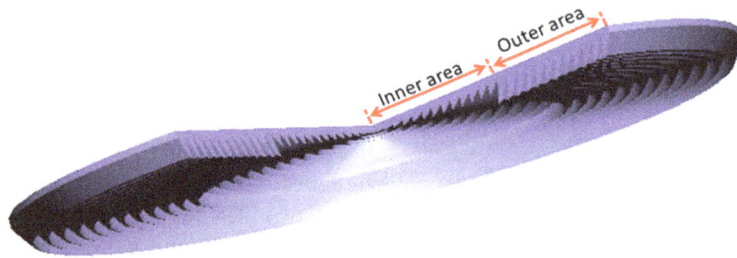

Figure 14. *The flat Fresnel lens in 3-D after finishing design process.*

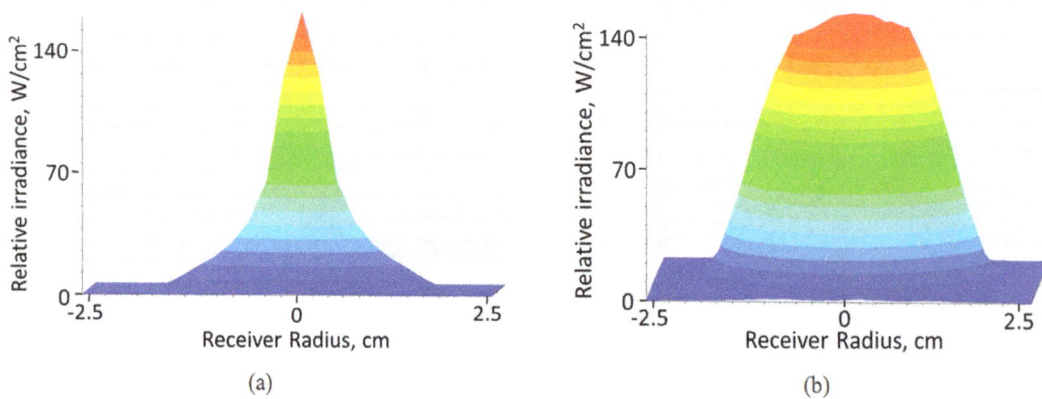

Figure 15. *Distribution of irradiance over the receiver: a) conventional Fresnel lens and b) our designed Fresnel lens.*

Figure 14 shows the flat Fresnel lens shape in 3-D. Lens shape consists of two parts: inner and outer areas. Inner area designed by using refraction phenomenon has been constructed by some smaller grooves. Meanwhile, outer area designed by using TIR has been constructed by some bigger grooves.

Furthermore, we have investigated the uniformity of irradiance distribution over receiver by using simulation in LightTools[TM]. The Fresnel lens collects sun light and distributes it over the receiver which has diameter 3mm. The uniformity has been described in Figure 15 (b). In this figure,the irradiance distribution over the receiver in our designed Fresnel lens is shown. It is quite uniform. There is no any hot spot over the receiver area, whereas, the conventional Fresnel lens has a hot spot in the center of lens.

5. Conclusions

A novel technique to design flat Fresnel lens for CPV system was presented to increase the concentration ratio, reduce optical loss, and improve uniformity of irradiance distribution over receiver. The simulation results have been identified that using our method to design Fresnel lens the concentration ratio can achieve 2500x while F-number keeping small (F-number=0.4). In addition, we can use this technique to design CPV system without SOE. Also, Fresnel lens designed by our technique collects and distributes uniform sunlight over receiver with any size. All these factors will be helpful to enhance the performance of CPV and reduce the cost of solar systemsof multi-junction solar cells.

Acknowledgements

This work was supported by a National Research Foundation of Korea (NRF) grant funded by the Korea government (MSIP) (No. 2014R1A2A1A11051888).

References

[1] Jui-Wen Pan, Jiun-Yang Huang, Chih-Ming Wang, Hwen-Fen Hong, Yi-Ping Liang, "High concentration and homogenized Fresnel lens without secondary optics element," Optics Communications, 2011, vol. 284, pp. 4284-4288.

[2] Mehrdad Khamooshi, Hana Salati, Fuat Egelioglu, Ali Hooshyar Faghiri, Judy Tarabishi, and Saeed Babadi, "A Review of Solar Photovoltaic Concentrators," International Journal of Photoenergy, 2014, vol. 2014, article ID 958521.

[3] E. Lorenzo and A. Luque, "Comparison of Fresnel lenses and parabolic mirrors as solar energy concentrators," Applied Optics, 1982, vol. 21, pp. 1851-1853.

[4] Lei Jing, Hua Liu, Hui-fu Zhao, Zhenwu Lu, Hongsheng Wu, He Wang, and Jialin Xu, "Design of novel compound Fresnel lens for High-performance photovoltaic concentrator," International Journal of Photoenergy, 2011, vol. 2012, article ID 630692.

[5] Stanislas Sanfo, Abdoulaye Ouedraogo, "Contribution to the Optical Design of A Concentrator with Uniform Flux for Photovoltaic Panels," Advances in Energy and Power, 2015, vol. 3, pp. 82-89.

[6] Ralf Leutz, Akio Suzuki, Atsushi Akisawa, Takao Kashiwagi "Design of A Nonimaging Fresnel Lens for Solar Concentrators," *Solar Energy*, 1999, Vol. 65, No. 6, pp. 379–387.

[7] Juan C. González, "Design and analysis of a curved cylindrical Fresnel lens that produces high irradiance uniformity on the solar cell," Applied Optics, 2009, Vol. 48, No. 11, pp. 2127-2132.

[8] N. Yamada and T. Nishikawa, "Evolutionary algorithm for optimization of nonimaging Fresnel lens geometry," Optics Express, 2010, Vol. 18.

[9] Juan C. Minano, Pablo Benitez, Pablo Zamora, Marina Buljan, Ruben Mohedano, and Asuncion Santamaria, "Free-form optics for Fresnel lens based photovoltaic concentrators" Optical Society of America, 2013, vol. 21, No. S3.

[10] Kwangsun Ryu, Jin-Geun Rhee, Kang-Min Park, Jeong Kim, "Concept and design of modular Fresnel lenses for concentration solar PV system," Solar energy, 2006, vol. 80, pp. 1580-1587.

[11] E. A. Katz, J. M. Gordon, D. Feuermann, "Effects of ultra-high flux and intensity distribution in multi-junction solar cells," Prog. Photovoltaics, 2006, vol. 14, pp. 297-303.

[12] http://www.fresneltech.com/.

[13] Ian Wallhead, Teresa Molina Jiménez, Jose Vicente García Ortiz, Ignacio Gonzalez Toledo, and Cristóbal Gonzalez Toledo, "Design of an efficient Fresnel-type lens utilizing double total internal reflection for solar energy collection," Optics express, 2012, vol. 20, No. S6.

15

Experiment and Simulation Study on Silicon Oil Immersion Cooling Densely-Packed Solar Cells Under High Concentration Ratio

Xue Kang[1], Yiping Wang[1, *], Ganchao Xin[2], Xusheng Shi[1]

[1]School of Chemical Engineering and Technology, Tianjin University, Tianjin, China

[2]Toppley (Zhuhai) Chemicals. Co., LTD. Guangdong, China

Email address:

wyp56@tju.edu.cn (Yiping Wang)

*Corresponding author

Abstract: In order to solve the heat dissipation problem of densely-packed solar cells in high concentration photovoltaic (HCPV) system, a new cooling method of using silicon oil directly immerse the solar cells was proposed. The heat transfer performance of silicon oil immersion cooling the densely-packed solar cells with and without fin structure was investigated through experiment and simulation methods. The results of heat transfer performance of solar cells without fin structure showed that the simulated data was consistent well with data of experiment and the temperature could be lowered down in the operation range of solar cell. Furthermore, the heat transfer performance of solar cells with fin structure was researched using the model under different silicon oil inlet temperatures, inlet flow rates and the flow pressure drop was measured. The results indicated that the solar cells temperature declined and distributed well with silicon oil inlet flow rate increasing but the solar cells temperature raised linearly with silicon oil inlet temperature increasing. The optimized parameters of cooling receiver with fin structure were that: height of fin was 14 mm, number of fin was 50 and the thickness of substrate was 1.5 mm, with which the large amount of heat of densely-packed solar cells under high concentration ratio could be well controlled and make sure the power generation of HCPV system was high efficient.

Keywords: High Concentration Ratio, Densely-Packed Solar Cells, Silicon Oil, Immersion Cooling, Fin Structure

1. Introduction

With the development of photovoltaic technology and solar cells, the high concentration photovoltaic (HCPV) played an important role in power generation system. However, the thermal management of solar cells is an important issue need to be dealt with. Significant researches have been dedicated to cool HCPV system as reviewed in literatures [1-5]. It has been shown that the low system thermal resistance was difficult to satisfy because of the presence of traditional wall resistance between solar cells and heat sinks with conventional cooling approaches.

A new cooling method of direct contact liquid immersion cooling was proposed [6] to cool the solar cells in CPV system.

With direct contact liquid immersion cooling, the bare solar cells were immersed in the circulating liquid directly. Thus, the traditional wall resistance between solar cells and heat sinks was eliminated to the boundary layer interface between the bulk liquid and the solar cells surface. Furthermore, the direct contact liquid immersion cooling provides an opportunity for heat to be taken away from both the upper and lower solar cells surface, which increased the heat transfer area by contrast with former cooling approaches. Our former study focus on liquid direct immersion cooling lower and medium CPV systems [7-13], furthermore, the attempt of using water immersion solar cells under 250X in dish CPV system had been done [14-15]. However, the water has some disadvantages and the proper immersion liquid was silicon oil [13, 16], meanwhile, the heat transfer performance of silicon oil direct immersion cooling solar cells under high concentration ratio has not been reported in literatures.

In this paper, silicon oil was chosen as immersion liquid to direct contact cooling the simulated solar cells model without fin structure. The heat transfer performance was investigated through numerical simulated and experimental methods, meanwhile, the CFD model was validated. Then the heat transfer performance of triple-junction solar cells model with fin structure was characterized by simulation method. Finally, the parameters of direct contact liquid immersion cooling receiver were optimized by simulation using the solar cells temperature and receiver pressure drop as evaluated targets.

2. Experimental

2.1. Experimental Setup and Procedure

Silicon oil was chosen as immersion liquid based on our former research. A stainless steel plate was designed to simulate the densely-packed solar cells and the electrical heated power of the plate simulated the solar concentration ratio under the premise that the power generation efficiency of triple-junction solar cells is 39.8% [17]. The liquid immersion receiver consisted of an electrical heating plate (EHP), two copper rods welded on EHP and flowing channel shaped by two iron plates and two teflon plates. The EHP consisted of four stainless plates with 40 mm length, 10 mm width, 0.28 mm thickness and 1 mm gap between each other. The flowing channel was 40 mm wide, 3.28 mm high and 200 mm long.

The silicon oil was pumped from the liquid tank and flew through the rotor flowmeter and injected in the liquid immersion receiver, and finally was driven back to the storage tank, in which there was a coil cooler to make sure the inlet temperature of silicon oil was fixed. The inlet flow rate of silicon oil was adjusted by a valve, and the temperature of simulated densely-packed solar cells was measured by K-type

thermocouples that welded on the EHP surface. The experimental setup was shown in Fig. 1.

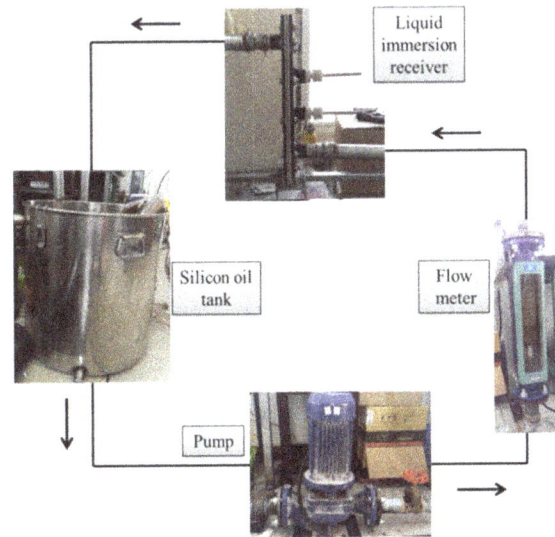

Fig. 1. *The experimental setup.*

2.2. CFD Models

The CFD model of liquid immersion receiver with simulated densely-packed solar cells without fin structure was designed, as Fig. 2a shown. The geometry dimension of receiver was 128 mm×24 mm×12 mm, the thickness of up and down flow channels was both 5.0 mm.

The CFD model of liquid immersion receiver with 64 triple-junction solar cells were densely-packed as 8×8, which were welded on the Aluminum substrate after insulation treatment. The geometry dimension of was 100 mm×100 mm×15.7 mm. Some fins were arranged on the back of substrates for heat dissipate enhancement, as shown in Fig. 2b.

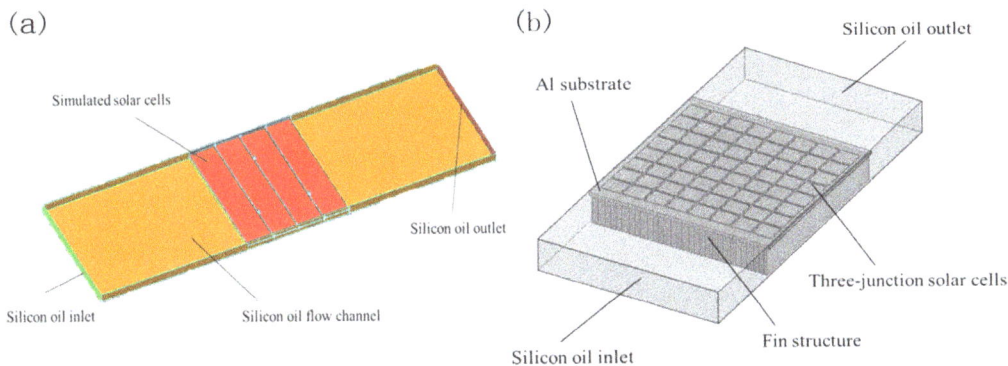

Fig. 2. *(a) The CFD model of liquid immersion receiver with simulated densely-packed solar cells without fin structure, (b) the CFD model of liquid immersion densely-packed solar cells with fin structure.*

3. Results and Discussion

3.1. Validation of Numerical Simulated Model

The temperature contour of simulated densely-packed solar cells was shown in Fig. 3a when inlet flow rate and

temperature of silicon oil was fixed at 4.2 m³/h and 50°C, respectively. It can be seen that the temperature of the main body was relatively uniform, the highest temperature was 70.31°C and the average temperature was 63.32°C. However, the temperature of the edge varied greatly, there was obvious temperature gradient along the flowing direction of silicon oil.

Because the flow condition of silicon oil was changing by the contact and detachment of silicon oil and simulated solar cells in the edge, the heat transfer was enhanced by the local fluid disturbance, leading to lower temperature in the edges. Hence, the edge effect was a main factor to consider when the liquid immersion receiver was developed. The comparison of simulated and experimental temperature under same condition was shown in Fig. 3b, which illustrated that the results kept well and there were only 2°C to 3°C temperature differences. Thus, it was reasonable and feasible to investigate the heat transfer performance of liquid immersion cooling densely-packed solar cells by numerical simulation method.

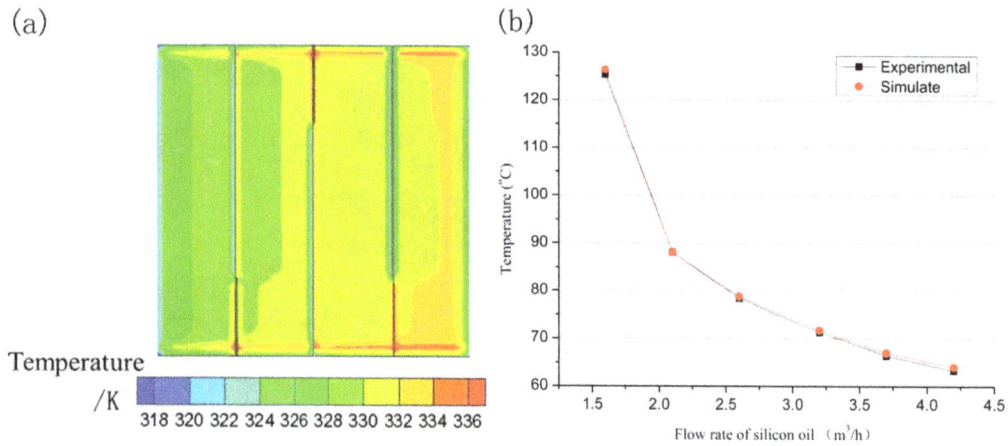

Fig. 3. (a) The temperature contour of simulated densely-packed solar cells, (b) the comparison of simulated and experimental results.

3.2. Heat Transfer Performance of Silicon Oil Immersion Cooling

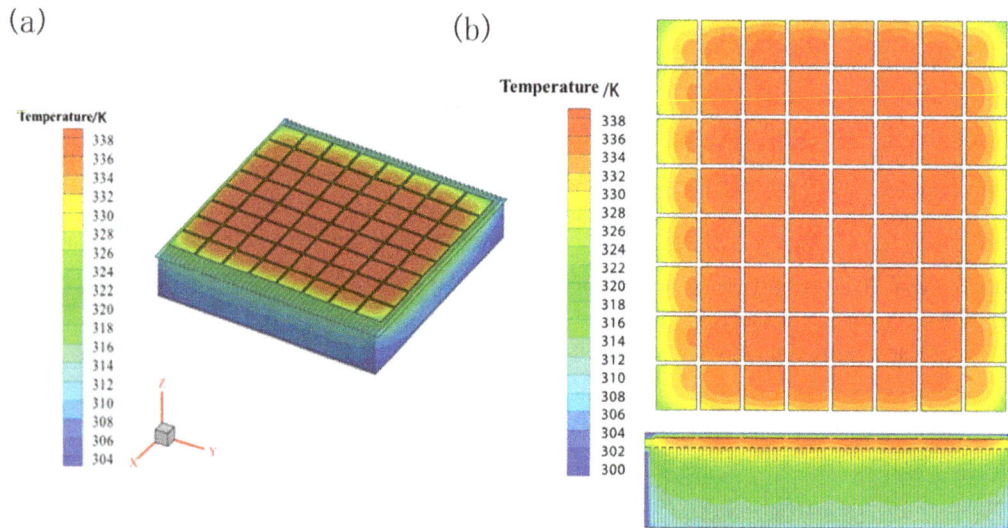

Fig. 4. (a) The temperature distribution of solar cells model, (b) the positive surface and axial temperature distribution of solar cells model.

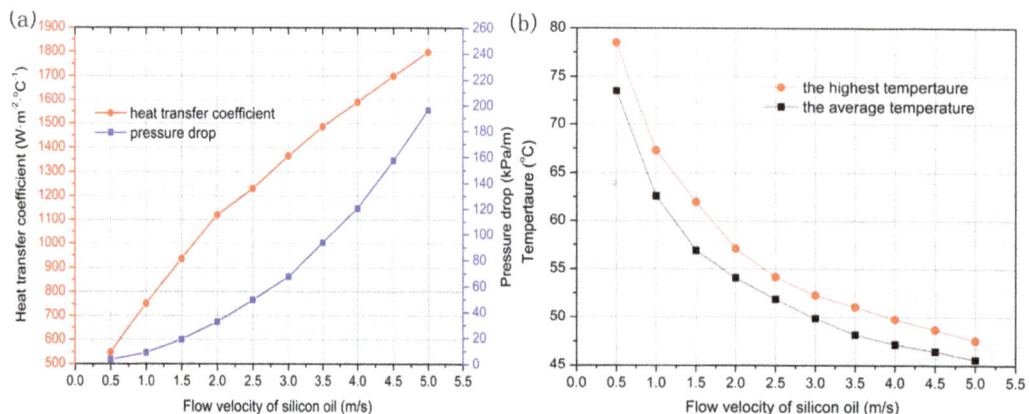

Fig. 5. The heat transfer performance of solar cells model with different silicon oil flow velocities.

The temperature distribution of densely-packed triple-junction solar cells with fins structure were shown in Fig. 4 when inlet flow velocity and temperature of silicon oil was 1.0 m/s and 25°C, respectively. The thickness of aluminum substrate was 1.5 mm, the height and amount of fins were 14 mm and 50, the thickness of liquid layer was 1.0 mm and the concentration ratio was 500X. The negative direction of X axis was silicon oil flowing direction. It can be seen that the highest temperature was lower than 70°C and the temperature distributed well uniform.

The varying trend of average temperature, highest temperature, heat transfer coefficient and pressure drop with silicon oil inlet flow velocity increasing was shown in Fig. 5. It was shown that the highest temperature was lower than 70°C, the average temperature was 63°C, the heat transfer coefficient was 750 W/(m² °C) and the pressure drop was lower than 10 kPa/m when inlet flow velocity and temperature of silicon oil was 1 m/s and 25°C. In order to further testify the heat transfer performance of silicon oil immersion cooling, the inlet flowing condition varied from 0.5 m/s to 5.0 m/s. The results showed that highest and average temperature decreased with inlet flow velocity increasing. The heat transfer coefficient and pressure drop increased with inlet flow velocity increasing. The direct contact of silicon oil and solar cells eliminated the traditional wall resistance and the heat transfer coefficient was improved from 451 W/(m² °C) at 0.5 m/s to 1800 W/(m² °C) at 5.0 m/s. However, the parasitic energy consumption increased with inlet velocity increasing, so the proper flow velocity of silicon oil was important for direct contact liquid immersion cooling receiver.

The varying trend of solar cells temperature with inlet temperature was shown in Fig. 6. It showed that the temperature of solar cells increased linearly with inlet temperature increasing. The temperature of solar cells increased from 26°C to 97°C when inlet temperature increased from -15°C to 65°C. The maximum temperature difference was within 4°C to 5°C under fixed inlet temperature, the reason of which was the liquid turbulence played a main factor in influencing the heat transfer. Thus, the lower inlet temperature was reasonable and better to obtain lower solar cells temperature.

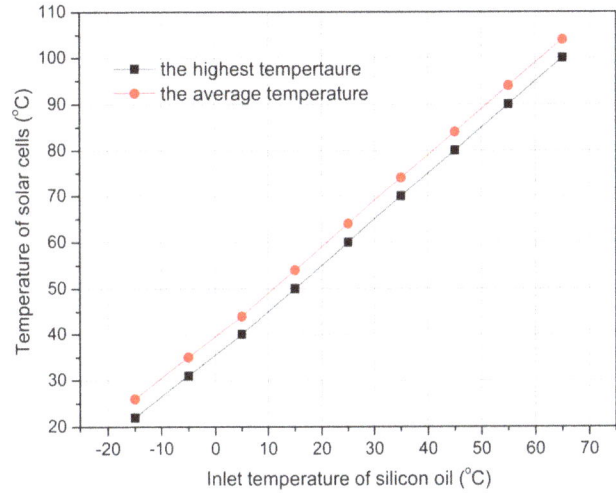

Fig. 6. *The temperature of solar cells model under different silicon oil inlet temperatures.*

3.3. Optimization of Silicon Oil Immersion Cooling Receiver

The thickness of liquid layer, fins structure and substrate thickness were main parameters of liquid immersion cooling receiver. The lower solar cells temperature and lower parasitic energy consumption were targets to optimize the parameters of receiver under concentration ratio of 500X.

3.3.1. The Effect of Liquid Layer Thickness

The thickness of liquid layer should be less than 6.3 mm when using silicon oil as immersion liquid according to our former research results [13]. The heat transfer performance and pressure drop were measured when liquid thickness ranged between 1 mm and 5 mm, the results were shown in Fig. 7. It can be seen that the temperature decreased with liquid layer thickness increasing and the thickness of the receiver increased with the liquid layer thickness increasing and the pressure drop decreased, which would decreased the parasitic energy consumption under fixed inlet flow velocity. Considering the results, the proper liquid layer thickness was 1 mm.

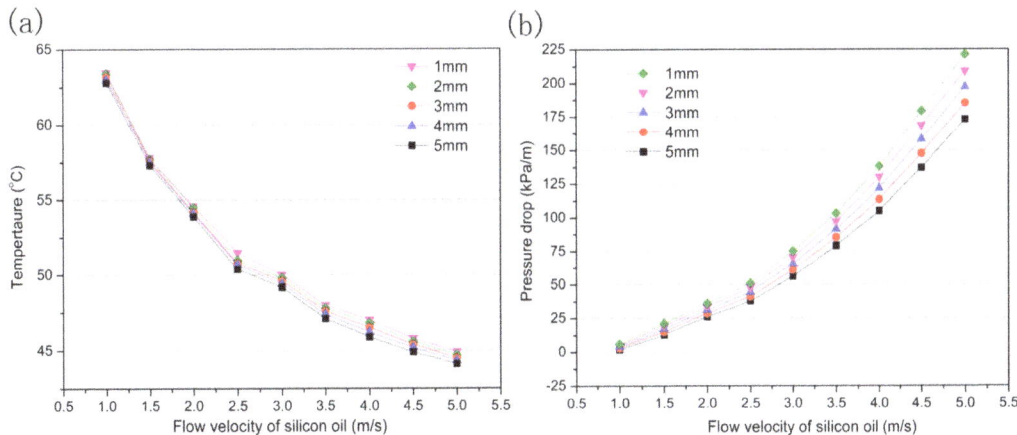

Fig. 7. *The solar cells temperature and pressure drop under different liquid thickness.*

3.3.2. The Effect of Height of Fins

The height of fins effect on solar cells temperature and pressure drop were evaluated under different inlet velocities of silicon oil. The height of fins ranged between 6 mm and 22 mm. The results were shown in Fig. 8.

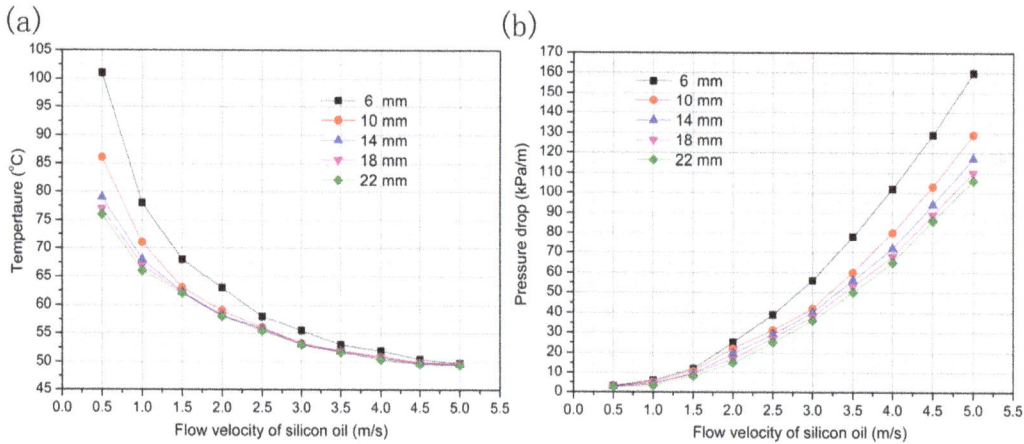

Fig. 8. The solar cells temperature and pressure drop under different height of fins.

It was shown that the solar cells temperature decreased from 100°C to 76°C with the fins height increased from 6 mm to 22 mm under inlet velocity was 0.5 m/s. The pressure drop decreased from 160 kPa/m to 107 kPa/m with the fins height increased from 6 mm to 22 mm under inlet velocity was 0.5 m/s. However, the decreased degree of temperature and pressure drop was not obvious when height of fins reached 18 mm and 22 mm, under which the velocity effect on solar cells temperature was little. Thus, the reasonable height of fins was 14 mm considering the solar cells temperature and pressure drop.

3.3.3. The Effect of Fins Amount

The effect of fins amount on temperature and pressure drop were measured under different inlet velocities. The fins amount ranged between 33 mm and 83 mm. The results were shown in Fig. 9. It was shown that the solar cells temperature decreased with fins amount increasing, the temperature decreased from 93°C to 48°C with fins amount increased from 33 mm to 83 mm under inlet velocity was 0.5 m/s/. The pressure drop increased with fin amount and silicon oil flow velocity increasing, but the pressure drop increased sharply when fins amount was higher than 50. The main reason was the heat transfer area increased with fins amount increasing but the liquid turbulence condition was also affected, which raised the parasitic energy consumption. Thus, the reasonable amount of fins was 50 considering the solar cells temperature and pressure drop.

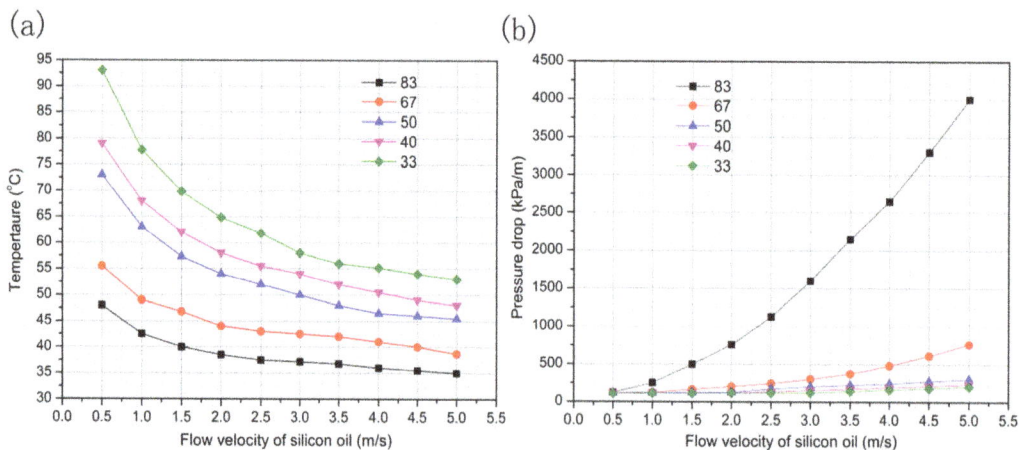

Fig. 9. The solar cells temperature and pressure drop under different amount of fins.

3.3.4. The Effect of Substrate Thickness

The effect of substrate thickness on temperature and pressure drop was measured under different inlet velocities. The substrate thickness ranged between 0.5 mm and 2.5 mm. The results were shown in Fig. 10. It can be seen that the solar cells temperature and pressure drop changed little with substrate thickness varying, which meant the main function of substrate was support for densely-packed solar cells and 1.5 mm was proper substrate thickness.

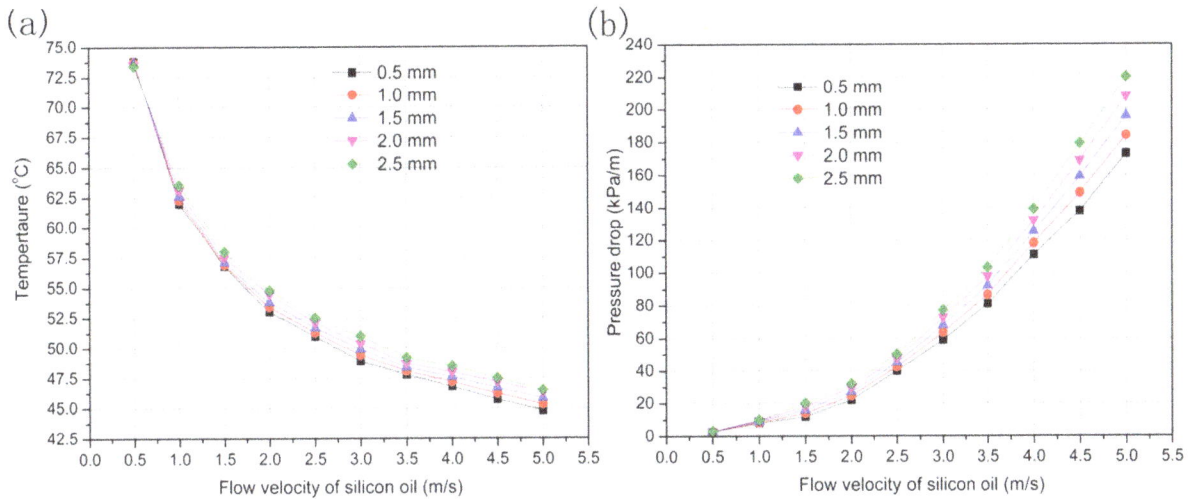

Fig. 10. The solar cells temperature and pressure drop under different substrate thickness.

Above all, the optimized parameters of liquid immersion cooling receiver were listed as follow: the thickness of liquid layer was 1.0 mm, the height and amount of fins were 14 mm and 50, the substrate thickness was 1.5 mm. The optimized silicon oil immersion cooling receiver would make sure the temperature of densely-packed solar cells distributed well uniform under 500X.

4. Conclusions

This paper mainly focused on the heat transfer performance of simulated densely-packed solar cells and optimization of liquid immersion cooing receiver. The conclusions were shown as follow.

1. The temperature could be controlled properly in the working range when using silicon oil to cool simulated solar cells model without fins structure under high concentration ratio. The results of simulated and experimental temperature of simulated densely-packed solar cells model were consisted well, which validated that the numerical simulation method was feasible to investigate the heat transfer performance of silicon oil direct immersion cooling solar cells.

2. The simulated results of densely-packed solar cells models with fins structure showed well temperature uniform. The temperature decreased and pressure drop increased with silicon oil inlet flow rate increasing. The model temperature increased linearly with silicon oil temperature increasing.

3. The direct contact liquid immersion cooling receiver was optimized using numerical simulated method. The optimized parameters were as follow: the thickness of liquid layer was 1 mm, the height and amount of fins was 14 mm and 50, the substrate thickness was 1.5 mm. The solar cells temperature distributed well uniform and relatively lower pressure drop when using the liquid immersion receiver for cooling densely-packed solar cells under 500X.

Acknowledgments

We acknowledge funding from the Chinese National Natural Science Foundation Project, Grant number 51478297, and the Programme of Introducing Talents of Discipline, Grant number B13011. Acknowledgements are also given to Dr. Yong Cui, Qunwu Huang in Green Energy Laboratory of Tianjin University for their contribution on language editing and result analysis to this paper.

References

[1] Royne A, Dey CJ, "Design of a jet impingement cooling device for densely packed PV cells under high concentration," Solar Energy, 81, pp. 1014-1024 (2007).

[2] Xu Z, Kleinstreuer C, "Computational Analysis of Nanofluid Cooling of High Concentration Photovoltaic Cells," Journal of Thermal Science and Engineering Applications, 6, (2014).

[3] Geng wg, Gao L, Shao M, Li Xy, "Numerical and experimental study on cooling high-concentration photovoltaic cells with oscillating heat pipe," International Journal of Low-Carbon Technologies, 7, pp. 168-173 (2012).

[4] Reeser A, Wang P, Hetsroni G, Bar-Cohen A, "Energy Efficient Two-Phase Microcooler Design for a Concentrated Photovoltaic Triple Junction Cell," Journal of Solar Energy Engineering, 136, (2014).

[5] Barrau J, Perona A, Dollet A, Rosell J, "Outdoor test of a hybrid jet impingement/micro-channel cooling device for densely packed concentrated photovoltaic cells," Solar Energy, 107, pp. 113-121 (2014).

[6] Wang Y, Fang Z, Zhu L, Huang Q, Zhang Y, Zhang Z, "The performance of silicon solar cells operated in liquids," Applied Energy, 86, pp. 1037-1042 (2009).

[7] Han X, Wang Y, Zhu L, "The performance and long-term stability of silicon concentrator solar cells immersed in dielectric liquids," Energy Conversion and Management, 66, pp. 189-198 (2013).

[8] Han X, Wang Y, Zhu L, Xiang H, Zhang H, "Reliability assessment of silicone coated silicon concentrator solar cells by accelerated aging tests for immersing in de-ionized water," Solar Energy, 85, pp. 2781-2788 (2011).

[9] Han X, Wang Y, Zhu L, Xiang H, Zhang H, "Mechanism study of the electrical performance change of silicon concentrator solar cells immersed in de-ionized water," Energy Conversion and Management, 53, pp. 1-10 (2012).

[10] Liu L, Zhu L, Wang Y, Huang Q, Sun Y, Yin Z, "Heat dissipation performance of silicon solar cells by direct dielectric liquid immersion under intensified illuminations," Solar Energy, 85, pp. 922-930 (2011).

[11] Sun Y, Wang Y, Zhu L, Yin B, Xiang H, Huang Q, "Direct liquid-immersion cooling of concentrator silicon solar cells in a linear concentrating photovoltaic receiver," Energy, 65, pp. 264-271 (2014).

[12] Xiang H, Wang Y, Zhu L, Han X, Sun Y, Zhao Z, "3D numerical simulation on heat transfer performance of a cylindrical liquid immersion solar receiver," Energy Conversion and Management, 64, pp. 97-105 (2012).

[13] Xin G, Wang Y, Sun Y, Huang Q, Zhu L, "Experimental study of liquid-immersion III–V multi-junction solar cells with dimethyl silicon oil under high concentrations," Energy Conversion and Management, 94, pp. 169-177 (2015).

[14] Zhu L, Boehm RF, Wang Y, Halford C, Sun Y, "Water immersion cooling of PV cells in a high concentration system," Solar Energy Materials and Solar Cells, 95, pp. 538-545 (2011).

[15] Zhu L, Wang Y, Fang Z, Sun Y, Huang Q, "An effective heat dissipation method for densely packed solar cells under high concentrations," Solar Energy Materials and Solar Cells, 94, pp. 133-140 (2010).

[16] Krauter S, "Increased electrical yield via water flow over the front of photovoltaic panels," Solar Energy Materials and Solar Cells, 82, pp. 131-137 (2004).

[17] Tianjin Lantian Solar Technology Limited Company, www.tjsolartech.com.

16

Simulation and Modelling of a Turbocharged Compression Ignition Engine

Brahim Menacer, Mostefa Bouchetara

Aeronautics and Systems Propelling Laboratory, Department of Mechanical Engineering, University of Sciences and the Technology of Oran, L.P 1505 El -Menaouer, Oran, Algeria

Email address:
acer.msn@hotmail.fr (B. Menacer), mbouchetara@hotmail.com (M. Bouchetara)

Abstract: The increase in fuel price is constraining car manufacturers to produce highly efficient engines with more regulations in terms of pollutant emissions. The increasing complexity of modern engines has rendered the prototyping phase long and expensive. This is where engine modeling becomes in the recent years extremely useful and can be used as an indispensable tool when developing new engine concepts. This study deals with the numerical simulation and performance prediction of a turbocharged diesel engine with direct injection. To predict the engine performances, we developed a computer program for simulating the operation of a turbocharged diesel engine, and used the commercial GT-Power software to validate the simulation results. In this work we carried out a comparative study of indicated mean effective pressure, mean effective pressure, power, torque and brake specific fuel consumption obtained by the analytical model for thermodynamic cycle simulation of a turbocharged diesel engine with the computer program developed in the language FORTRAN and those with the GT-Power software. The language FORTRAN program developed is currently used in the course of modeling and simulation of engine performance.

Keywords: One Zone Model, Ignition Compression Engine, Heat Transfer, Friction, Turbocharged Diesel Engine, GT-Power

1. Introduction

To predict the behavior of the internal combustion engine and trends of design and operating parameters, the thermodynamic modeling is a useful and effective tool. In recent years, numerical simulation has contributed enormously towards new evaluation of internal combustion engines. The numerical simulation based on mathematical modeling of diesel engine processes has long been used as an aid for design engineers to develop new design concepts. The literature offers various modeling approaches for internal combustion engines. In the development of engine models, we may distinguish three major steps: pure thermodynamic models based on first and second law analysis used since 1950, empirical models introduced in early 1970s and physically based nonlinear models for engine simulation and control design introduced over the last forty years [1], [2]. In this paper we propose a thermodynamic zero-dimensional, one-zone computational model based on the step-by-step filling and emptying method proposed by Watson and jonata [3] and valid for turbocharged diesel engine. The developed simple model should offer a reasonably accurate prediction of in-cycle engine state evolution. For comprehensive reviews on diesel engine modeling see the literature [4], [5]. Combustion in the diesel engine is a complex process because it involves several simultaneous physical and chemical phenomena which are not well understood until now. During the combustion, fuel is injected into the combustion chamber where the compressed air is at elevated temperature (self-ignition temperature). The fine droplets of fuel come in contact with hot air and commence to burn from the outer surface of the droplets.

The objective of the present work is to predict and analyze the performance of a turbocharged compression ignition engine using a developed computational thermodynamic model. This simulation model predicts in-cylinder temperatures and pressures as function of the crank angle, with the application of modified Vibe function for combustion model and Woschni correlation modified by Hohenberg for heat transfer at cylinder walls. It also takes into consideration the effects of heat losses and temperature dependent specific heats. The simulation code can provide useful information on the behavior under steady state operating conditions of main performance characteristics of the turbocharged diesel engine studied, such as instantaneous cylinder pressure $pcyl$, instantaneous cylinder temperature $Tcyl$, fuel burning rate $\frac{dm_{fb}}{dt}$, indicated mean effective pressure $imep$, friction mean effective pressure $fmep$, mean effective pressure $bmep$, indicated work Wi, brake power $bpower$, friction power, effective torque, indicated efficiency R_{ind}, effective efficiency R_{eff} and brake specific fuel consumption $bsfc$.

In this work, we will compare the effects on performance characteristics mentioned above of changing values for example; engine speed, crank angle, injection timing and compression ratio, respectively predicted by the developed simulation program and the selected known commercial GT-Power software.

2. Assumptions of the Developed Simulation Model

The expansion phase which both inlet and exhaust valve are closed represent the significant part of engine cycle, for it is in this period the power is developed by the engine. To simulate the engine performance, there are eight combustion models are proposed in the literature [5]. They can be classified into two main categories, single-zone combustion and multi-zone combustion models. In the engine cycle simulation model, we chose the single zone model proposed by Watson and al. which gives a heat of combustion satisfactory for calculations of the thermodynamics cycle, especially of high speed direct injection diesel engine. The model adopted by Lyn assumes that the heat takes place over a period of 40°CA to 50°CA.

The assumptions that have been made in developing the in-cylinder model for the direct injection diesel engine are:

1. The pressure and temperature of cylinder charge are assumed to be uniform throughout the cylinder and vary with crank angle.
2. The unburned mixture at any instant is composed of air and residual gases without chemical reaction.
3. No gas leakage through the valves and piston rings so that the mass remains constant.
4. The heat transfer region is limited by the cylinder head, the bottom surface of the piston and the instantaneous cylinder wall.
5. The temperature of the surfaces mentioned above is constant during the cycle.
6. The rate of heat transfer of gases to the wall is calculated from the temperature of the combustion gases and the wall. The heat transfer of gas-wall is changing rapidly due to the motion of the gas during the piston motion and to the geometry of the combustion chamber. The correlation of Hohenberg is used to calculate the rate of heat transfer cylinder [6].
7. We consider uniform crank speed (steady state engine).

3. Governing Equations of the Developed Simulation Model

3.1. Cylinder Volume

The cylinder volume at each crank angle position or the instantaneous total volume of gases in the cylinder can be obtained by knowing the geometrical engine parameters and using the equation of crank-slider mechanism. It is given as follows, Figure 1 [7]:

$$V_{cyl}(t) = V_m + \frac{\pi D^2}{4}\left(l + \frac{S}{2} - \left(\frac{S}{2}\cos(\omega t) + \sqrt{l^2 - \frac{S^2}{4}\sin^2(\omega t)}\right)\right)(1)$$

The derivative of equation (1) is written as follows:

$$\frac{dV_{cyl}}{dt} = \frac{\pi D^2}{4}\left(\frac{S}{2}\sin(\omega t) + \frac{\frac{S^2}{8}\sin(2\omega t)}{\sqrt{l^2 - \frac{S^2}{4}\sin^2(\omega t)}}\right) \quad (2)$$

t: Time corresponding to crank angle measured with respect to the top dead center [s]

ω: Engine speed [rad/s]

V_m: Clearance volume

r_c: Compression ratio ($r_c = (V_d + V_m)/V_m$)

l: Connecting rod length [m]

r: Crank radius [m]

S: Piston stroke [m], with ($S = 2.r$)

D: Cylinder bore [m]

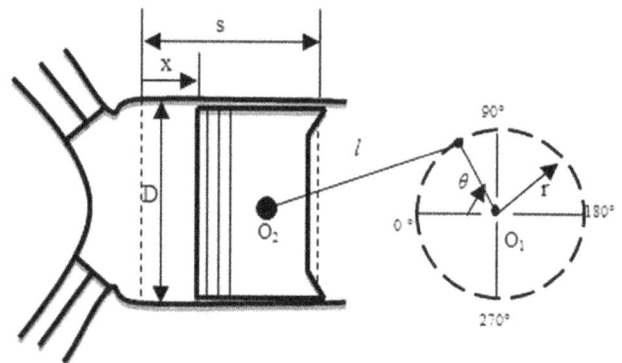

Figure 1. Crank-slider mechanism and cylinder geometry

3.2. Fuel Burning Rate

Fuel burning rate in internal combustion engines are generally governed by functions based on the law of normal distribution of continuous random variables.

There are two empirical models to determine the fuel burning rate: the simple Vibe law and the modified or double Vibe function following the Watson and al. model. In this simulation, we chose the single zone combustion model proposed by Watson and al. This correlation developed from experimental tests carried out on engines with different characteristics in different operating regimes [8].This model reproduces in two combustion phases; the first is the faster combustion process, said the premixed combustion and the second is the diffusion combustion which is slower and represents the main combustion phase. Figure 2 illustrates schematically the thermal balance of the single-zone cylinder model.

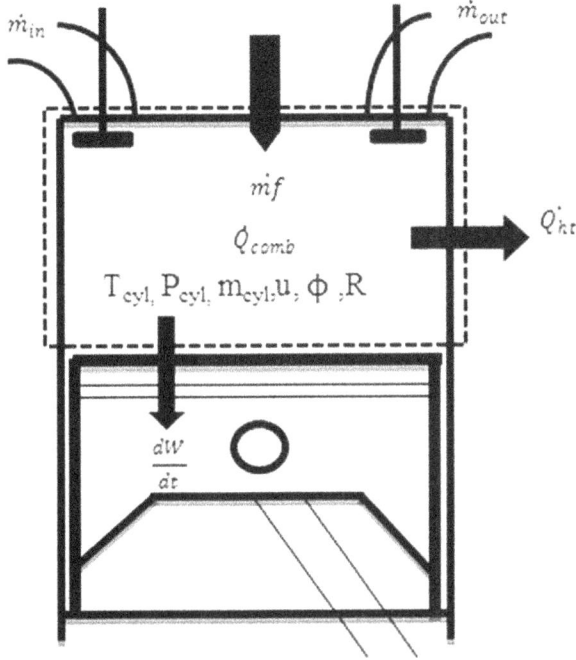

Figure 2. Thermal balance of the single zone model (--- system boundary)

During combustion, the amount of heat release $\dot{Q}_{comb}$ is assumed proportional to the burned fuel mass:

$$\frac{dQ_{comb}}{dt} = \frac{dm_{fb}}{dt} h_{for} \qquad (3)$$

$$\frac{dm_{fb}}{dt} = \frac{dm_{fb}^*}{dt} \frac{m_f}{\Delta t_{comb}} \qquad (4)$$

The combustion process is described using an empirical model, the single zone model obtained by Watson and al. [3]:

$$\frac{dm_{fb}^*}{dt} = \beta \left(\frac{dm_{fb}}{dt}\right)_p + (1 - \beta)\left(\frac{dm_{fb}}{dt}\right)_d \qquad (5)$$

$\frac{dQ_{comb}}{dt}$: Rate of heat release during combustion [kJ/s]

$\frac{dm_{fb}}{dt}$: Burned fuel mass rate [kg/ s]

h_{for}: Enthalpy of formation of the fuel [kJ/kg]

$\frac{dm_{fb}^*}{dt}$: Normalized burned fuel mass rate

m_f: Injected fuel mass per cycle [kg/cycle]

$\left(\frac{dm_{fb}}{dt}\right)_p$: Normalized fuel burning rate in the premixed combustion

$\left(\frac{dm_{fb}}{dt}\right)_d$: Normalized fuel burning rate in the diffusion combustion

β : Fraction of the fuel injected into the cylinder and participated in the premixed combustion phase. It depends on

the ignition delay τ_{id} described by Arrhenius formula [9] and the equivalence ratio ϕ.

$$\beta = 1 - \beta_1 \phi^{\beta_2}/\tau_{id}^{\beta_3} \qquad (6)$$

β_1, β_2 β_3 : Empirical constants for fuel fraction in the premixed combustion ($\beta_1 = 0.9, \beta_2 = 0.35, \beta_3 = 0.40$)

ϕ: Fuel-air equivalence ratio

The equivalence ratio ϕ is defined as:

$$\phi = \left(\frac{m_{fb}}{m_a}\right)/\phi_s \qquad (7)$$

m_a: Mass air participating in fuel combustion [kg]

ϕ_s: Stoichiometric fuel-air ratio

In diesel engine, in which quality governing of mixture is used, the equivalence ratio varies greatly depending on the load (from (5) and over at small loads to 1.40…1.10 at full load) [10].

The fuel burned mass m_{fb} is written as follows:

$$m_{fb} = \frac{m_{cyl}\phi_s\phi}{1+\phi_s\phi} \qquad (8)$$

From the equations (7) and (8), one obtains the state equation of the equivalence ratio [11]:

$$\frac{d\phi}{dt} = \left(\frac{1+\phi_s\phi}{m_{cyl}}\right)\left(\frac{1+\phi_s\phi}{\phi_s}\frac{dm_{fb}}{dt} - \phi\frac{dm_{cyl}}{dt}\right) \qquad (9)$$

The ignition delay τ_{id} in [ms] is the period between injection time and ignition time and it calculated by Arrhenius formula:

$$\tau_{id} = k_1 \bar{p}_{cyl}^{k_2} e^{\left(\frac{k_3}{\bar{T}_{cyl}}\right)} \qquad (10)$$

$\bar{p}_{cyl}$ and $\bar{T}_{cyl}$: Average values of the pressure and temperature in the cylinder when the piston is at the top dead center.

$k_1 = 0,0405$; $k_2 = 0,757$; $k_3 = 5473$: These coefficients are experimentally determined on rapid compression engines and valid for the cetane number between 45 and 50, [11].

3.2.1. Fuel Burning Rate During the Premixed Combustion

The normalized fuel burning rate in the premixed combustion is [9], [10]:

$$\left(\frac{dm_{fb}}{dt}\right)_p = C_{1p}C_{2p}t_{norm}^{(C_{1p}-1)}(1 - t_{norm}^{C_{1p}})^{(C_{2p}-1)} \qquad (11)$$

$$t_{norm} = \frac{t-t_{inj}}{\Delta t_{comb}} = (\theta - \theta_{inj})/\Delta\theta_{comb} \qquad (12)$$

t_{norm} : Normalized time vary between 0 (ignition beginning or injection time) and 1 (combustion end)

$\Delta t_{comb}, \Delta\theta_{comb}$: Combustion duration [s, °CA]

t_{inj}, θ_{inj}: Injection time and angle [s, °CA]

t, θ: Actual time and angle [s, °CA]

C_{1p}, C_{2p} : Constants model of the premixed combustion

$$C_{1p} = 2 + 1.25 \times 10^{-8}(\tau_{id}N)^{2.4}$$

$$C_{2p} = 5000$$

3.2.2. Fuel Burning Rate During the Diffusion Combustion

The fuel burning rate in the diffusion combustion is calculated as [10], [11]:

$$\left(\frac{dm_{fb}}{dt}\right)_d = C_{3d}C_{4d}t_{norm}^{(C_{4d}-1)} \times e^{(-C_{3d}t_{norm}^{C_{4d}})} \qquad (13)$$

C_{3d}, C_{4d}: Constants of the diffusion combustion model, then;

$$C_{3d} = 14.2/\phi_{tot}^{0.644}$$

$$C_{4d} = 0.79C_{3d}^{0.25}$$

3.3. Heat Transfer in the Cylinder

Heat transfer affects engine performance and efficiency. The heat transfer model takes into account the forced convection between the gases trapped into the cylinder and the cylinder wall. The heat transfer by conduction and radiation in the engine block are much less important than the heat transfer by convection [6]. The instantaneous convective heat transfer rate from the in-cylinder gas to cylinder wall $\dot{Q}_{ht}$ is calculated by [11]:

$$\frac{dQ_{ht}}{dt} = A_{cyl}h_t(T_{cyl} - T_{wall}) \qquad (14)$$

T_{wall}: Temperature walls of the combustion chamber (bounded by the cylinder head, piston head and the cylinder liner). From the results of Rakapoulos and al. [12], T_{wall} is assumed constant.

The instantaneous heat exchange area A_{cyl} in [m^2] can be expressed roughly by the following relation:

$$A_{cyl} = \left(\alpha_p + \alpha_{cyl,h}\right)\frac{\pi D^2}{4} + \pi D \frac{S}{2}\left(\frac{l}{r} + 1 - \cos \omega t - \sqrt{\left(\frac{l}{r}\right)^2 - \sin^2(\omega t)}\right) \qquad (15)$$

$\alpha_p, \alpha_{cyl,h}$: Coefficient shape respectively of the piston and cylinder head (for flat area $\alpha = 2$ and for no flat area $\alpha > 2$).

The global heat transfer coefficient in the cylinder can be estimated by the empirical correlation of Hohenberg which is a simplification of the Woschni correlation; it presents the advantage to be simpler of use and is the most adequate among all available relations to compute the heat transfer rate through cylinder walls for diesel engine [13].

The heat transfer coefficient h_t in [kW/K .m^2] at a given piston position, according to Hohenberg's correlation [6] is:

$$h_t(t) = k_{hoh} \, p_{cyl}^{0.8} V_{cyl}^{-0.06} T_{cyl}^{-0.4} (\bar{v}_{pis} + 1.4)^{0.8} \qquad (16)$$

where;

p_{cyl} is the cylinder pressure and V_{cyl} the in cylinder gas volume at each crank angle position.

k_{hoh}: Constant of Hohenberg which characterize the engine, (k_{hoh}=130)

The mean piston speed $\bar{v}_{pis}$ [m/s], is equal to:

$$\bar{v}_{pis} = 2 \times S \times N \qquad (17)$$

N: Engine speed [rpm]

3.4. Energy Balance Equations

In the filling and empting method, only the law of conservation energy is considered. The energy balance of the engine for a control volume constituted by the cylinder gasses is established over a complete cycle:

$$\frac{dU}{dt} = \frac{dW}{dt} + \frac{dQ}{dt} \qquad (18)$$

U is the internal energy, W is the external work and $\dot{Q}$ is the total heat release during the combustion.

The internal energy U per unit mass of gas is calculated from a polynomial interpolation deduced from the calculation results of the combustion products at equilibrium for a reaction between air and fuel C_nH_{2n}. The polynomial interpolation is a continuous function of temperature and equivalence ratio. It is valid for a temperature range T between 250 °K and 2400 °K and equivalence ratio ϕ between 0 and 1.6. To determine the change in internal energy, we use the expressions of Krieger and Borman [14]:

$$\frac{dU}{dT} = \left(\frac{dA}{dT} - \frac{dB}{dT}\phi\right)/(1 + \phi_s\phi) \qquad (19)$$

$$\frac{dA}{dT} = C_0 + C_1T + C_2T^2 - C_3T^3 + C_4T^4 \qquad (20)$$

$$\frac{dB}{dT} = -C_5 - C_6T + C_7T^2 - C_8T^3 \qquad (21)$$

$\frac{dA}{dT}, \frac{dB}{dT}$: Interpolation polynomial of Krieger and Borman
$C_0, C_1, C_2, C_3, C_4, C_5, C_6, C_7, C_8$: Krieger and Borman constants

The work rate is calculated from the cylinder pressure and the change in cylinder volume:

$$\frac{dW}{dt} = -p_{cyl}\frac{dV_{cyl}}{dt} \qquad (22)$$

The total heat release $\dot{Q}$ during the combustion is divided in four main terms:

$$\frac{dQ}{dt} = \frac{dQ_{in}}{dt} + \frac{dQ_{comb}}{dt} - \frac{dQ_{out}}{dt} - \frac{dQ_{ht}}{dt} \qquad (23)$$

with;

$$\begin{cases} \frac{dQ_{in}}{dt} = C_p\dot{m}_{in}T_a \\ \frac{dQ_{out}}{dt} = C_p\dot{m}_{out}T_{cyl} \\ \frac{dQ_{comb}}{dt} = \dot{m}_{fb}Q_{LHV} \end{cases} \qquad (24)$$

$\frac{dQ_{ht}}{dt}$: Rate of the convective heat transfer from gas to cylinder walls [kW]

$\frac{dQ_{in}}{dt}$ and $\frac{dQ_{out}}{dt}$: Inlet and outlet enthalpy flows in the open system [kW]

$\dot{m}_{in}$: Mass flow through the intake valve [kg/s]

$\dot{m}_{out}$: Mass flow through the exhaust valve [kg/s]

Q_{LHV}: Lower heating value of fuel [kJ/kg]

C_p : Specific heat at constant pressure [kJ/kg.K]

C_v: Specific heat at constant volume [kJ/kg.K]

The rate of change of mass inside the cylinder is evaluated from mass conservation, and is as follows:

$$\frac{dm_{cyl}}{dt} = \dot{m}_f + \dot{m}_{in} - \dot{m}_{out} \tag{25}$$

From the energy balance, we can deduce the temperature of gases in the cylinder $\dot{T}_{cyl}$ [15]:

$$\frac{dT_{cyl}}{dt} = [(\frac{dQ_{ht}}{dt} + \Sigma \left(h_0 \frac{dm}{dt}\right)_{in} - \Sigma \left(h_0 \frac{dm}{dt}\right)_{out} + \frac{dQ_{comb}}{dt} - u\frac{dm_{cyl}}{dt})\frac{1}{m_{cyl}} - \frac{RT_{cyl}}{V_{cyl}}\frac{dV_{cyl}}{dt} - \frac{\partial u}{\partial \phi}\frac{d\phi}{dt}]/\left(\frac{\partial u}{\partial T_{cyl}}\right) \tag{26}$$

In equation (26), many terms will be zero in some control volumes all or some of the time. For examples:

$\frac{dV_{cyl}}{dt}$ is zero for the manifolds,

$\left(h_0 \frac{dm}{dt}\right)_{in}$ and $\left(h_0 \frac{dm}{dt}\right)_{out}$ are zero for the cylinder,

$\frac{dm_{fb}}{dt}$ is zero the manifolds,

$u\frac{dm_{cyl}}{dt}$ is zero for the cylinder except for mass addition of fuel during combustion,

$\frac{dQ_{ht}}{dt}$ is neglected for the inlet manifolds,

$\frac{\partial u}{\partial \phi}$ is zero for the cylinder except during combustion (when fuel is added, hence ф changes),

Specific enthalpies $(h_0)_{in}$ and $(h_0)_{out}$ (except the specific enthalpy of formation h_{for}) are constant values.

By application of the first Law of thermodynamics for the cylinder gas, the equation (26) became:

$$\frac{dT_{cyl}}{dt} = \frac{1}{m_{cyl}C_v}\left(\frac{dQ}{dt} - p_{cyl}\frac{dV_{cyl}}{dt}\right) \tag{27}$$

The state equation of ideal gas is given by:

$$p_{cyl}V_{cyl} = m_{cyl}RT_{cyl} \tag{28}$$

R : Gas constant [kJ/kg.K]

Rearranging equations (23), (26), (27), (28); the state equation for cylinder pressure finally becomes:

$$\frac{dp_{cyl}}{dt} = \frac{\gamma}{V_{cyl}}\left[RT_{in}\dot{m}_{in} - RT_{cyl}\dot{m}_{out} - p_{cyl}\dot{V}_{cyl}\right] + \frac{\gamma-1}{V_{cyl}}\left[\dot{m}_{bf}Q_{LHV} - \dot{Q}_{ht}\right] \tag{29}$$

γ: Specific heat ratio ($\gamma = C_p/C_v$)

To evaluate the differential equation (26) or (29), all terms of the right side must be found. The most adapted numerical solution method for these equations is the Runge-Kutta method.

3.5. Engine Performance Parameters

3.5.1. Indicated Mean Effective Pressure, Friction Mean Effective Pressure and Mean Effective Pressure

The indicated mean effective pressure $imep$ is an important parameter because it is the potential output of the thermodynamic cycle:

$$imep = W_i/V_d \tag{30}$$

Where;

$$Wi = \Delta t.\Sigma_{t_0}^{t_n} p(t_i)\frac{dV_{cyl}(t_i)}{dt} \tag{31}$$

$p(t_i)$: Cylinder pressure at time position corresponding to crank angle position

Δt : Time interval, $\Delta t = \frac{\Delta\theta}{\omega}$

Wi : Indicated Work per cycle [kJ]

V_d: Displacement volume [m³], $V_d = \pi D^2 S/4$

Friction losses not only affect the performance, but also increase the size of the cooling system, and they often represent a good criterion of engine design. The indicated mean effective pressure lost to overcome friction due to gas pressure behind the rings, to wall tension rings and to piston and rings.

The model proposed by Chen and Flynn [19], [20] demonstrate that the value of the mean friction pressure $fmep$ [bar], be composed of a mean value C and additive terms correlated with the maximal cycle pressure p_{max} and the mean piston speed $\bar{v}_{pis}$. The mean value C, supposed constant, depends on the engine type and represents a constant base pressure which is to be overcome first. The term depending on $\bar{v}_{pis}$, reflect the friction losses in the cylinder (piston-shirt). The maximal cycle pressure p_{max} characterizes the losses in the mechanism piston-rod-crankshaft. So the friction mean effective pressure is calculated by [19]:

$$fmep = C + (0.005p_{max}) + 0.162\bar{v}_{pis} \tag{32}$$

p_{max}: Maximal cycle pressure [bar]

For direct injection diesel engine $C = 0.130$ bar

The mean effective pressure is used to evaluate the performance of an internal combustion engine. It is defined as the pressure acting on the piston during the expansion stroke to produce the same amount of work as the real power cycle. The mean effective pressure $bmep$ is the difference between $imep$ and $fmep$:

$$bmep = imep - fmep \tag{33}$$

3.5.2. Work Done

The work done Wd is given by:

$$Wd = bmepV_dN_{cyls} \tag{34}$$

N_{cyls} : Cylinder number

3.5.3. Effective and Indicated Power

The effective power $bpower$ for 4-stroke engine is:

$$bpower = bmep V_d N_{cyls} N/2 \qquad (35)$$

The indicated power *ipower* is expressed as:

$$ipower = imep V_d N_{cyls} N/2 \qquad (36)$$

The friction power *fpower* is evaluated as:

$$fpower = ipower - bpower \qquad (37)$$

3.5.4. Effective Torque
The effective torque is given by:

$$torque = bpower/(2\pi N) \qquad (38)$$

3.5.5. Indicated, Effective and Mechanical Efficiency
Indicated efficiency R_{ind} is equal to:

$$R_{ind} = Wi/Q_{comb} \qquad (39)$$

Q_{comb} : Heat release during combustion [kJ]
The effective efficiency R_{eff} is given by:

$$R_{eff} = Wd/Q_{comb} \qquad (40)$$

The mechanical efficiency R_{mec} is evaluated as:

$$R_{mec} = Wd/Wi \qquad (41)$$

3.5.6. Brake Specific Fuel Consumption
The brake specific fuel consumption *bsfc* [g/kW.h] equal:

$$bsfc = \dot{m}_f/bpower \qquad (42)$$

4. Simulation Program of Supercharged Diesel Engines

4.1. Computing Steps of the Developed Simulation Program

The calculation of the thermodynamic cycle according to the basic equations mentioned above requires an algorithm for solving the differential equations for a large number of equations describing the initial and boundary conditions, the kinematics of the crank mechanism, the engine geometry, the fuel and kinetic data.

It is therefore wise to choose a modular form of the computer program. The developed power cycle simulation program includes a main program as an organizational routine, but which incorporates a few technical calculations, and also several subroutines. The computer program calculates in discrete crank angle incremental steps from the start of the compression, combustion and expansion stroke.

The program configuration allows through subroutines to improve the clarity of the program and its flexibility. The

basis of any power cycle simulation is above all the knowledge of the combustion process. This can be described using the modified Wiebe function including parameters such as the combustion time and the fraction of the fuel injected into the cylinder.

For the closed cycle period, Watson recommended the following engine calculation crank angle steps: 10 °CA before ignition, 1° CA at fuel injection timing, 2° CA between ignition and combustion end, and finally 10 ° CA for expansion.

The computer simulation program includes the following parts:

- *Input engine, turbocharger and intercooler data*

Engine geometry (D , S , l , r ....), Engine constant (N, ϕ, C_r ..), Turbocharger constant ($\pi c, \pi t, pamb, Tamb,$ m , ICE , $p_{out,tur}$, $T_{out,tur}, p_{out,man}$, $T_{out,man}$) and polynomial coefficient of thermodynamic properties of species.

- *Calculation of intercooler and turbocharger thermodynamic parameters*

Compressor outlet pressure p_c , compressor outlet temperature T_c , compressor outlet masse flow rate $\dot{m}_c$, intercooler outlet pressure p_{ic} , intercooler outlet temperature T_{ic} , intercooler outlet masse flow rate $\dot{m}_{ic}$, turbine outlet pressure p_t , turbine outlet temperature T_t , turbine outlet masse flow rate $\dot{m}_t$.

- *Calculation of engine performance parameters*

- Calculation of the initial thermodynamic data (calorific value of the mixture, state variables to close the inlet valve, compression ratio C_r).

- Calculation of the piston kinematic and heat transfer areas.

-Main program for calculating the thermodynamic cycle parameters of compression, combustion and expansion stroke.

-Numerical solution of the differential equation (the first law of thermodynamics) with the Runge-Kutta method.

-Calculation of the specific heat (specific heat at constant pressure C_p and specific heat at constant volume C_v).

-Calculation of the combustion heat, the heat through walls and the gas inside and outside the open system.

- Calculation of main engine performance parameters mentioned above.

- *Output of Data block*

Instantaneous cylinder pressure p_{cyl} , instantaneous cylinder temperature T_{cyl} , indicated mean effective pressure *imep*, friction mean effective pressure *fmep*, mean effective pressure *bmep*, indicated power *ipower*, friction power *fpower*, brake power *bpower*.

The computer simulation steps of a turbocharged diesel engine are given by the flowchart in Figures 3.

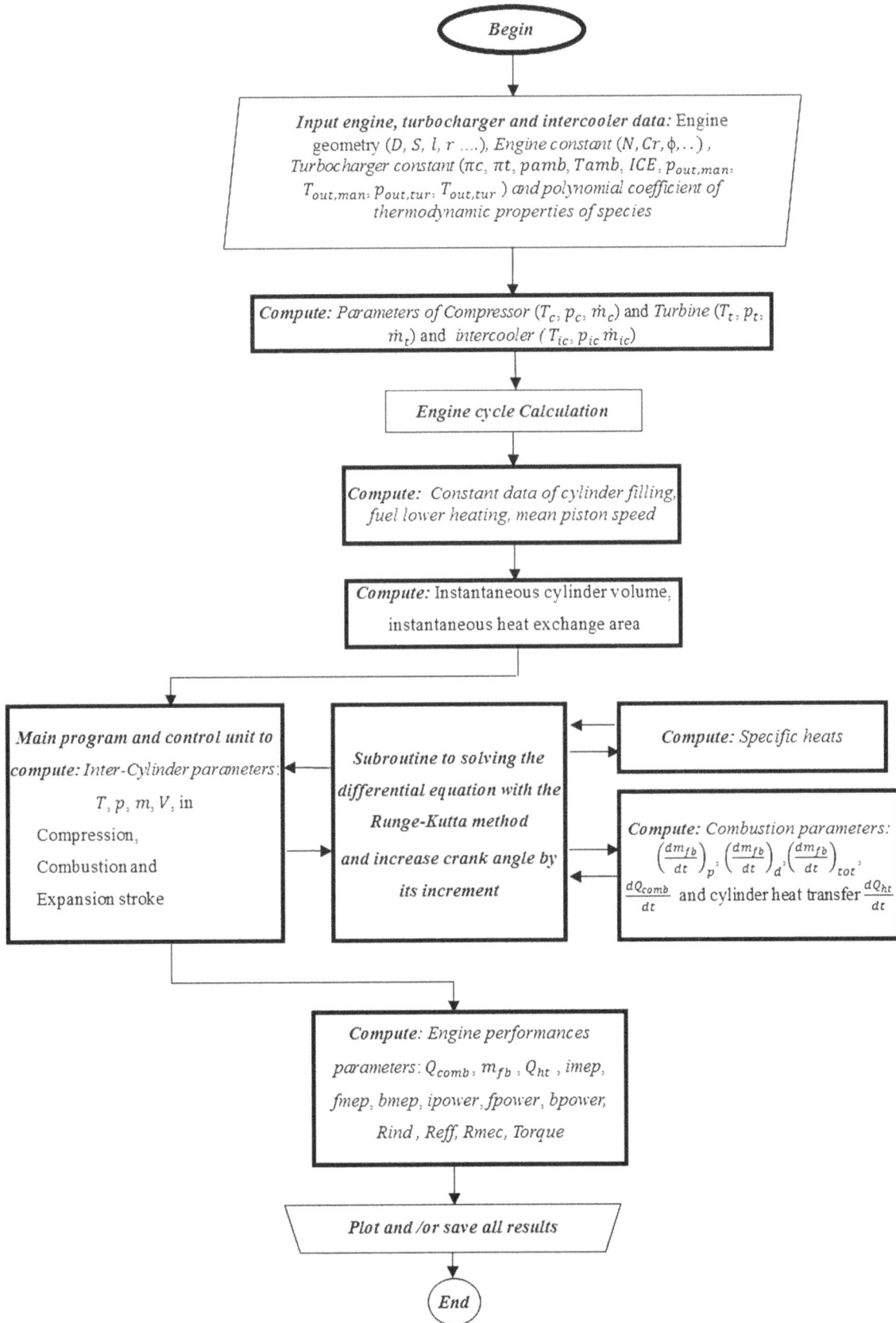

Figure 3. *Schematic Flowchart of steady state engine cycle computer program*

4.2. Simulation with the GT-Power Software

The GT-Power is a powerful tool for the simulation of internal combustion engines for vehicles, and systems of energy production. Among its advantages is the facility of use and modeling. GT-Power is designed for steady state and transient simulation and analysis of the power control of the engine. The diesel engine combustion can be modeled using two functions Wiebe [21]. GT-Power is an object-based code, including template library for engine components (pipes, cylinders, crankshaft, compressors, valves, etc...). Figure 4 shows the model of a turbocharged diesel engine with 6 cylinders and intercooler made with GT-Power. In the modeling technique, the engine, turbocharger, intercooler, fuel injection system, intake and exhaust system are considered as components interconnected in series.

4.2.1. Injection System

The simple injection system is used to inject fluid into cylinder and used for direct-injection diesel engines. Table1 shows the parameters of the injection system.

Table 1. Injection system parameters [21]

Injectors parameters	units	values
Injection pressure	[bar]	1000
Start of injection bTDC	[°CA]	15° BTDC
Number of holes per nozzle	[-]	8
Nozzle hole diameter	[mm]	0.25

4.2.2. Inlet Manifold and Exhaust Manifold

In the intake manifold, the thermal transfers are negligible in the gas-wall interface. This hypothesis is acceptable since the collector's temperature is near to the one of gases that it contains.

The variation of the mass in the intake manifold depends on the compressor mass flow and the flow through of valves when they are open. In the modeling view, the line of exhaust manifold of the engine is composed in three volumes. The cylinders are grouped by three and emerge on two independent manifold, component two thermodynamic systems opened of identical volumes. A third volume smaller assures the junction with the wheel of the turbine.

4.2.3. Turbocharger

Turbocharging the internal combustion engine is an efficient way to increase the power and torque output. The turbocharger consists of an axial compressor linked with a turbine by a shaft. The compressor is powered by the turbine which is driven by exhaust gas. In this way, energy of the exhaust gas is used to increase the pressure in the intake manifold via the turbocharger. As a result more air can be added into the cylinders allowing increasing the amount of fuel to be burned compared to a naturally aspirated engine [22].

4.2.4. Heat Exchanger or Intercooler

The heat exchanger can be assimilated to an intermediate volume between the compressor and the intake manifold. It comes to solve a system of differential equations supplementary identical to the manifold. It appeared to assimilate the heat exchanger as a non-dimensional organ (one supposes that it doesn't accumulate any gas).

5. Results and Discussions

For this investigation, the specifications of the selected turbocharged direct injection diesel engine are presented in the following Table.

Table 2. Engine specifications [21]

Engine parameters	Units	Values
Bore	mm	120.0
Stroke	mm	175.0
Displacement volume	cm^3	1978.2
Connecting Rod Length	mm	300.0
Compression ratio	-	16.0
Inlet valve diameter	mm	60
Exhaust valve diameter	mm	38
IVO	°CA	314
IVC	°CA	-118
EVO	°CA	100
EVC	°CA	400
Injection timing	°CA	15° BTDC
Fuel system	-	Direct injection
Firing order	-	1-5-3-6-2-4

In this paper the simulation result of engine performance are fuel burning rate for premixed, diffusion and total combustion, normalized burned fuel mass in premixed, diffusion and total combustion, cylinder pressure, cylinder temperature, friction pressure, mean effective pressure, friction power, effective power, effective torque, mechanical efficiency, effective efficiency, brake specific fuel consumption. All those performance characteristic magnitudes are presented as a function of engine load and engine speed.

Figures 5 and 6 show the evolution of the fuel burning rate and the normalized burned fuel mass for premixed, diffusion and total combustion at a speed of 1400 rpm and full load.

Figure 7 presents the cylinder pressure trace for half and full load at N = 1400 rpm, and for an injection advance of 15° before TDC (15° BTDC) obtained using the developed simulation model and GT-Power. The irregularities shown in the diagram may be due to the residual gas portion in the combustion chamber and also the vaporization of diesel fuel in the cylinder walls causing a cooling effect which affects the pressure and temperature. We can recognize a good agreement between both pressure curves. This indicates that the developed simulation model is correct.

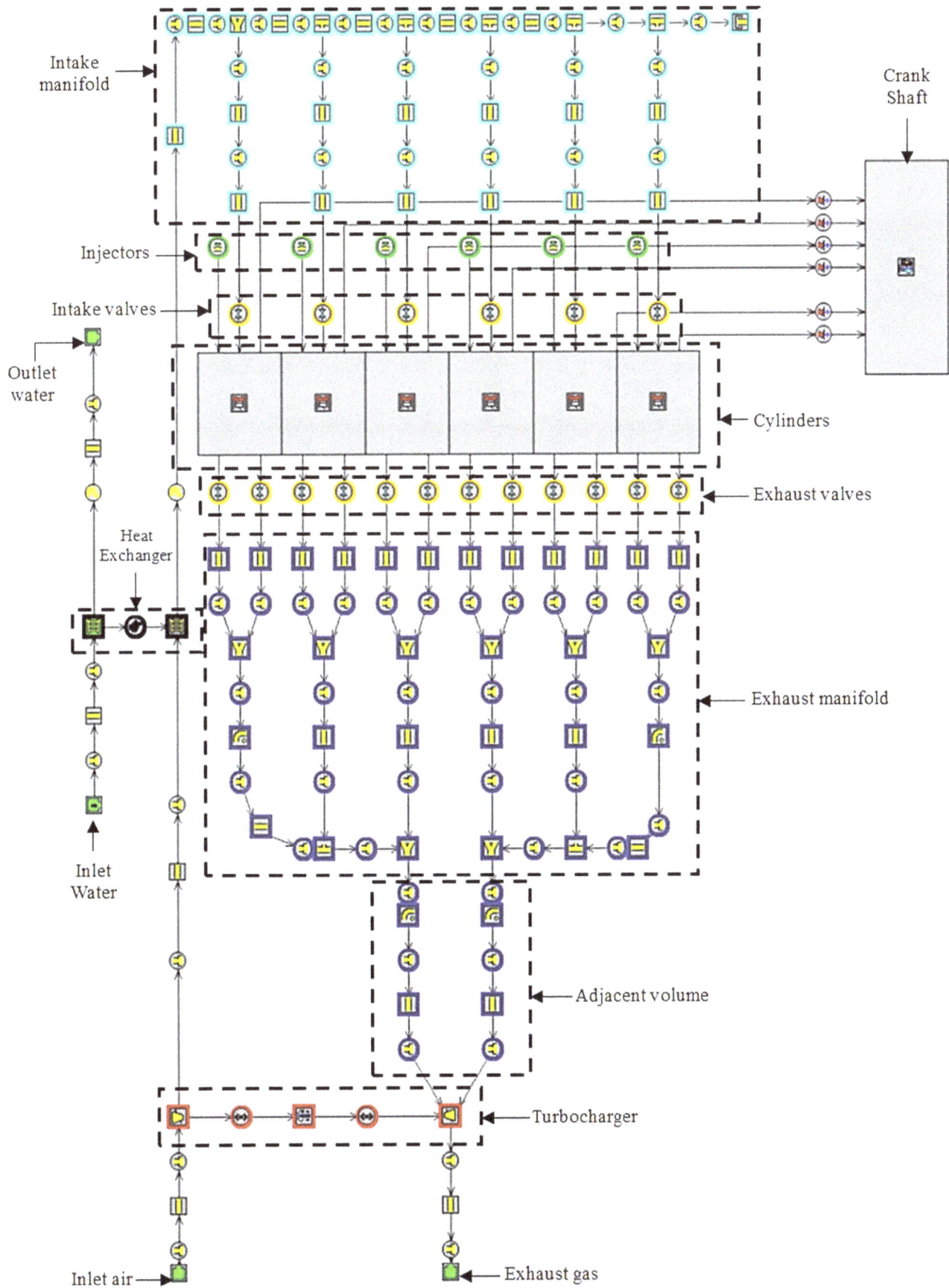

Figure 4. *Developed model of the 6-cylinders turbocharged engine using the GT-Power software*

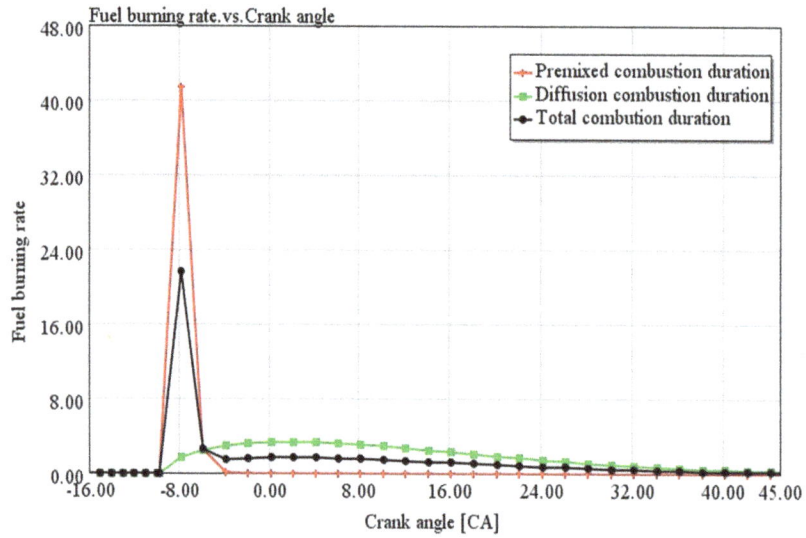

Figure 5. *Fuel burning rate for premixed, diffusion and total combustion*

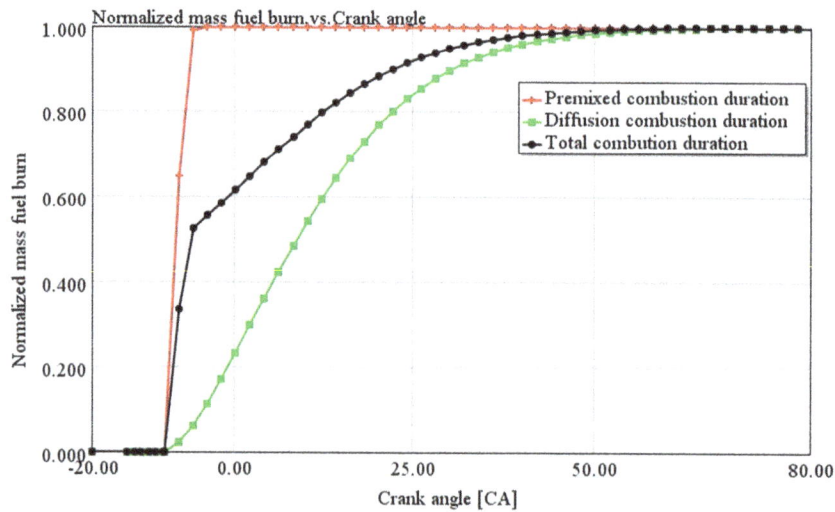

Figure 6. *Normalized burned fuel mass in premixed, diffusion and total combustion*

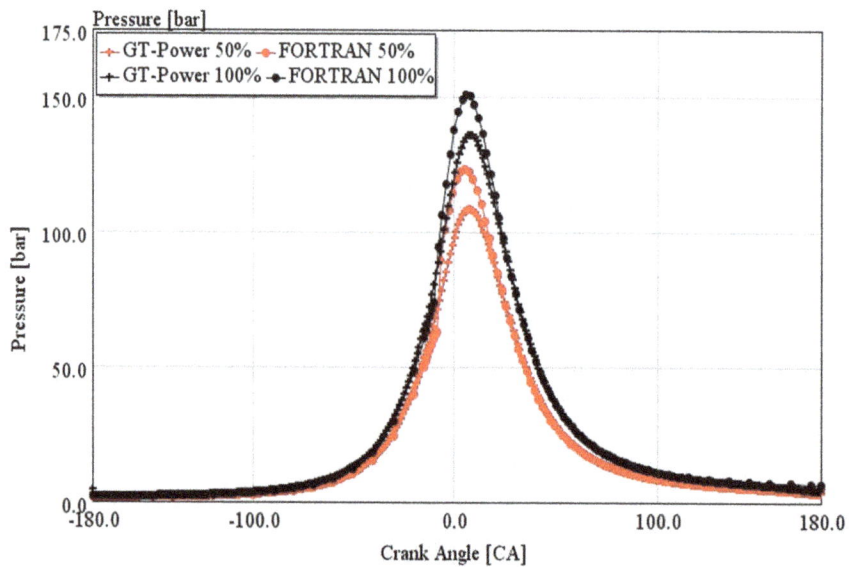

Figure 7. *Comparison of cylinder pressure at N=1400 rpm and different loads*

Figure 8 shows the variation of the temperature cylinder with two simulation models for 50% and 100% load up on crank angle and for an engine speed of 1400 rpm and advance injection of 15° BTDC. The appearance of both curves is almost identical.

The evolution of the friction power as a function of engine speed for different loads is shown in Figure 9. The friction power increases parabolically with an engine speed due to friction losses of moving parts and pumping losses, which correspond to the polynomial function found by Hendricks and Sorenson [23]. The friction Power also increases with the engine load due to increase of lateral forces on the piston. We can observe that the engine speed and load has the major influence over the engine friction. The friction powers according to the GT-Power model are smaller than those of the developed simulation and this is due to the choice of adopted friction model, Chen-Flynn friction model. For a speed of 2100 rpm, the middle gap between the two results, developed simulation model and GT-Power, is lower than 8%.

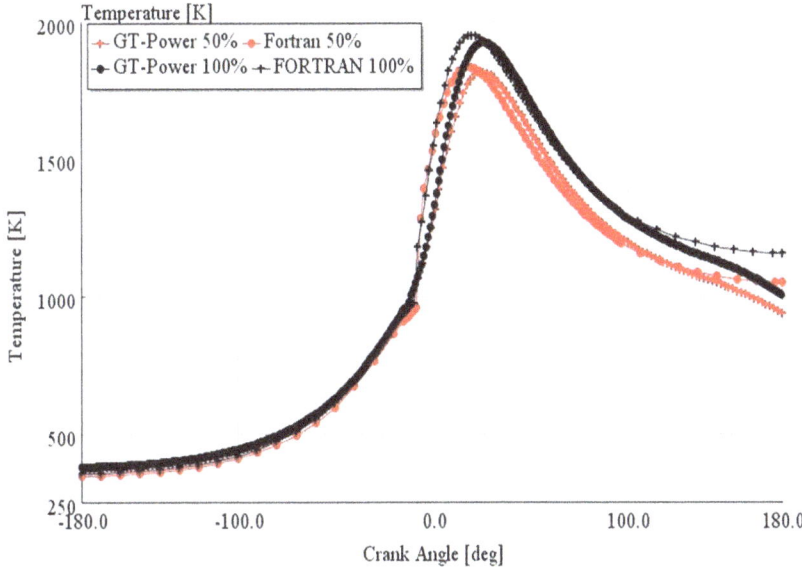

Figure 8. *Temperature cylinder for N=1400 rpm*

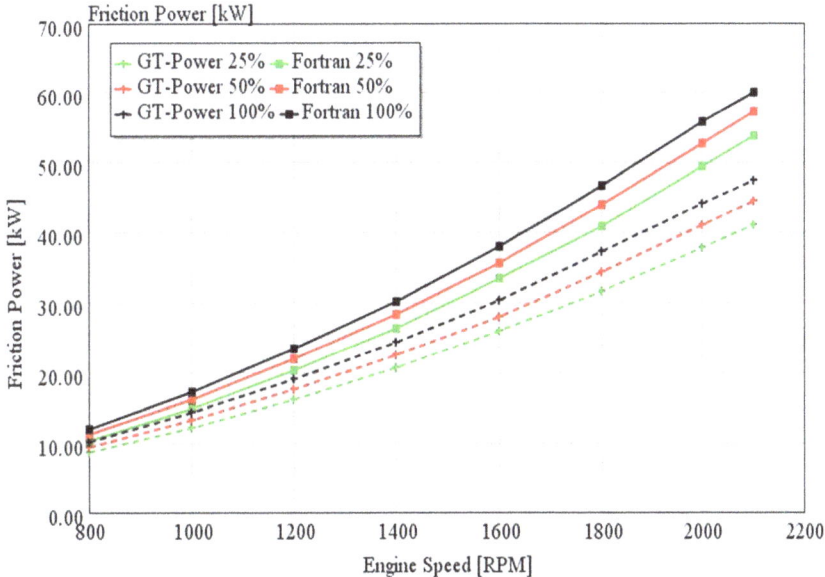

Figure 9. *Friction Power at full load and partial loads*

Figure 10 shows the change of brake power engine upon engine speed for different loads. The effective powers increase with the engine speed and loads. The gap between the two results, the developed simulation model and with GT-Power grows with the engine speed and loads, but the middle gap is lower than 8%.

The variation of the brake efficiency and the mechanical efficiency with the engine speed and for different engine loads are shown by Figures 11.*a* and *b*. There is some difference in results between the developed simulation model

and the GT-Power model due to the pressure losses with the   model.
engine speed, and essentially to the chosen friction losses

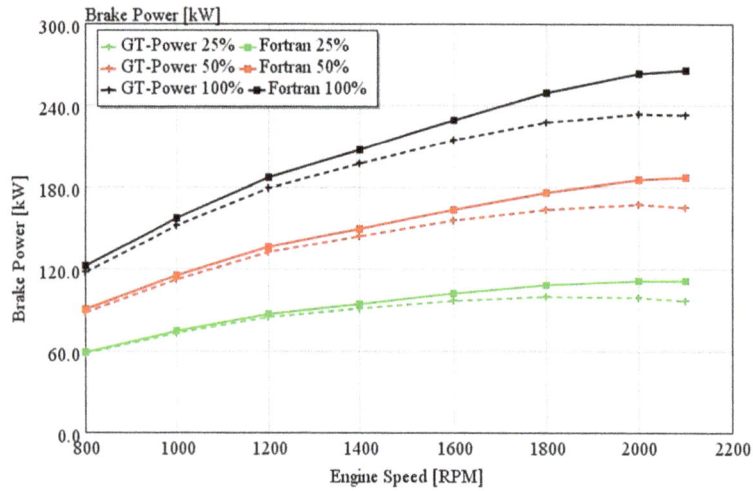

Figure 10. Brake power at full load and partial loads

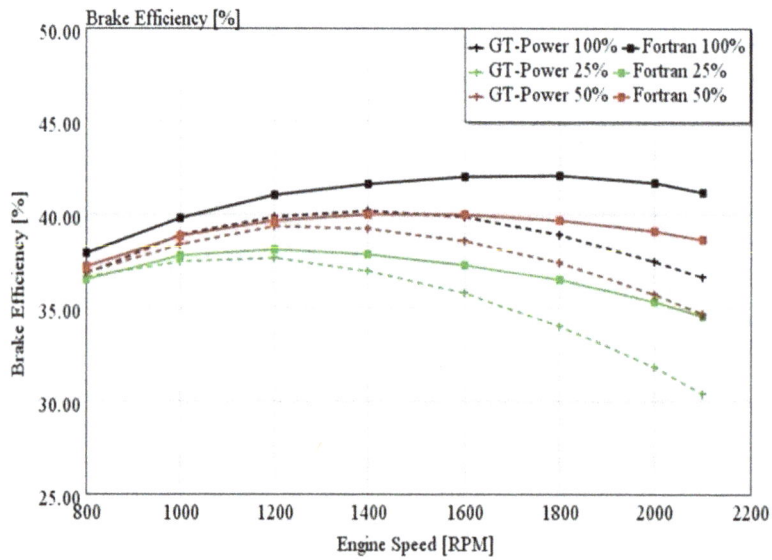

Figure 11-a. Brake efficiency at full and partial load

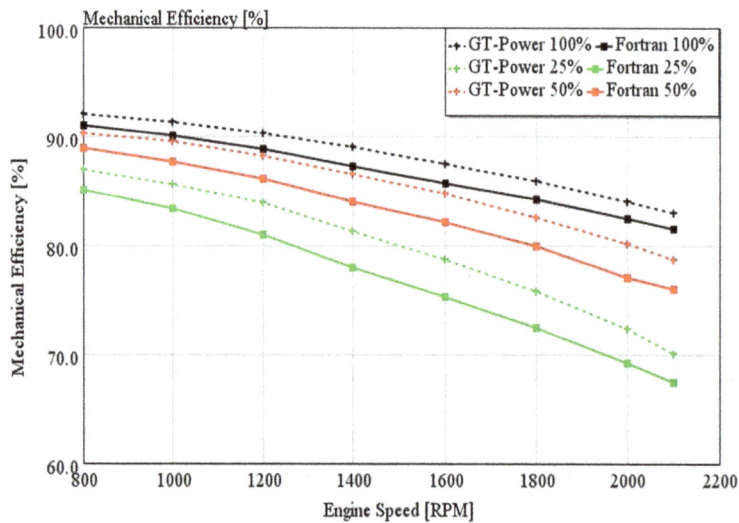

Figure 11-b. Mechanical efficiency at full and partial load

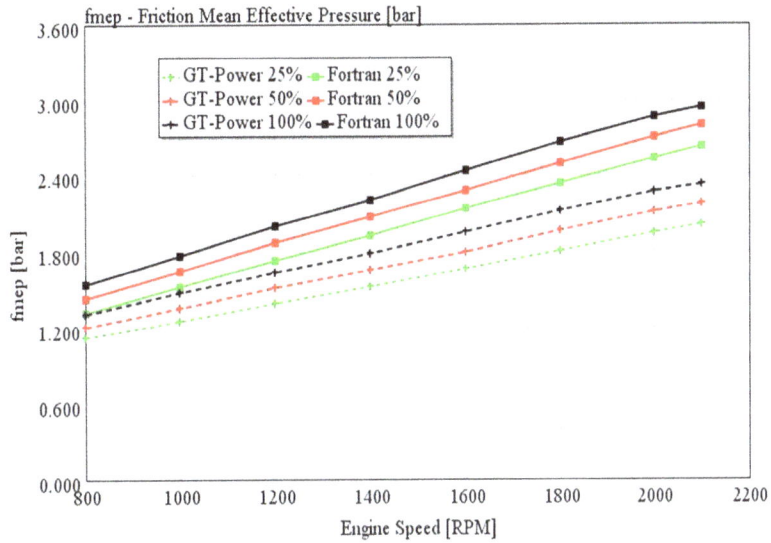

Figure 12-a. *Friction pressure at full and partial load*

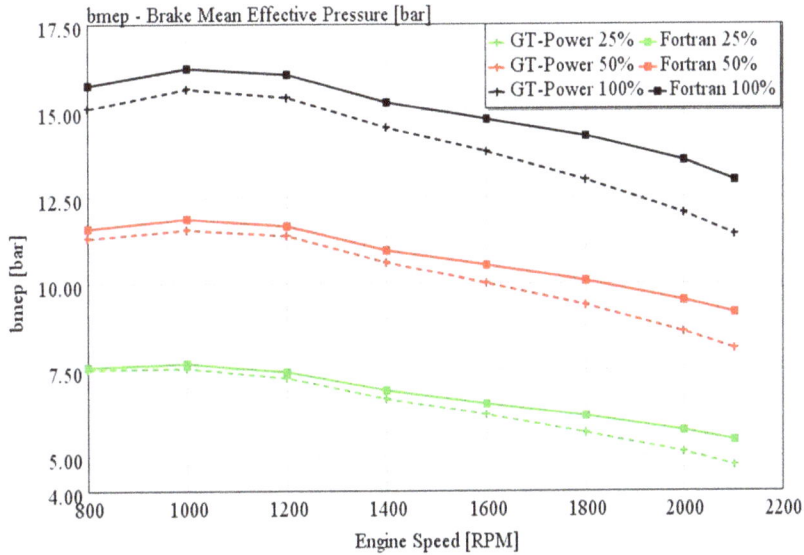

Figure 12-b. *Mean effective pressure at full and partial load*

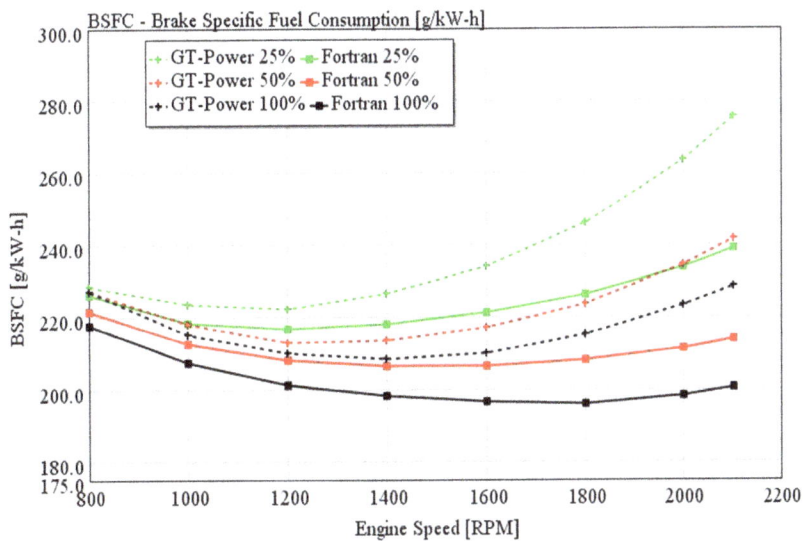

Figure 12-c. *Brake specific fuel consumption at full and partial load*

In Figures 12.*a* and *b*, we can see that the friction pressure increases linearly and the mean effective pressure drops with the engine speed, regardless of the engine load.

The reciprocal behavior is shown under the brake specific fuel consumption, figure 12.*c*. We notice that with simulation model; the calculated brake specific fuel consumption is lower than with GT-Power model; however the global tendency is nearly the same. At a given engine speed, the higher the load grows, the specific consumption decreases. This means that at full load diesel engine operates in an economic way. The minimal specific fuel consumption decreases with increasing load and moves toward higher engine speeds.

Another engine parameter is the injection timing which plays an important role in combustion process. If the injection is too early majority of the combustion takes place in the compression stroke causing high compression work and hot losses. If the injection is retarded then majority of the combustion takes place in the expansion stroke causing a loss of expansion, hence correct injection timing is required to achieve minimum best timing.

Figure 13 shows the effect of the injection time on the brake efficiency at full load. The maximal value of brake efficiency is at 15° BTDC. Figure 13 confirms the observations mentioned above.

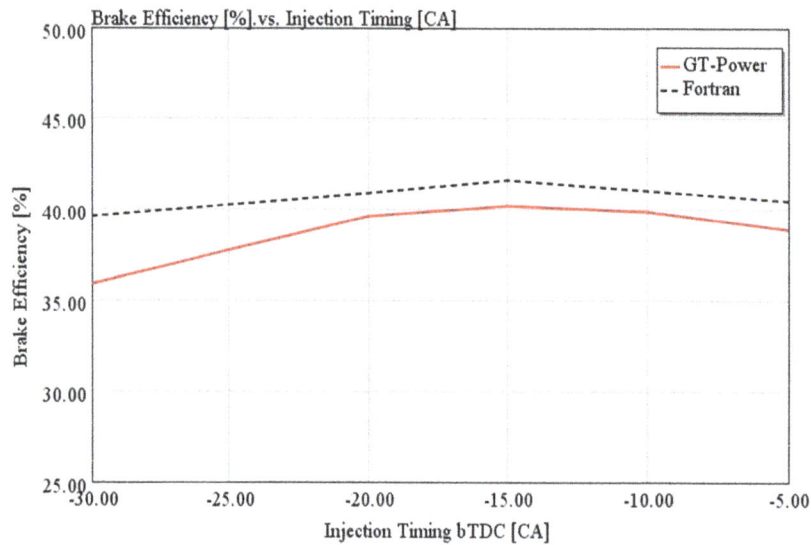

Figure 13. *Brake efficiency at different injection timing*

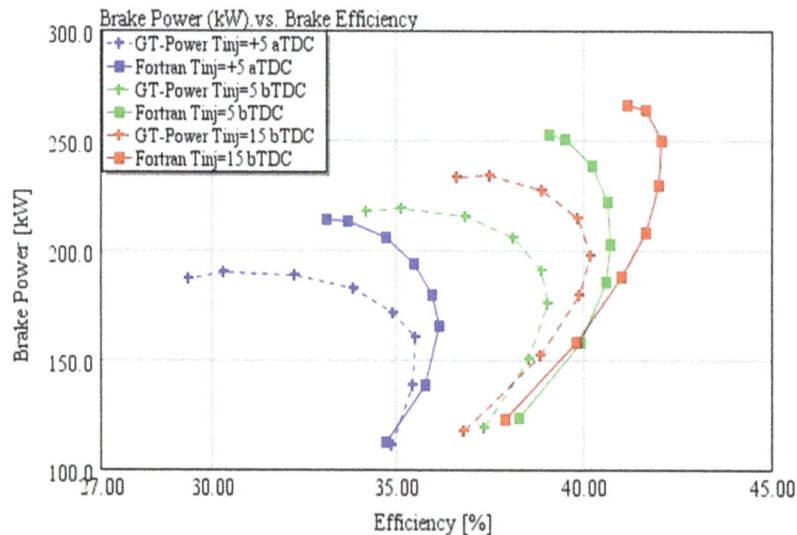

Figure 14. *Influence of injection timing on brake power and effective efficiency at full load*

As show the figure 14, the optimum effective efficiency and maximal brake power increase both when the injection advance increases. If we change the injection advance from 5 ° to 15° BDC, then the optimum effective efficiency increase of 5%, and the maximum brake power of 6%.

The influence of the compression ratio on the indicated mean effective pressure at the constant engine speed and for different injection times is presented in Figure 15. With the increase of the compression ratio results a higher maximum cycle pressure, which should not exceed the structural limits of the engine. The higher the compression ratio is, the larger is the indicated mean effective pressure. The highest indicated mean effective pressure values are achieved at 15°

BTDC. We note that for an injection advance of 30 ° BTDC the mean indicated pressure decreases slightly with the increase of compression ratio. For a constant compression ratio, the indicated mean effective pressure decreases with increasing of injection advance.

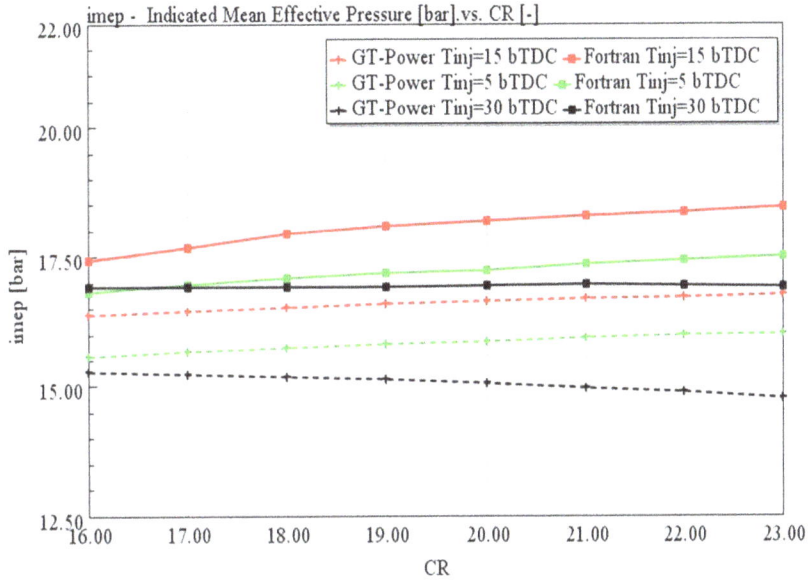

Figure 15. *Influence of compression ratio and injection time on the indicated mean effective pressure (N= 1400 rpm)*

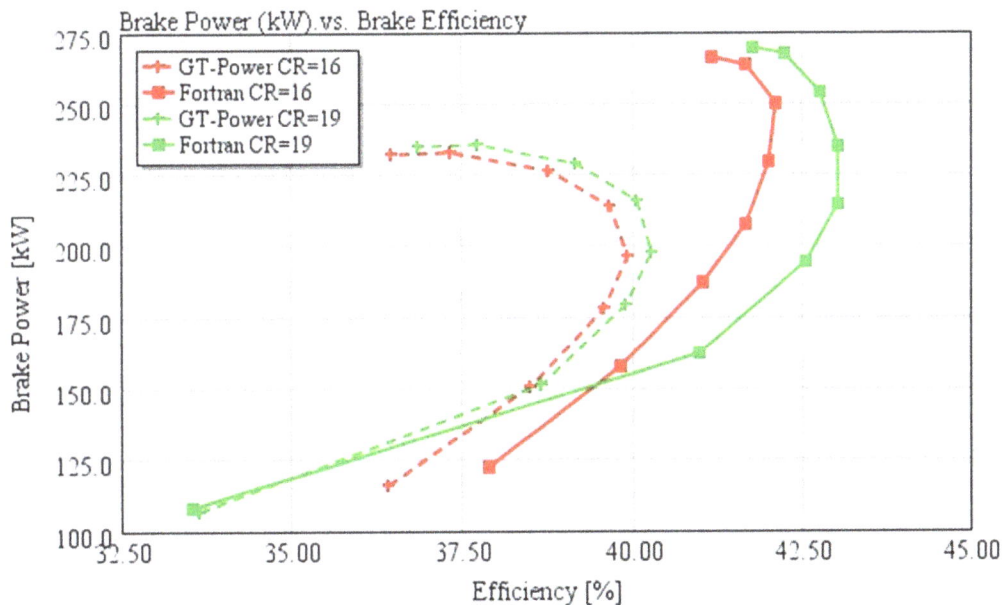

Figure 16. *Compression ratio influence, full load, injection timing 15° BTDC*

Figure 16 shows another presentation form of the influence of the compression ratio on the brake power at full load, for an injection advance of 15° BTDC, and engine speed between 800 and 2100 rpm. The effective efficiency increases with increase of the brake power until its maximum value, after it begins to decrease until the point corresponding of a maximal value of the brake power. It is also valid for the brake power. When the compression ratio is increased from 16.0 to 19.0, the maximum effective efficiency improved by 7%, and the maximum brake power by 15%, this is valid for

the developed simulation program and GT-Power.

6. Conclusion

The present work is a contribution to investigate the performance characteristics of a turbocharged diesel engine using a new developed thermodynamic numerical simulation model. Despite its dependence on empirical models and data, the developed simulation program is characterized by its simplicity and the capability to predict the main performance of turbocharged direct injection diesel engines. Modeling is based on the first law of thermodynamics in its simple form. The simulated power cycle was a closed cycle. Using the one zone combustion modeling concept, we have been analyzed the sensitivity of different engine parameters such as engine speed, engine load, compression ratio, injection time upon performance of a turbocharged diesel engine. The comparative study showed that the results obtained with the developed simulation program are sufficiently similar to those with the used commercial software GT-Power. The results from the present numerical simulation model could be used to improve the design and control strategy of the engine in terms of performance. In future work, we will try to develop a semi-empirical model describing more accurately the real operating conditions of a turbocharged diesel engine, incorporating for example real values for timing, injection rate, combustion rates, real engine geometry, heat transfer in combustion chamber, input and the exhaust manifold, and engine-turbocharger interaction. Due to the increasing complexity of combustion processes, we must try to find other alternatives for a numerical modeling of the combustion process and all engine heat losses. This is an important condition for a good agreement of the power cycle calculation with measurements on the engine test bench. The present study leaves open many possibilities for future researches based upon the engine modeling.

References

[1]   P. A. Hazell and J. O. Flower; Sampled-data theory applied to the modeling and control analysis of compression ignition engines – Part 1;International Journal of Control, pp. 549–562, 1971.

[2]   D. Descieux and M. Feidt; One zone thermodynamic model simulation of an ignition compression engine, Applied Thermal Engineering, vol. 27, pp 1457–1466, 2007.

[3]   N. Watson, A. D. Pilley and M. Marzouk; a combustion correlation for diesel engine simulation. SAE Technical Paper, (800029), 1980.

[4]   O. Grondin, R. K. Stobart, H. Chafouk and J. Maquet; Modelling the compression ignition engine for control; review and future trends, SAE Paper, 2004-01-0423, 2004.

[5]   A.Nafis; Evaluation of different types of friction to improve the performance of a direct injection diesel engine; Journal of Energy and Environment, 2004, pp. 33-47.

[6]   G. F. Hohenberg; Advanced approaches for heat transfer calculations, SAE Paper (1979), N°. 790825.

[7]   P. Gunter, Christian Schwarz, Gunnar Stiesch and Frank Otto; simulation combustion and pollutant formation for engine-development; springer-verlag berlin Heidelberg, 2006.

[8]   A. Sanli, A. N. Ozsezen, I. Kilicaslan and M. Canakci; The influence of engine speed and load on the heat transfer between gases and in-cylinder walls at fired and motored conditions of an IDI diesel engine; Applied Thermal Engineering, vol. 28, No. 11-12, pp. 1395-1404, 2008.

[9]   Robert Bosch GmbH, Automotive Handbook, 8th ed.; Plochingen: John Wiley and Sons, 2011.

[10]   A. Sakhrieh; Computational thermodynamic analysis of compression ignition engine; International Communications in Heat and Mass Transfer; No. 37, PP. 299–303, 2010.

[11]   J. B. Heywood; Internal combustion engine fundamentals, McGraw-Hill, Newyork, 1988.

[12]   C.D. Rakopoulos, D.C. Rakopoulos, G.C. Mavropoulos and E.G. Giakoumis; Experimental and theoretical study of the short-term response temperature transientsin the cylinder walls of a diesel engine at various operating conditions; Applied Thermal Engineering;2004,pp. 679-702.

[13]   R. Stone; Introduction to Internal Combustion Engines, Society of Automotive Engineers, Warrendale, 2d edition, 1995.

[14]   R. B. Krieger and G. L. Borman; The computation of apparent heat release for internal combustion engines, Proceedings of Diesel Gas Power, ASME (1966), N°. 66-WA/DGP-4.

[15]   N. Watson and M. S. Janota; Turbocharging the internal combustion engine; The Macmillan Press, New York, 1982.

[16]   T.K. Gogoi and D.C. Baruah; A cycle simulation model for predicting the performance of a diesel engine fuelled by diesel and biodiesel blends; Energy, N° 35.2010, pp. 1317–1323.

[17]   Kumar SV and Minkowycz WJ; Numerical Simulation of the thermodynamic fluid flow and heat transfer processes in a diesel engine; Numerical Heat Transfer, Part A.1990.pp.17-143.

[18]   Rakopoulos CD and Giakoumis EG; Simulation and exergy analysis of transient diesel engine operation; Energy; N° 22, 1997.pp. 875–885.

[19]   Z. Bazari and S. H. Chan; Diesel engine thermodynamic simulation: current status and future developments; The Mechanical Engineer 1992, pp.25-32.

[20]   S. H. Chan, Y. He and J. H. Sun; Prediction of transient nitric oxide in diesel exhaust; Proceedings of the Institution of Mechanical Engineers, Part D: Journal of Automobile Engineering 1999, pp. 213- 327.

[21]   Gamma Technologies, GT-Power User'sManual, GT-Suite Version 7.0, 2009.

[22]   J. Galindo, F. J. Arnau, A. Tiseira and P. Piqueras; Solution of the Turbocompressor Boundary Condition for One-Dimensional Gas-Dynamic Codes; Mathematical and Computer Modelling, vol. 52, No. 7-8, pp. 1288-1297, 2010.

[23]   E, Hendricks and S.C, Sorenson; Mean value modeling of spark ignition engines, SAE technical paper series, 900616, 1990.

17

Performance Characteristics of Automotive Air Conditioning System with Refrigerant R134a and Its Alternatives

Abdalla Gomaa

Refrigeration and Air Conditioning Technology Department, Faculty of Industrial Education, Helwan University, Cairo, Egypt

Email address:

gomaa@metcohvac.com, abdallagomaa@hotmail.com

Abstract: In this paper, the thermal performance characteristics of automotive air conditioning are carried out. Experimental analysis of R134a automotive air conditioning system with variable speed compressor is investigated. The purpose is to present a clear view on the effect of compressor speed, and condensing temperature on the thermal characteristics of automotive air conditioning. This study is exteneded theoritcally to cover more alternatives of the current R134a due to its impact of the Global Warming Potential GWP. The possibility of using low-GWP refrigerants of R152a, R1234yf, and R1234ze as alternatives to R134a in automotive air conditioning has been assessed. The refrigerants are investigated over a wide range of condensing temperature, evaporating temperature and refrigerant mass flow rate. The assessment is accomplished with cooling capacity, compressor power, coefficient of performance, pressure ratio, and condenser load. The results indicated that, the refrigerant R1234yf is much more environmentally accepted and has the best thermal performance among all investigated refrigerants.

Keywords: Automotive Air Conditioning, Variable Speed Compressor, R134a Alternatives

1. Introduction

Although the automotive industry has been using HFC134a (R134a) as a standard replacement for CFC12 since 1994 for its zero ozone depletion potential (ODP), this refrigerant has a very high Global Warming Potential (GWP = 1430). HFC134a contributes to global warming because of its fluorine content. Ozone depletion and total climate change depends on both global warming potential and ozone depletion potential. So, there is a need to find out alternatives of R134a under Kyoto protocol and Montreal protocol. The European Union issued a directive requiring for all car manufacturers to begin using a new refrigerant with a global warming potential (GWP) of less than 150 on all cars built for sale in the European Union by 2017, all cars assembled for sale in the European Union must be charged with an alternative refrigerant of R134a.

Over the last several years, much research and development effort has been focused on potential refrigerants possessing low Global Warming Potentials (GWPs). The evaluation of an automotive air conditioning system of R134a with a variable capacity compressor was studied by

J.M. Saiz Jabardo, et al,[1]. They developed a computer simulation model which includes a variable capacity compressor and a thermostatic expansion valve in addition to the evaporator and micro channel parallel flow condenser. Effects of design parameters on system performance of compressor speed, return air to the evaporator and condensing air temperatures have been experimentally simulated by means of developed model.

Comparative performance of an automotive air conditioning system of R134a using fixed and variable capacity compressors was studied by *Alpaslan Alkan, and Murat Hosoz,*[2]. They concluded that the operation with the variable speed compressor usually yields a higher COP than the operation with the fixed speed compressor in expense of a lower cooling capacity.

Jitendra Verma et al, [3] carried out a review of alternative to R134a refrigerant. They stated that, R152a is almost a straight drop-in substitute for R134a. The molecule is similar to R134a except that two hydrogen atoms are substituted for two fluorine atoms. It has similar operating characteristics to R134a but cools even better. An environmental benefit of R152a is that it has a global warming rating of 10 times less than R134a. *Ghodbane,* [4] simulated the performance of automotive air conditioning systems with several hydrocarbons. He determined that the systems with R152a and R270 yield a better performance than the one with R134a. In addition, a comparative assessment of the performance of a secondary loop system using these refrigerants was provided

E. Navarro, et al, [5] presented a comparative study between R1234yf, R134a and R290 for an open piston compressor of automotive air conditioning at different operating conditions. The test matrix comprised two compressor speeds, evaporation temperatures and condensation temperatures. They concluded that R290 has shown a significant improvement in compressor and volumetric efficiencies while R1234yf improves its efficiencies compared to R134a for pressure ratios higher than 8.

J. Navarro-Esbri, et al, [6] carried out an experimental analysis of R1234yf as a drop-in replacement for R134a in a vapor compression system. The experimental tests were carried out varying the condensing temperature, the evaporating temperature, the superheating degree, the compressor speed, and the internal heat exchanger use. Comparisons are made taking refrigerant R134a as baseline and the results show that the cooling capacity obtained with R1234yf is about 9% lower than that obtained with R134a. *Claudio Zilio, et al,* [7] studied experimentally an automotive air conditioning system equipped with variable displacement compressor. They concluded that, the R1234yf systems present lower performance than the R134a system at a given cooling capacity. *Yohan Lee and Dongsoo Jung,* [8] carried out a brief performance comparison of R1234yf and R134a in a bench tester for automobile applications. They concluded that the coefficient of performance and cooling capacity of R1234yf were 2.7% and 4.0% lower than that of R134a respectively.

Gustavo Pottker, and Pega Hrnjak, [9] studied the effect of condenser subcooling on the performance of an air conditioning system operating with R134a and R1234yf. It was concluded that the COP of the system operating with R1234yf can benefit more from the condenser subcooling than that with R134a due differences in thermodynamic properties.

In the present study, the thermal performance of R134a automotive air conditioning is carried out experimentally and theoretically. The study is assessed over wider range of compressor speed, condensing temperature and evaporating temperature. This study is extended to cover possible alternatives of R134a with low GWP of 150 or less according to Europe union recommendation. The low Global Warming Potential GWP refrigerants of hydrofluorocarbon- HFC-152a (R152a), and a very low GWP refrigerant of hydrofluoro-olefins of HFO-1234yf (R1234yf) and HFO-1234ze (R1234ze) are concerned in this investigation. The properties

of these Refrigerants are listed in table (1). The possibility of using R152a, R1234yf, and R1234ze, as alternatives to R134a in automotive air conditioning has been investigated using Engineering Equation Solver (EES, 2013). This investigation is done with standard parameters such as cooling capacity, compressor power, coefficient of performance (COP), pressure ratio and condenser heat load.

2. Experimental Test Rig

The experimental setup of automotive air conditioning system is shown in Figs. (1-a, 1-b) which is consists of R134a refrigeration system with variable speed compressor, condenser, thermostatic expansion valve and evaporator. The compressor is belt driven by a three-phase 1.5 kW electric motor energized through a frequency inverter, which allows the operation of the compressors at the required speed. It contains auxiliary equipment of a liquid receiver/filter-drier, flowmeter, and thermostat.

The experimental system contains two air ducts in which the evaporator and condenser have been inserted. The duct containing the evaporator has a cross-section area of 0.0504 m^2 and a length of 1.2 m. This duct has an axial fan driven by a DC motor and an electric heater with a maximum capacity of 1.8 kW. The air flow rate passing through the evaporator can be maintained at the required value by varying the voltage across the fan motor via a voltage regulator. Furthermore, the required air temperature at the evaporator inlet can be achieved by varying the voltage across the heater via another voltage regulator. On the other hand, the duct containing the condenser has a cross-section area of 0.187 m^2 and a length of 1.2 m. This duct contains a condenser axial fan driven by DC motors and another electric heater with a maximum capacity of 3 kW. The condenser air flow rate can be varied by adjusting the voltage across the fan motors. Moreover, the temperature of the air stream entering the condenser can be kept at the required value by varying the voltage across the heater. The reading of the measuring instruments of temperature, air velocity, refrigerant flow rate are recorded after the experiment reach steady state condition which in most cases takes time about 45 minutes.

Table 1. Details of refrigerants properties.

Item	R134a	R152a	R1234yf	R1234ze
Chemical formula	$C_2H_2F_4$	$C_2H_4F_2$	$C_3H_2F_4$	$C_3H_2F_4$
Molecular weight (kg/kmol)	102	66	114	114
ASHRAE safety classification	A1	A2	A2L	A2L
ODP	0	0	0	0
100-year GWP	1430	140	4	6
Critical temperature (k)	374.21	386.26	367.85	382.51
Critical pressure (kPa)	4059	4580	3382	3636
Boiling point (°C)	-26.1	-24.0	-30	−19

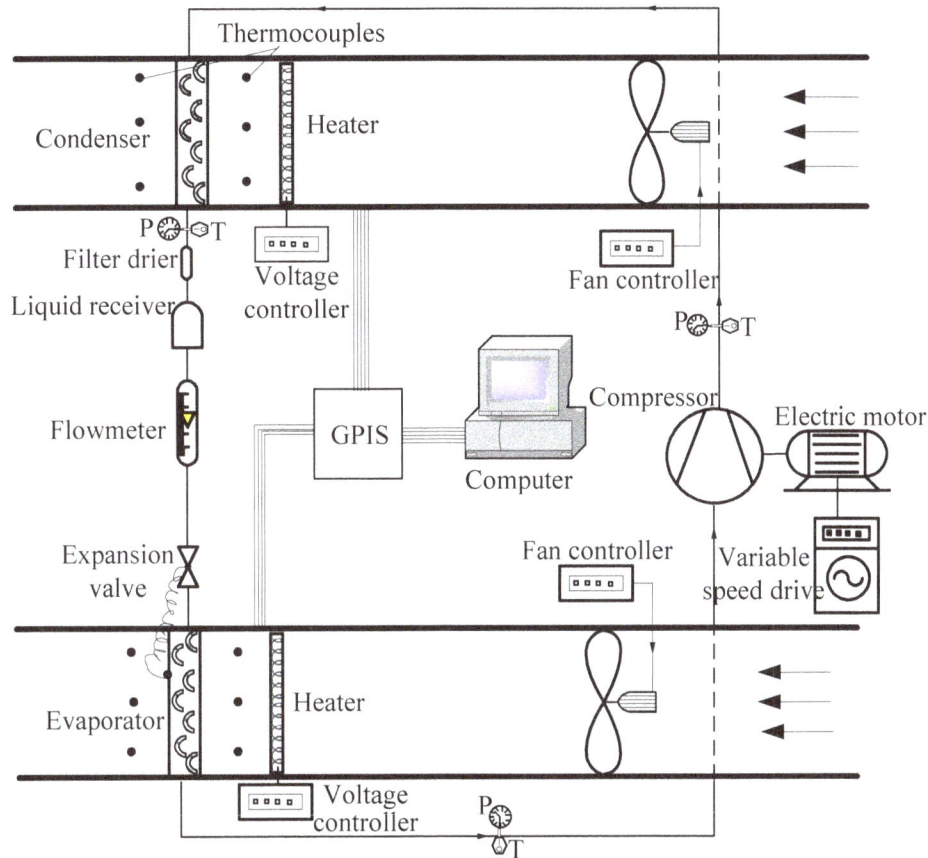

Fig. (1a). Schematic of experimental test rig.

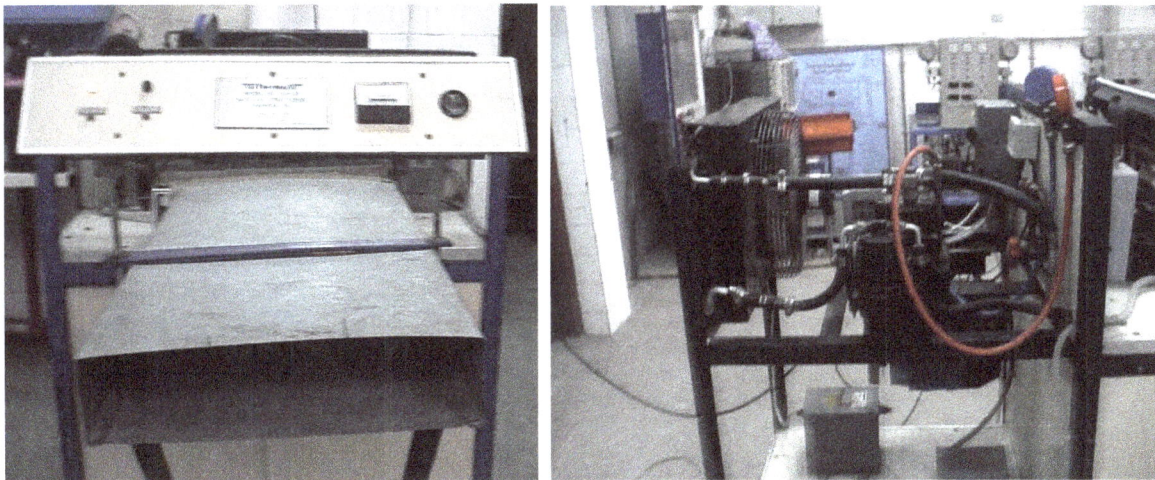

Fig. (1b). Photographs of the experimental test rig.

3. Measuring Techniques

The temperatures of the air side are measured using pre-calibrated K-type thermocouples. Two points of the K-type thermocouple probes with accuracy of 0.5 °C are placed on the upstream air and four thermocouples probes are placed on downstream of the test section to measure the air temperatures for both the evaporator and condenser respectively. All thermocouples are connected via a data acquisition system with accuracy of ± 0.1%. The relative humidity of the air upstream and downstream is measured by a humidity meter with accuracy of ± 1%. The air velocity profile through the duct section is identified according to ASHRAE recommendations by hot wire anemometer with an accuracy of ± 0. 1% of full scale. The refrigerant flow rate is measured by using refrigerant flow meter with an accuracy of ± 1%. Refrigerant pressure gauges with an accuracy of ± 0.5 % are fixed on high pressure and low pressure sides to measure the pressure before and after of the evaporator and condenser respectively.

4. Measurements Uncertainties

The experimental error analysis indicates the implication of error of the measured parameters on the uncertainty of the results. The uncertainty analysis of the various calculated parameters are estimated according to Holman JP, [10]. Given W_1, W_2, W_3, ..., W_n uncertainties in the independent variables (X_1, X_2, X_3, ...X_n) and W_R is the uncertainty in the result at the same odds. Then the uncertainty in the result can be given as;

$$W_R = \left[\left(\frac{\partial R}{\partial X_1} W_1 \right)^2 + \left(\frac{\partial R}{\partial X_2} W_2 \right)^2 + \left(\frac{\partial R}{\partial X_3} W_3 \right)^2 + ... + \left(\frac{\partial R}{\partial X_n} W_n \right)^2 \right]^{\frac{1}{2}} \qquad (1)$$

The uncertainties of the calculated experimental parameters are given in Table 2.

Table 2. *Range of uncertainties of calculated parameters.*

	Q_{eva} (%)	Q_{cond} (%)	W (%)	COP (%)
Minimum uncertainty	± 2	± 2.1	± 2.4	± 4
Maximum uncertainty	±3.6	±3.8	± 4.9	± 6.5

5. Data Reduction

The refrigeration pressure-enthalpy diagram of the automotive air conditioning cycle is illustrated in Fig. (2). Thermodynamically, The cooling capacity (Q_{eva}) is given by,[11]

$$Q_{eva} = \dot{m}_{ref} \left(h_1 - h_4 \right) \qquad (2)$$

The power required to drive the compressor is given by:

$$W_{com} = \dot{m}_{ref} \left(h_2 - h_1 \right) \qquad (3)$$

Actual specific enthalpy of the superheated vapor refrigerant at the compressor outlet (h_2) can be calculated as follows:

$$h_2 = h_1 + \frac{(h_{2,is} - h_1)}{\eta_{is,com}} \qquad (4)$$

The isentropic compressor efficiency ($\eta_{is,com}$) is taken as 0.65, M. Fatouh, et al [12].

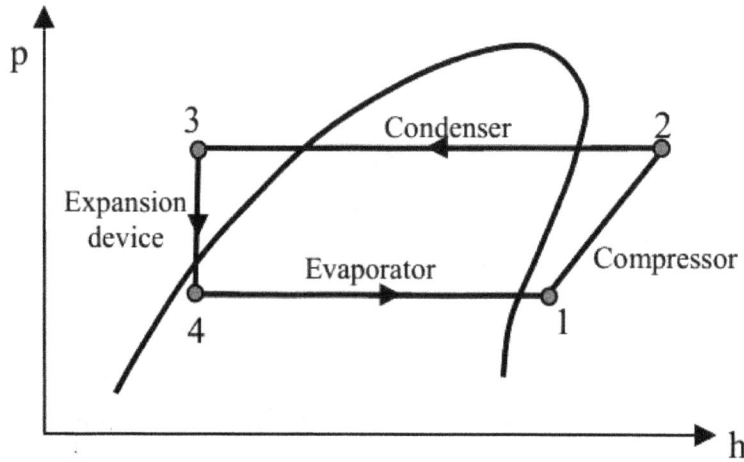

Fig. (2). Pressure-enthalpy diagram of the air conditioning refrigeration cycle.

Condenser heat load can be written as;

$$Q_{con} = \dot{m}_{ref} \left(h_2 - h_3 \right) \qquad (5)$$

The coefficient of performance (COP) is defined as the ratio of the cooling capacity to the compressor power, i.e.

$$COP = \frac{Q_{eva}}{W} \qquad (6)$$

Pressure ratio (*PR*) is defined as the ratio between condenser and evaporator pressures in which they depend mainly on the condensation and evaporation temperatures, respectively,

$$PR = \frac{P_{con}}{P_{eva}} \qquad (7)$$

The pervious equations are used to develop a computer program using Engineering Equation Solver (EES, 2013), [13]. Input parameters are refrigerant type, evaporating temperature, condensing temperature, refrigerant mass flow rate, evaporator specifications and condenser specifications. Output data are pressure ratio, refrigerant cooling capacity, condenser heat load, input power and coefficient of performance, for the existing system.

6. Results and Discussion

In this investigation, the results are comprised in two categories, the performance of R134a system with variable speed compressor and the performance of low global warming potential refrigerants as an alternative of R134a in automotive air conditioning.

6.1. Effect of Compressor Speed (RPM)

In automotive air conditioning, the compressors are usually belt-driven by the engine and therefore the compressor speed varies according to crankshaft RPM which led to a varying in mass flow rate through the air conditioning cycle. Experimentally the effect of compressor speed on the cooling capacity of R134a for different condensing temperature is illustrated in Fig. (3). For a certain condensing temperature, the cooling capacity increases as the compressor RPM increases. This is due to the increase of refrigerant mass flow rate with the increase in compressor speed. The ambient air temperature affects directly the cooling capacity. When the condensing temperature increased by 5°C, the cooling capacity decreased by 9%, while the COP decreased by 27% as illustrated in Fig. (4). The lower compressor RPM led to a higher in COP which can revealed that the increasing of compressor speed produce more friction power and hence more heat release at the compressor is occurred. A lower COP means a lower energy efficiency system and hence more global worming potential at a given duty.

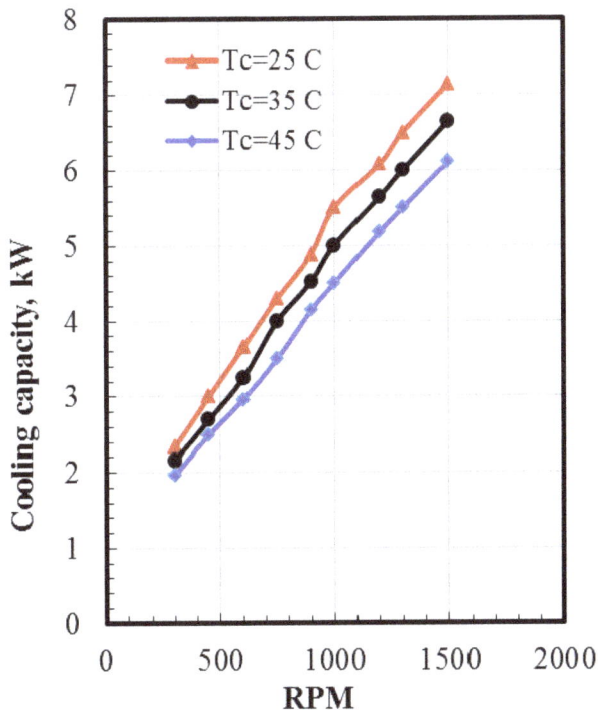

Fig. (4). *COP versus RPM of R134a. (Experimental results).*

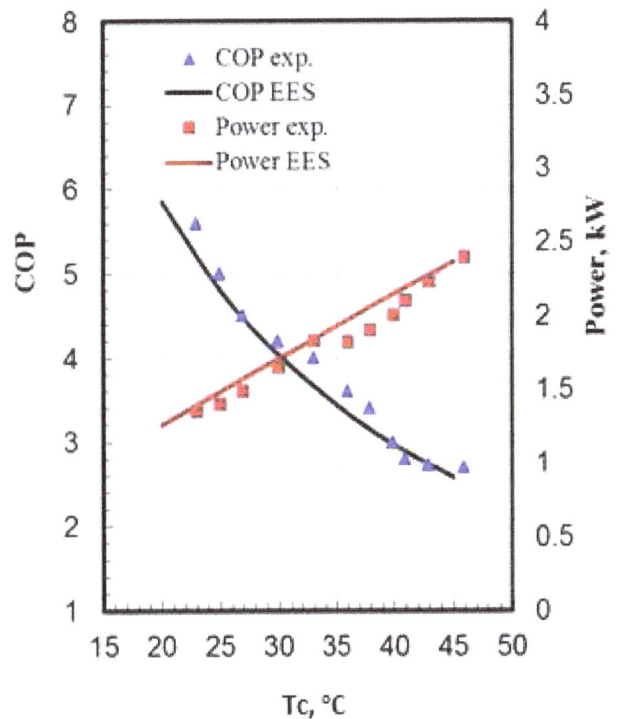

Fig. (5). *Validation of experimental and theoretical results of R134a for COP and input power.*

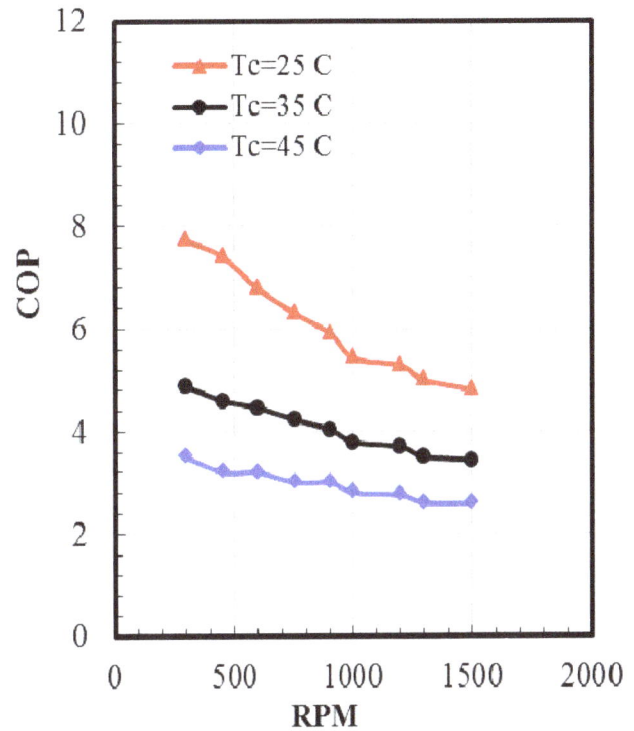

Fig. (3). *Cooling capacity versus RPM of R134a. (Experimental results).*

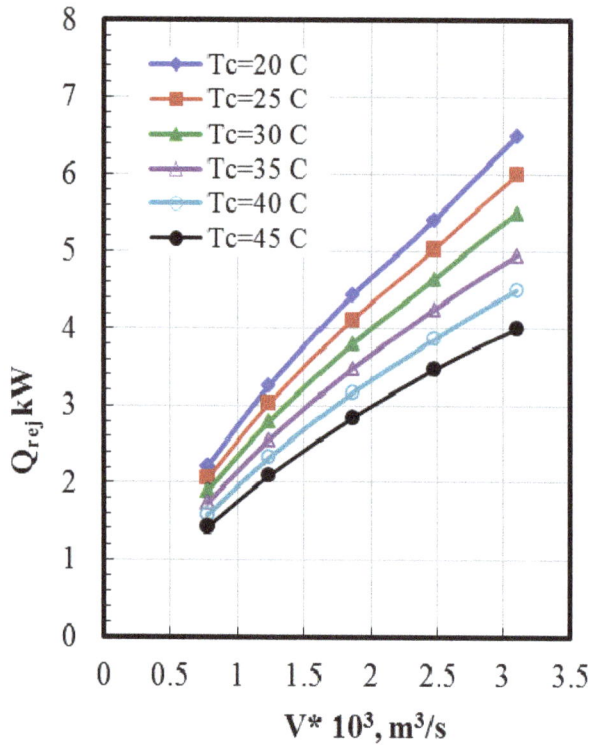

Fig. (6). Condenser heat rejection versus refrigerant flow rate of R134a.

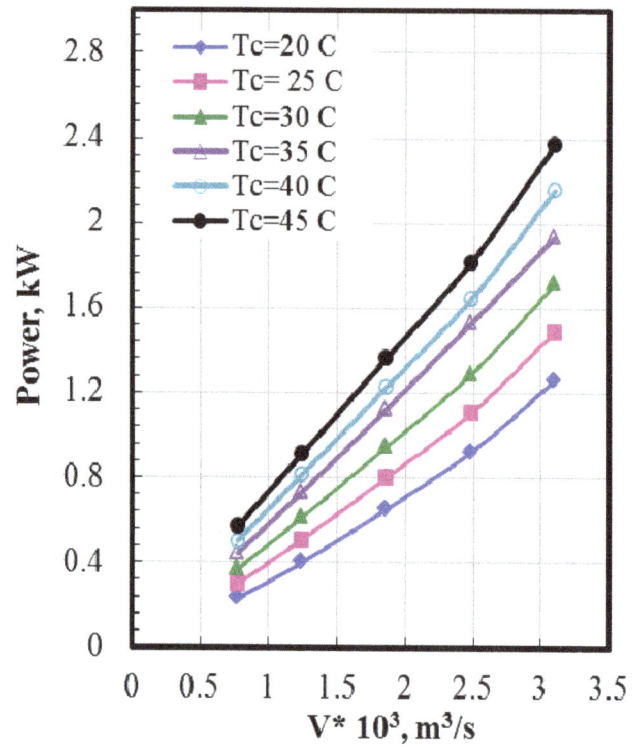

Fig. (7). Compressor power versus refrigerant flow rate of R134a.

The analysis of the automotive air conditioning refrigeration cycle is extended using Engineering Equation Solver (EES, 2013) in order to study a wide range of design parameters in addition to a study the performance of low Global Warming Potential (GWP) alternatives refrigerants as low as 150 according to EU recommendation. Validation between the experimental and theoretical (EES) results is needed. Figure (5) shows the COP and input power validation of the theoretical EES results with the present experimental results. The results show that simulated EES results are comparable to the experimental results.

6.2. Effect of R134a Condensing Temperature

The effect of condensing temperature on the compressor power and heat rejection from the condenser of R134a system is illustrated in Fig. (6) and Fig (7) respectively. It evident that the increase of condensing temperature is led to increase in compressor consumption power furthermore increase in condenser load at the same refrigerant flow rate. As the refrigerant flow rate increase which is happened as a reason of compressor speed, both the compressor power and condenser load increase on the other hand, the increase in cooling capacity is evidenced. Figure (8) indicates the trend of the COP with the refrigerant flow rate of R134a system at different condensing temperature. For all condensing temperature, the COP increase with the increase of the volume flow rate. It is noted that to obtain enhanced COP, the condensing temperature should maintain as low as possible.

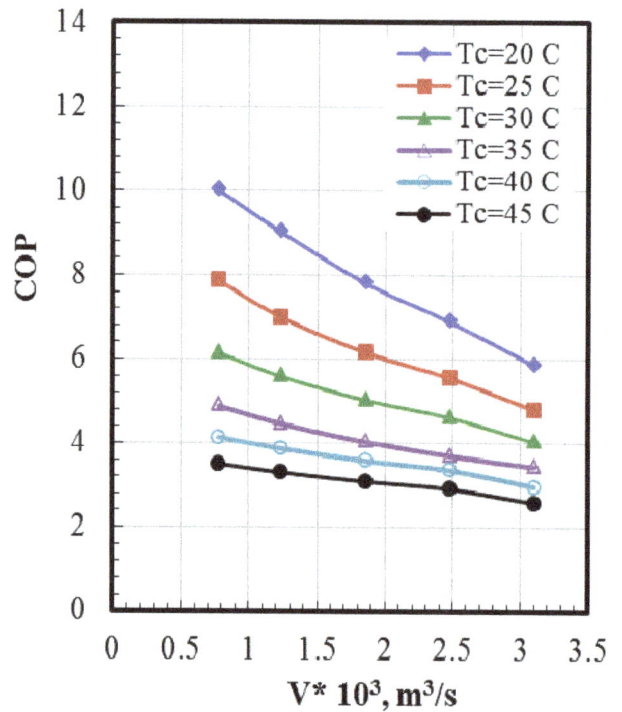

Fig. (8). COP versus refrigerant flow rate of R134a, EES results.

6.3. Performance of Low GWP Refrigerants

The selection of an alternative refrigerant of the automotive air conditioning system receives special significance not only with zero ozone depletion potential but also with the global warming potential (GWP $\leq$ 150), furthermore a low energy demand by the system which is seen as an essential criterion. In particular, the refrigerants R152a, R1234yf and R1234ze automotive air conditioning systems are the most possible as an alternative with R134a. An environmental benefit of R152a is that it has a global warming rating of 140, which is 10 times less than R134a. The refrigerant R1234yf (HFO-1234yf) is the first in a new class of refrigerants acquiring a global warming potential (GWP = 4) rating 335 times less than that of R134a. It was developed to meet the European directive that went into effect in 2011 requiring that all new car platforms for sale in Europe use a refrigerant in its air conditioning system with a GWP below 150. The refrigerant R1234ze is a hydrofluoroolefin and it was developed as a "fourth generation" refrigerant to replace R134a in automotive air conditioning which it has zero ozone-depletion potential and a low global warming potential (GWP = 6), [14].

6.4. Effect of Condensing Temperature

Figure (9) shows the cooling capacity of R152a, R124yf and R1234ze with particular reference to R134a for a typical evaporation temperature of 10 °C and 0.0031 m^3/s refrigerant volume flow rate. The cooling capacity of R134a is higher than that of R152a, R1234yf and R1234ze by 3.5%, 3.8%, and 19% respectively at the same operating condition. The effect of the condensing temperature on the compressor power consumption is illustrated in Fig. (10). The condensing temperature is ranged from 20°C to 45°C which is most practically applicable. This figure indicates that, the compressor power consumption of R134a is higher than that of R152a, R1234yf and R1234ze by 8.5%, 1.6%, and 28% respectively at the same operating condition (T$_e$ = 10°C and V = 0.0031 m^3/s).

Figure (11) shows the COP variation with the condensing temperature of R134a, R152a, R124yf and R1234ze. The COP of R1234yf system is lower than that of R134a by 2.2% this is confirmed with *Yohan Lee and Dongsoo Jung*, [8]. The COP of R1234ze and R152a is higher than that of R134a by 10.8% and 5.6% respectively.

The effect of evaporating temperature on the coefficient of performance of R134a, R152a, R124yf and R1234ze is illustrated if Fig. (12). The evaporating temperature affects the COP positively. It is evident that the coefficient of performance of all investigated refrigerants increases when the evaporating temperature increases. This is due to the increase in cooling effect and the decrease in compressor power. For all evaporating temperature, the COP of R1234ze is the highest among the other refrigerants. It confirmed that the system performance of R1234yf is the most closely to the performance of R134a system in which it is more environmentally sustainable refrigerant for automobiles which has a 99.7% better GWP score than R134a.

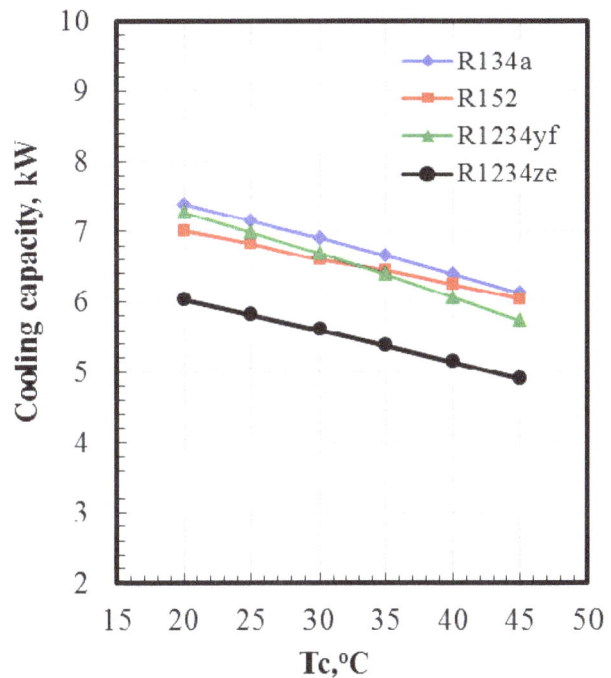

Fig. (9). *Cooling capacity versus condensing temperature at T$_e$=10°C*

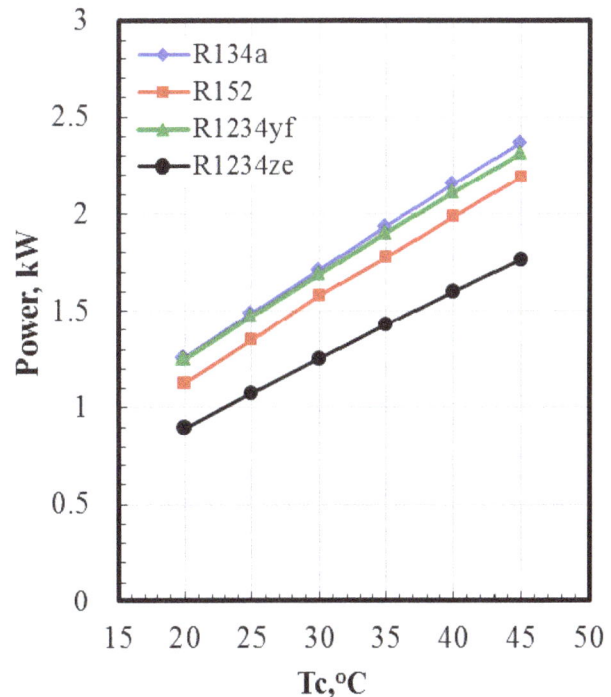

Fig. (10). *Compressor power versus condensing temperature at T$_e$= 10°C.*

6.5. Effect of Refrigerant Mass Flow Rate

The performance of the low GWP refrigerants of R152a, R1234yf and R1234ze are stated in this section in which the representations of cooling capacity, compressor power, pressure ratio, and COP with the refrigerant volume flow rate are shown in Figs. (13 to 16) for a typical evaporation and condensing temperatures of 10°C and 35°C respectively.

When the change in car speed is occurred which is due to change of fuel consumption, this led to a change in compressor RPM and hence variation in refrigerant flow rate is achieved. As the refrigerant flow rate increase, the cooling capacity, compressor power and pressure ratio increase also. The pressure ratio of R134a is higher that of R152a, R1234yf, R1234ze by 2%, 7.3%, 4.4% respectively. It can be seen that the R152a has a higher COP and lower values of pressure ratio and compressor input power.

It is evident that the coefficient of performance of all investigated refrigerants decreases when the refrigerant volume flow rate increases. At all values of refrigerant flow rate, the highest coefficient of performance is obtained for R1234ze among all investigated refrigerants. From the environmental and thermal performance point of view, the refrigerant R1234yf has the best thermal performance among all investigated refrigerants and the automakers would not have to make significant modifications in production lines or in automotive system designs to accommodate this refrigerant.

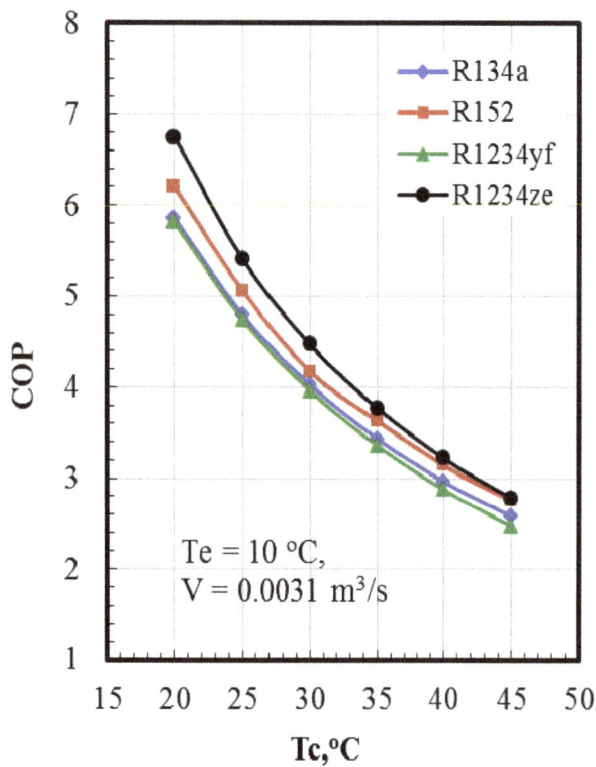

Fig. (12). COP versus evaporating temperature for Tc = 35ºC and V = 0.0031 m³/s.

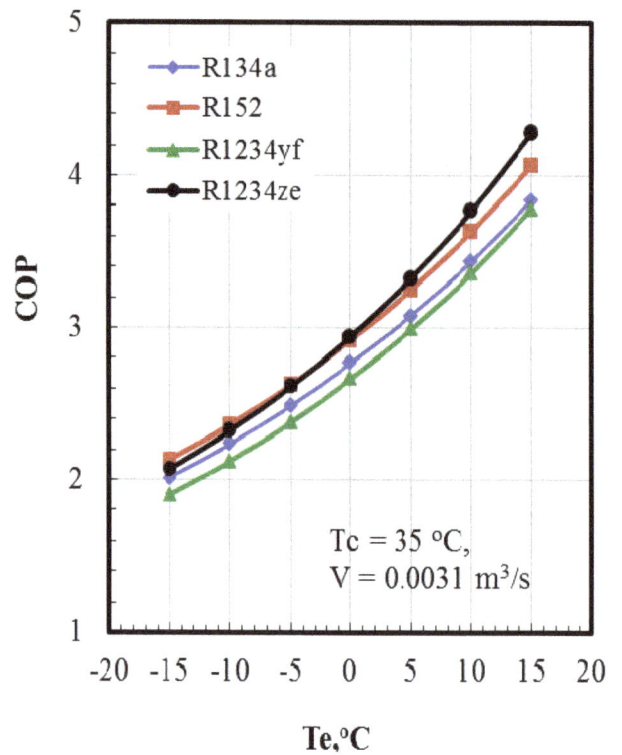

Fig. (11). COP versus condensing temperature for Te = 10ºC and V = 0.0031 m³/s.

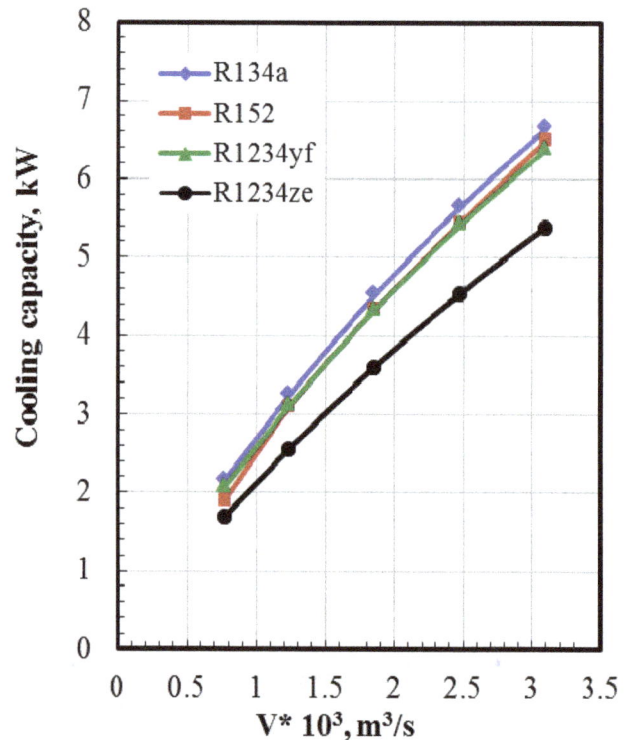

Fig. (13). Cooling capacity versus refrigerant flow rate at Tc = 35ºC and Te = 10ºC.

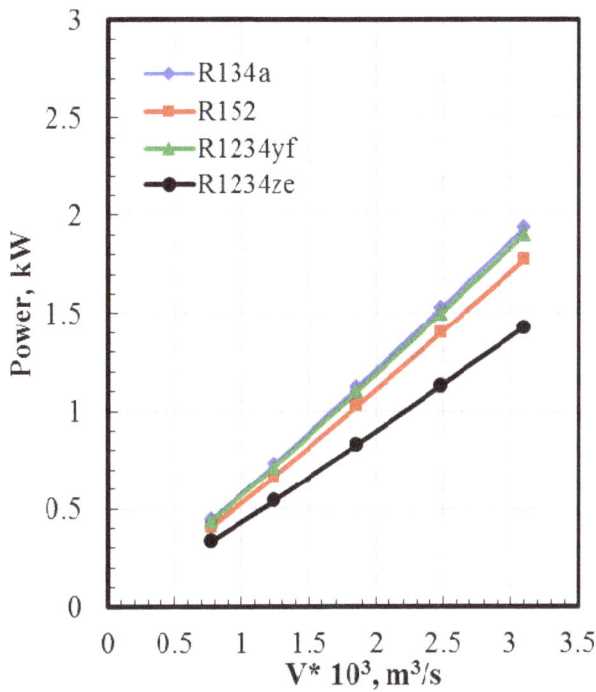

Fig. (14). Compressor power versus refrigerant flow rate at $T_c = 35^oC$, $T_e = 10^oC$.

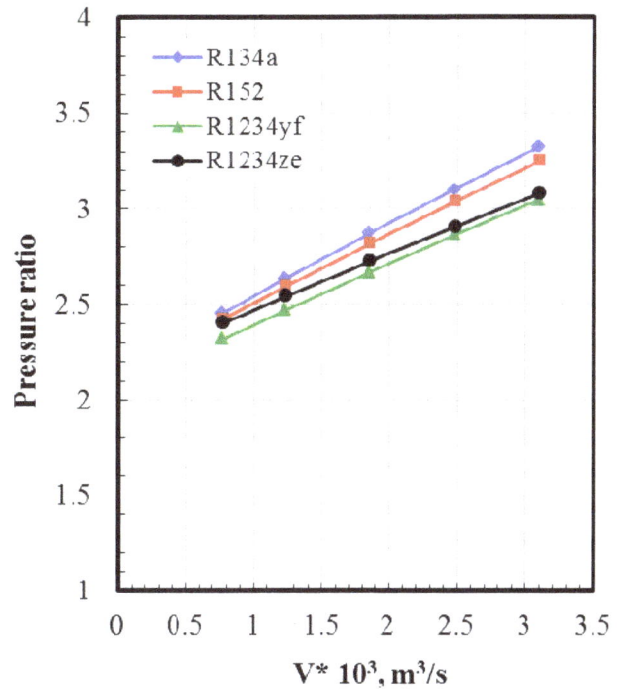

Fig. (16). Pressure ratio versus refrigerant flow rate at $T_c = 35^oC$ and $T_e = 10^oC$.

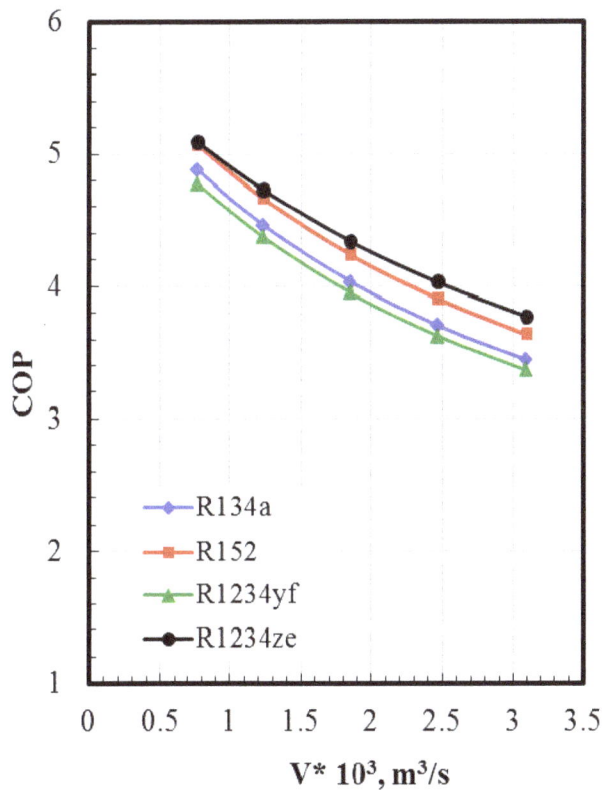

Fig. (15). COP versus refrigerant flow rate at Tc = 35°C and Te = 10°C.

7. Conclusion

The thermal performance of R134a automotive air conditioning is carried out experimentally and theoretically. The performance of R134a alternatives (R152a, R1234yf, and R1234ze) which are characterized by low GWP of less than 150 is presented. The study is assessed over wider range of compressor speed, condensing temperature and evaporating temperature. This investigation is done with standard parameters such as cooling capacity, compressor power, coefficient of performance, pressure ratio and condenser heat load and the main conclusion are;

- The increase in compressor speed (RPM) produces a lower values of COP for all values of condensing temperature.
- When the condensing temperature increased by 5°C, the cooling capacity decreased by 9%, while the COP decreased by 27%.
- For all values of condensing and evaporating temperature, the highest coefficient of performance is obtained for R1234ze among all investigated refrigerants.
- The performance of refrigerant R1234yf is the most similar to refrigerant R134a in all parameters.
- From the environmental and thermal performance point of view, the refrigerant R1234yf has the best thermal performance among all investigated refrigerants.
- For all investigated refrigerant, the increase of condensing temperature led to increase in compressor consumption power furthermore increase in condenser load at the same refrigerant flow rate.

Nomenclature

COP	coefficient of performance	(-)
h	specific enthalpy	(kJ/kg)
$\dot{m}$	mass flow rate	(kg/s)
W	Compressor power	(kW)
RPM	Revolution per minute	(min^{-1})
P	pressure	(N/m^2)
PR	pressure ratio	(-)
Q	heat transfer rate	(kW)
η_{is}	Isentropic efficiency	(-)

Subscripts

eva	Evaporator
com	Compressor
con	Condenser
ref	Refrigerant

References

[1]   J.M. Saiz Jabardo, W. Gonzales Mamani, M.R. Ianella, (2002), Modeling and experimental evaluation of an automotive air conditioning system with a variable capacity compressor, International Journal of Refrigeration, 25 (2002) 1157–1172

[2]   Alpaslan Alkan, and Murat Hosoz (2010), Comparative performance of an automotive air conditioning system using fixed and variable capacity compressors. International Journal of Refrigeration, 33 (2010) 487–495.

[3]   Jitendra Kumar Verma, Ankit Satsangi, Vishal Chaturani, (2013) A Review of Alternative to R134a (CH3CH2F) Refrigerant, International Journal of Emerging Technology and Advanced Engineering, 3, 1, (2013), 300-304

[4]   E. Navarro, I.O. Martınez-Galvan, J. Nohales, J. Gonzalvez-Macia (2013), Comparative experimental study of an open piston compressor working with R-1234yf, R-134a and R-290, International Journal of Refrigeration, 36 (2013) 768–775

[5]   J. Navarro-Esbrı, J.M. Mendoza-Miranda , A. Mota-Babiloni , A. Barraga-Cervera , J.M. Belman-Flores, (2013), Experimental analysis of R1234yf as a drop-in replacement for R134a in a vapor compression system, International Journal of Refrigeration, 36 (2013) 870-880

[6]   Ghodbane, M., 1999. An investigation of R152a and hydrocarbon refrigerants in mobile air conditioning. SAE, technical paper, 1999-01-0874.

[7]   Claudio Zilio, Steven Brown, Giovanni Schiochet, Alberto Cavallini, (2011). The refrigerant R1234yf in air conditioning systems. Journal of Energy, 36, (10), (2011), 6110–6120

[8]   Yohan Lee and a Dongsoo Jung, (2012), A brief performance comparison of R1234yf and R134a in a bench tester for automobile applications. Applied Thermal Engineering, 35, (2012), 240–242

[9]   Gustavo Pottker, and Pega Hrnjak, (2015), Experimental investigation of the effect of condenser subcooling in R134a and R1234yf air conditioning systems with and without internal heat exchanger. International Journal of Refrigeration, 50 (2015) 104–113

[10] Holman JP. 2001. Experimental Method for Engineers. Seventh editions, McGraw-Hill Book Company, New York.

[11] Bolaji BO. Experimental study of R152a and R32 to replace R134a in a domestic refrigerator. Journal of Energy, 35, 2010, 3793-3798.

[12] M. Fatouh, A. Ibrahem Eid, F. Nabil 2009. Performance of water- to - water vapor compression refrigeration system using R22 alternatives, part I: system simulation, Engineering Research Journal 124 (December 2009) M19- M43

[13] EES, 2013 Engineering Equation Solver. F-Chart Software, Middleton, WI, USA (2013)

[14] Honeywell Sells Novel Low-Global-Warming Blowing Agent to European Customers, Honeywell press release, Oct. 7, 2008.

18

Thermal Analysis for an Ultra High Temperature Gas-Cooled Reactor with Pebble Type Fuels

Motoo Fumizawa, Naoya Uchiyama, Takahiro Nakayama

Department of Mechanical Engineering, Shonan Institute of Technology, Fujisawa, Kanagawa, Japan

Email Address:
fumizawa@mech.shonan-it.ac.jp (M. Fumizawa), psn-man.lv.030@softbank.ne.jp (N. Uchiyama), 2991taka@gmail.com (T. Nakayama)

Abstract: This study presents a predictive thermal-hydraulic analysis with packed spheres in a nuclear gas-cooled reactor core. The predictive analysis considering the effects of high power density and the some porosity value were applied as a design condition for an Ultra High Temperature Reactor (UHTR). The thermal-hydraulic computer code was developed and identified as PEBTEMP. The highest outlet coolant temperature of 1316 °C was achieved in the case of an UHTREX at LASL, which was a small scale UHTR using hollow-rod as a fuel element. In the present study, the fuel was changed to a pebble type, a porous media. Several calculation based on HTGR-GT300 through GT600 were 4.8 w/cm^3 through 9.6 w/cm^3, respectively. As a result, the relation between the fuel temperature and the power density was obtained under the different system pressure and coolant outlet temperature. Finally, available design conditions are selected.

Keywords: Thermal Hydraulics, Ultra High Temperature Reactor (UHTR), Pressure Drop, Porosity and Pebble Type Fuel

1. Introduction

Very high temperature gas-cooled reactor project is energetically developing the design study to establish 1,000 °C as a coolant outlet temperature and to realize the hydrogen production [1-2], where GIF is the Generation IV International Forum. For a long time, a fundamental design study has been carried out in the field of the high temperature gas-cooled reactor i.e. HTGR [3-8], which showed that a coolant outlet temperature was around 900 °C. The interest of HTGR is increasing in many countries as a promising energy future option. There are currently two research reactors of THGR type that are being operated in Japan and China. The inherent safety of HTGR is due to the large heat capacity and negative temperature reactivity coefficient. The high temperature heat supply can achieve more effective utilization of nuclear energy. For example, high temperature heat supply can provide for hydrogen production, which is expected as an alternative energy source for oil. Also, outstanding thermal efficiency will be achieved at about 900 °C with a Brayton-cycle gas turbine plant.

However, the highest outlet coolant temperature of 1316 °C had been achieved by UHTREX as shown in Figure 1, in Los Alamos Scientific Laboratory at the end of 1960's [3-4]. It was a small scale Ultra High Temperature Nuclear Reactor

(UHTR). The coolant outlet temperature would be higher than 1000 °C in the UHTR. The UHTREX adopted the hollow rod type fuel; the highest fuel temperature was 1,582 °C, which indicated that the value was over the current design limit. According to the handy calculation, it was derived that the pebble type fuel was superior to the hollow type in the field of fuel surface heat transfer condition [9].In the present study, the fuels have changed to the pebble type so called the porous media. In order to compare the present pebble bed reactor and UHTREX, a calculation based on HTGR-GT300 was carried out in the similar conditions with UHTREX i.e. the inlet coolant temperature of 871°C, system pressure of 3.45 MPa and power density of 1.3 w/cm^3. The main advantage of the pebble bed reactor (PBR) is that high outlet coolant temperature can be achieved due to its large cooling surface and high heat transfer coefficient that have the possibility to get high thermal efficiency. Besides, the fuel loading and discharging procedures are simplified; the PBR system makes it possible that the frequent load and discharge are easier than the other reactor system loaded block type fuel without reactor shutdown. This report presents thermal-hydraulic calculated results for a concept design PBR system of 300MWth of the modular HTGR-GT300 with the pebble types of fuel element as shown in Figure 2. A calculation for comparison with UHTREX have been carried out and presented as well.

2. Reactor Description

2.1. Concept of Modular HTGR-GT300, GT600 and GT600

A concept of pebble-bed type HTGR are shown in Figures 2 and 3 with the main nuclear and thermal-hydraulic specifications presented in Table 1. In the case that the thermal power is 300MW (GT-300), the average power density changes to 4.8 MW/m³. The coolant gas enters from the outer shell of the primary coolant coaxial tube to the pressure vessel at temperature of 550°C and pressure of 6 MPa, follows the peripheral region of side reflectors up to the top and goes downward through the reactor active core. The outlet coolant goes out through the inner shell of primary coolant tube at temperature of 900°C. The cylindrical core is formed by the blocks of graphite reflector with the height of 9.4m and the diameter of 2.91m. There exist holes in the reflector that some of them used for control rod channels and the others used for boron ball insertion in case of an accident. In the case that the thermal powers are 450MW (GT450) and 600MW (GT600), the average power densities change to 7.2 MW/m³ and 9.6 MW/m³, respectively.

2.2. Fuel Element

The two types of pebble fuel elements, consisting of fuel and moderator, are shown in Figure 4. One is a solid type where radius of inner graphite r_{co}=0, and the other is a shell type fuel element. The fuel compacts are a mixture of coated particles [9].

3. Thermlhydraulic Analysis

3.1. PEPTEMP Code

A one-dimensional thermal-hydraulic computer code was developed that was named PEPTEMP [5] as shown in Figure 5. The code solves for the temperature of fuel element, coolant gas and core pressure drop using assumed power, power distribution, inlet and outlet temperature, the system pressure, fuel size and fuel type as input data.

The options for fuel type are of the pebble type; the multi holes block type and the pin-in-block type. The power distribution for cases of cosine and exponential is available., The users can calculate for the other distributions by preparing the input file.

The maximum fuel temperature will be calculated in PEPTEMP as follows:

$$T_{max}(z) = T_{in} + \Delta T_{cl}(z) + \Delta T_{film}(z) + \Delta T_{sl}(z) + \Delta T_{com} \quad (1)$$

where $T_{max}(z)$: fuel temperature at the center of fuel element i.e. the maximum fuel temperature; ΔT_{cl}: gas temperature increment from inlet to height z; T_{in}: gas inlet temperature; $\Delta T_{film}(z)$: temperature difference between fuel element surface and coolant gas at z; $\Delta T_{sl}(z)$: temperature difference between

fuel matrix outer surface and fuel element surface; $\Delta T_{com}(z)$: temperature difference between fuel matrix outer surface and fuel center; q''': power density; A_f: fuel element surface area; z: axial distance from the top of the core; C_p: coolant heat capacity.

3.2. Temperature Difference in the Spherical Fuel Element

Figure 4 shows fuel configuration of the solid type and the shell type fuel element. In the solid type, ΔT_{com} is given as follows

$$\Delta T_{com}(z) = T_{co} - T_c = \frac{q'''(z)r_c^2}{6\lambda_c} \quad (2)$$

In the case of the shell type fuel element, ΔT_{com} can be calculated by the following expression;

$$\Delta T_{com}(z) = T_{co} - T_c = \frac{q'''(z)}{6\lambda_c}\left(r_c^2 - 3r_{co}^2 + \frac{2r_{co}^3}{r_c}\right) \quad (3)$$

3.3. Film Temperature Difference

The film temperature differences are calculated as follows;

$$\Delta T_{film} = T_s - T_{ch} = \frac{q'''(z)r_c^3}{3r_s^2 h} \quad (4)$$

3.4. Heat Transfer Coefficient

Heat transfer coefficient h in Equation (4) is calculated using the following correlation [10]:

$$h = 0.68\rho v_s C_p \, \text{Re}^{-0.3} \, \text{Pr}^{-0.66} \quad (5)$$

$$\text{Re} = \frac{\rho v_s d}{(1-\varepsilon)\mu} \quad (6)$$

where, ρ: coolant density; vs.: coolant velocity; Re: Reynolds number; Pr: Prandtl number; ε: Porosity; d: fuel element diameter and μ: viscosity of fluid.

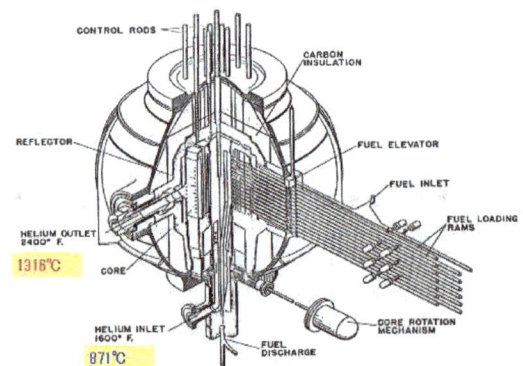

Fig. 1. Reactor structure of UHTREX, quoted from reference [3].

Table 1. *Major nuclear and thermal-hydraulic specification.*

Thermal power (MW)	300 / 450 / 600
Coolant	Helium
Inlet coolant temperature (oC)	550
Outlet coolant temperature (oC)	900 (900 – 1650 oC)
Coolant Pressure (MPa)	6.0 (1 – 15 MPa)
Total coolant flow rate (kg/s)	172.1 / 258.2 / 344.2
Core coolant flow rate (kg/s)	165.2 / 247.8 / 320.8
Core diameter (m)	2.91
Core height (m)	9.4
Core fuel porosity ($-$)	0.39 (0.26 – 0.50)
Average power density (MW/m3)	4.8 / 7.2 / 9.6
Fuel type (for standard case)	6 cm diameter pebble

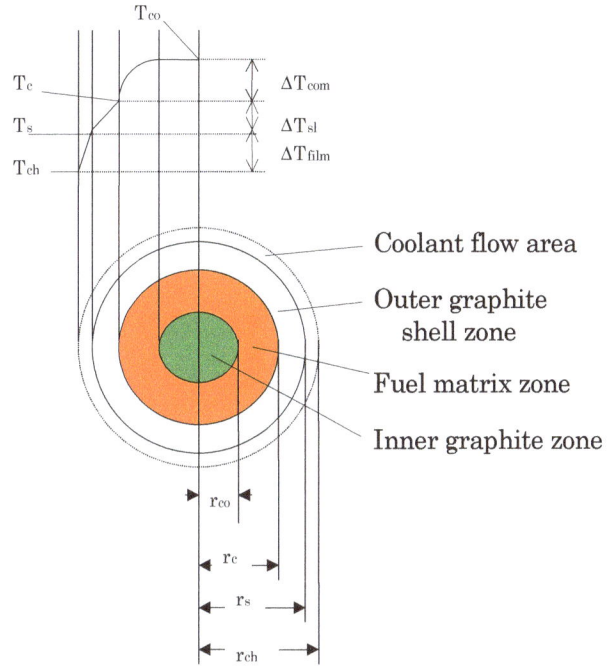

Fig. 4. *Relation of shell type fuel element and temperature difference, in the case that no inner graphite zone is called the solid type, i.e., $r_{co}=0$.*

The concept of pebble bed type HTGR

Parameter study
considered as follows:
1. Thermal power
2. Porosity
3. System pressure
4. Outlet coolant temperature
5. Solid , shell type fuel

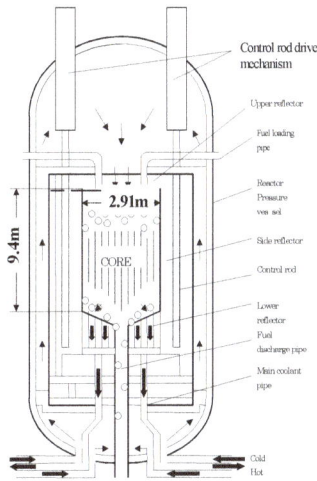

Fig. 2. *A concept of pebble bed reactor of HTGR –GT300.*

Figure   Core arrangement plane view

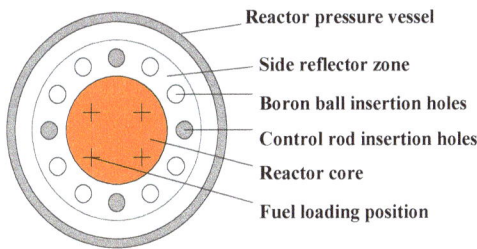

- Reactor pressure vessel
- Side reflector zone
- Boron ball insertion holes
- Control rod insertion holes
- Reactor core
- Fuel loading position

Four percent Coolant flows Control rod insertion holes.
W_{eff}: effective coolant flow rate that has dimensionless value due to the normalization by the total coolant flow rate. Maximum W_{eff} is 0.96.

Fig. 3. *Core arrangement plane view.*

Analysis method

Fuel Temperature analysis code for High Temperature gas-cooled reactor
PEBTEMP

Sphere fuel
Unit cell model
(porosity=0.39)

Thermal conduction Eq.

The option of thermal power distribution are as follows:
(1) Cosine
(2) exponential
i.e. same fuel center temperature in axial

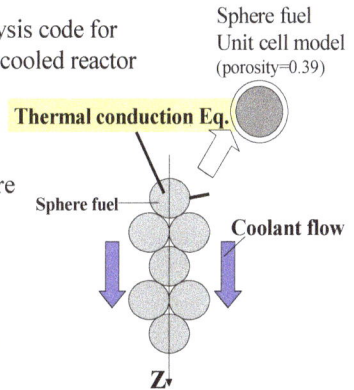

Fig. 5. *Analysis method of thermal-hydraulic computer code PEPTEMP.*

Parameter	Fuel element type		
	Hollow-rod	Multi-hole	Pebble-bed
Heat transfer area of fuel (m²)	777	1073	3814
Heat transfer coefficient (W/m·K)	2101	2801	4037

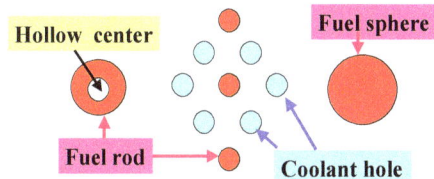

Fig. 6. *Heat transfer performances in the small size reactor design by handy evaluation.*

Table 2. Analysis cases and fuel maximum temperature in the 300MW of thermal power i.e.GT300.

Case	P(MPa)	Tout	ε=0.26	ε=0.39	ε=0.40	ε=0.50
A01	1	900	983.7	1015.6	1019	1060.3
A02	1	1150	1209	1236.4	1239	1274.3
A03	1	1400	1446.7	1468.9	1471	1502
A04	1	1650	1688.9	1707.4	1709.6	1736.6
A05	5	900	983.7	1015.6	1019	1060.2
A06	5	1150	1209	1236.4	1239	1274.2
A07	5	1400	1446.7	1468.8	1471	1501.9
A08	5	1650	1688.8	1707.4	1709.6	1736.6
A09	6	900	983.7	1015.6	1019	1053.8
A10	6	1150	1209	1236.3	1239	1274.2
A11	6	1400	1446.7	1468.8	1471	1501.9
A12	6	1650	1688.8	1707.4	1709.6	1736.6
A13	10	900	983.7	1015.6	1019	1060.2
A14	10	1150	1209	1236.3	1238.9	1274.2
A15	10	1400	1446.7	1468.8	1470.9	1501.9
A16	10	1650	1688.8	1707.4	1709.6	1736.6
A17	15	900	983.7	1015.6	1019	1060.1
A18	15	1150	1208.9	1236.3	1238.9	1274.2
A19	15	1400	1446.7	1468.8	1470.9	1501.9
A20	15	1650	1688.8	1707.4	1709.6	1736.6

3.5. Pressure Drop

Pressure drop through the core expresses by the following correlation [6]:

$$\Delta P = 6.986 \frac{(1-\varepsilon)^{n+1}}{\varepsilon^3} \text{Re}_p^{-n} \rho v_s^2 \frac{H}{d} K + \Delta P_a \quad (7)$$

$$n=0.22 \quad (8)$$

$$K = 1 - 0.26(1+n+3\frac{1-\varepsilon}{\varepsilon})\frac{d}{R} \quad (9)$$

$$\text{Re}_p = \frac{\rho v_s d}{\eta} \quad (10)$$

where, H: core height; R: core radius and ΔP_a: acceleration pressure drop.

3.6. Effective Flow Rate Consideration

As many blocks of graphite form the reflector, there exist gaps by which the coolant flow may pass through [11]. Actually, only one portion of coolant passes through the reactor core from the top to the bottom. This portion is called effective flow rate and can be calculated iteratively in the code. The empirical equation used in this code is as follows [11]:

$$W_{eff} = 0.98 - 0.012\Delta P \quad (11)$$

where, W_{eff}: effective coolant flow rate that has dimensionless value due to the normalization by the total coolant flow rate.

ΔP: pressure drop through the core

4. Calculation Results

4.1. Handy Calculation Results for Small Scale HTGR

Before the main calculation, we have done the prediction study of the comparison of key factors of heat transfer in the small scale HTGR with three types of different fuel elements. They are the hollow-rod [3], the multi-hole [1,8] and the pebble-bed [10]. The small reactor thermal data are as follows; thermal power of 50MW, power density of 2.5 MW/m^3 and inlet/ outlet coolant temperature of 395 °C/ 850 °C, respectively. Figure 6 shows the results of heat transfer area in the core and heat transfer coefficient on the fuel surface. Heat transfer area of the pebble-bed is 5 times larger than that of the hollow-rod. Heat transfer coefficient of the pebble-bed is twice larger than that of the hollow-rod. Therefore heat transfer performance of pebble-bed is superior to other types of fuel elements.

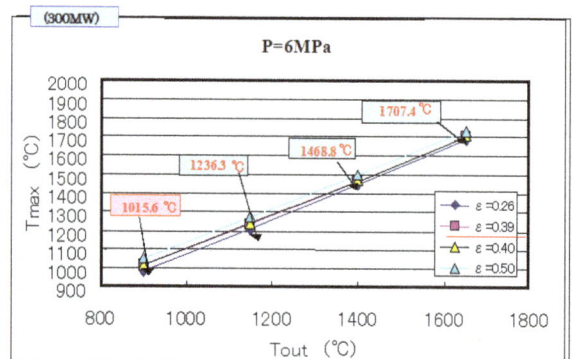

Fig. 7. Dependence of maximum fuel temperature on outlet coolant temperature for GT300 with different porosity and W_{eff}=1.

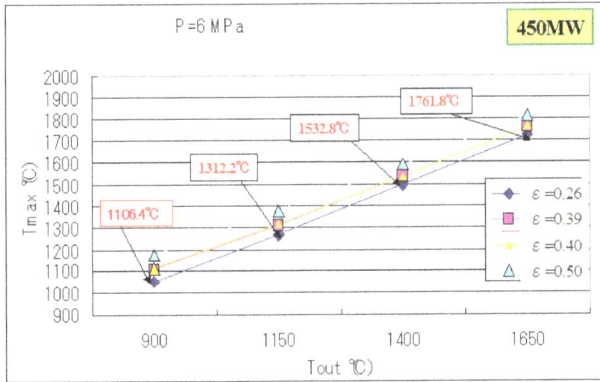

Fig. 8. *Dependence of maximum fuel temperature on outlet coolant temperature for GT450 with different porosity and W_{eff}=1.*

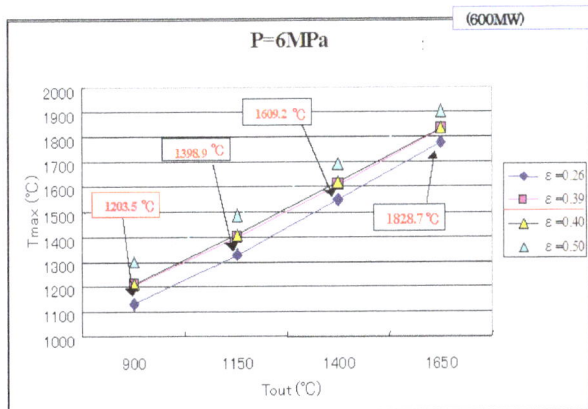

Fig. 9. *Dependence of maximum fuel temperature on outlet coolant temperature for GT600 with different porosity and W_{eff}=1.*

4.2. Temperature Calculation for HTGR-GT300 to GT600

Table 2 shows the 20 analysis cases and fuel maximum temperature in the 300MW of thermal power i.e.GT300. The system pressure ranges from 1 MPa through 15 MPa. The system pressure does not have any effect on the fuel maximum temperature. Thus we focus our intension to 6 MPa of system pressure [1]. Figure 7, 8 and 9 show the dependence of maximum fuel temperature on outlet coolant temperature for GT300, GT450 and GT-600 with different porosity and W_{eff}=1. The maximum fuel temperature for GT600 is 168 °C higher than that for GT300 where the outlet coolant temperature is 900 °C and the porosity is 0.39. The maximum fuel temperature for GT-600 is 163 °C higher than that for GT-300, where the outlet coolant temperature is 1150 °C and the porosity is 0.39. The high porosity leads to low fuel maximum temperature.

4.3. Pressure Drop Calculation for HTGR-GT300 to GT600

Table 3 shows the 20 analysis cases and pressure drop (ΔP) in the core of 300MW of thermal power. The system pressure ranges from 1 MPa through 15 MPa. The high system pressure

leads to low-pressure drop in the core. In the case of 6 MPa of system pressure, the ΔP changes from 40.2 kPa to 16.7 kPa, where the T_{out} increases from 900 °C to 1150 °C. The ΔP changes from 16.7 kPa to 6.3 kPa, where the porosity increases from 0.39 to 0.50 with 1150 °C of T_{out}. The high porosity leads to low-pressure drop. In the case of 15 MPa of system pressure, the ΔP changes from 16.7 kPa to 6.7 kPa, where the T_{out} increases from 900 °C to 1150 °C. The ΔP changes from 6.7 kPa to 2.6 kPa, where the porosity increases from 0.39 to 0.50 with 1150 °C of T_{out}.

Table 4 shows the 20 analysis cases and pressure drop (ΔP) in the core of 450MW of thermal power. The system pressure ranges from 1 MPa through 15 MPa. The high system pressure leads to low-pressure drop in the core. In the case of 6 MPa of system pressure, the ΔP changes from 86.2 kPa to 35.3 kPa, where the T_{out} increases from 900 °C to 1150 °C. The ΔP changes from 35.3 kPa to 13.7 kPa, where the porosity increases from 0.39 to 0.50 with 1150 °C of T_{out}. The high porosity leads to low-pressure drop. In the case of 15 MPa of system pressure, the ΔP changes from 35.3 kPa to 14.7 kPa, where the T_{out} increases from 900 °C to 1150 °C. The ΔP changes from 14.7 kPa to 5.5 kPa, where the porosity increases from 0.39 to 0.50 with 1150 °C of T_{out}.

Table 5 shows the 20 analysis cases and pressure drop (ΔP) in the core of 600MW of thermal power. The system pressure ranges from 1 MPa through 15 MPa. The high system pressure leads to low-pressure drop in the core. In the case of 6 MPa of system pressure, the ΔP changes from 147 kPa to 60.8 kPa, where the T_{out} increases from 900 °C to 1150 °C. The ΔP changes from 60.8 kPa to 23.5 kPa, where the porosity increases from 0.39 to 0.50 with 1150 °C of T_{out}. The high porosity leads to low-pressure drop. In the case of 15 MPa of system pressure, the ΔP changes from 59.8 kPa to 24.5 kPa, where the T_{out} increases from 900 °C to 1150 °C. The ΔP changes from 124.5 kPa to 9.4 kPa, where the porosity increases from 0.39 to 0.50 with 1150 °C of T_{out}.

Fig. 10. *The procedures to evaluate the pressure drop considering the available effective coolant flow rate.*

Table 3. Analysis cases and pressure drop in the GT300.

Case	P(MPa)	Tout	ε =0.26	ε =0.39	ε =0.40	ε =0.50
A01	1	900	9.80E+02	2.35E+02	2.16E+02	9.11E+01
A02	1	1150	4.12E+02	9.80E+01	8.92E+01	3.72E+01
A03	1	1400	2.35E+02	5.78E+01	5.19E+01	2.16E+01
A04	1	1650	1.67E+02	3.92E+01	3.63E+01	1.47E+01
A05	5	900	1.96E+02	4.80E+01	4.41E+01	1.86E+01
A06	5	1150	8.23E+01	1.96E+01	1.76E+01	7.55E+00
A07	5	1400	4.80E+01	1.18E+01	1.08E+01	4.41E+00
A08	5	1650	2.74E+01	7.94E+00	7.15E+00	3.04E+00
A09	6	900	1.67E+02	4.02E+01	3.63E+01	1.57E+01
A10	6	1150	6.86E+01	1.67E+01	1.47E+01	6.27E+00
A11	6	1400	4.02E+01	9.60E+00	8.72E+00	3.72E+00
A12	6	1650	2.74E+01	6.57E+00	5.98E+00	2.55E+00
A13	10	900	9.80E+01	2.45E+01	2.16E+01	9.21E+00
A14	10	1150	4.12E+01	9.80E+00	9.02E+00	3.82E+00
A15	10	1400	2.45E+01	5.78E+00	5.29E+00	2.25E+00
A16	10	1650	1.67E+01	3.92E+00	3.63E+00	1.57E+00
A17	15	900	6.76E+01	1.67E+01	1.47E+01	6.17E+00
A18	15	1150	2.74E+01	6.66E+00	6.08E+00	2.55E+00
A19	15	1400	1.57E+01	3.82E+00	3.53E+00	1.47E+00
A20	15	1650	1.08E+01	2.65E+00	2.45E+00	9.80E-01

Table 4. Analysis cases and pressure drop in the GT450.

Case	P(MPa)	Tout	ε =0.26	ε =0.39	ε =0.40	ε =0.50
A01	1	900	2.16E+03	5.19E+02	4.70E+02	1.96E+02
A02	1	1150	8.82E+02	2.16E+02	1.96E+02	8.13E+01
A03	1	1400	5.10E+02	1.27E+02	1.08E+02	4.70E+01
A04	1	1650	3.53E+02	8.43E+01	7.64E+01	3.23E+01
A05	5	900	4.31E+02	1.08E+02	9.41E+01	3.92E+01
A06	5	1150	1.76E+02	6.32E+03	3.82E+01	1.67E+01
A07	5	1400	1.08E+02	2.45E+01	2.25E+01	9.51E+00
A08	5	1650	5.88E+01	1.67E+01	1.57E+01	6.47E+00
A09	6	900	3.63E+02	8.62E+01	7.84E+01	3.33E+01
A10	6	1150	1.47E+02	3.53E+01	3.23E+01	1.37E+01
A11	6	1400	8.62E+01	2.06E+01	1.86E+01	7.94E+00
A12	6	1650	5.88E+01	1.37E+01	1.27E+01	5.39E+00
A13	10	900	2.16E+02	5.19E+01	4.70E+01	1.96E+01
A14	10	1150	8.92E+01	2.16E+01	1.96E+01	8.13E+00
A15	10	1400	5.19E+01	1.27E+01	1.18E+01	4.70E+00
A16	10	1650	3.53E+01	8.53E+00	7.74E+00	3.23E+00
A17	15	900	1.47E+02	3.53E+01	3.14E+01	1.37E+01
A18	15	1150	5.98E+01	1.47E+01	1.27E+01	5.49E+00
A19	15	1400	3.43E+01	8.33E+00	7.55E+00	3.14E+00
A20	15	1650	2.35E+01	5.68E+00	5.19E+00	2.16E+00

Table 5. Analysis cases and pressure drop in the GT600.

Case	P(MPa)	Tout	ε =0.26	ε =0.39	ε =0.40	ε =0.50
A01	1	900	3.72E+03	8.92E+02	8.13E+02	3.43E+02
A02	1	1150	1.57E+03	3.63E+02	3.33E+02	1.37E+02
A03	1	1400	8.82E+02	2.16E+02	1.96E+02	8.13E+01
A04	1	1650	6.08E+02	1.47E+02	1.27E+02	5.59E+01
A05	5	900	7.45E+02	1.76E+02	1.67E+02	6.86E+01
A06	5	1150	3.04E+02	7.35E+01	6.66E+01	2.84E+01
A07	5	1400	1.76E+02	4.21E+01	3.92E+01	1.67E+01
A08	5	1650	9.80E+01	2.94E+01	2.65E+01	1.08E+01
A09	6	900	6.17E+02	1.47E+02	1.37E+02	5.68E+01
A10	6	1150	2.55E+02	6.08E+01	5.59E+01	2.35E+01
A11	6	1400	1.47E+02	3.53E+01	3.23E+01	1.37E+01
A12	6	1650	9.80E+01	2.45E+01	2.25E+01	9.31E+00
A13	10	900	3.72E+02	9.02E+01	8.13E+01	3.43E+01
A14	10	1150	1.57E+02	3.72E+01	3.33E+01	1.37E+01
A15	10	1400	8.92E+01	2.16E+01	1.96E+01	8.13E+00
A16	10	1650	6.08E+01	1.47E+01	1.37E+01	5.59E+00
A17	15	900	2.55E+02	5.98E+01	5.49E+01	2.25E+01
A18	15	1150	9.80E+01	2.45E+01	2.25E+01	9.41E+00
A19	15	1400	5.98E+01	1.47E+01	1.27E+01	5.49E+00
A20	15	1650	4.12E+01	9.80E+00	8.92E+00	3.72E+00

Table 6. The available analysis cases in the GT300 considering the design limits.

Case	P(MPa)	Tout	ε=0.26	ε=0.39	ε=0.40	ε=0.50
A01	1	900	983.7	1016	1019	1060
A02	1	1150	1209	1236	1239	1274
A05	5	900	983.7	1015.6	1019	1060.2
A06	5	1150	1209	1236.4	1239	1274.2
A09	6	900	983.7	1016	1019	1054
A10	6	1150	1209	1236	1239	1274
A13	10	900	983.7	1015.6	1019	1060.2
A14	10	1150	1209	1236.3	1238.9	1274.2
A17	15	900	983.7	1016	1019	1060
A18	15	1150	1208.9	1236	1239	1274

Case	P(MPa)	Tout	ε=0.26	ε=0.39	ε=0.40	ε=0.50
A03	1	1400				2.16E+01
A04	1	1650				1.47E+01
A05	5	900				1.86E+01
A06	5	1150		1.96E+01	1.76E+01	7.55E+00
A09	6	900				1.57E+01
A10	6	1150		1.67E+01	1.47E+01	6.27E+00
A13	10	900			2.16E+01	9.21E+00
A14	10	1150		9.80E+00	9.02E+00	3.82E+00
A17	15	900		1.67E+01	1.47E+01	6.17E+00
A18	15	1150		6.66E+00	6.08E+00	2.55E+00

Design Limit of maximum fuel temperature(℃) and pressure drop (kPa) in 300MW thermal power

Design Limit

Tmax ≦ 1300℃

ΔPmax ≦23.3kPa

Table 7. The available analysis cases in the GT600 considering the design limits.

Case	P(MPa)	Tout	ε=0.26	ε=0.39	ε=0.40	ε=0.50
A01	1	900	1130.3	1203.5	1210.5	1297.9
A05	5	900	1130.3	1203.5	1210.5	1297.9
A09	6	900	1130.3	1203.5	1210.4	1297.9
A13	10	900	1130.3	1203.4	1210.4	1297.8
A17	15	900	1130.3	1203.4	1210.4	1297.8

Case	P(MPa)	Tout	ε=0.26	ε=0.39	ε=0.40	ε=0.50
A07	5	1400				1.67E+01
A08	5	1650				1.08E+01
A10	6	1150				2.35E+01
A11	6	1400				1.37E+01
A12	6	1650			2.25E+01	9.31E+00
A14	10	1150				1.37E+01
A15	10	1400		2.16E+01	1.96E+01	8.13E+00
A16	10	1650		1.47E+01	1.37E+01	5.59E+00
A17	15	900				2.25E+01
A18	15	1150			2.25E+01	9.41E+00
A19	15	1400		1.47E+01	1.27E+01	5.49E+00
A20	15	1650		9.80E+00	8.92E+00	3.72E+00

Design Limit of maximum fuel temperature(℃) and pressure drop (kPa) in 600MW thermal power

Design Limit:

Tmax ≦ 1300℃

ΔPmax ≦23.3kPa

4.4. Consideration of Available Reactor Core Design

The traditional design limits in Japan suggested that the maximum fuel temperature should be lower than 1300 °C. From the engineering judgments, the effective coolant flow rate should be higher than 70 %. Thus the design limit of the pressure drop in the core is 23.3 kPa, according to the eq. (11) as shown in Figure 10. Table 6 and 7 show the available analysis cases of GT300 and GT600. The available analysis cases in GT300 are case A05, A06, A09, A10, A13, A14, A17, and A18. It means that 1150 °C of the outlet coolant temperature is available. On the contrary, the available analysis case in GT600 is the case A17. It means that 900 °C of the outlet coolant temperature is available.

5. Conclusions

The followings can be concluded:
1. High porosity leads to low fuel maximum temperature.
2. High system pressure leads to low-pressure drop in the core.
3. High porosity leads to low-pressure drop.
4. The available analysis cases in 300MW of thermal power are 8 cases, which indicates that the outlet coolant temperature is lower than 1150 °C.
5. On the contrary, the available analysis case in 600MW of thermal power is only 1 case, which indicate that the outlet coolant temperature is up to 900 °C.

Nomenclature

A_f: fuel element surface area; (m^2)
C_p: coolant heat capacity; (J/kgK)
H: core height; (m)
h : heat transfer coefficient; (W/m^{2}K)
q''': power density; (W/m^3)
R: core radius; (m)
Re: Reynolds number
$T_f(z)$: fuel temperature at the center of fuel element, i.e., the maximum fuel temperature; (°C)
T_{in}: gas inlet temperature; (°C)
T_{out}: gas outlet temperature; (°C)
W_{eff}: effective coolant flow rate, dimensionless value due to the normalization
z: axial distance from the top of the core; (m)
ΔP: pressure drop through the core (kPa)
ΔP_a: acceleration pressure drop; ((kPa)
ΔT_{cl}: gas temperature increment from inlet to height z; (°C)
$\Delta T_{com}(z)$: temperature difference between fuel matrix outer surface and fuel center; (°C)

$\Delta T_{film}(z)$: temperature difference between fuel element surface and coolant gas at z; (°C)
$\Delta T_{sl}(z)$: temperature difference between fuel matrix outer surface and fuel element surface; (°C)

References

[1] A Technology Roadmap for Generation IV Nuclear Energy Systems, GIF-002-00, Generation IV International Forum (2002), http://gif.inel.gov/roadmap/

[2] Shusaku SHIOZAWA et al., "The HTTR Project as the World Leader of HTGR Research and Development", J. of AESJ Vol.47, pp. 342-349 (2005)

[3] B.M. HOGLUND,:"UHTREX Operation Near", Vol.9, pp. 1, Power Reactor Technology (1966)

[4] "UHTREX: Alive and Running with Coolant at 2400 oF", Nuclear News (1969)

[5] Progress Report – Pebble Bed Reactor Program, NYO-9071, US Atomic Energy Commission (1960)

[6] M.M.El-Wakil, Nuclear Energy Conversion, Thomas Y. Crowell Company Inc., USA (1982)

[7] M.M El-Wakil, Nuclear Heat Transport, International Textbook Company, USA (1971)

[8] Motoo Fumizawa et al., "Effective Coolant Flow Rate of Flange Type Fuel Element for Very High Temperature Gas-Cooled Reactor ", J. of AESJ Vol.31, pp 828-836 (1989)

[9] Motoo Fumizawa et al., "Preliminary Study for Analysis of Gas-cooled Reactor with Sphere Fuel Element ", AESJ Spring MTG, I66 (2000)

[10] Fumizawa,M.;Nuclear Reactors (ISBN 978-953-51-0967 -9), Edited by Amir Zacarias Mesquita, InTech, pp.177-191 (2013)

[11] Motoo Fumizawa et al., "The Conceptual Design of High Temperature Engineering Test Reactor Upgraded through Utilizing Pebble-in-block Fuel", JAERI-M 89-222 (1989)

19

Methods of Temperature Measurement in Magnetic Components

Abakar Mahamat Tahir[1, 2], Amir Moungache[1, 3], Mahamat Barka[1], Jean-Jacques Rousseau[3], Dominique Ligot[2], Pascal Bevilacqua[2]

[1]Faculté Des Sciences Exactes Et Appliquées de l'Université De Ndjaména, Département De Technologie, Ndjaména, Tchad
[2]Laboratoire Ampère Insa-Lyon 20, Avenue Al. Einstein, Villeurbanne Cedex, France
[3]Laboratoire LT2C (ex DIOM 2), Rue Du Dr Paul Michelon, Etienne Cedex, France

Email address:

abakarmt@gmail.com (M. T. Abakar), amirmab@yahoo.fr (A. Moungache), mahamat.barka@gmail.com (M. Barka),
Jean.Jacques.Rousseau@univ-st-etienne.fr (J. J. Rousseau), Dominique.ligot@insa-lyon.fr (D. Ligot),
Pascal.BEVILACQUA@insa-lyon.fr (P. Bevilacqua)

Abstract: In this paper, we reveal the different temperatures recorded in selected magnetic components and, we show the methods of determination and measurement used. As magnetic components have characteristics that strongly depend on the level of local temperature, it is essential to take into consideration the temperature and its influence on the magnetic and electrical characteristics of the component. From the modeling objectives, we describe the constitution, the originality and the main functions of the device used to determine the temperature of the material, its winding, and connection. Finally the result of deepened tests will be presented, allowing to determining parameters of our model. Our work focuses on the development of thermal models capable of determining the working temperature of the chosen magnetic component at given points. It aims to help develop a methodology for designing thermal models of magnetic components; an approach that will be validated through a practical demonstration.

Keywords: Thermal Model, Temperature, Magnetic Component, Thermocouple, Coil, Magnetic Material

1. Introduction

In order to determine the elements of the model, to reject certain parameters or validate the model, it is essential to carry out the measurement of the temperature in different parts of the magnetic component used in power electronics, that's to say (i.e.) the temperatures of the magnetic material, the temperature in the coiling, and the temperature of the connection. The model should also accurately measure the losses occurring in the magnetic material and the windings. To avoid damaging the magnetic component under test, we chose an indirect measure of average temperatures obtained from a bench test. Particular attention has also been paid to the extent of losses to minimize sources of error. In the first part, we present the principles of measurement of temperature used. In the second part, we describe the apparatus used. We specify its constitution, its originality, and its main functions, and finally we assess the accuracy of measurements.

2. Measurement of Temperature

2.1. Assumptions

We first recall our assumptions about thermal modeling of magnetic components to justify the measurement principles used that lead to the measure of average temperature [1, 2, 3, and 10]. Temperatures are assumed to be uniform in the material and in the different windings. Thus, a component made up of a magnetic circuit and the two windings will be defined by three figures of temperatures. Thus, we have to measure the average temperature of these elements in the one hand and validate the model in the other. For the same reasons, it is essential to measure the ambient temperature and the temperature of the connection.

Since we seek to design a model that can predict the temperature with only a few degrees difference to the most, we need to have a means of characterization to measure the temperature with equal to or greater accuracy, under either static (set temperature) or transient conditions. In addition to this, the temperature measuring device will minimally disturb the functioning of the component to be characterized.

2.2. Temperature Measurement Method

Two solutions are possible
- A direct method with a probe, a thermocouple, an IR sensor or another temperature sensor is frequently used. [4, 5, 11]. The direct method is difficult to implement, however. In fact, the geometric dimensions of magnetic components are sometimes too short to accurately allow fixing certain types of probes. Furthermore, the direct method requires priming the device under test (DUT) and possibly modifying it (drilling a hole to insert a temperature probe).
- An Indirect measurement that allows determining the average temperature of the magnetic material and the different windings. This approach typically used for thermal measurements in power electronics is more convenient than the direct method, since it does not require any modification of the component [6, 7, and 8]. It is based on the measurement of a quantity whose value is a function of temperature such as the variation of the threshold voltage of a diode in relation with the temperature. As for the magnetic components, their saturation is representative of the temperature of the magnetic material. In addition to that, the resistance of the coil reliably reflects the average temperature of the copper. We have chosen this approach since it has many advantages.

2.3. Measuring the Temperature of the Magnetic Material

Determining the average temperature of the magnetic material is the same as measuring the saturation induction. Indeed, as shown in Figure 1, the saturation magnetization is a function of temperature. This feature naturally depends on the type of material under study.

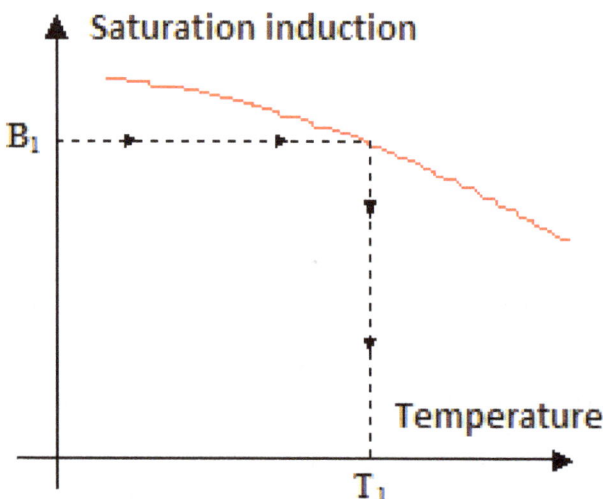

Figure 1. Features of saturation induction $B_{sat} = f(T)$ of a material.

A first phase of identification is needed. It is to measure the saturation induction (saturation induction or induction for a given value of the applied field) for some values of the temperature. Typically a few points are sufficient between 20 °C and 120 °C due to the appearance of this feature. This characterization is to be performed once for a given material. There is a correlation between measuring the average temperature of the magnetic material and measuring the saturation induction using a flux meter. Measuring the saturation average is a non-destructive approach, simple to implement, and requires no instrumentation of the device under test since it has two coils.

2.4. Measuring the Temperature of the Resistance of the Winding

With the same approach, the average temperature of a coil is given by measuring the continuous winding resistance as shown in Figure 2.

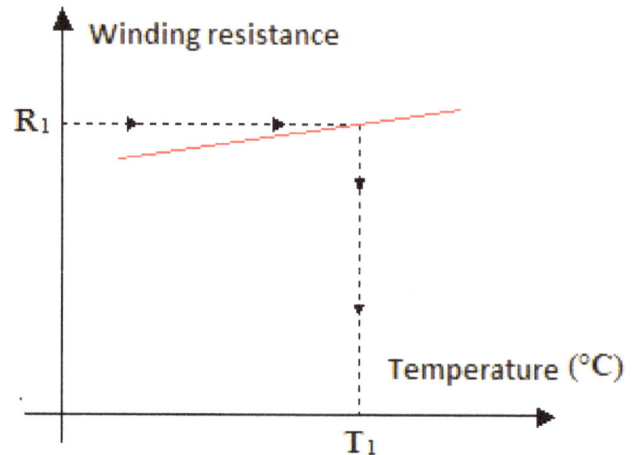

Figure 2. Evolution of the resistance of a winding according to the temperature.

The variation of a winding resistance versus temperature is given by: $R = R_0(1 + \alpha T)$ where R_0 is the resistance at 0°C and T the temperature. This variation is linear. The copper resistance approximately varies up to 30% when the temperature varies between 20°C and 125°C which requires an accurate measurement process of the winding resistance. This method with 4 connection wires provides satisfactory results with minimal complexity. For accuracy purposes it is better to proceed with the statement of the characteristic $R_{(Temperature)}$ of the pure copper rather than using α, the coefficient of the temperature.

2.5. Measuring the Temperature of the Connecting Terminals

We have shown in the previous section that the terminals of the magnetic component played an important role in the transfer of heat from the component outward. Typically, the coil is soldered to a printed circuit board, which is a much more effective heat sink as the tracks are wide. The temperature of the connecting terminals is not equal to the ambient temperature and the identification of the model parameters requires measuring this temperature. As an indirect measurement proves difficult, using a thermocouple is a solution that seems simple to be implemented. However, the validity and accuracy of the measurement are to be watched closely. Make sure that the thermocouple does not affect the heat transfer. Measures with the thermocouple seem to be complicated. Thus the following precautions are taken in order to not disturb the measurement accuracy. Thermocouples are considered to be accurate within ± 1 °C but great precautions are needed [9,10].

- They may pick up electrical noise
- They need to be calibrated to a "cold junction".
- Their attachment to the DUT can be complicated.
- The wires of the thermocouple or disc type thermocouples may make the device to be measured considerably colder
- The wires of the thermocouples may obstruct or disturb the airstream.

3. Equipment Characterization

3.1. Incorporation

The thermal characterization of a magnetic component is to operate in conditions as close as possible to the nominal operation conditions (same constraints: current, voltage, frequency...) and periodically meet different temperatures. Temperature measurements should not disturb the trial or alter the magnetic component. The measurement time should be very short compared to the thermal time constants of the component (a few tenths of a second for all measures) [2, 4]. The following diagram (Figure 3) specifies the constitution of the panel that includes the following four subsets:

- A power supply that allows exciting the sample under test with the conventional forms of wave power electronics or by using a quasi - sinusoidal source;
- A device for measuring the saturation induction flow meter consisting of an integrator;
- A magnetizing device which allows access to the temperature of the magnetic material.
- Measuring device 4 for measuring the resistance of the windings and determine the operating temperature (or more).

3.2. A Switch Card

The equipment completed by a system for measuring the temperature of the connection consists of a thermocouple and an associated packer. This delivers 0 - 10V signal to the capture card.

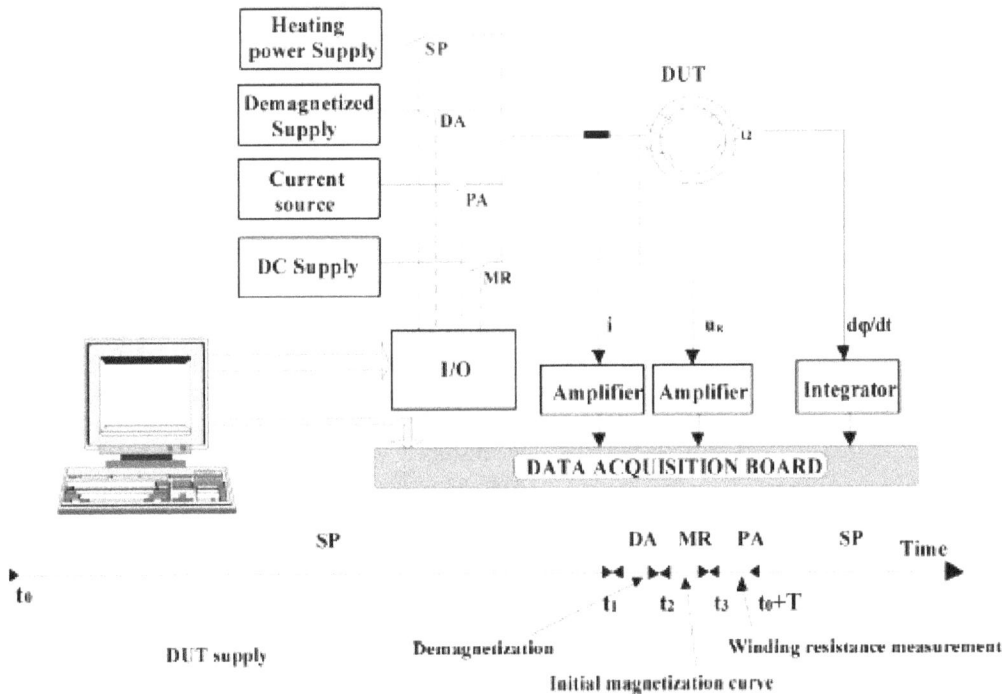

Figure 3. Schematic diagram thermal equipment and operating timing.

At regular intervals (for example every minute), the temperature measurement is carried out in the following order:

- Disconnection of the power source.
- Measuring of the average temperature of the magnetic material which involves the demagnetization of the material.

- The connection of the excitation source and recording of the initial magnetization curve from magnetization to saturation.
- Measuring of the average winding temperature by commissioning a DC current source, and measuring the resistance of the winding. This process is repeated if necessary for different windings.

- The unit is controlled by a computer with a capture card for analog quantities measures and a map of digital inputs/outputs for the control of different elements. It also enables the processing and data storage.

3.3. The Power Source

It is typically composed of a switching power supply that provides power to the sample under test by the usual wave electronic forms of power. For accuracy purposes mentioned above, this power source can also be replaced by a sinusoidal voltage source or a DC power supply based on the tests to be performed and the expected extent of the losses dissipated in the component accuracy. The start and stop of this source in all cases are controlled by the PC via inputs / outputs digital card.

3.4. The Device for Measuring the Saturation Induction

It comprises a demagnetization circuit and a flow meter integrator:

The demagnetizing circuit: The demagnetization is obtained using an oscillating capacitor discharge through the coil. This is controlled by the PC via the digital input output card assembly mainly composed of an LC circuit. The capacitor is charged at constant voltage of adjustable amplitude, a potentiometer is used to control the charging voltage E of the capacitor, which adjusts the amplitude of max ($I_M = E\sqrt{\dfrac{C}{L}}$).The oscillating frequency of the discharge is adjusted by a set of capacitors (C_1 and between 22 uF). A switch allows you to choose different values of the capacitor C to adjust the frequency of the oscillating discharge ($f = \dfrac{1}{2.\pi.\sqrt{LC}}$).

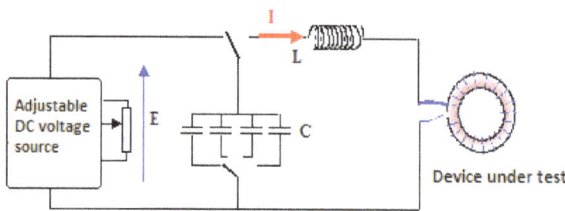

C ranging from 1 µF to 22 µF and $L \approx 100mH$ source voltage 0 -150V

Figure 4. *Circuit demagnetization.*

3.5. The Fluxmeter

Recording the characteristic $\phi(i)$ and B(H) is obtained by the simultaneous acquisition of the current in the primary winding and the integral of the induced electromotive force (emf) across the terminals of an auxiliary winding (the emf is proportional to the derivative of the flux $e_2 = -n_2 \dfrac{d\varphi}{dt}$). The knowledge of the characteristics of the test sample (effective area A_e, the effective length, number of turns of winding exciter winding n_1 and n_2 measurement) $\phi(i)$ or B (H) provides the magnetic characteristics) from the measurements. The magnetic field is given by the equation:

$H(t) = \dfrac{n_1.i_1(t)}{l_e}$. The flow is obtained by the integration of the analog voltage output as knowing of the numbers by coil $\varphi(t) = -\int \dfrac{e_2(t)}{n_2} dt$. The flux meter is equipped with programmable integrator whose time constants are chosen so that to give output signals whose amplitude is as close as possible to the full scale. The average induction is obtained from: $B(t) = \dfrac{\varphi(t)}{Ae}$

Figure 5 below describes the constitution of the flow meter integrator including the devices shown below:
- A demagnetization;
- An excitation circuit comprising a function generator driven by the PC and a power amplifier;
- A flow measurement performed by a programmable integrator;
- An exciter current measurement I_1 photo current is obtained by means of a non-inductive shunt.

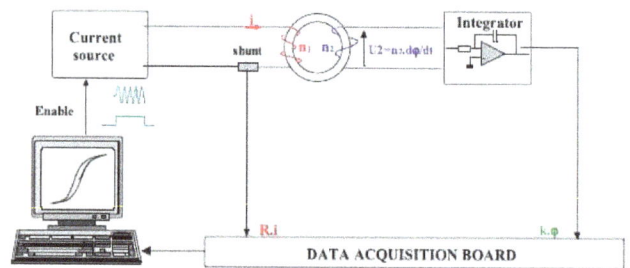

Figure 5. *Integratorfluxmeter.*

Grabber card Keithley DAS 1802 Keithley inputs and outputs PIO12 Wavetek Model 184 generator triggered - Amplifier Sodilec 36V- 12A

The material is demagnetized before processing the measurement, (Figure 6). Next, a measurement is performed without injecting current (I(t) = 0) in the exciter coil which can take into account the drift of the analog integrator. Finally, we proceed to the statement of the initial magnetization curve. To overcome the problems of the drift, the results of the first acquisition at zero current is subtracted from those of the second measurement.

Measuring time (demagnetization - measurement drift and measurement signal: 0.2s).

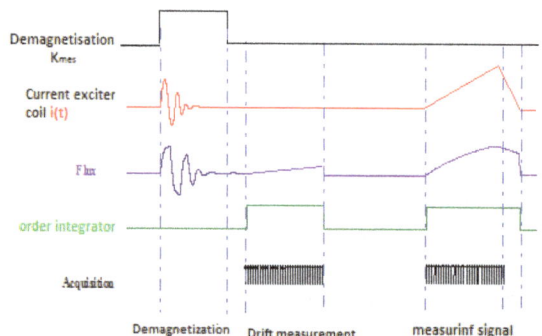

Figure 6. *Sequencing steps to the statement of the initial magnetization curve.*

Figure 7 below shows the results of measurements made using the flow meter integrator.

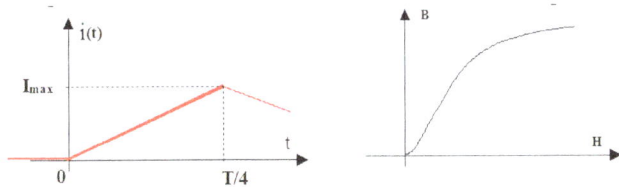

Figure 7. First magnetization curve.

3.6. Apparatus for Measuring the Resistance of the Winding

Determine the average winding temperature that can be achieved in another way by measuring the DC resistance of the coil floor by four voltmetersto access the value of the resistance method with sufficient accuracy.A source of energy injects a continuous current in the winding to determine the strength;and four valuesare available (0.5 - 1 - 2.5 and 5A).

Two programmable instrumentationsamplifiers are used: one for measuring the current (voltage across a shunt), the other for measuring the voltage across the winding enables operation in full-scale (0 -10V).

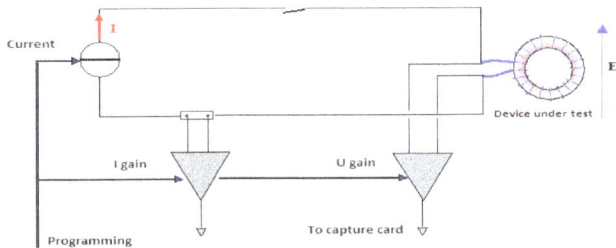

Figure 8. Apparatus for measuring the resistance of the coil.

3.7. Switch Card

A switch card can be connected sequentially to the sample under test with the followingdifferent subsets of the bench characterization: power supply, circuit demagnetization, flow meter integrator device for measuring the resistance of the winding.

To control the various elements (demagnetization heating, etc.), 8 logical signal generators are managed using 8 relays. This card is controlled from a digital input -output card (PIOcard).

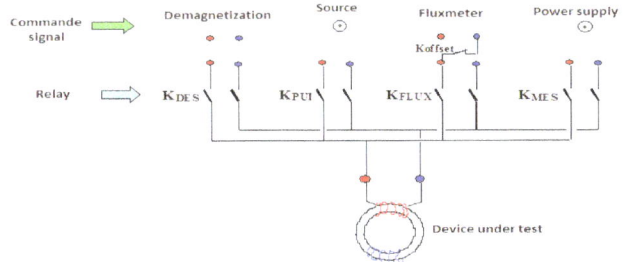

Figure 9. Simplified Block Diagram of the switch board.

This card allows you to connect the magnetic component under test that is electrically isolated, and delivers control signals to it. The typical measurement cycle is indicated in Figure 10 below:

From t_0 to t_1 (relay closed K_{PUI}), the component under test is powered by the power source, the duration of this first phase covers about 99 % of the period T. As the component works in conditions very close to its nominal operating conditions, the temperature will rise.

At time t_1,the phase demagnetization of the material (relay closed K_{DES})starts.

From t_2 to t_3, there is the first magnetization characteristic to derive the average temperature of the magnetic material (K_{FLUX} closed relay).

From t_3 (K_{MES} relay closed),the resistance of the winding is measured, thus determining the temperature of the coilat time $t_0 + T$ (K_{PUI} relay closed).Then, a new measurement cyclestarts.

With heating for 60s (for example), we have about 800 ms for measurements (demagnetization, first magnetization, and resistance measurement).

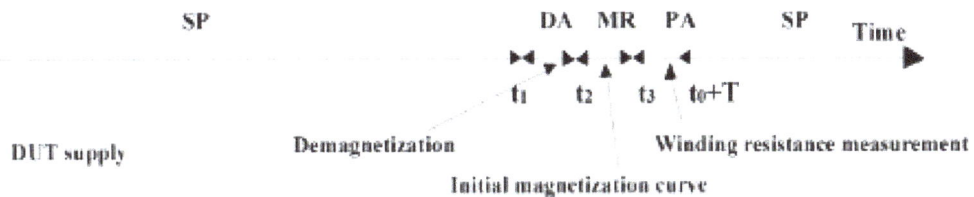

Figure 10. Measuring cycle.

Conventionally, several tens of points of measurement are indicated.The length between two points is greater than a minute while the time required for measures does not exceed a few tenths of a second. Of course, it is all about sizes that depend mainly on the thermal time constants of the studied component.

All control signals are provided by the PC, and the switch board ensures the format and the galvanic isolation.

3.8. Software Test Drive

The software for the control of the tester was developed using the software called test point. This software is to develop and uses test programs, measurement, and data acquisition. All acquisitions are managed via a PC with an analog card and a digital map. This program also allows the measurement acquisition, processing and display, and data storage. Various parameters for the test are provided by the operator, these steps include:

- The choice of parameters running the test (number of points of measurement, frequency...);
- The parameters for the acquisition of the initial magnetization curve (time constant of the integrator, sampling frequency, the value of shunt used for current measurement, the sample data: number of turns of the windings, section and effective length);
- The semi -automatic determination of the gain of the instrumentation amplifier according to the current selected for measuring the winding resistance (manual selection of current 0.5, 1.25, 2.5 and 5A);
- Then the acquisition proceeds automatically until the full display of curves giving the evolution of different temperatures as a function of time.

4. Validation of Measuring Bench

Various tests were carried out in order to validate the measurements obtained using the device described above. We were particularly interested in:

- Checking that the measurement (relevant to the sampling time) did not affect the heating of the component under test.
- Analyzing the validity of the temperature measurement using the connection of a thermocouple.
- studying the repeatability and the accuracy of the measurements of the average temperatures of the winding and the magnetic material.

4.1. Influence Measures on the Heating of the Component Under Test

Any measurement device would typically modify the circuit in which it takes place. In these circumstances, it is important that the perturbation induced by the device should be negligible. This notice applies to any measurement process and naturally so to measuring temperatures. This is why it was essential to ensure that the average temperature measuring of the winding and the magnetic material had a negligible impact on the heating of the component under test. Sampling the temperature every minute with 800ms holding time unlikely leads to significant disturbances. For this reason, it was necessary to check.

We carried out several types of tests to check the validity of this approach: for an 800ms holding time, the device under test is not supplied. Then it is excited by different sources for performing the measurements. To ensure that the absence of power during the measurement does not significantly disrupt the heating of the component, we observed the evolution of temperatures by varying the sampling time. The figures below show the results for two tests, one corresponding to one minute and the otherto five minutes as sampling times.

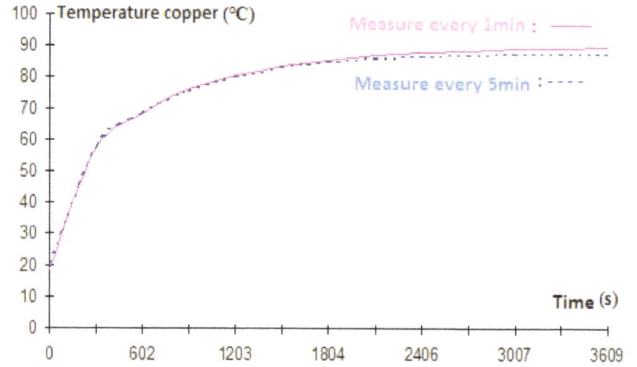

Figure 11. *Influence of the sampling time of the copper temperature.*

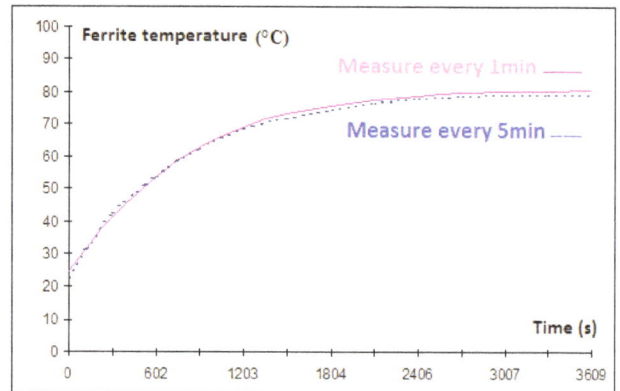

Figure 12. *Influence of the sampling time of the ferrite temperature.*

The tests do not show significant temperature differences between copper and ferrite. The maximum gap between their temperatures reached only 2 (two) degrees. Therefore, it can be concluded that for a sufficiently long holding time (more than 1 minute) the influence of the measurement process is negligible.

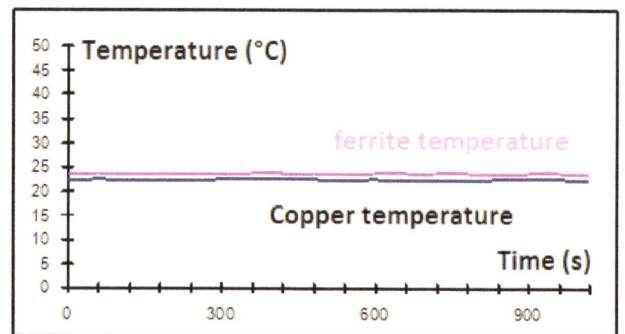

Figure 13. *Influence of the measurement of the temperature of the component under test (Ferrite temperatureand copper temperature).*

We also verified that the measure doesn't lead to additional heating of the component under test, neither heating due to the current demagnetization, nor excitement due to the measurement of the winding resistance. For this reason, every single minute, we recorded the unchanged temperature of every component in the category. Figure 14 below shows a high temperature stability of the tested component. The measured temperature of the copper fluctuates between 22.3 °C and 22.6 °C while that of the ferrite varies between 23.5 and 23.9 °C. As assessed above, this confirms that the temperature of the component under test doesn't change significantly when it is sampled every minute.

4.2. Temperature Measuring Means for Connection to a Thermocouple

Measuring the temperature of the connection using a thermocouple is a major challenge. Many tests have been made to obtain, first a reproducible measurement, and then a measurement as close as possible to reality.

The reproducibility of the measurement depends mainly on how the thermocouple is linked to the connection. Figure 16 below shows what seems the most appropriate way to do it. The tip of the thermocouple is covered with a drop of tin. It is not just the welding thatis important. Particular care should be taken of the position of the thermocouple. The use of a magnifying glass helps to position the tip of the thermocouple as near as possible to the conductor in order to ensure a good replication of the measurement.

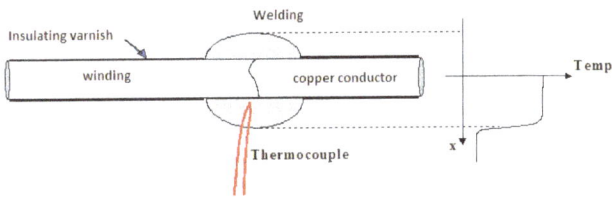

Figure 14. Measurement of the temperature in connection with a thermocouple.

In order to assess the accuracy of the measurements, we compared the results of thermocouple measurement with those obtained by an indirect method. The system is described in Figure 15below. A coil of about 30 turns is driven by a direct current I, whose intensity is adjusted to achieve a significant heating (up to 120 °C). The winding temperature was measured:

- Straightforwardly bythethermocouple placed in the center;
- Indirectly by measuring the resistance of a part of the winding.

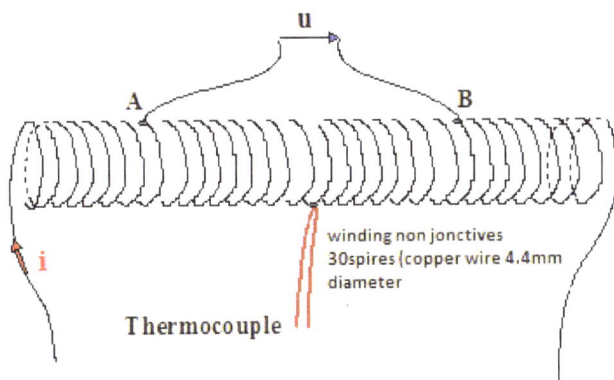

Figure 15. A comparison between the direct measurement and the indirect measurement of the winding temperature.

Given the geometric dimensions, it is reasonable to consider that the coil has a uniform temperature between the two measuring points A and B. Thus, the comparison of both measurements is legitimate.

Figure 16 below is entirely representative of the many comparisons made. We canobserve that the temperature indicated by the thermocouple is about 10% lower than the one obtained by the indirect method.

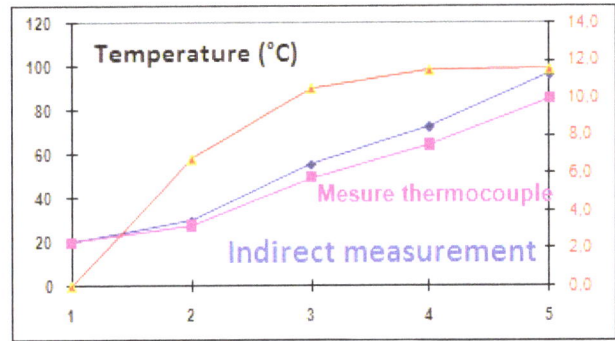

Figure 16. Comparison of direct measurement and indirect measurement.

It is difficult to explain this difference. Several hypotheses are plausible:

- the thermocouple acts as a thermal shunt. This contributes to locally reduce the temperature of the connection;
- as the thermocouple is not centrally located, it does not measure the temperature of the copper, but a slightly lower temperature taking into account the temperature gradient between the center of the copper and the ambient air;
- The thermocouple is not solidly welded (rather it is a cold junction).A thin layer of air surrounding the active part helps to locally reduce the temperature measured very locally.

Our aim was to improve measurement accuracy but the many trials carried out did not provide results with better than 10% accuracy. This type of precision will only be found in the determination of the elements of an equivalent circuit.

Precision and reproducibility of the measurement temperature of the winding and the magnetic material.

To characterize the reproducibility of measurements achieved through the use of the heat bench, we carried out the same test over three years. This test (DC$_{test}$) consists in feeding the sample under test by a high magnitude direct current to raise the temperature changes of the copper and the ferrite over a given time. Figure 20 below corresponding to the results obtained during the last three years shows good replication of the measurements.

Regarding the accuracy of measurements, we placed in a thermal temperature controlled enclosure, the sample previously characterized. For three different temperatures (62 °C, 91 °C and 120 °C)programmed into the heating chamber, we set a direct measurement for the internal temperature, and an in direct measurement for both the winding temperature and the temperature of the magnetic material. These measurements were carried out in steady state which allowed the temperature of the component to be uniform. About thirty measures were carried out each time.

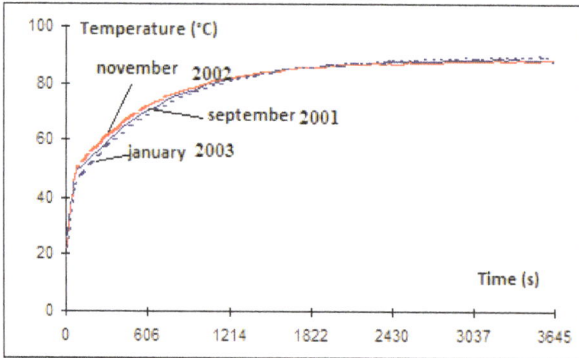

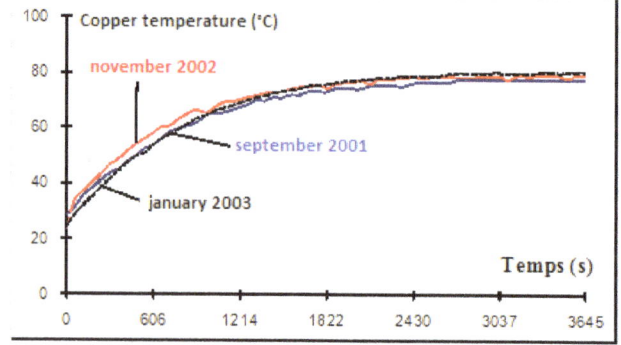

Figure 17. Measurement repeatability.

Figures 17, 18 and 19 indicate the temperature distribution of the copper and that of the magnetic material measured by the indirect method previously described

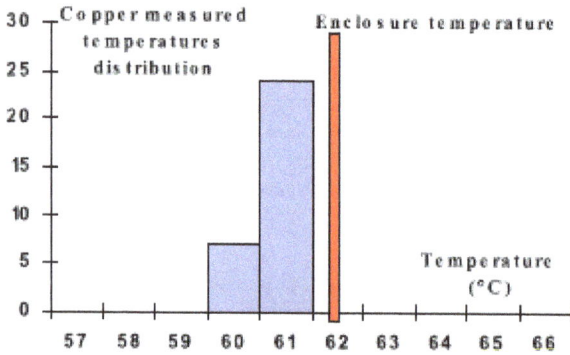

Figure 18. Measured temperatures distribution at $T_{Enclosure}$ = 62 °C.

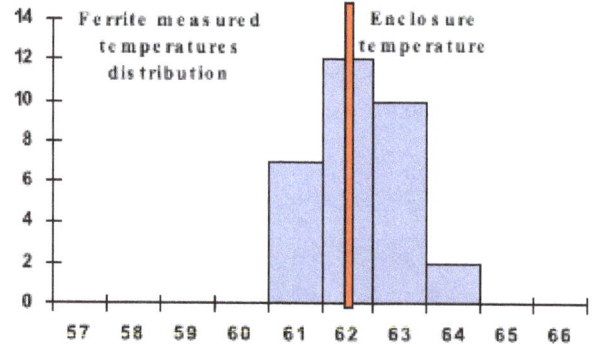

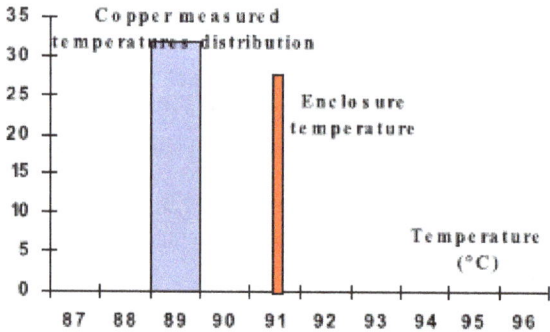

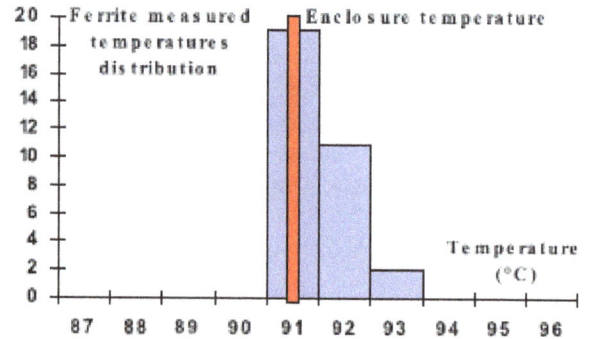

Figure 19. Measured temperatures distribution at $T_{Enclosure}$ = 91 °C.

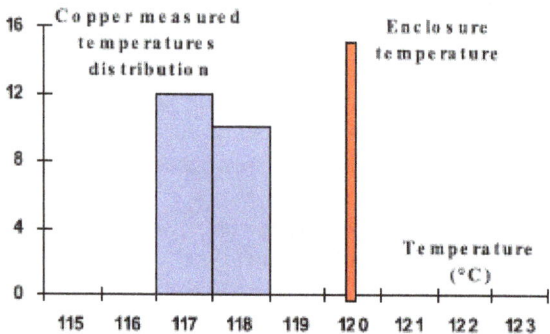

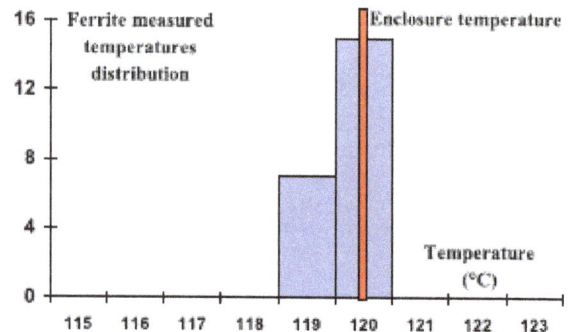

Figure 20. Measured temperatures Distribution at $T_{Enclosure}$ = 120 °C.

The above figures show that the accuracy of the measurement effected is very acceptable, as the variation is only about $\pm 2°C$. Only the winding temperature in the higher temperature conditions is stained by a larger error of about$\pm 3°C$.

5. Conclusion

In order to measure the temperature of the component under test, a vital step into the determination of the elements of the models, we developed a thermal characterization bench. To avoid the modification of the components, we mainly adopted the indirect measuring method. It consists in determining the temperature of the winding and that of the magnetic material from measurements of resistance and initial magnetization curve. Only, the determination of the temperature of connection requires a direct thermocouple measurement. For each of these methods we verified the validity of the approach and the accuracy of the results. So we can state and support that we have set up an appropriate bench for the thermal characterization of magnetic components in permanent or transient states. This bench was also used to validate several proposed models.

The quality of a model depends largely on the accuracy of losses and temperature measurements. We have developed a methodology to accurately determine the various losses in a magnetic component in a frequency range extending from DC to maximum operating frequency in several hundred kHz. Accuracy of measurement had been a constant concern in our approach.

References

[1] Abakar M. T. Modélisation thermique des composants magnétiques utilisés en électronique de puissance. Thèse de Doctorat, INSA de Lyon, 2003, 114 p.

[2] Abakar M. T., J.J ROUSSEAU, BEVILACQUA P., and MAHMOUD Y.Thermal measurementequipment for magnetic component.Eur.Phys.J.appl.Phys, 2005, 30: 57-64.

[3] Abakar M. T. et al. Measuring methods of losses in magnetic component. Pelagia Research Library. Advances in Applied Science Research, 2014, 5(1):156-168.

[4] M. T Abakar et al. Magnetic components Thermal modeling. Part 1: Theorical Issues. IREMOS ISSN 1974 - 9821 Vol 2 N°1 Fev 2009.

[5] Choi B., PEARCE J. A., WELCH A. J. Modeling infrared temperature measurements: comparaison of experimental results with simulation. Conference title: Laser-Tissue Interaction XI: Photochemical, photo thermal and photomechanical, San Diego CA, USA, 2000, 3914: 48 - 53.

[6] Rousseauj. J. Modélisation des composants magnétiques. Habilitation à Diriger des Recherches. INSA de Lyon, 1996, 104 p.

[7] Joiner B., Adams V. Measurement and simulation of junction to board thermal resistance and its application in thermal modeling. Annual IEEE Semiconductor Thermal Measurement and Management Symposium, San Diego CA, USA, 1999, p 212-220.

[8] Thottuvelil V. J., WILSON T. G., OWEN H. A. High-frequency measurement techniques for magnetic cores. IEEE PESC 1985, p. 412-425.

[9] Imre. T. G., CRONJE W. A., VAN WYK. J. D., FERREIRA J. A. Loss Modeling and Thermal Measurement in Planar Inductors - A Case Study. IEEE Transactions on Industry Applications, November/December 2002, 38 (6): 634 – 647.

[10] Petersson, A. (Ericsson), Conway, P. (Murata Power Solutions) and Boylan, J. (TDK Innoveta). Thermal Measurements of Power Converters – How and Why? EPSMA. P14. 2009.

[11] Alaa, H. Magnetic components modeling including thermal effects for DC-DC converters virtual prototyping. Electric power. University Claude Bernard - Lyon I, 2014; p26-43.

20

Study on the Experiments and the Numerical Analysis of Exchange Flow Behavior in the Unstably Stratified Field

Motoo Fumizawa, Yoshiharu Saito, Naoya Uchiyama, Takahiro Nakayama

Department of Mechanical Engineering, Shonan Institute of Technology Fujisawa, Kanagawa, Japan

Email address:

fumizawa@mech.shonan-it.ac.jp (M. Fumizawa), sutere0port96@gmail.com (Y. Saito), psn-man.lv.030@softbank.ne.jp (N. Uchiyama), 2991taka@gmail.com (T. Nakayama)

Abstract: In the flow mechanism of unstably stratified field, occurs after Rayleigh-Taylor instability. Buoyancy-driven exchange flows were investigated the helium-air flow in the vertical narrow pathway between upper air chamber and lower helium chamber. Exchange flows may occur following the opening of a window for ventilation, when fire breaks out in a room, as well as when a pipe ruptures in a high temperature gas-cooled nuclear reactor. The numerical analysis and experiment in this paper was carried out in a test chamber filled with helium and the flow was visualized using the smoke wire method. The flow behavior was recorded by a high-speed camera combined with a computer system. The image of the flow was transferred to digital data, and the flow velocity was measured by PTV and PIV software. The mass fraction in the test chamber was measured using electronic balance. The detected data was arranged by the densimetric Froude number of the exchange flow rate derived from the dimensional analysis. A method of mass increment was developed and applied to measure the exchange flow rate. As the result, it is revealed that three dimensional structure of counter current exchange flow in the narrow flow path such as rotation and circulation flows by the optical system, mass inclement method and numerical analysis of moving particle method as well as HSMAC method.

Keywords: Buoyancy, Exchange Flow, Helium, Moving Particle Method, Clockwise Flow

1. Introduction

In the flow mechanism of unstably stratified field, occurs after Rayleigh-Taylor instability. Buoyancy-driven exchange flows were investigated the helium-air flow in the vertical narrow pathway between upper air chamber and lower helium chamber. Exchange flows may occur following the opening of a window for ventilation, the outbreak of fire in a room or over an escalator in an underground shopping center, as well as when a pipe ruptures in a modular high temperature gas-cooled nuclear reactor. The fuel loading pipe is located in an inclined position in a pebble bed reactor such as the Modular reactor [1,2,3] and AVR [4].

In safety studies of High Temperature Gas-Cooled Reactor (HTGR), the failure of a standpipe at the top of the reactor vessel or a fuel loading pipe may be one of the most critical design-based accidents. Once the pipe ruptures, helium immediately rushes up through the breach. Once the pressure between the inside and outside of the pressure vessel has balanced, helium flows upward and air flows downward through the breach into the pressure vessel. This means that buoyancy-driven exchange flow occurs through the breach, caused by the density difference of the gases in the unstably stratified field. Since an air stream corrodes graphite structures in the reactor, it is important to evaluate and reduce the air ingress flow rate when a standpipe rupture occurs.

Studies have been performed on the exchange flow of two fluids with different densities through vertical short tubes. Epstein[5] studied the exchange flow of water and brine through various vertical tubes, experimentally and theoretically. Mercer et al.[6] experimentally studied an exchange flow through inclined tubes with water and brine. The latter experiments were carried out in the range of 3.5 $< L/D < 18$ and 0 deg $< \theta < 90$ deg, and indicated that the length-to-diameter ratio L/D, and the inclination angle θ of the tube are the important parameters for the exchange flow rate. Most of these studies were performed on the exchange flow using a relatively small difference in the densities of the two fluids (up to 10 per cent). However, in the case of a HTGR standpipe rupture, the density of the outside gas is at least three times larger than that of the gas inside the pressure vessel. Few studies have been performed so far using such a large density difference. Kang et al.[7] studied experimentally the exchange flow through a round tube with a partition plate. Although one may assume that the partitioned plate, a kind of obstacle in the tube, would reduce the exchange flow rate, Kang found that the exchange flow rate was increased by the partition plate because of separation of the upward and downward flows.

The main objectives of the present study are to investigate the behavior of the exchange flow in the vertical narrow pathway then to evaluate the exchange flow rate using mass increment with the helium-air system. The following methods are applied the present study.

(1) Optical system of the Mach-Zehnder interferometer
(2) Numerical analysis
(3) Mass increment method

2. Optical System of Mach- Zehnder Interferometer

2.1. Experimental Apparatus and Procedure

The optical system of the Mach-Zehnder interferometer, MZC-60S to visualize the exchange flow is shown in Figure 1. After being rejoined behind the splitter, the test and reference laser beams interfere, and the pattern of interference fringe appears on the screen. If the density of the test section is homogeneous, the interference fringes are parallel and equidistant [8,9]. If the density is not homogeneous, the interference fringes are curved. Inhomogeneity in the test section produces a certain disturbance of the non-flow fringe pattern. The digital camera and high-speed camera using a D-file can be attached to the interferometer.

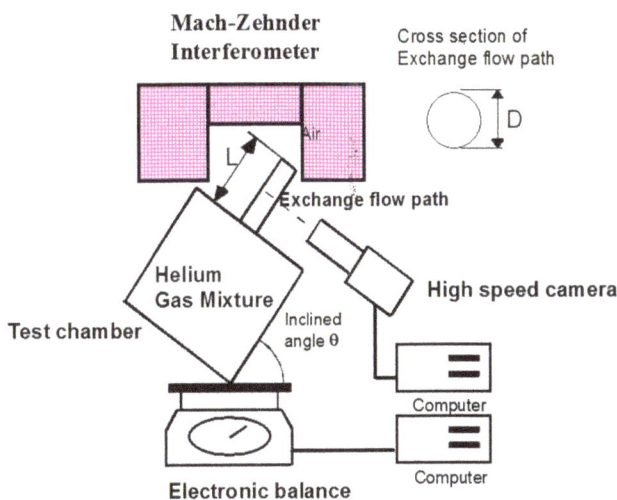

Fig. 1. Experimental apparatus of optical system of Mach-Zehnder.

2.2. Results of Optical System

To investigate the flow pattern, the exchange flows are visualized by the Mach-Zehnder interferometer. The photo in Figure 2 shows the typical interferogram of the fringes for the vertical narrow tube (L/D=5). The upward helium plume and the downward air plume break intermittently through the passage and swing from left to right in the lateral direction. The period of the swing is~2s.It is clearly observed that the up flow of helium and the down flow of air do not take place smoothly, because they interact strongly with each other. The flow is fluctuated and unstable.

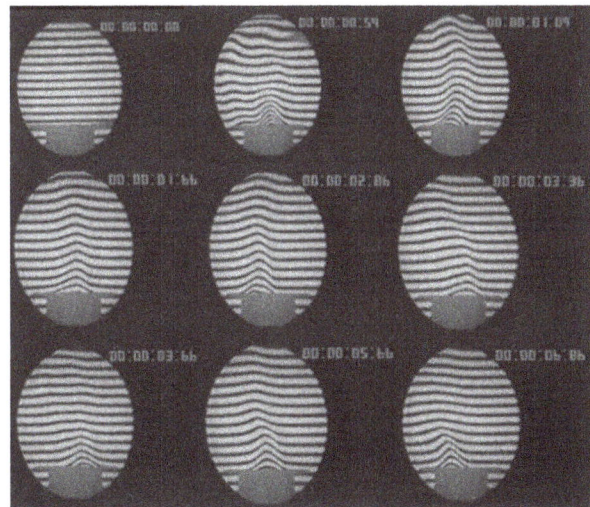

Fig. 2. Visualization result of Mach-Zehnder method upper part of vertical narrow tube.

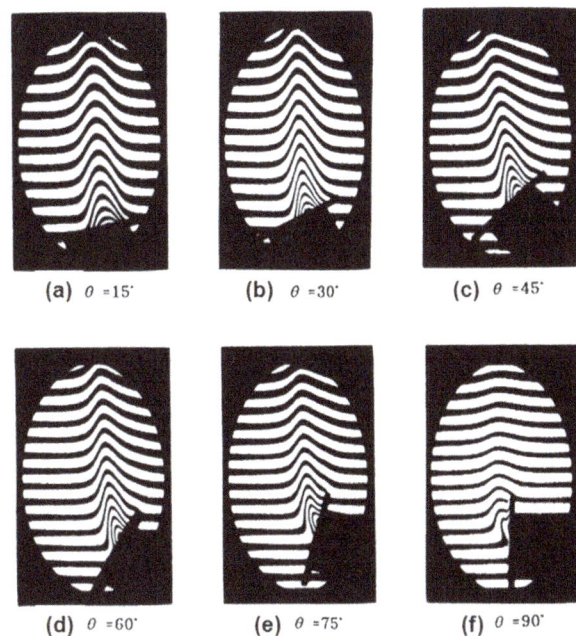

Fig. 3. Typical interference fringes for an inclined narrow tube (L/D=5).

Figure 3 shows the typical interference fringes for an inclined long tube (L/D=5). The curved interference fringes indicate that the lighter helium flows in the upper passage of the tube. The straight fringes indicate that the heavier air flows in the bottom of the tube. It is clearly visualized that the exchange flows take place smoothly and in a stable manner in the separated passages of the tube. This leads to less resistance for the exchange flow in the inclined tubes compared to the vertical ones. In the case of a 30 deg. angle, the curvature of the interference fringes is larger than that at other angles, indicating that the exchange flow rate and the densimetric Froude number are the largest at 30 deg.

3. Numerical Analysis

3.1. Moving Particle Method

Figure 4 and 5 show cross sectional view and bird-eye view of numerical boundary conditions of moving particle method of 3D coordinate, respectively. Table 1 shows the calculation condition of moving particle method. Lagrange method is adopted calculation program of moving particle method. Therefore, the numbers of particles are around two million. The calculation program code is adopted Particleworks3.01 and possessor is composed of TESLA C2070 [10].

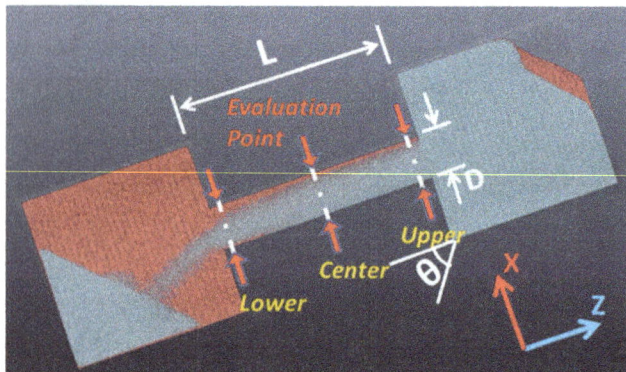

Fig. 4. *Cross sectional view of numerical boundary conditions of moving particle method.*

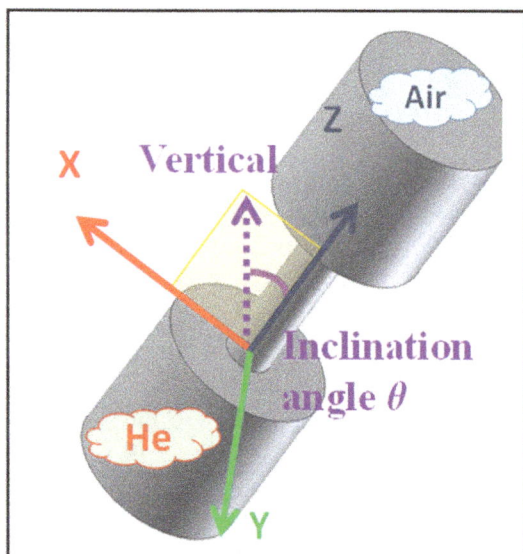

Fig. 5. *Bird -eye view of numerical boundary conditions of moving particle method.*

Table 1. *Calculation condition of moving particle method.*

Number of particle N	1~2million
Inclination angle θ	0°,15°,30°,45°,90°
Particle diameter a	0.815~1.0mm
Flow path length L	100mm
Diameter flow path d	20mm
Container size h=D	80mm,110mm

Figure 6 shows unstably and asymmetric flow patterns became dominant in the vertical narrow pipe. It is interesting that the horizontal cut view, in the figure, reveals that the clockwise upward flow was observed, which means spiral upward flow pattern. Figure 7 shows that the upward plume behaves like the snake movement in the upper chamber. Numerical calculation of moving particle method predicts that the clockwise movement of the plume and snake behavior of plume and the behavior is consistent with the result that Mach-Zehnder method shows random behavior of the plume from left to right.

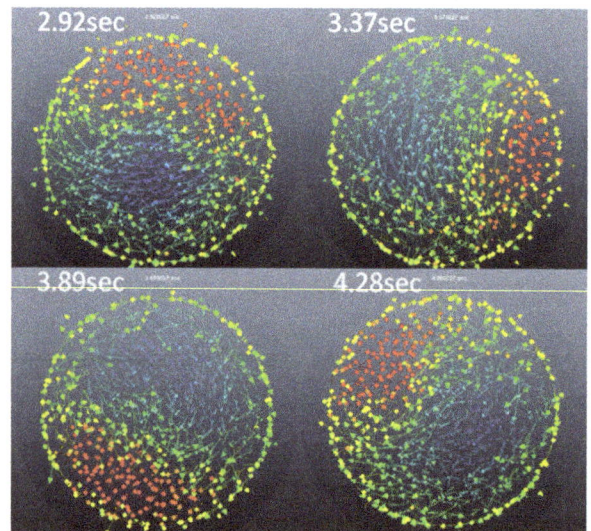

Fig. 6. *Horizontal view of exchange flow in the narrow tube.*

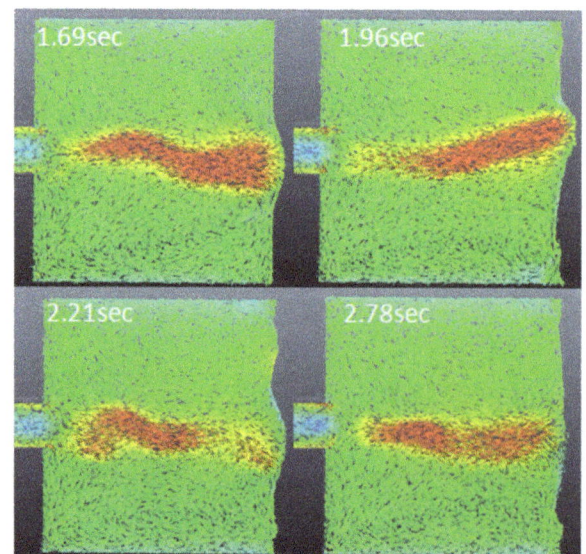

Fig. 7. *Vertical view of light plume behavior upper chamber on the narrow tube.*

3.2. HSMAC Method

Two dimensional unsteady code of HSMAC method is adopted to the buoyancy-driven exchange flow system[2]. In the code, FORTRAN program is described basic equation of mass, momentum and energy. Analysis coordinate is shown in Figure 8. The left part is test chamber filled with helium gas and the right part is outside region filled with air. Typical calculation result is shown in Figure 9, where is the narrow channel, between the left and right. The exchange flow occurs with vortex in the narrow channel. Therefore, the center flow rate Q_1 is larger than the right edge flow rate Q_2, as shown in Figure 10.

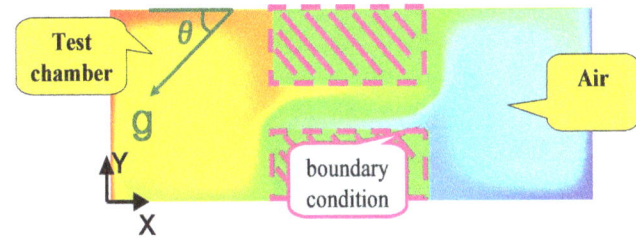

Fig. 8. *Analytical coordinate and conditions.*

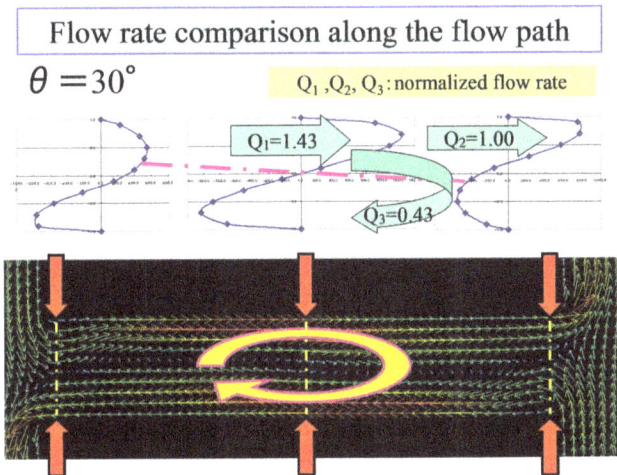

Fig. 9. *Analytical result of narrow flow path of Exchange flow.*

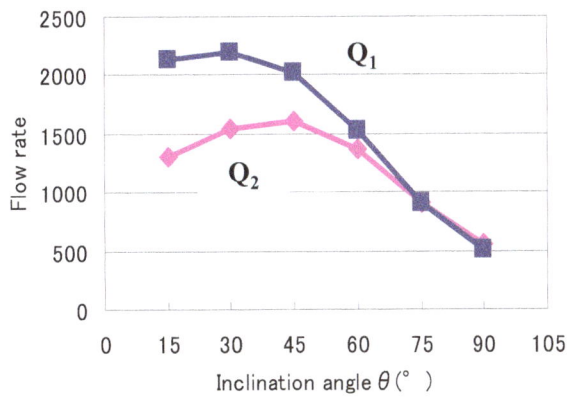

Fig. 10. *Relation of Exchange flow rate and inclination angle of analytical results.*

4. Method of Mass Increment

4.1. Experimental Apparatus and Procedure

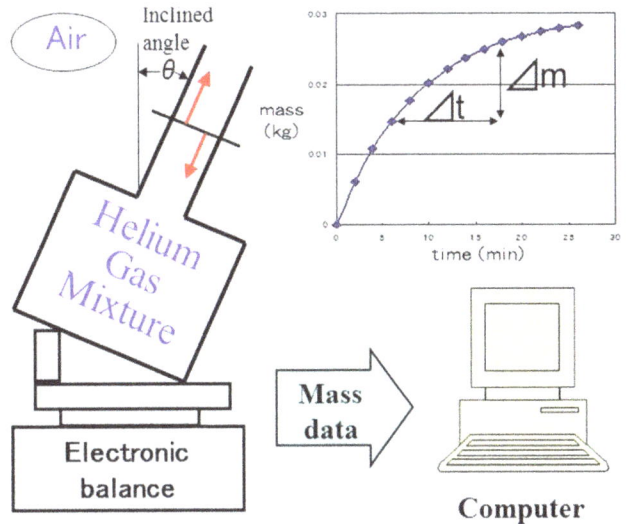

Fig. 11. *Sketch of the mass increment apparatus.*

The mass increment method was used for the investigations. Figure 11 shows a rough sketch of the apparatus. It consists of a test chamber, an electronic balance and a personal computer for data acquisition. The experimental procedure are as follows. Air enters the test chamber and the mass of the gas mixture in the test chamber increases. The mass increment Δm is automatically measured by the high accurate electronic balance. The density increment of the gas mixture $\Delta \rho_L = \Delta m/V$ is calculated From mass increment data,. The density increment signifies the difference between densities of the gas mixture from the density of pure helium in the test chamber. Subsequently, the volumetric exchange flow rate is evaluated using the following equation:

$$Q = \frac{V}{\rho_H - \rho_L} \cdot \frac{d(\Delta \rho_L)}{dt} \quad (1)$$

The densimetric Froude number is defined by the following equation derived from the dimensional analysis suggested by Keulegan [11]:

$$Fr = \frac{Q}{A} \sqrt{\frac{\rho}{gD\Delta\rho}} \quad (2)$$

In the above equations, V is the volume of the test chamber, ρ_H the density of air, ρ_L the density of the gas mixture in the test chamber, $\triangle\rho_L$ ($=\rho_H$ -ρ_{He}) = the density increment of the gas mixture, t the elapsed time, U(=Q/A) the exchange-velocity, ρ($=\rho_H$ +ρ_L)/2, D the diameter and g the acceleration of gravity. The experiments are performed under atmospheric pressure and room temperature using the vertical and inclined round tubes, and using a vertical annular tube. The density of the gas mixture is close to that of helium in the present experiment. The sizes of the tubes are as follows.

The diameter of the round tube D is 20 mm, which is much smaller than that of the test chamber. The inclination angle θ ranges from 15 to 90 deg and the height L ranges from 0.5 to 200 mm.

4.2. Comparison between Two Experimental Methods

It is already known that the densimetric Froude number is regarded as constant within a time duration when the gas in the upward flow is assumed to be helium [12]. Figure 12 shows the relationship between Fr and inclination angle θ with L/D as a parameter. For inclined tubes, Fr is larger than that for vertical tubes. (i.e. L/D = 10). The densimetric Froude number reaches the maximum at 60 deg for the orifice and 30 deg for the long tube. It is found that the angle for the maximum Fr decreases with the increase of L/D in the helium-air system.

Comparison between Smoke wire method (SW) and Mass increment method (MI)

$$Fr = \frac{Q}{A} \sqrt{\frac{\rho}{g \, D \, \Delta\rho}} \qquad \textbf{L/D=10}$$

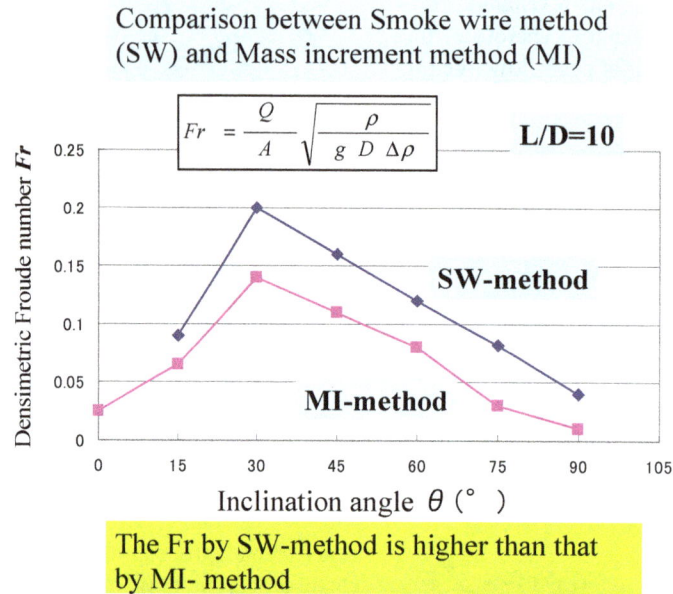

The Fr by SW-method is higher than that by MI- method

Fig. 12. Relation of Densimetric Froude number and inclination angle in both experiments.

5. Conclusion

It is revealed that three dimensional structure of counter current exchange flow in the narrow flow path such as rotation and circulation flows by the optical system, mass inclement method and numerical analysis of moving particle method as well as HSMAC method.

(1) Numerical calculation of moving particle method predicts the clockwise movement of the plume behavior flow occurs in the narrow flow path

(2) Numerical calculation of HSMAC method predicts the circulation flow of the tube length direction occurs in the narrow flow path.

(3) Numerical calculation of moving particle method predicts snake behavior of plume and the behavior is consistent with the result that Mach-Zehnder method shows random behavior of the plume from left to right.

(4) (4) Densimetric Froude number Fr by SW-method is higher than that by MI- method.

Acknowledgements

The authors are deeply indebted to Mr. Shuhei Ohkawa, working Sanwa-koki Co. Ltd., for his unfailing interest and helpful input to this study.

References

[1] Fumizawa,M.; Nuclear Reactors (ISBN 978-953-51-0018-8) , Edited by Amir Zacarias Mesquita, InTech, pp.47-56 (2012)

[2] Fumizawa,M.; Proc. HT2005 ASME Summer Heat Transfer Conference, HT2005 -72131, Track 1-7-1 , pp.1-7 (2005).

[3] Fumizawa,M.;Nuclear Reactors (ISBN 978-953-51-0967-9) , Edited by Amir Zacarias Mesquita, InTech, pp.177-191 (2013)

[4] M.M.El-Wakil, Nuclear Energy Conversion, Thomas Y. Crowell Company Inc., USA (1982)

[5] Epstein,M., Trans. ASME J. Heat Transfer, pp.885 -893 (1988)

[6] Mercer, A. and Thompson.H., J. Br. Nucl. Energy Soc., 14, pp.327-340 (1975),

[7] Kang,T. et al., NURETH-5, pp.541-546 (1992)

[8] Merzkirch.W., "Flow Visualization", Academic Press (1974)

[9] Feng,J. et al., Chemical Engineering Journal, Volume 86, pp.243-250 (2002)

[10] Prometech software web-site, http://www.prometech.co.jp/ (2015)

[11] Keulegan, G. H., U. S. N. B. S. Report 5831 (1958)

[12] Fumizawa,M. et. al., J. At. Energy Soc. Japan, Vol.31, pp.1127-1128 (1989)

21

Optical Fiber Daylighting System Featuring Alignment-Free

Ngoc Hai Vu, Seoyong Shin[*]

Department of Information and Communication Engineering, Myongji University, Yongin, South Korea

Email address:

anh_haicntn@yahoo.com (N. H. Vu), sshin@mju.ac.kr (S. Shin)

[*]Corresponding author

Abstract: We present a cost-effective optical fiber daylighting system composed of prism and compound parabolic concentrator (P-CPC). Our simulation results demonstrate an optical efficiency of up to 89% when the concentration ratio of the P-CPC is fixed at 100. We have also used a simulation to determine an optimal geometric structure of P-CPCs. Because of the simplicity of the P-CPC structure, a lower-cost mass production process is possible. Our quest for an optimal structure has also shown that P-CPC has high tolerance for input angle of sunlight. The high tolerance allows replacing a highly dual precise active sun-tracking system with a single sun-tracking system as a cost-effective solution. Therefore, our results provide an important breakthrough for the commercialization of optical fiber daylighting systems that are faced with challenges related to high cost.

Keywords: Compound Parabolic Concentrator, Plastic Optical Fiber, Daylighting

1. Introduction

As a type of green energy, solar energy has been attracting increasing attention in recent years. Common ways of harvesting solar energy include photovoltaics (PV), solar thermal, and daylighting [1]. So far, conversion efficiency of solar cell still is challenge and it is difficult to do this with the cost effectiveness necessary to make solar generated electricity a commercial reality. In certain instances, however, solar energy can be made more competitive by applying it directly to the end use [2]. One of the direct applications of solar energy is daylighting. Daylight is used to illuminate building interiors to affect the indoor environment, health, lighting quality, and energy efficiency [3]–[5]. In sustainable buildings, daylighting can provide energy reductions through the use of electric light controls, and it can reduce the dependence on artificial lighting, which cannot fulfill the needs of the human body [6] .

Daylighting involves collecting natural sunlight for interior illumination. For illumination inside buildings, the collected sunlight is typically guided through a duct or a fiber bundle. [2], [6]–[12]. In building integration, one of the most important features of the remote light transportation is the wiring method and the wiring method is expected to be as simple as that of electrical wires[12]. However, the light ducts have their difficulties for wiring so that daylight transportation through optical fibers is considered as the best approach so far [12]. Only optical fibers are suitable for this requirement. Optical fiber daylighting technology is one of the most efficient solutions for the delivery of natural light to a space in a building where daylight is limited. Optical fiber daylighting systems are composed of three main components: the sunlight collector with a sun tracking mechanism, optical fibers, and luminaires that distribute light in the required space. To facilitate coupling with the fiber bundle, an optical concentrator must be used to concentrate the sunlight [1]. Therefore, optical concentrators play a crucial role in harvesting solar energy. Through the research and development of many public and private groups, two basic collector designs have proven to be the most effective and reliable. The first strategy uses optical lenses to refract and concentrate sunlight into optical fibers; the second design captures incoming light by reflection from parabolic mirrors[2], [6]–[10], [12]. However, both designs suffer from the non-uniformity of the light beam over the end-face of the optical fibers, and additional secondary optics are needed to homogenize the sunlight and increase the optical fiber coupling efficiency and the tracking tolerance [11]. Figure 1

(a) shows the typical mechanism for a optical daylighting system using Fresnel lenses. Numerous designs related to the traditional concentrator have been proposed, which can provide a considerable concentration ratio, but requires a sophisticated alignment between primary concentrator, second optics and optical fiber. They also require acuracy dual-axis tracker and typically a large space.

In the field of concentrated solar energy applications, solid dielectric compound parabolic concentrators (CPCs) recently have been one of the best choices because of simple structure and high efficiency. A solid dielectric CPC concentrates light via reflection and refraction by incorporating a solid dielectric refractive material into the CPC structure as shown in Figure 1 (b). The total internal reflection within the the solid CPC has a high reflectance and therefore may lead to a higher optical efficiency. Mallick et al. [13], [14] have investigated an asymmetric CPC consisting of two different parabolas using a transparent dielectric material. Winston et al. [15]–[17] stated that a solid dielectric CPC has an increased angular acceptance and reduced optical loss compared with its non-dielectric counterpart. However, important characteristics the solid dictric CPC are low concentration ratio and the very high non-homogeneity in the spatial flux distribution produced at the exit aperture. These features is not suite for optcial fiber daylighting system that requires high concentration ratio and uniformly irradiation at exit aperture for optical fiber coupling.

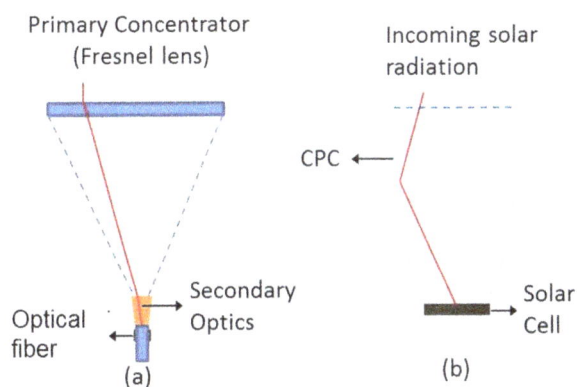

Figure 1. *Physical layout of (a) one optical fiber daylighting system using Fresnel lens and (b) a solid dielectric CPC using for CPV applications.*

In this study, we introduce an optical fiber daylighting system using a combination of prism and solid dielectric CPC which can achieve high concentration ratio and uniform distribution of solar irradiation inside the optical fibers. This proposed system remains some advantages of CPCs such as low fabrication cost, alignment-free, thus facilitates the viable commercialization of cost effective mass-produced systems. To our knowledge, the optical fiber daylighting system using prism-CPC combination described in this study is the first system that can show cost effective potential when manufactured in volume.

The remainder of the paper is organized in the following manner: Section 2 describes the design concept and model

principle of daylighting systems using combination of prism solid dielectric CPC (P-CPC). A detailed description of optical fiber coupling is also discussed in this part. In Section 3, the optical fiber daylighting system based on P-CPC is modeled in LightTools™ software (Synopsys Inc., California, USA) to evaluate the performance of such a system. We also optimize all of parameters that affect on the optical efficiency and angular tolerence of system. Finally, brief concluding remarks and possibilities for future work are included in Section 4.

2. Design Concept and Model Principle

This part introduces the conceptual design and working principle of a combination of Prism-compound parabolic concentrator (P-CPC). The foundation of idea is based on a CPC that was used for many different applications, ranging from high-energy physics to solar energy collection. To modify the CPCs for our purpose, we recall the theory of conventional CPCs. A symmetrical CPC as shown in Figure 2 (a) consists of two identical parabolic reflectors that funnel radiation from the aperture to the absorber [18]. The right-hand side and the left-hand side parabolas are axisymmetric. The focuses of two parabolas form the base of the CPC, as shown in Figure 2 (b). When a sun ray beam is parallel to the main axis of parabolic rim, it will be focused on the focus of parabola as illustrated in Figure 2 (c). A solid dielectric CPC is filled with dielectric materials such as poly-methyl methacrylate (PMMA) [19]. When the incidence angles of the incoming rays are smaller than acceptance angle, the rays would undergo total internal reflection or mirror reflection to reach the base of CPC [17].

In this study, we propose a new aspect of using solid dielectric CPC that utilize the imaging optics property of CPC - a non-imaging optics device. Figure 3 (a) shows the physical layout of our proposed solar concentration device based on combination of prism-CPC. The prism is placed at the top of CPC which can change the direction of incoming solar ray. With appropriate prism angle, the direct sunlight refracts at two edge of prism and divided in two separate beams that are parallel to the axe of 2 parabolic rim of CPC. After reflection at the wall of CPC two beams focus at focal point of parabola as shown in Figure 3 (b).

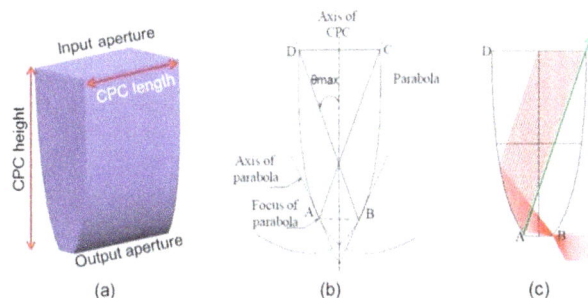

Figure 2. *(a) 3D view of a solid dielectric CPC; (b) A symmetrical CPC with parameters and (b) ray tracing of sunlight beam that parallel to axis of parabolic rim.*

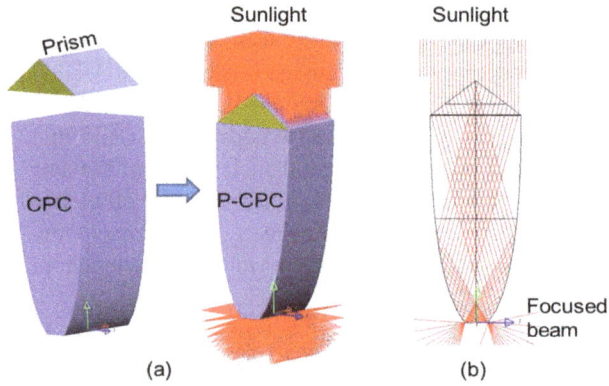

Figure 3. *(a) Physical layout of P-CPC and (b) mechanism of sunlight concentration by ray tracing.*

Figure 4 shows the method to calculate the structrure of prism based on accepctane angle of CPC. The incident angle Θ_i of sunlight ray at the edge of prism is equal to angle α of prism ($\Theta_i = \alpha$). The refracted ray should have direction of parabolic rim of CPC (acceptance angle of CPC: Θ_{acc}). Base on Snells law, the relation between Θ_i and Θ_{acc} is shown in Equation 1.

$$\sin \Theta_i = n_1 \sin(\theta_i - \theta_{acc}) \qquad (1)$$

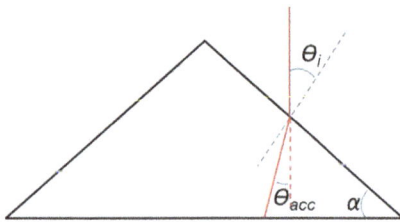

Figure 4. *The relation between prism structure and acceptance angle of CPC.*

The optical fiber consists of a core, cladding, and an external protective coating. The light travels inside the core, while the cladding, which has a lower refractive index, provides internal reflection at the boundary of the core. The optical fibers used in daylighting and solar thermal applications for the transmission of sunlight need to transmit a broad spectrum. One of the most significant features of sunlight transportation is the wiring, and the wiring must be as simple as electrical wiring. Therefore, only optical fibers can fulfill this requirement. Optical fibers were utilized to deliver sunlight to the interior with small losses. Silica optical fibers (SOFs) are known to be good light-transmission media and have the best resistance to heating; however, SOFs are expensive. Plastic optical fibers (POFs) have substantially higher attenuation coefficients than SOFs, but POFs are preferred in daylighting systems due to their lower cost, tighter minimum bend radius, ease of installation and durability for complex wiring in buildings [3], [20]. The light can be transferred over long distances without visible changing of the input color because the POFs are made with PMMA, which has attenuation minima of 64, 73 and 130 dB/km, occurring at 520, 570 and 650 nm,

respectively. These wavelengths indicate that the PMMA fibers will transmit green, yellow and red light particularly well. The POF parameters are listed in Table 1.

Table 1. *POF parameters for design and simulation.*

Parameters	
Attenuation	0.45 dB/m
Core/Cladding Diameter	1.960/2.0 mm
Refractive Index: Core/Cladding	1.492/1.402
Minimum Bend Radius	50 mm
Spectral Trans. Range	380–750 nm

The exit port from the P-CPC concentrator has a rectangular shape, so a ribbon configuration of optical fiber is proposed for the optical fiber coupling. We remove apart of optical fiber ribbon and connects to P-CPC by a index matching gel as describe in Fig. 5. (a,b). The light propagates by reflection in POFs to reach the interior for illumination. Fig. 5 (c) shows the optical fiber coupling mechanism using ray tracing method. Rays that exceed the critical angle, as defined by Snell's law, propagate via total internal reflection (TIR) within the waveguide to the exit aperture. Otherwise it will be gone out as loss. For proposed system, the effective sunlight collecting area is calculated by product of CPC length and input size D. The output are two end faces of POF. Therefore, the geometric concentration ratio of system C_R as shown in Equation 2. For simply, C_R is ratio of CPC width and POF diameter.

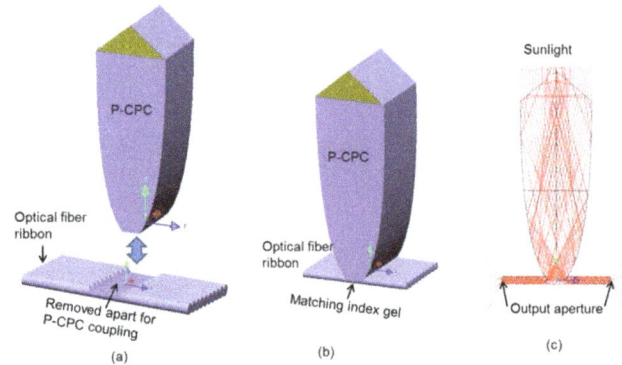

Figure 5. *(a) Optical fibers arranged in ribbon configuration and removed apart for coupling purpose (b) Physical layout of P-CPC coupled with optical fiber ribbon; (b) ray tracing analysis.*

$$C_R = \frac{CPC\ length \times CPC\ width}{Total\ area\ of\ POF\ endfaces} = \frac{D}{2d} \qquad (2)$$

The design of daylighting system powered by renewable solar energy was presented in this section. The sunlight concentrator researched in this study is composed of a prism, CPC. The concentrator is supposed to be equipped with a tracking system to collect sunlight in the normal direction [4]. A ribbon optical fiber is attached at the output aperture of CPC to collect the focused sunlight beams. The components of the optical system, design parameters and their effects on the optical performance are discussed in more detail in following section.

3. Optical Analysis and Performance

Optical modelling plays a crucial role in the efficiency evaluation of an optical system. Commercial optical modeling software, LightTools™, was used to design and simulate the geometrical structure of daylighting system based on P-CPC [21]. In the designed system, one of the most common optical plastic, poly-methyl methacrylate (PMMA) with refractive index of $n_{PMMA} = 1.518$ is selected for prism, CPC and POFs[11]. To evaluate the losses in the system, in simulation model, we inserted three luminous flux receivers as shown in Figure 6.

The optical efficiency, which is simply defined as the ratio of the output luminous flux to the input luminous flux, is a function of the reflection and absorption losses (Equation 3).

$$\eta = \frac{Flux \ on \ Receiver \ 2 + Flux \ on \ Receiver \ 3}{Flux \ on \ Receiver \ 1} \qquad (3)$$

The efficiency of system depends on shape of CPC and system concentration ratio. Section 3.1 and 3.2 will discussed these problems in details.

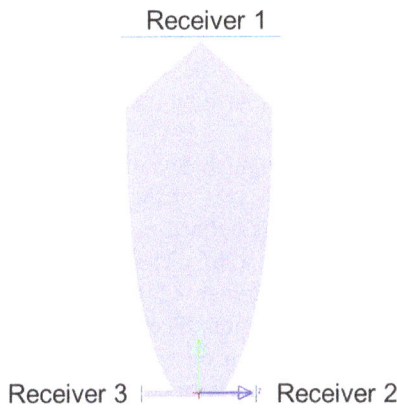

Figure 6. Illustration of the simulation structure for efficiency analysis.

3.1. Optimization the Shape of CPC

Loss mechannism is illustrated by ray tracing as shown in Figure 7. The Fresnel reflection losses occur at the boundaries where the light passes from one region to another with different refractive indices. In this proposed system, the Fresnel losses occur at the surface of prism. The Fresnel loss at the conjunction between POFs and waveguide can be reduced to below 0.1% by filling the matching index and then it can be ignored in comparison with other losses. The leak at the parabolic wall of CPC and bottom surface of slab waveguide are also imprortant and they can significantly affect the final efficiency of the system. This kind of losses cause by some ray can not satisfy the TIR condition inside the P-CPC concentrator. These losses depend on the shape of CPC. This could be prevented when the solid CPC and slab waveguide has a mirror coating. However the mirror coating usually has a lower reflectance than the total internal reflection, so the coating may have positive or negative effect on the optical performance of a P-CPC.

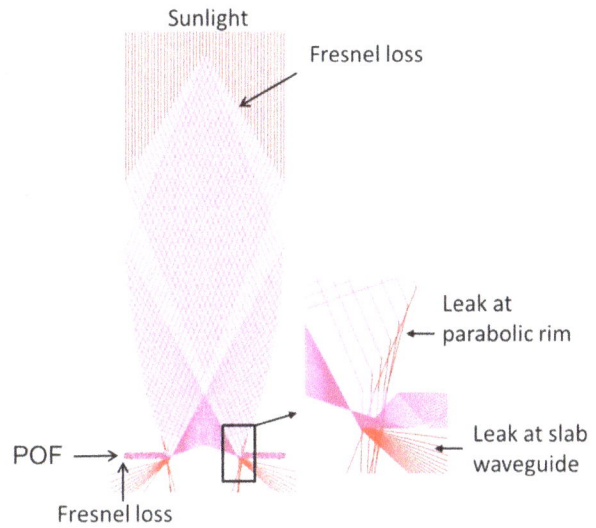

Figure 7. Losses mechanism of system.

In order to quantify the effect of CPC shape on the efficiency of the system, a parametric analysis is carried out whilst varying its CPC concentration ratio C_{CPC}. The C_{CPC} is defined by ratio between input apature size and output apature size of CPC. Figure 8 shows the variation of P-CPC concentration system with different C_{CPC}. We fix the input aperture of CPC (input size of system) $D = 200$ mm. The diameter of POF is fixed at $d = 2$ mm. Base on Equation 2, the concentration ratio of system archieves $C_R = 50$.

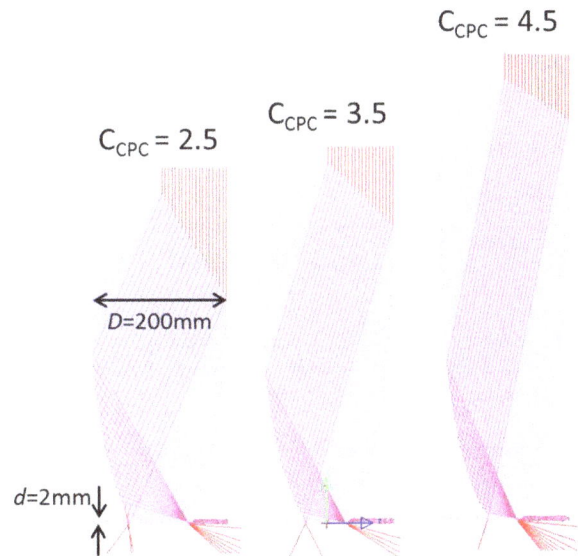

Figure 8. P-CPC concentration system shape with different shape of CPC.

To optimization the structure of P-CPC to get highest efficiency, we carried out simulation with several different CPC concentration ratio C_{CPC} in the range of 2 to 8 mm in increments of 0.5 mm. Figure 9 shows that $C_{CPC} = 3.5$ is the optimal size for structure of CPC to obtain the highest efficiency of 89%.

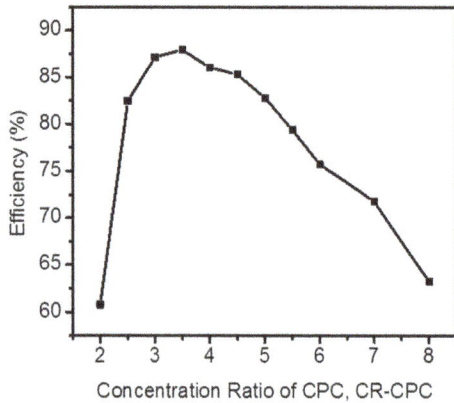

Figure 9. *Variation of optical efficiency at diffirent concentration ratio C_{R-CPC} of CPC.*

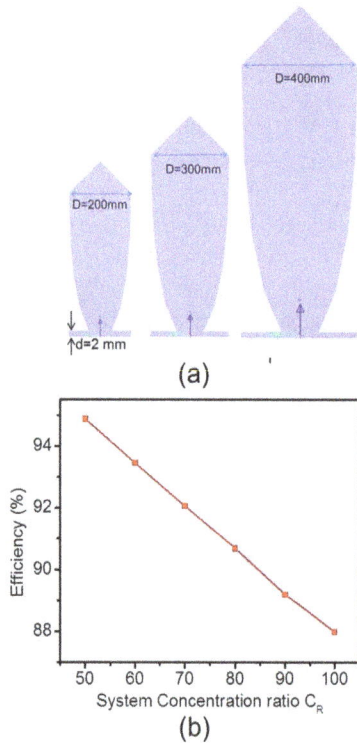

(a)

(b)

Figure 10. *(a) The variation of P-CPC shape with different concentration ratio C_R. (b) The variation of optical efficiency at different system concentration ratio.*

3.2. The Dependence of Efficiency on the System Concentration ratio C_R

In this proposed system, the prism attached at the top of CPC to direct sunlight beam as shown in Figure 3. However, the prism has an inherent disadvantage that is the dispersion of the solar spectrum. For optical fiber daylighting systems, the dispersion of solar spectrum range is very important since it leads to an essential decrease of the optical efficiency and concentration ratio of the systems. The sunlight is dispersed at the focal point of P-CPC due to the wavelength dependence of refractive index of prism material. The focused area also defines how big the sun image will be at the focal point, and it affect on the coupling between CPC

and POFs. Larger core of POFs will capture more focused sunlight but decreases the concentration ratio of system.

We used ray tracing in LightTools^{TM} to analyze the dependence of efficiency on concentration system that is directed by the dispersion phenomanon. The sunlight source used in the analysis is in the range of 300 - 750 nm. POF diameter is fixed at $d = 2$ mm and P-CPC with input aperture D vary from 200 mm to 400 mm in step of 40 mm. It means the concentration ratio C_R decreases from 100 to 50 in step of 10. Figure 10 (a) illustrates the P-CPC structure with some different input aperture width $D = 200$ mm, 300 mm, 400 mm, respectively. Figure 10 (b) shows the simulated optical efficiency at different sytem concentration ratio. It can be seen that because of dispersion, system efficiency η is almost linearly decreased with the increase of C_R. The lower concentration ratio can provide higher optical efficiency but also reduce sunlight capturing area.

3.3. Tolerance of the System

For proper operation of the proposed daylighting concentration system, direct sunlight should always be paralle to the main axis of P-CPC. This is a difficult task since the position of the Sun is always changing, and this led us to use a Sun tracking system. The required accuracy of the Sun tracking system is determined by the solar concentrating collector's angle of tolerance [22]–[25]. The tolerance of the system is the acceptable angular deviation of the sunlight direction from the two main axes of the system, within the allowable efficiency loss. It is defined as the angle where the efficiency drops by 10% [26]. The acceptance angle determines the required accuracy of the tracking system mounted upon the concentrator. The dependence of the optical efficiency of the system on angular deviation along the North-South (NS) and East-West (ES) direction are very different because the system is not symmetric. We examined the efficiency with different angular deviations of the sunlight direction along the NS and EW directions. The alignment of system along NS and EW direction was shown in Figure 11.

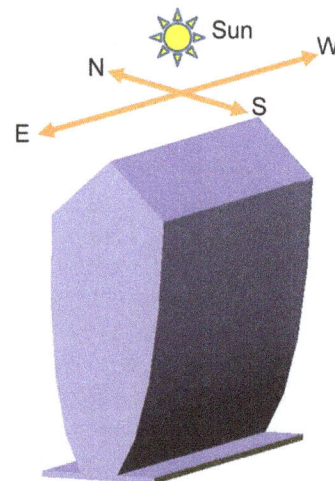

Figure 11. *The alignment of system along NS and EW direction.*

Figure 12 (a) shows the optical efficiency output relative to angular deviation along the EW direction, and Figure 12 (b) shows a graph of effciency versus angular deviation along the NS-direction. The simulation results show that the tolerance is more than ±6 degrees along the EW-direction, which is far larger than that for the EW-direction (±0.5°). This indicates that by using a P-CPC concentrator, the acceptance angle

along the EW-direction can be greatly increased without sacrificing too much optical efficiency. Therefore, the proposed system uses P-CPC instead of a conventional lens or parabolic mirror as the concentrator can lower the accuracy requirements along the EW-direction, and this reduces the cost of the tracking system.

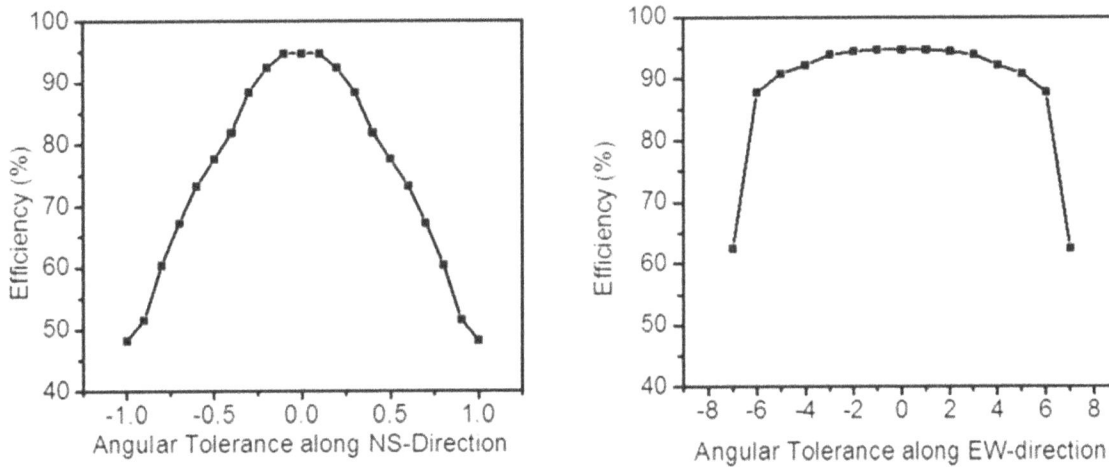

Figure 12. Variation of optical efficiency of concentrator at different angular deviations along the (a) EW-direction and (b) NS-direction.

Table 2. Average illuminance at different times of the day and calculated luminous flux at output for the proposed system.

Time	Solar Altitude (°)	Sunlight Illuminance (lux)	Luminous Flux on the Surface Concentrator (lm)	Luminous Flux at the Output of Concentrator (lm)
6 AM	8	20,000	4000	3800
7 AM	19	40,000	8000	7600
8 AM	31	60,000	12000	11400
9 AM	43	80,000	16000	15200
10 AM	54	100,000	20000	19000
11 AM	65	105,000	21000	19950
12 PM	74	110,000	22000	20900

3.4. Daylighting

The illuminance from the sunlight was measured at different times of the day. The site of application was located at 127° longitude and 37.5° latitude. Here we will look at the illuminance on a summer day as an example: The highest solar elevation (zenith) angle at the site is 76°, and the time is set for 12:30 PM. To achieve direct sunlight, we assume that the daylighting system has a Sun tracking device which rotates the concentrator module toward the Sun all day. The area of the sunlight collector is 0.2 m² if we assume that length of P-CPC is 1 m and width is 0.2 m. The measured illuminances of the input flux at the surface of the concentrator and of the luminous flux at the output concentrator are listed in Table 2.

4. Conclusion

An optical fiber daylighting system using P-CPCs has been designed and discussed with the purpose of saving the energy. To explore the practical performance of the proposed system,

a sample optical system was modeled and simulated with LightTools™. The simulation results indicate that 89% of optical efficiency was achieved at $C_R = 100$ for the proposed concentrator system. In addition, the tolerance (acceptance angles) in the NS- and the EW-directions also were analyzed. By using a P-CPC, an acceptance angle of ±6° was achieved in the EW-direction. This allows us to use a lower accuracy sun tracking system, such as a passive Sun tracking system along the EW-direction as a cost effective solution. This study is the first to use a combination of prism and compound parabolic concentrator for an daylighting system. It shows great potential for the commercial and industrial scale daylighting application. In the future, we will try to implement experimentation under real conditions to verifying the accuracy of the simulation and the commercial viability of the system.

Acknowledgements

This work was supported by a National Research Foundation of Korea (NRF) grant funded by the Korea government (MSIP) (No. 2014R1A2A1A11051888).

References

[1] T.-C. Teng and W.-C. Lai, "Planar solar concentrator featuring alignment-free total-internal-reflection collectors and an innovative compound tracker," *Opt. Express*, vol. 22, no. S7, p. A1818, 2014.

[2] M. A. Duguay and R. M. Edgar, "Lighting with sunlight using sun tracking concentrators," *Appl. Opt.*, vol. 16, no. 5, p. 1444, 1977.

[3] I. Ullah and S. Shin, "Highly concentrated optical fiber-based daylighting systems for multi-floor office buildings," *Energy Build.*, vol. 72, pp. 246–261, 2014.

[4] C.-H. Tsuei, W.-S. Sun, and C.-C. Kuo, "Hybrid sunlight/LED illumination and renewable solar energy saving concepts for indoor lighting.," *Opt. Express*, vol. 18 Suppl 4, no. November, pp. A640–A653, 2010.

[5] A. J.-W. Whang, Y.-Y. Chen, S.-H. Yang, P.-H. Pan, K.-H. Chou, Y.-C. Lee, Z.-Y. Lee, C.-A. Chen, and C.-N. Chen, "Natural light illumination system.," *Appl. Opt.*, vol. 49, no. 35, pp. 6789–6801, 2010.

[6] I. Ullah and S. Shin, "Highly concentrated optical fiber-based daylighting systems for multi-floor office buildings," *Energy Build.*, vol. 72, pp. 246–261, 2014.

[7] R. Núñez, I. Antón, and G. Sala, "Hybrid lighting-CPV, a new efficient concept combining illumination with CPV," *AIP Conf. Proc.*, vol. 1477, no. 4, pp. 221–224, 2012.

[8] W. T. Xie, Y. J. Dai, R. Z. Wang, and K. Sumathy, "Concentrated solar energy applications using Fresnel lenses: A review," *Renew. Sustain. Energy Rev.*, vol. 15, no. 6, pp. 2588–2606, 2011.

[9] H. Abdul-Rahman and C. Wang, "Limitations in current day lighting related solar concentration devices : A critical review," *Int. J. Phys. Sci. Vol.*, vol. 5, no. 18, pp. 2730–2756, 2010.

[10] B. Bouchet and M. Fontoynont, "Day-lighting of underground spaces: Design rules," *Energy Build.*, vol. 23, no. 3, pp. 293–298, 1996.

[11] N. Vu and S. Shin, "A Large Scale Daylighting System Based on a Stepped Thickness Waveguide," *Energies*, vol. 9, no. 2, p. 71, 2016.

[12] E. Ghisi and J. A. Tinker, "Evaluating the potential for energy savings on lighting by integrating fibre optics in buildings," *Build. Environ.*, vol. 41, no. 12, pp. 1611–1621, 2006.

[13] T. K. Mallick and P. C. Eames, "Design and fabrication of low concentrating second generation PRIDE concentrator," *Sol. Energy Mater. Sol. Cells*, vol. 91, no. 7, pp. 597–608, 2007.

[14] T. K. Mallick, P. C. Eames, and B. Norton, "Non-concentrating and asymmetric compound parabolic concentrating building facade integrated photovoltaics: An experimental comparison," *Sol. Energy*, vol. 80, no. 7, pp. 834–849, 2006.

[15] R. Winston, "NONIMAGING OPTICS Limits to Concentration," 2008.

[16] R. Winston and J. M. Gordon, "Planar concentrators near the étendue limit.," *Opt. Lett.*, vol. 30, no. 19, pp. 2617–2619, 2005.

[17] R. Winston and W. Zhang, "Pushing concentration of stationary solar concentrators to the limit," *Opt. Express*, vol. 18, no. 9, pp. A64–72, 2010.

[18] A. Rabl, "Comparison of solar concentrators," *Sol. Energy*, vol. 18, no. 2, pp. 93–111, 1976.

[19] N. B. Goodman, R. Ignatius, L. Wharton, and R. Winston, "Solid-dielectric compound parabolic concentrators: on their use with photovoltaic devices.," *Appl. Opt.*, vol. 15, no. 10, pp. 2434–6, 1976.

[20] H. J. Han, S. B. Riffat, S. H. Lim, and S. J. Oh, "Fiber optic solar lighting: Functional competitiveness and potential," *Sol. Energy*, vol. 94, pp. 86–101, 2013.

[21] O. Selimoglu and R. Turan, "Exploration of the horizontally staggered light guides for high concentration CPV applications," *Opt. Express*, vol. 20, no. 17, p. 19137, 2012.

[22] C. Lee, P. Chou, C. Chiang, and C. Lin, "Sun tracking systems: A review," *Sensors*, vol. 9, no. 5, pp. 3875–90, 2009.

[23] M. T. A. Khan, S. M. S. Tanzil, R. Rahman, and S. M. S. Alam, "Design and construction of an automatic solar tracking system," *ICECE 2010 - 6th Int. Conf. Electr. Comput. Eng.*, no. December, pp. 326–329, 2010.

[24] J. Song, Y. Zhu, Z. Jin, and Y. Yang, "Daylighting system via fibers based on two-stage sun-tracking model," *Sol. Energy*, vol. 108, pp. 331–339, 2014.

[25] H. Mousazadeh, A. Keyhani, A. Javadi, H. Mobli, K. Abrinia, and A. Sharifi, "A review of principle and sun-tracking methods for maximizing solar systems output," *Renew. Sustain. Energy Rev.*, vol. 13, no. 8, pp. 1800–1818, 2009.

[26] P. Xie, H. Lin, Y. Liu, and B. Li, "Total internal reflection-based planar waveguide solar concentrator with symmetric air prisms as couplers," *Opt. Express*, vol. 22, no. S6, p. A1389, 2014.

22

Prandtl Number Effect of Mixed Convection Heat and Mass Transfer in a Triangular Enclosure with Heated Circular Obstacle

Sayeda Fahmida Ferdousi[1, 2, *], **Md. Abdul Alim**[2], **Raju Chowdhury**[1, 2]

[1]Department of Natural Science, Stamford University Bangladesh, Dhaka, Bangladesh
[2]Department of Mathematics, Bangladesh University of Engineering & Technology, Dhaka, Bangladesh

Email address:
fahmida.buet@gamil.com (S. F. Ferdousi)
[*]Corresponding author

Abstract: The effect of Prandtl number of mixed convection heat and mass transfer in a triangular enclosure with heated and concentrated circular obstacle is analyzed by solving mass, momentum, energy and concentration balance equations. The left lower middle and right upper middle walls are kept at low temperature and concentration. All others wall are assumed to be adiabatic. The lower wall is moving in the +x direction and all others walls are maintained at no-slip condition. Moreover, Galerkin Weighted Residuals finite element method is applied to solve the governing equations. The study is performed for different values of Prandtl number, Richardson number and buoyancy ratio. A simple transformation is employed to transfer the governing equations into a dimensionless form. The result shows that at high *Pr* heat transfer rate increase rapidly and at low *Pr* it increases linearly with the increase of *Ri*. However, buoyancy ratio and Lewis number plays an important role for the flow, temperature and concentration fields.

Keywords: Mixed Convection, Circular Obstacle, Heat and Mass Transfer, Sliding Wall, Triangular Cavity

1. Introduction

Mixed convection heat and mass transfer in a lid-driven closed cavity have received a considerable attention in the past decades [1-4]. This problem is often encountered in a variety of engineering applications such as in the cooling of electronic devices, food processing and nuclear reactors etc. Numerous studies have been performed on the heat and mass transfer in enclosure.

Combined heat transfer in triangular enclosure has been investigated by increasing number of researchers. Triangular enclosures can be used in the roofs of the buildings or electronic heaters [5-8]. Koca et al. [9] analyzed the effect of Prandtl number on natural convection in triangular enclosures with localized heating from below. Teamah et al. [10] studied numerical simulation of double diffusive natural convective flow in an inclined rectangular enclosure in the

presence of magnetic field and heat source. Xu et al [11] investigated numerical simulation of double diffusive mixed convection in an open enclosure with different cylinder locations. Rahman et al. [12] studied natural convection on heat and mass transfer in a curvilinear triangular cavity. Hasanuzzaman et al. [13] analyzed the effects of Lewis number on heat and mass transfer in a triangular cavity. Comparison of flow and heat transfer characteristics in a lid-driven cavity between flexible and modified geometry of a heated bottom wall was investigated by K. Khanafer [14]. Ching et al. [15] studied finite element simulation of mixed convection heat and mass transfer in a right triangular enclosure. Finite element simulation of MHD combined convection through a triangular wavy channel was investigated by Parvin and Hossain [16]. Rahman et al. [17] investigated numerical study on the conjugate effect of joule heating and magneto-hydrodynamics mixed convection in an obstructed lid-driven square cavity. Al-Salem et al [18] studied effects of moving lid direction on MHD mixed convection in a linearly heated cavity. They found that direction of lid is more effective on heat transfer and fluid flow in the case of mixed convection than it is the case in forced convection. Al-Amiri et al [19] studied effect of sinusoidal wavy bottom surface on mixed convection heat transfer in a lid-driven cavity. Recently, Chowdhury et al. [20] has analyzed natural convection in a porous triangular enclosure with a circular obstacle in presence of heat generation.

On the basis of the above literature review, it appears that the mixed convection heat and mass transfer in enclosures was widely studied under different circumstances. However, little attention has been paid on such a problem. The main objective of this paper is to examine the effect of buoyancy ratio, Lewis number, Richardson number and the moving wall on the heat and mass transport in the cavity.

2. Mathematical Model

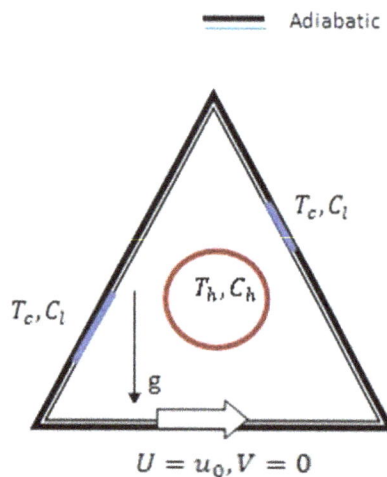

Fig. 1. Schematic diagram of the triangular cavity.

The physical system considered in the present study is presented in the fig. 1. It is a two dimensional triangular enclosure with a heated obstacle in the middle of the triangular. The length and height of the enclosure are depicted by L and H respectively. The bottom wall is moves to the positive x-direction. The circular obstacle has the high temperature and high concentration. An external flow with low temperature and low concentration enters the middle-lower of the left wall and middle-upper to the right wall. All other walls are assumed to be adiabatic. The diameter of the circular obstacle is kept at 0.4 in this work. The inlet size of both the wall is equal to one fifth of the enclosure length $(L/5)$. All solid boundaries are assumed to be rigid and no-slip walls.

3. Governing Equations

The governing equations are based on the conservation of mass, momentum, energy and species concentration. The density in the buoyancy force term of the y-momentum equation follows the Boussinesq approximation. The detailed dimensionless governing equations can be written as following:

$$\frac{\partial U}{\partial X} + \frac{\partial V}{\partial Y} = 0 \tag{1}$$

$$U\frac{\partial U}{\partial X} + V\frac{\partial U}{\partial Y} = -\frac{\partial P}{\partial X} + \frac{1}{Re}\left(\frac{\partial^2 U}{\partial X^2} + \frac{\partial^2 U}{\partial Y^2}\right) \tag{2}$$

$$U\frac{\partial V}{\partial X} + V\frac{\partial V}{\partial Y} = -\frac{\partial P}{\partial Y} + \frac{1}{Re}\left(\frac{\partial^2 V}{\partial X^2} + \frac{\partial^2 V}{\partial Y^2}\right) + Ri(\theta + Br\,C) \tag{3}$$

$$U\frac{\partial \theta}{\partial X} + V\frac{\partial \theta}{\partial Y} = \frac{1}{Re\,Pr}\left(\frac{\partial^2 \theta}{\partial X^2} + \frac{\partial^2 \theta}{\partial Y^2}\right) \tag{4}$$

$$U\frac{\partial C}{\partial X} + V\frac{\partial C}{\partial Y} = \frac{1}{LeRePr}\left(\frac{\partial^2 C}{\partial X^2} + \frac{\partial^2 C}{\partial Y^2}\right) \tag{5}$$

where the dimensionless variables are defined as:

$$X = \frac{x}{L}, Y = \frac{y}{L}, U = \frac{u}{u_0}, V = \frac{v}{u_0}, P = \frac{(p+\rho gy)}{\rho u_0^2}, \theta = \frac{T-T_c}{T_H-T_c}, C = \frac{C-C_L}{C_H-C_L} \tag{6}$$

and the five dimensionless parameters, Reynolds number Re, Prandtl number Pr, Richardson number Ri, Lewis number Le and buoyancy ratio Br, are:

$$Pr = \frac{v}{\alpha}, Re = \frac{Lu_0}{v}, Ri = \frac{g\beta_T(T_H-T_L)\,L}{u_0^2}, Br = \frac{\beta_C(C_H-C_L)}{\beta_T(T_H-T_C)} \text{ and } Le = \frac{\alpha}{D} \tag{7}$$

The Richardson number Ri, characterizes the relative importance of natural to forced convections. In this study, Ri is set to be 0.1,1,5 and 10 representing the natural convection to forced convection. Lewis number Le, reflects the mass transfer exchange in the whole enclosure. The buoyancy ratio Br, are also studied with different values to consider the effect of double diffusive mixed convection. Prandtl number Pr, is taken 0.1 for air and 7.0 for water. The dimensionless boundary conditions are specified as follows:
On the middle-lower left wall:

$$U = V = 0, \theta = C = 0 \tag{8a}$$

On the middle-upper right wall:

$$U = V = 0, \theta = C = 0 \tag{8b}$$

On the heated obstacle:

$$U = V = 0, \theta = C = 1 \tag{8c}$$

On the bottom wall:

$$U = 1, V = 0 \tag{8d}$$

On all solid cavity walls:

$$U = V = 0, \frac{\partial \theta}{\partial n} = \frac{\partial C}{\partial n} = 0 \qquad (8e)$$

where n is the non-dimensional distances either X or Y direction acting normal to the surface.

The average nusselt number and sherwood numbers evaluated along the circular obstacle of the cavity based on the dimensionless quantities may be expressed as

$$Nu_{av} = -\frac{1}{2\pi} \int_0^{2\pi} \left(\frac{\partial \theta}{\partial n} \right) d\varphi \qquad (9)$$

and

$$Sh_{av} = -\frac{1}{2\pi} \int_0^{2\pi} \left(\frac{\partial C}{\partial n} \right) d\varphi \qquad (10)$$

Table 1. Grid sensitivity check at $Re = 100, Ri = 0.1, Le = 10$ and $Br = 20$.

Pr	Element	7649	1666	1159	781
0.1	Nu_{av}	8.2867	8.3209	8.26817	8.43853
	Sh_{av}	0.08453	0.08261	0.08201	0.08351
7	Nu_{av}	12.43819	12.6645	12.56452	13.0009
	Sh_{av}	0.00118	0.00118	0.00117	0.00119

4. Numerical Treatment

Finite element method has been used to solve Eqs. $(2) - (5)$ with the boundary conditions of Eqs. $7(a) - 7(e)$ numerically. The numerical method used in this study is based on the finite element method to discretize the governing equations (Eqs. $(2) - (5)$) and the set of algebraic equations are solved using a stationary non-linear together with direct linear system solver. In the case of all dependent variables, the relative tolerance for the error criteria was 10^{-4}. Non-uniform triangular grid system is engaged in this study. The Nusselt and Sherwood numbers for different grid sizes (781 to 7649) are presented to develop an understanding of the grid fineness that is necessary for accurate numerical simulation (Table 1). There is considerable change in Nusselt number for different values of Prandtl number with different grid sizes and no considerable change found in Sherwood number for different values of Prandtl number but changes occurs for different grid size.

5. Program Validation

Simulations were carried out to study the double-diffusive mixed convection in an open enclosure with different cylinder locations [11]. Here an external flow with low temperature and concentration enters the lower-left wall and there is an exit at the upper-right wall. Heated and concentrated cylinder is placed on the middle of the enclosure. Rest of the walls are assumed to be adiabatic. All solid boundaries are assumed to be rigid and no-slip condition. Figs. 2 and 3 show the comparison of the predicted results of isoconcentration, streamline and isotherm by this program with the result from Xu et al. [11].

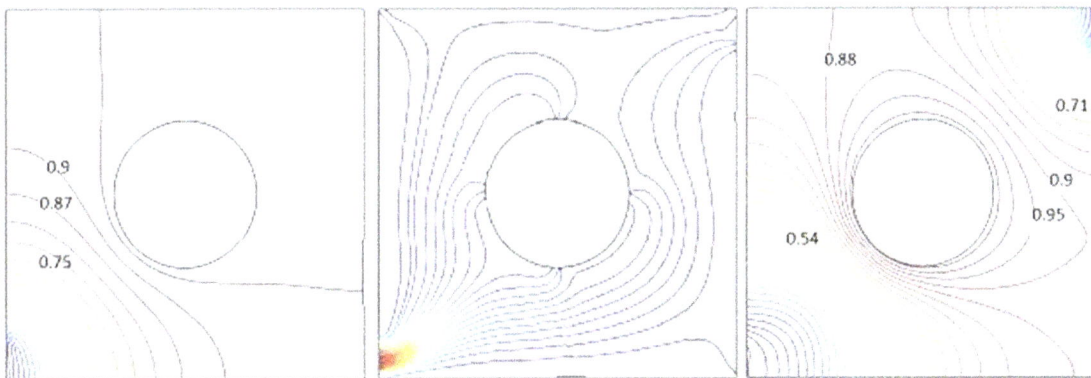

Fig. 2. Isoconcentration, streamline and isotherm at $Pr = 0.7, Ri = 1.0, Br = 1.0$ and $Le = 0.1$ (present).

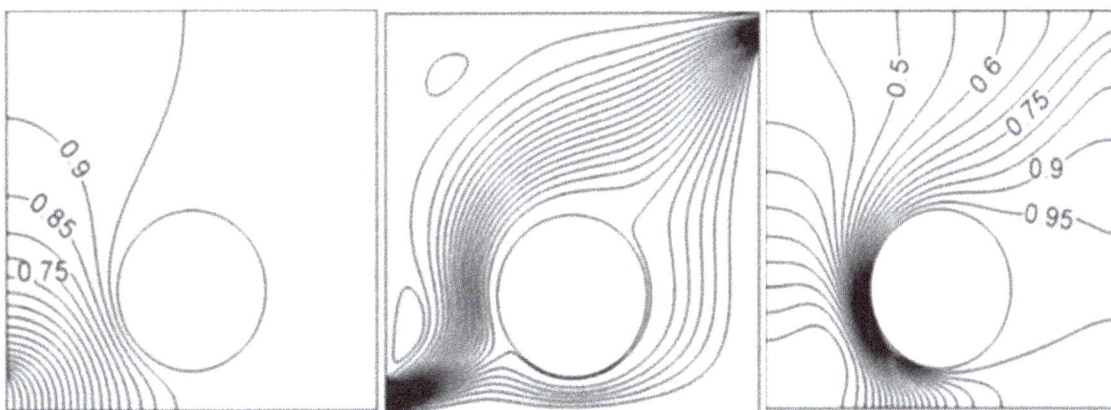

Fig. 3. Isoconcentration, streamline and isotherm at $Pr = 0.7, Ri = 1.0, Br = 1.0$ and $Le = 0.1$ (Xu et al. [11]).

6. Results and Discussion

A numerical simulation has been done by using finite element method and has investigated the mixed convection heat and mass transfer in a lid-driven triangular enclosure having a heated obstacle. Two cases were considered and tested according to different values of Prandtl number $i.e.\,0.1$ for air and 7 for water respectively. Here the governing parameter affecting heat transfer is the Richardson number Ri. For $Ri < 1$, the flow and heat transfer is dominated by forced convection, for $Ri > 1$, it is dominated by natural convection and for $Ri = 1$, it is a mixed regime. Computations are chosen for different parameters such as Prandtl number, $Pr(0.1$ and $7)$, the Reynolds number, $Re(= 100)$, buoyancy ratio $(10 < Br < 50)$, Lewis number $(0 < Le < 100)$ and Richardson number $(0.1 < Ri < 10)$.

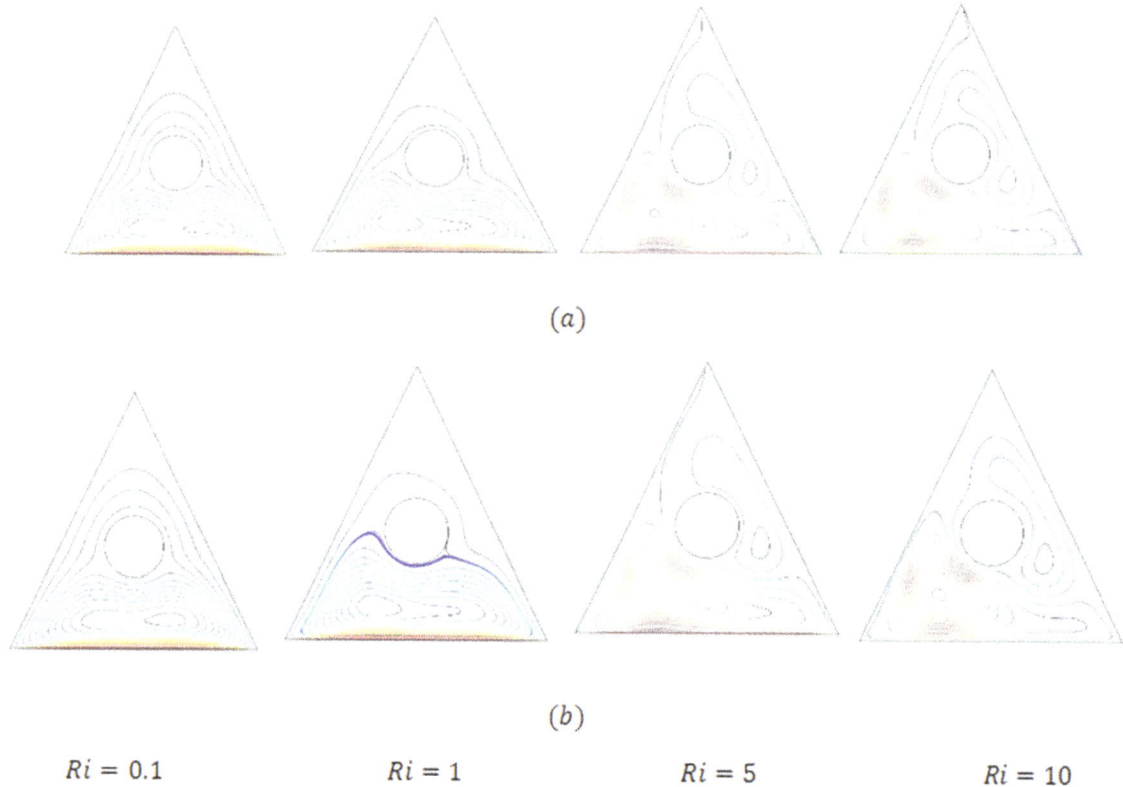

(a)

(b)

$Ri = 0.1$ $Ri = 1$ $Ri = 5$ $Ri = 10$

Fig. 4. *Effect of Richardson number on streamlines for (a) case 1 and (b) case 2 with $Br = 10$.*

Fig. 4. shows the effect of Richardson number on streamlines for two different cases as case 1 (for air with $Pr = 0.1$) and case 2 (for water with $Pr = 7$) at $Br = 10$. Here the lid moves to the $+x$ direction and fluid inside the cavity impinges to the left corner of the triangular cavity with the increase of Ri and it circulates in clockwise direction for both cases. However, at $Pr = 7$ the top corner of the triangular cavity becomes stagnant and there is no other circulation cell inside the cavity with the increases of Richardson number. As seen from the figures, effects of Lewis number become insignificant on the flow field inside the cavity. Again fig. 7. shows the effect of buoyancy ratio on streamlines for case1 and case 2 with Ri=5. These shows that with the increase of buoyancy ratio circulation can be formed within the right wall included the circular obstacle. At low Prandtl number no circulation can be twisted but at high Prandtl number circulation becomes twisted and and another circulation found at the right corner of the triangular cavity. Also Lewis number is not effective for both cases.

Effect of Richardson number on isotherm are presented in fig. 5. for both case. The cavity is heated due to the circular obstacle in the middle but is suppressed by moving lid in the $+x$ direction. Isotherms are clustered near the left lower middle and the right upper middle wall respectively. At low Prandtl number with the increase of Ri, heat can flows all over the cavity. But at high Prandtl number the right corner of the cavity become stagnant and no heat can flows in that region. Also heat flows around the circular obstacle and reaches slowly left to top corner of the cavity with the increase of Ri at High Prandtl number. Fig. 8. shows the effect of buoyancy ratio on isotherms for both cases. These figures shows that what happens if Br varies from 10 to 50. Here we observed that with increase of buoyancy ratio heat moves from top to right wall i.e. it moves in clockwise direction at high Prandtl number. Whereas at low Prandtl number the changes in isotherms are negligible.

Isoconcentration contours are shown in fig. 6. and fig. 9. Fig. 6. illustrates that at low Richardson number there is no change in isoconcentration for both cases, but at high Ri there is a little bit change. Fig. 9. shows that with the increase of buoyancy ratio isoconcentration are distributed from circular obstacle to the right inclined wall.

Heat and mass transfer are presented graphically in fig. 10. It shows that with increase of Richardson number Nusselt number increases. But at low Prandtl number this increasing graph is negligible whereas at high Prandtl number this increasing graph is noticeable. On the other hand at high Prandlt number Sherwood number become constant with the increase of Lewis number. But at high Prandtl number Sherwood number decreases with the increase of Lewis number.

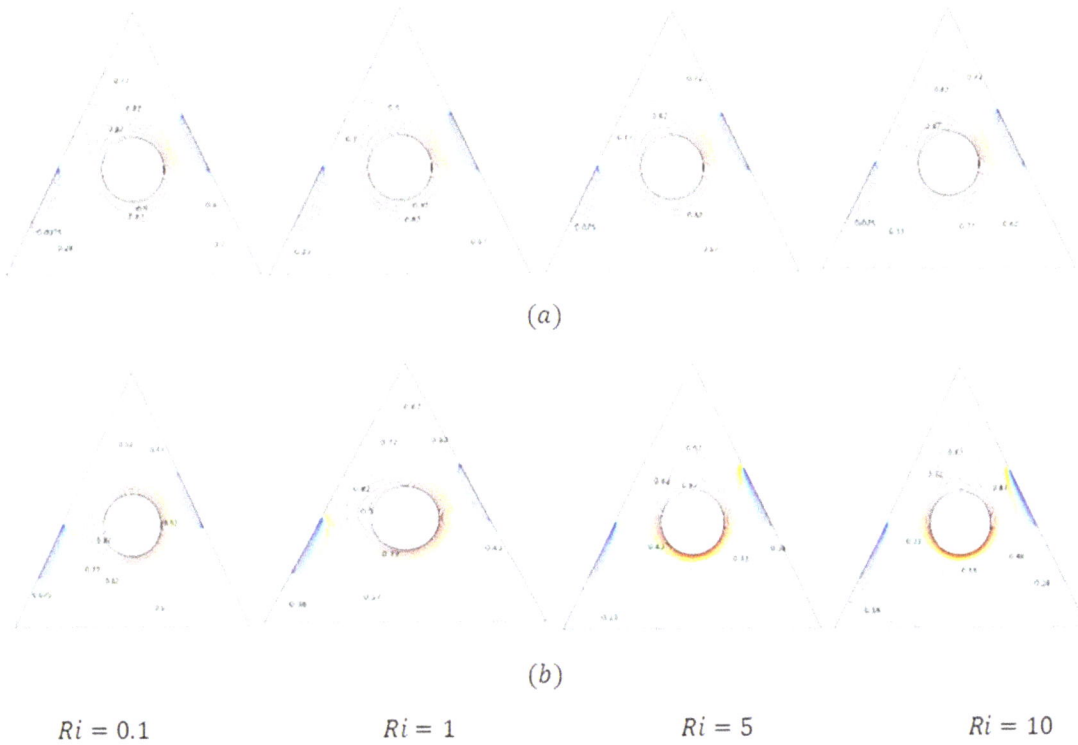

(a)

(b)

$Ri = 0.1$ $Ri = 1$ $Ri = 5$ $Ri = 10$

Fig. 5. *Effect of Richardson number on isotherm for (a) case 1 and (b) case 2 with Br = 10.*

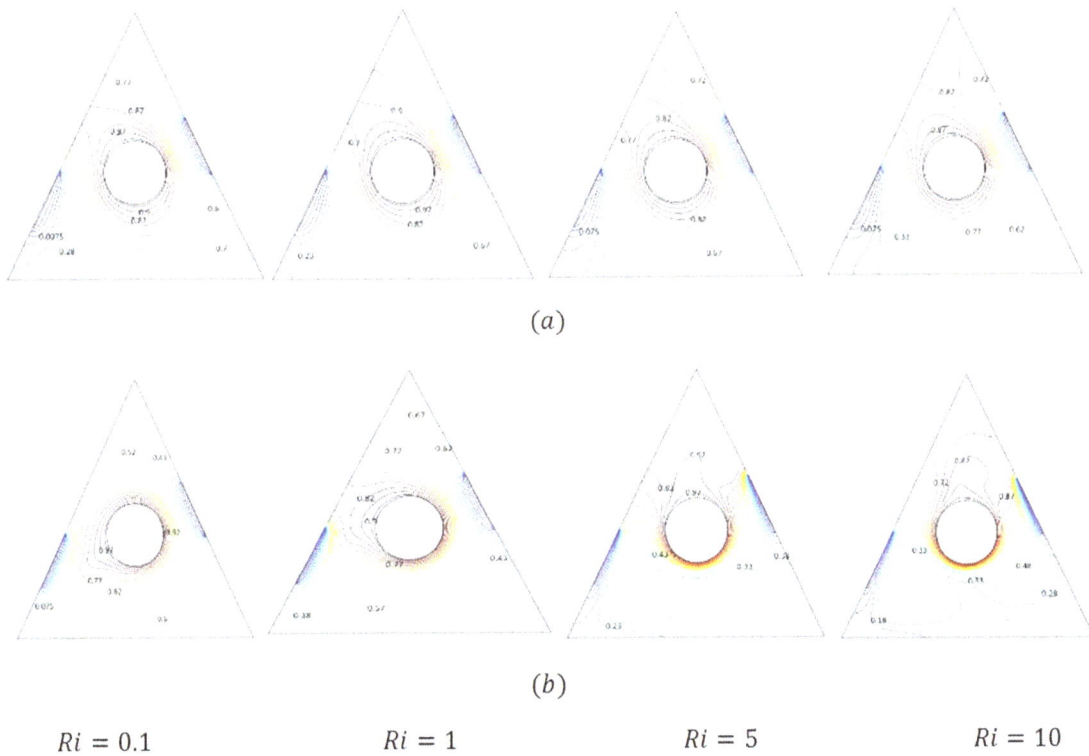

(a)

(b)

$Ri = 0.1$ $Ri = 1$ $Ri = 5$ $Ri = 10$

Fig. 6. *Effect of Richardson number on isoconcentration for (a) case 1 and (b) case 2 with Br = 10.*

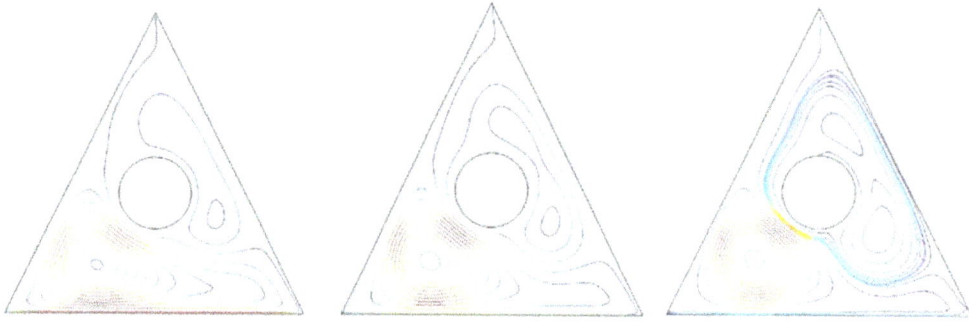

(a)

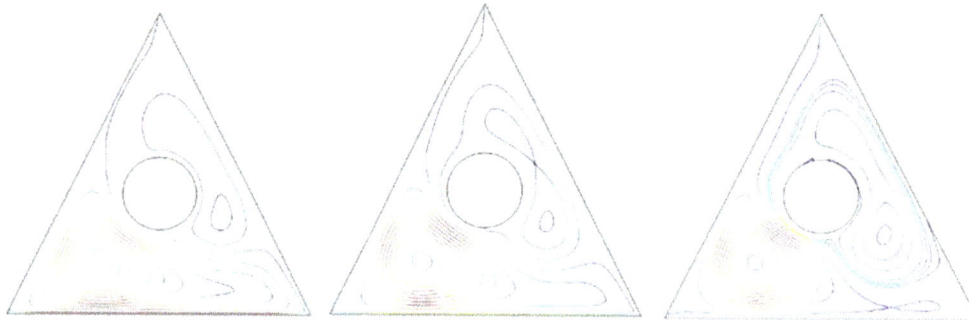

(b)

$Br = 10$ $Br = 20$ $Br = 50$

Fig. 7. *Effect of buoyancy ratio on streamlines for (a) case 1 and (b) case 2 with Ri = 5.*

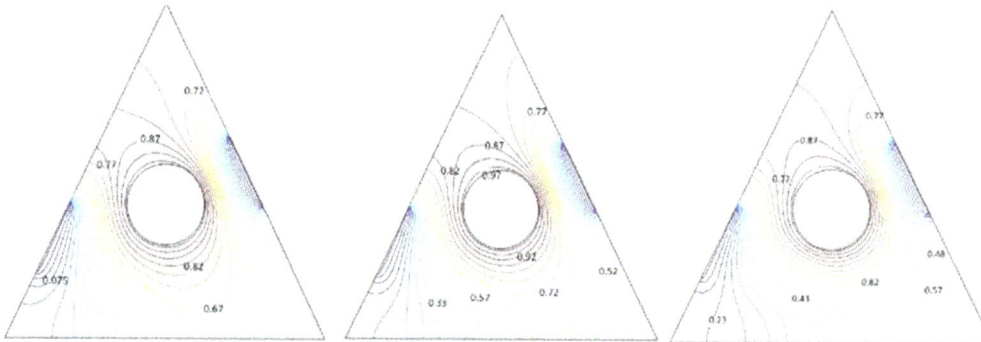

(a)

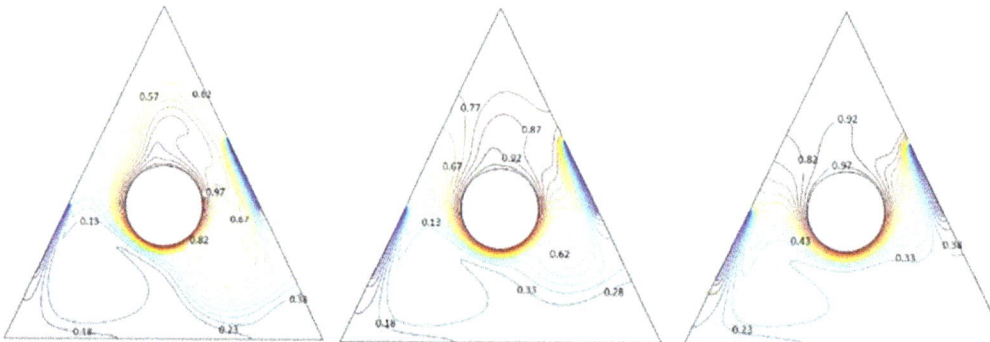

(b)

$Br = 10$ $Br = 20$ $Br = 50$

Fig. 8. *Effect of buoyancy ratio on isotherms for (a) case 1 and (b) case 2 with Ri = 5.*

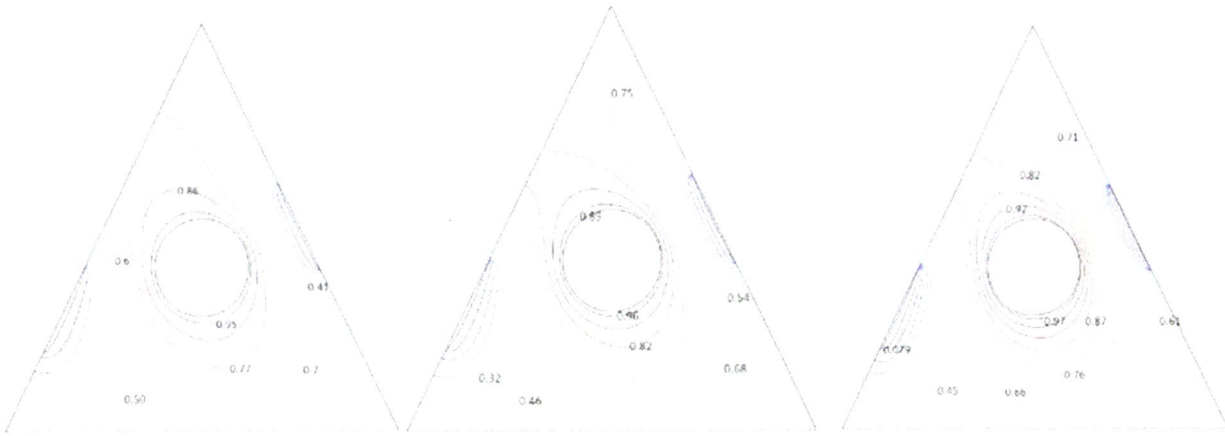

(a)

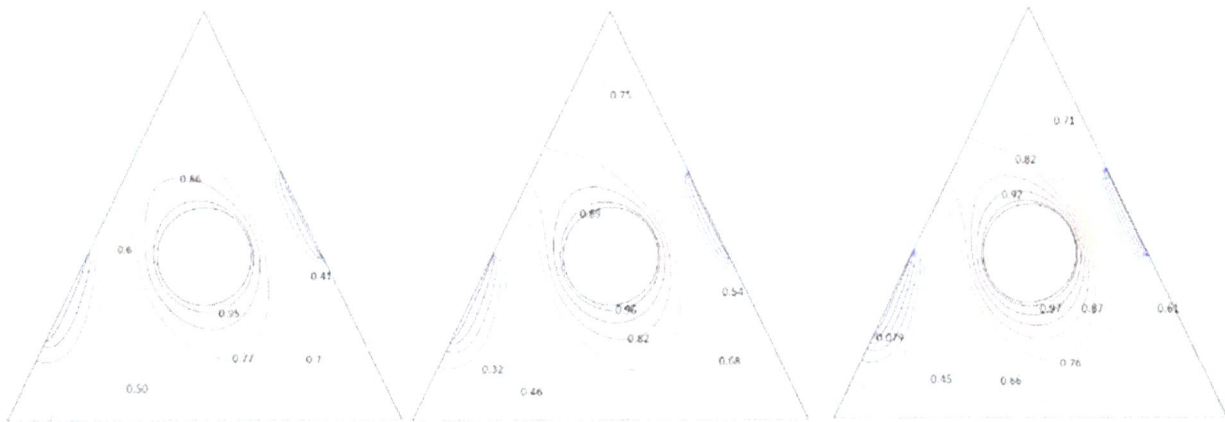

(b)

$Br = 10$ $Br = 20$ $Br = 50$

Fig. 9. *Effect of buoyancy ratio on isoconcentration for (a) case 1 and (b) case 2 with Ri = 5.*

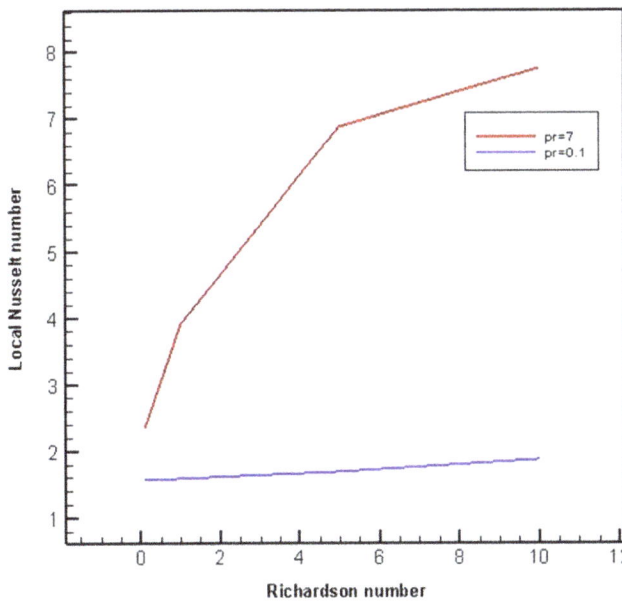

(a)

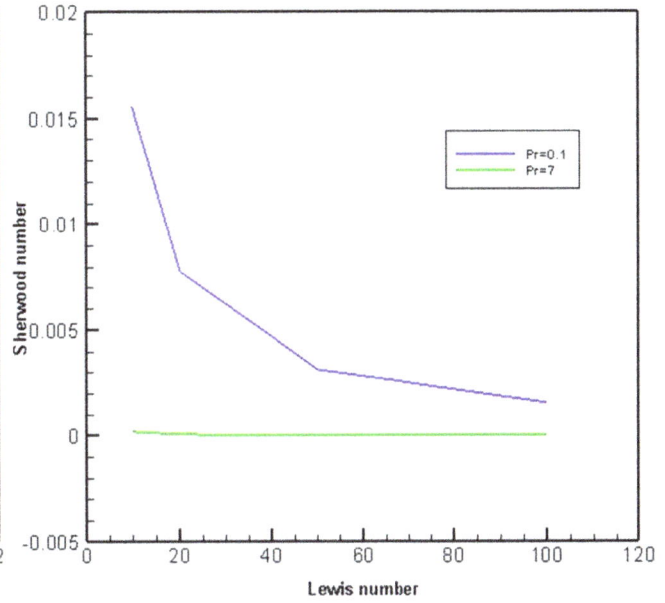

(c)

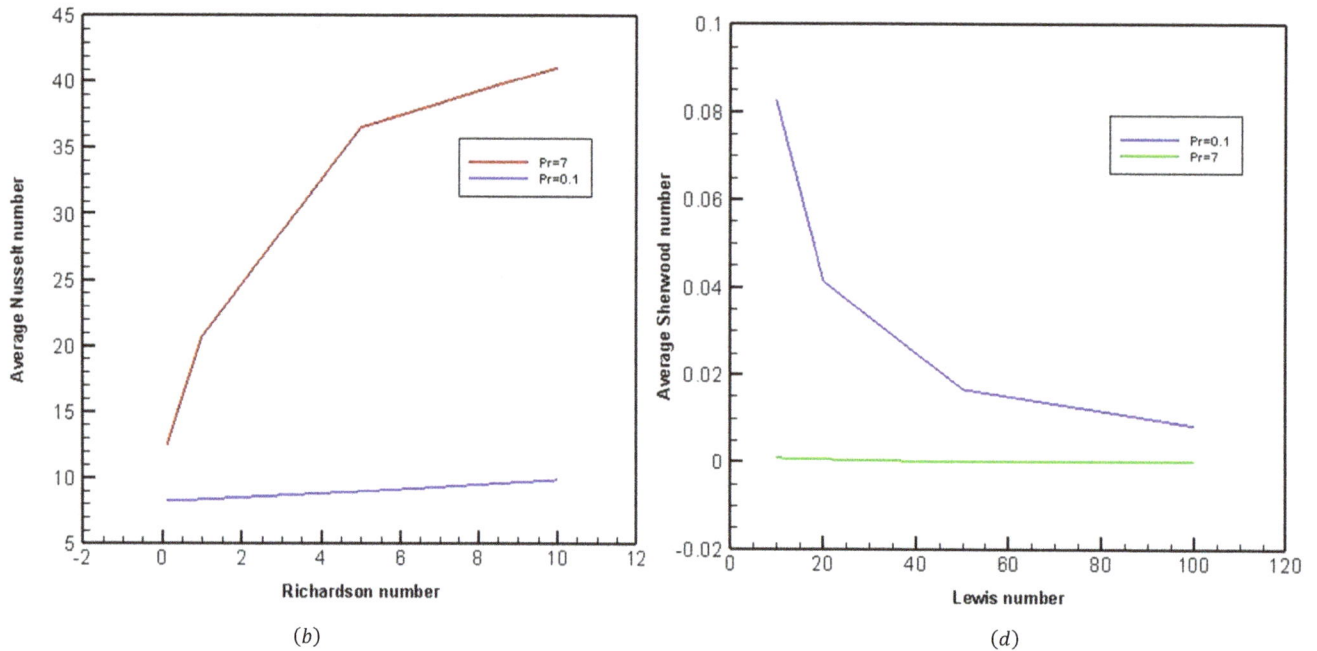

(b)   (d)

Fig. 10. *(a) Local Nusselt number versus Ri (b) Average Nusselt number versus Ri (c) Sherwood number versus Le (d) Average Sherwood number versus Le for both cases.*

7. Conclusion

A computational work has been performed to study the effects of Richardson number and buoyancy ratio on heat and mass transfer in a triangular cavity with heated obstacle and right and left inlet cold wall for different values of Prandtl number. Some important findings can be listed as follows:

Heat transfer increases with the increase of Richardson number. At high Prandtl number heat transfer rate increases rapidly with the increase of Richardson number. But at low Prandtl number heat transfer rate increases linearly with the increase of Richardson number.

Sherwood number decreases with the increase of Lewis number. At low Prandtl number Sherwood number decrease rapidly with the increase of Lewis number. But at high Prandtl number Sherwood number moves constantly with the increase of Lewis number.

Buoyancy Ratio plays an important role on temperature distribution. Also it plays important role for high Prandtl number than low Prandtl number.

Lewis number becomes insignificant on the flow field and temperature distribution. However, Lewis number is an effective parameter on isoconcentration.

Nomenclature

Br	buoyancy ratio
h	heat transfer coefficient of the fluid, $w/(m^2.k)$
c	mass concentration, kg/m^3
c_H	concentration at the circular obstacle, kg/m^3
c_L	concentration at the inlet, kg/m^3
C	dimension less mass concentration, $C = \frac{c-c_L}{c_H-c_L}$
D	mass diffusivity, m^2/s

L	cavity length
Le	Lewis number, $Le = \frac{v}{D}$
p	pressure, N/m^2
P	dimensionless pressure, $P = \frac{(p+\rho gy)}{\rho u_0{}^2}$
g	acceleration of gravity, m/s^2
Re	Reynolds number, $Re = \frac{Lv_0}{v}$
Ri	Richardson number, $Ri = \frac{g\beta_T(T_H-T_L)\,L}{u_0{}^2}$
Pr	Prandtl number, $Pr = \frac{v}{\alpha}$
Nu	Nusselt number
Sh	local Sherwood number
Nu_{av}	average Nusselt number
Sh_{av}	average Sherwood number
n	outward normal direction
u	velocity components in x direction, m/s
v	velocity components in y direction, m/s
u_0	lid velocity
U	dimensionless velocity components in X direction
V	dimensionless velocity components in Y direction
x, y	dimensional coordinates, m
X, Y	dimensionless coordinates
T	local temperature
T_H	hot wall temperature
T_L	cold wall temperature
Greek symbols	
α	thermal diffusivity
β_T	thermal expansion coefficient
β_C	compositional expansion coefficient
v	kinematic viscosity
θ	non-dimensional temperature
ρ	density

References

[1] M. Morzynski, C. O. Popiel, Laminar heat transfer in a two-dimensional cavity covered by a moving wall, Numer. Heat Transfer 12 (1988) 265–273.

[2] M. K. Moallemi, K. S. Jang, Prandtl number effects on laminar mixed convection heat transfer in a lid-driven cavity, Int. J. Heat Mass Transfer 35 (1992) 1881–1892.

[3] R. Iwatsu, J. M. Hyun, K. Kuwahara, Mixed convection in a driven cavity with a stable vertical temperature gradient, Int. J. Heat Mass Transfer 36 (1993) 1601–1608.

[4] A. K. Prasad, J. R. Koseff, Combined forced and natural convection heat transfer in a deep lid-driven cavity flow, Int. J. Heat Fluid Flow 17 (1996) 460–467.

[5] H. Asan, L. Namli, Laminar natural convection in a pitched roof of triangular cross-section: summer day boundary conditions, Energy and Buildings 33 (2000) 69–73.

[6] G. A. Holtzman, R. W. Hill, K. S. Ball, Laminar natural convection in isosceles triangular enclosures heated from below and symmetrically cooled from above, J. Heat Transfer 122 (2000) 485–491.

[7] V. A. Akinsete, T. A. Coleman, Heat transfer by steady laminar free convection in triangular enclosures, Int. J. Heat Mass Transfer 25 (1982) 991–998.

[8] H. Salmun, Convection patterns in a triangular domain, Int. J. Heat Mass Transfer 38 (1995) 351–362.

[9] A. Koca, H. F. Oztop, Y. Varol, The effects of Prandtl number on natural convection in triangular enclosure with localized heating from below, Int. Commun. Heat Mass Transfer 34 (2007) 511-519.

[10] M. A. Teamah, A. F. Elsafty, E. Z. Massoud, Numerical simulation of double-diffusive natural convective flow in an inclined rectangular enclosure in the presence of magnetic field and heat source, Int. J. Therm. Sci. 52 (2012) 161–175.

[11] H. T. Xu, Z. Y. Wang, F. Karimi, M. Yang, Y. W. Zhang, Numerical simulation of double diffusive mixed convection in an open enclosure with different cylinder locations, Int. Commun. Heat Mass Transfer 52 (2014) 33-45.

[12] M. M. Rahman, H. F. Oztop, A. Ahsan, J. Orfi, Natural convection on heat and mass transfer in a curvilinear triangular cavity, Int. J. Heat Mass Transfer 55 (2012) 6250–6259.

[13] M. Hasanuzzamzn, M. M. Raahman, H. F. Oztop, N. A. Rahim, R. Saidur, Effects of Lewis number on heat and mass transfer in a triangular cavity, Int. Commun. Heat Mass Transfer 39 (2012) 1213-1219.

[14] Khalil Khanafer, Comparison of flow and heat transfer characteristics in a lid-driven cavity between flexible and modified geometry of a heated bottom wall, Int. J. Heat Mass Transfer 78 (2014) 1032–1041.

[15] Y. C. Ching, H. F. Oztop, M. M. Rahman, M. R. Islam, A. Ahsan, Finite element simulation of mixed convection heat and mass transfer in a right triangular enclosure, Int. Commun. Heat Mass Transfer 39 (2012) 689-696.

[16] S. Parvin, N. F. Hossain, Finite element simulation of MHD combined convection through a triangular wavy channel, Int. Commun. Heat Mass Transfer 39 (2012) 811-817.

[17] M. M. Rahman, M. A. Alim, M. M. A. Sarker, Numerical study on the conjugate effect of joule heating and magneto-hydrodynamics mixed convection in an obstructed lid-driven square cavity, Int. Commun. Heat Mass Transfer 37 (2010) 524-534.

[18] K. Al-Salem, H. F. Öztop, I. Pop, Y. Varol, " Effects of moving lid direction on MHD mixed convection in a linearly heated cavity", Int. J. Heat Mass Transfer 55 (2012) 1103–1112.

[19] A. Al-Amiri, K. Khanafer, J. Bull, I. Pop, "Effect of sinusoidal wavy bottom surface on mixed convection heat transfer in a lid-driven cavity", Int. J. Heat Mass Transfer 50 (2007) 1771–1780.

[20] R. Chowdhury, M. A. H Khan, M. N. A. Siddiki, Natural convection in porous triangular enclosure with a circular obstacle in presence of heat generation, Amer. J. App. Math. 3 (2015) 51-58.

[21] S. V. Patankar, Numerical Heat Transfer and Fluid Flow, Hemisphere McGraw-Hill, Washington DC, 1980.

23

Energy Management Strategies for Hybrid PV/Diesel Energy Systems: Simulation and Experimental Validation

Gabin Koucoi[1], Daniel Yamegueu[1, *], Quoc-Tuan Tran[2], Yézouma Couliblay[1], Hervé Buttin[2]

[1]Laboratory for Solar Energy and Energy Savings (LESEE), International Institute for Water and Environmental Engineering (2IE),
  Ouagadougou, Burkina Faso
[2]Smart Grid Laboratory (LSEI), National Solar Energy Institute (CEA/INES), Bourget Du Lac, France

Email address:

dan.yamegueu@gmail.com (D. Yamegueu)

Abstract: Hybrid photovoltaic-diesel systems are becoming more and more attractive for rural electrification in sub-Saharan Africa region. In this paper, some energy management strategies for a photovoltaic-diesel system without battery storage have been theoretically and experimentally studied. The proposed strategies are respectively based on active power control of inverters and controllable loads to ensure security operation for the system and maximize the solar energy penetration. Simulations and experiments have been performed under two different climate conditions and have been applied to an African rural load profile. All the energy management strategies developed have been implemented with the Matlab environment. The obtained results have shown the effectiveness of the proposed strategies to avoid power reserve to the diesel generator, to increase solar energy fraction, to reduce CO_2 emissions, and to ensure the system's frequency and voltage stability.

Keywords: Hybrid PV/Diesel System, Simulation, Experimentation, Energy Management, Off-Grid Electrification

1. Introduction

At a worldwide level, between 1.7 and 2.0 billion people do not have access to public grid-based electricity. The majority of this population lives in rural areas [1, 2]. In fact, grid extension is often considered as the first option for the electrification of these areas. However, electrification by grid extension requires a large investment that generally does not coincide with disperse populations with low energy demands and low income. Despite many efforts to extend the existing grids to rural areas, most remote areas will not be reached within a foreseeable future. The electrification rate of sub-Saharan Africa is one of the lowest among developing countries. The situation is more catastrophic in rural areas where less than 18% of population have access to electricity [3]. Furthermore, the progressive electrification of such rural areas has mainly been achieved through stand-alone energy production units and widely by Diesel generators (DG). However, the volatile prices of fossil fuels, their high cost of maintenance coupled to their greenhouse gas emissions make this option costly and unsustainable [4, 5]. Fortunately, these regions have abundant solar energy potential ranging from 4 to 6 kWh/m²/day which is enough to develop solar energy power generation systems [4, 5]. As shown by many authors, PV/Diesel hybrid systems can be more reliable and cost effective than stand-alone PV or Diesel systems. It could be a great opportunity for rural electrification that could trigger social and economic development in rural areas [4-6].

However most of these hybrid energy systems included a battery storage system. Due to manifold drawbacks of batteries, namely their high investment and replacement costs, their short lifetime (between 3 and 5 years maximum in rural areas for solar lead-acid batteries which are commonly used for PV systems), and their chemical pollution of the environment (batteries are released into nature after their use because of the lack of recycling infrastructure for batteries in developing countries), hybrid PV/Diesel systems without batteries for storage for rural electrification are being considered more often [4, 7]. Furthermore, in a hybrid PV/Diesel system without battery, an energy management system is mandatory to increase their reliability and their optimal operation. Many studies have been published and demonstrated that there is a potential for hybrid systems based on PV/Diesel system but very few include experimental verification of the system behavior [5, 8, 9]. In this paper, energy management strategies based respectively on inverter active power control and on controllable loads using are proposed and compared to inverter ON/OFF control strategy from simulation and experimental analysis. First, the control strategies studied are described followed by the models of the system' components used for simulation. Second, the experimental setup and case of study considered are displayed. Then, the results from simulation and experimental validation are presented and discussed. Finally, the conclusion is made.

2. Description of Energy Management Strategies for Hybrid PV/Diesel Energy Systems

The hybrid PV/Diesel energy system studied in this work includes a diesel generator coupled with a PV inverter to supply a load profile demand. In the following subsection, the different energy management strategies studied are described.

2.1. Energy Management Strategy Based on Inverter ON/OFF Control

This strategy called "Inverter ON/OFF control" aims to control the hybrid PV/Diesel system by switching ON or OFF the inverter according to climate conditions and load variation. Figure 1 shows the flowchart of this control strategy. As displayed in Figure 1, the inverter is switch OFF when the remaining power to be supplied by the diesel generator is less than its minimal output power value (P_{Diesel_min}), the inverter is set OFF to avoid DG to operate

under this limit.

Based on the PV array power ($P_{pv}(t)$) measurement the inverter is kept connected to the system whenever the DG output power is higher than it minimal output power value.

$X_{Inverter}$ =0 means that the inverter is disconnected from the system. When $X_{Inverter}$ = 1, the inverter is connected and the load will be fed by both inverter and diesel generator.

This control strategy based on PV inverter switching is the simplest approach to control PV energy penetration on a hybrid system. However this approach could lead to a waste of a significant amount of the PV energy produced.

To overcome this drawback, we proposed in what follows two new strategies, which aim to achieve more PV energy integration in the energy system and to guarantee reliable operation of the whole system.

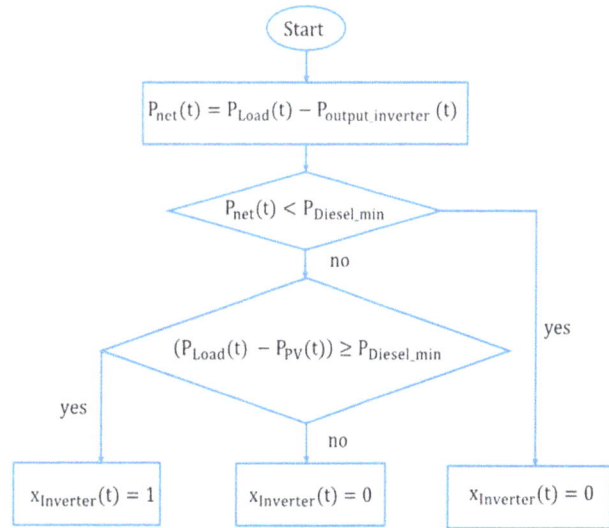

Figure 1. Flowchart of inverter ON/OFF control strategy.

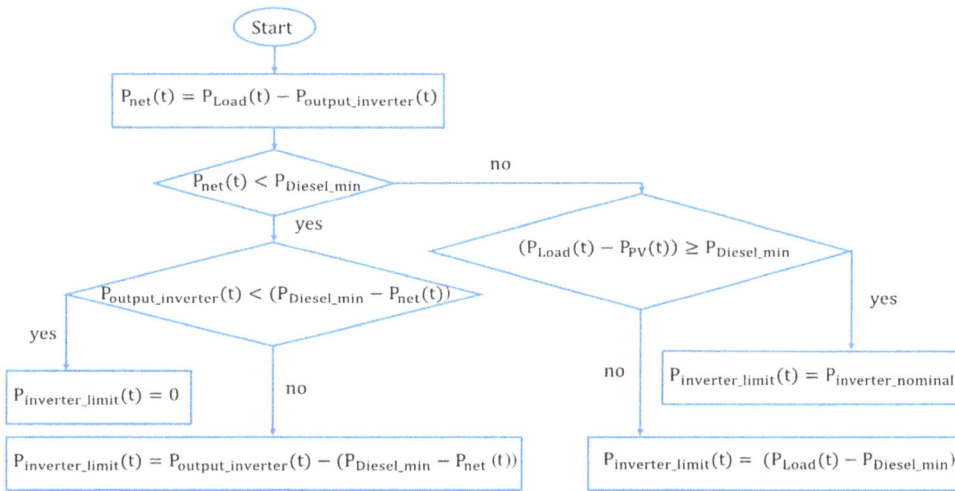

Figure 2. Flowchart of inverter active power control.

2.2. Energy Management Strategy Based on Inverter Active Power Control

Whatever climate conditions and load variations, the hybrid PV/Diesel system is expected to deliver energy

without interruption at any time and ensure a stability of the system. This energy management strategy shown on Figure 2 aims to ensure reliable operation of the system, to maximize the solar energy penetration by the inverter's active power control and to avoid operation of the DG under its minimal set point value especially during high PV generation and low loads periods.

According to the algorithm shown on Figure 2, one has:

- If Pnet (t) < P_{Diesel_min}, two scenarii can occur:

If the PV energy generated by the inverter ($P_{output\_inverter}$ (t)) is less than the power that must be cut [P_{Diesel_min} - P_{net} (t)] in order to keep output power of DG P_{Diesel} (t) higher than their minimal set point value, all the PV energy generated is cut off. In this situation, the DG alone will supply the load.

$$P_{inverter\_limit} (t) = 0 \qquad (1)$$

If $P_{output\_inverter}$(t) is higher than [P_{Diesel_min} - P_{net}(t)], the PV energy generated from inverter is reduced. In this case:

$$P_{inverter\_limit}(t) = P_{output\_inverter} (t) - [P_{Diesel\_min} - P_{net} (t)] \qquad (2)$$

- If P_{net} (t) ≥ P_{Diesel_min}, two situations can occur:

If $[P_{Load}(t) - P_{PV}(t)]$ is higher than P_{Diesel_min}, the inverter output power limit is set at its nominal output value.

$$P_{inverter\_limit}(t) = P_{inverter\_nominal} \qquad (3)$$

This step aims to increase the PV power penetration when the PV energy estimated is higher than the output power from the inverter at that time.

If $[P_{Load}(t) - P_{PV}(t)]$ is less than P_{Diesel_min}, the inverter active power limit is evaluated as:

$$P_{inverter\_limit}(t) = P_{Load} - P_{Diesel\_min} \qquad (4)$$

2.3. Energy Management Strategy 3: Control with Controllable loads

In this strategy, whenever the PV energy is higher than the load demand or the DG is in a situation to operate under its minimal output value, the controllable loads are connected to the system to consume the surplus of PV energy generated and consequently allow the DG to operate up its minimal output set point. The flowchart of the control strategy developed in this case is displayed in Figure 3.

For the performance evaluation of the energy management strategies proposed, the mathematical model of each component of the hybrid PV/Diesel system considered for the simulation is necessary. These models are presented in subsequent sections.

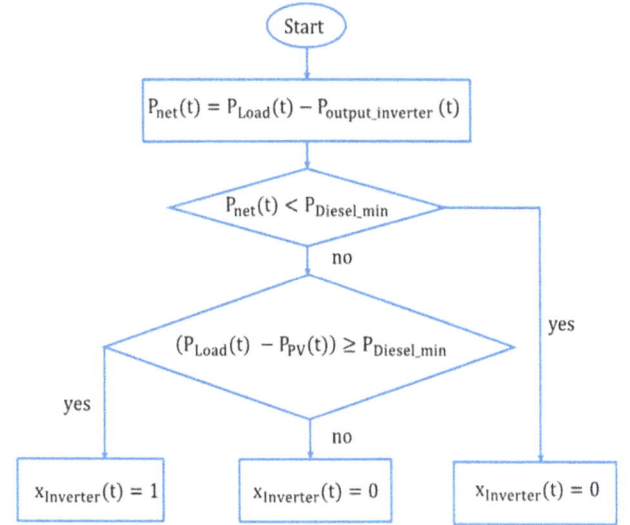

Figure 3. Flowchart of control strategy with controllable loads.

3. Mathematical Models of Hybrid PV/Diesel System Components

3.1. PV generator (PV Array + Inverter)

It is well known that a PV array produces DC current. However, many electrical appliances use AC current. Therefore, the DC current is, very often, converted into AC current. The inverter is the device that converts the power flow from the DC current to the AC current. The AC output power of the inverter is expressed as it follows [10, 11]:

$$P_{output\_inverter} = P_{PV\_STC} * \left(\frac{G}{G_{STC}}\right) * \left[1 + K_t * \left(T_{cell} - T_{cell\_STC}\right)\right] * \eta_{inverter} \qquad (5)$$

Where:

$P_{output\_inverter}$ (kW) is the output power of the inverter,

P_{pv_STC} (kW) is the output power of the PV array under standard test conditions (STC),

G (W/m²) is the local global solar radiation,

G_{STC} (W/m²) is global solar radiation under STC conditions. K_t (°C⁻¹) is the temperature coefficient,

T_{cell_STC} (°C) is cell temperature under standard test conditions,

$\eta_{inverter}$ is the inverter efficiency.

T_{cell} (°C) is the PV cell operation temperature (°C). This latter can be expressed by:

$$T_{cell} = T_{amb} + \left(\frac{G}{G_{NOCT}}\right) * \left(NOCT - T_{amb_{NOCT}}\right) \qquad (6)$$

T_{amb} (°C) is the ambient temperature, NOCT is the

nominal cell operating temperature (°C) which is measured under G_{NOCT} = 800 W/m² and T_{amb_NOCT} = 20°C. NOCT is given by manufacturers

3.2. Diesel Generators Model

A Diesel generator is widely modeled by its hourly fuel consumption. This later is deeply linked to the output power produced and the rated power of the DG. In this study the following linear approximation has been used [4, 5].

$$f(t) = a * P_{Diesel}(t) + b * P_{Diesel\_nominal} \qquad (7)$$

Where f(t) is the hourly fuel consumption in L/h;

$P_{Diesel}(t)$ is the output power of the DG;

$P_{Diesel\_nominal}$ is the rated power of diesel generator.

a and b are constants parameters of diesel generators.

4. Experimental Setup and Case Study

4.1. Experimental Setup

The experimental validation of the energy management strategies developed is performed under the microgrid platform at the National Institute for Solar Energy (CEA-INES) in France (see Figure 4).

The microgrid experimental platform includes:

- One Diesel generator (400/230 V, 50 Hz, 35 kW)
- A PV emulator; it is used to simulate PV production for different climate conditions. It's maximal output power is 10 kWp,
- One SMA inverter (25 kW),
- A loads bank of resistors of 88 kVA as maximal active power; that is used to simulate the load profile considered.
- A variable electronic load with active power ranging from 0 to 10 kW used as controllable loads.

The platform monitoring, control-command and data acquisition is made with the SCADA unit developed with Labview software. The communication protocol used is the standard TCP/IP Modbus protocol. All the energy management strategies developed are implemented with the Matlab environment. For the tests, 5 minutes has been chosen as sampling period and corresponds to 1 hour in real conditions. Thus, the daily load profile has been run over 2 hours during the experiments.

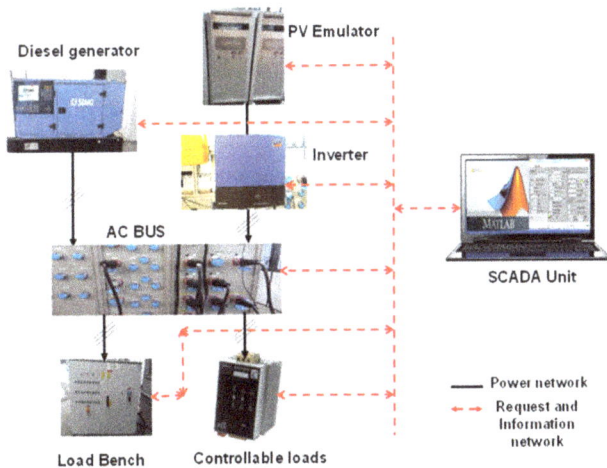

Figure 4. *Microgrid experimental platform at INES.*

4.2. Case of Study: Load Profile and Climate Data

The load profile that is taken account in this study and shown in Figure 5 is typical of sub-Saharan Africa rural areas. The daily energy demand is 187 kWh with 17 kW as the load peak value.

For the purpose of evaluating the performance of the proposed control strategies, two different climate conditions are considered (one for a cloudy day and the second for a sunny day) for simulation and experiment analysis.

Figures 6 and 7 display the global solar radiation and ambient temperature profiles for respectively cloudy and

sunny days. The parameters of simulations and experimental analysis are presented in Table 1.

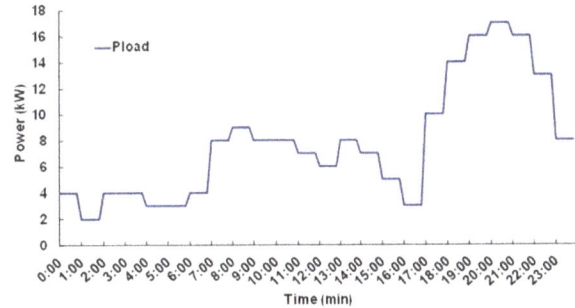

Figure 5. *Daily rural areas load profile.*

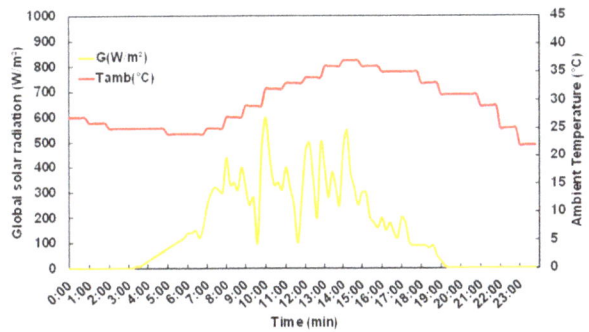

Figure 6. *Global solar radiation and ambient temperature profiles of a cloudy day.*

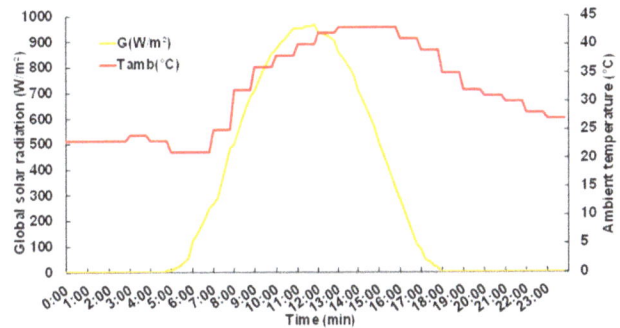

Figure 7. *Global solar radiation and ambient temperature profile of a sunny day.*

Table 1. *Simulation and experimental parameters.*

Simulation and experimental parameters	Values
Nominal power of Diesel generator (kW)	35
Minimal output power of DG set point (kW)	1
a: Diesel generator fuel factor (L/kWh)	0.246
b: Diesel generator fuel factor (L/kWh)	0.084
Nominal power of inverter (kW)	25
NOCT (°C)	45
Kt (°C)	0.004
Carbon emission factor (kg/kWh)	0.34

5. Results and Discussion

To assess the performance of the control strategies studies, both simulation and experimental results are presented and discussed.

5.1. Simulation Results

5.1.1. Energy Flow Balance of the Hybrid PV/Diesel System with Inverter ON/OFF Control Strategy

Figure 8 shows the energy flow balance for a cloudy day with inverter ON/OFF control strategy. It can be noticed that at any time, the load demand has been supplied by the diesel generator alone or by both diesel and PV generators. The minimal output power from DG during this day is obtained at 16 h and equal to 1.5 kW. This value is higher than the minimal output set point assumed to 1 kW. In this case, no PV energy curtailment has been observed and all the PV energy produced and generated through the inverter is consumed.

On the sunny day, especially from 9h to 14h30, 15h to 15h 30 and 16h to 16h15 as shown in Figure 9, when the remaining power to be supplied by the DG is lower than its minimal output set point, the PV inverter is disconnected from the system. No PV energy has been generated by the inverter to the system in this period. The load is supplied only by the diesel generator. It can be observed that an important amount of PV energy has been wasted under this strategy.

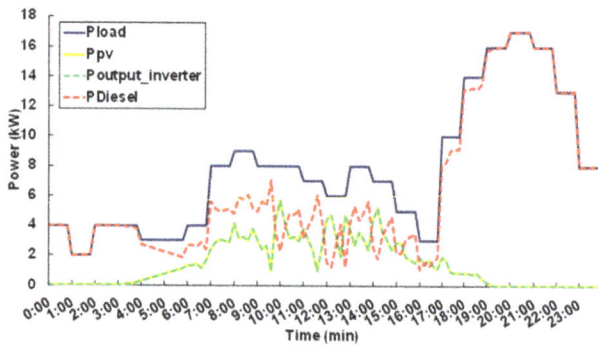

Figure 8. *Simulation of energy flow balance on cloudy day.*

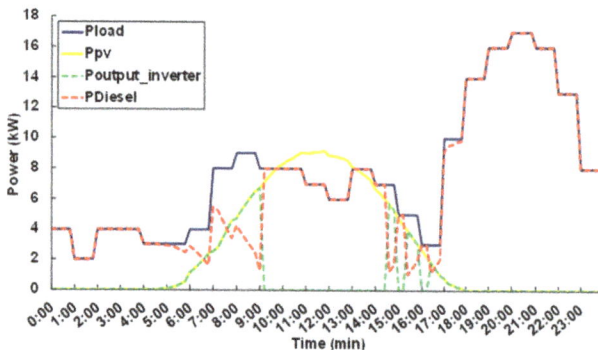

Figure 9. *Simulation of energy flow balance in sunny day under inverter ON/OFF control strategy.*

5.1.2. Energy Flow Balance Under inverter Active Power Control Strategy

On the cloudy day, the energy flow balance for this strategy leads to the same results as those of the inverter ON/OFF control strategy presented in figure 8.

On the sunny day, the system's operation under inverter active power control strategy proposed is shown in figure 10. Comparing to the inverter ON/OFF control strategy, from 9h to 14h30, 15h to 15 30 and 16h to 16h15, the PV energy

produced through the inverter is controlled as well as possible in order to keep the output power of diesel generator at least equal to the minimal set point.

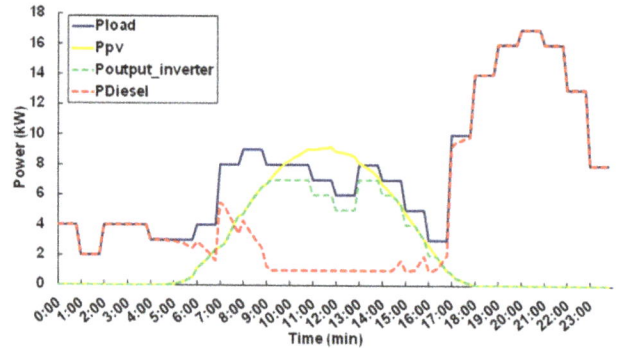

Figure 10. *Simulation of energy flow balance on sunny day under active power control strategy.*

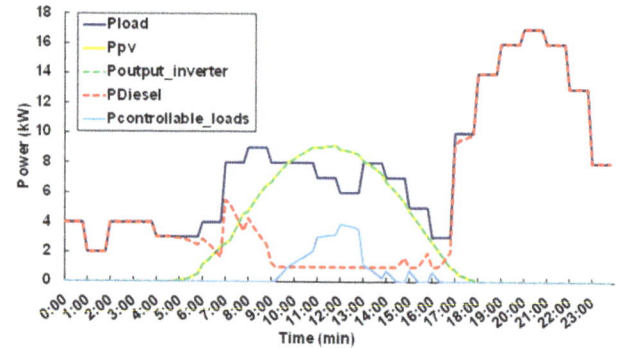

Figure 11. *Simulation of energy flow balance on sunny day under control strategy with controllable loads.*

5.1.3. Energy Flow Balance Under Strategy with Controllable Loads

As described in sub-section 2.3, the controllable loads have been used especially to consume the surplus of PV energy and then avoid the operation of DG under its minimal output power set point. Under this strategy, since the minimal DG output power observed (1.5 kW) on the cloudy day is higher than the minimal set point value, no controllable load is needed for the system management. Figure 11 presents the daily operation of the hybrid system under this strategy on the sunny day. The effectiveness of the control approach can be observed. During the period for high PV energy and low load demand, the controllable loads power need is evaluated and connected to the hybrid energy system. The Diesel generator at this instance runs at least at its minimal output power set point.

From the simulation results obtained, the impact for each of the control strategies on solar energy fraction and on carbon emission are investigated and compared in the following sub-section.

5.1.4. Effect of Control Strategies on Solar Energy Fraction

In order to investigate the effect of each strategy on solar energy penetration in the energy system, the solar energy fraction (F_{PV}) is selected as criteria. It is defined as the ratio of solar energy consumed to the total energy production of the hybrid system. The equation for solar energy fraction evaluation is shown as follows in [12].

$$F_{PV} = \frac{E_{PV}}{E_{PV} + E_{Diesel}} * 100 \qquad (8)$$

On the sunny day and as presented in Figure 12, the control strategy with controllable loads offers the high value (34%) of solar energy fraction. This is the ideal approach to avoid solar energy curtailment and to ensure reliable operation to the whole system.

However, for control strategies without controllable loads, it is clear that the strategy proposed, which is based on active power control achieves more solar energy in the system than inverter ON/OFF control strategy. Almost 65% and 71% of the solar energy is wasted by using ON/OFF control strategy comparing respectively to inverter active power control strategy and control strategy with controllable loads-(Figure 12).

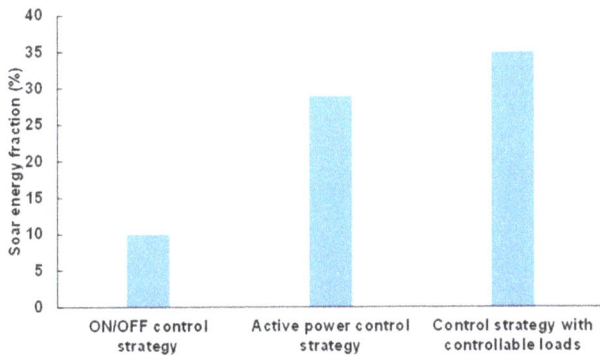

Figure 12. Solar energy fraction on sunny day for the control strategies studied.

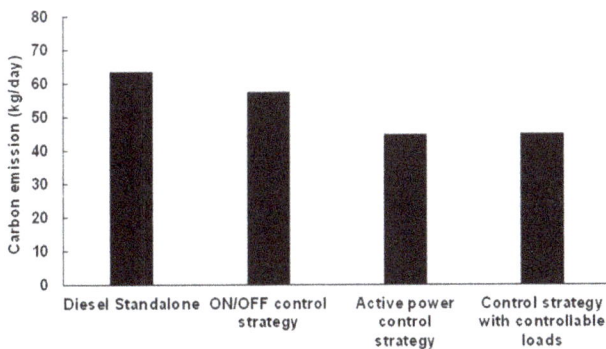

Figure 13. Carbon emission level.

5.1.5. Effect of Control Strategies on Carbon Emissions Level

The most important environmental benefit of the PV system is its carbon emission reduction resulting in a cleaner and less polluted environment that has a great impact on people's lives [6]. The impact of the control strategies proposed in this study is evaluated and compared to diesel standalone system. Figure 13 shows that the lower amount of carbon (almost the same value: 45 kg/day) is generated under the inverter control active power strategy and control strategy with controllable loads. Respectively 22% and 30% of carbon has been eliminated under inverter active power control strategy, control strategy with controllable loads

comparing to inverter ON/OFF control strategy and the diesel standalone system. It appears clearly that the control strategies proposed based on the controllable loads and on the inverter active power control contribute to reduce the carbon emission.

5.2. Experimental Results

Due to the advantages and the effectiveness of the proposed strategies comparing to inverter ON/OFF control strategy based on the simulation results previously presented, their experimental validation has been performed in this sub-section. First, the energy flow balance under these strategies for the cloudy and sunny day are presented and discussed. Finally, the power quality of the electricity produced has been monitored by measuring the voltage and frequency.

5.2.1. Energy Flow Balance from Experimental Analysis
  i.   Inverter active power control strategy

On the cloudy day, the experimental results obtained have shown the dynamic of the DG to supply the load at any time (Figure 14). From 9h35 to 10h35 when more fluctuations of PV energy is observed, the DG supplies the load as well as possible without interruption. This result highlights the ability of the DG to offer a better dynamic and ensure reliable operation of the whole system. A similar results was obtained from the simulation as presented previously on figure 8 for the cloudy day.

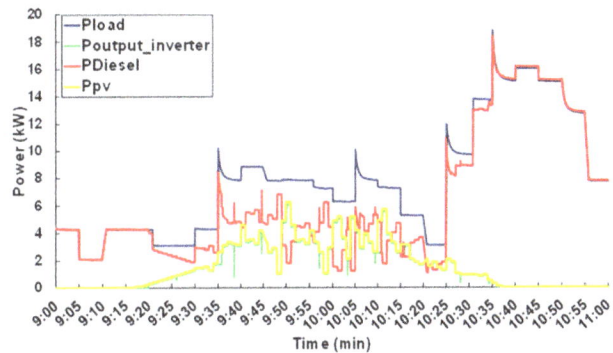

Figure 14. Experimentation of energy flow balance on cloudy day under inverter active power control.

As displayed in Figure 15, the control strategy based on inverter active power control proposed correctly manages the energy in hybrid PV/Diesel as expected from the simulation results. The inverter output power has been controlled from 14h to 14h30 in order to ensure better operation to the system especially when PV energy (P_{pv}) becomes more than the load demand (P_{load}).

  ii.   Control strategy with controllable loads

The operation of the hybrid PV/Diesel energy system with the control strategy based on the controllable loads is presented in figure 16 for the sunny day. The experimental results have demonstrated the feasibility of controllable loads use to avoid reserve power on the diesel generator and at the same time to consume the total PV energy produced by the system.

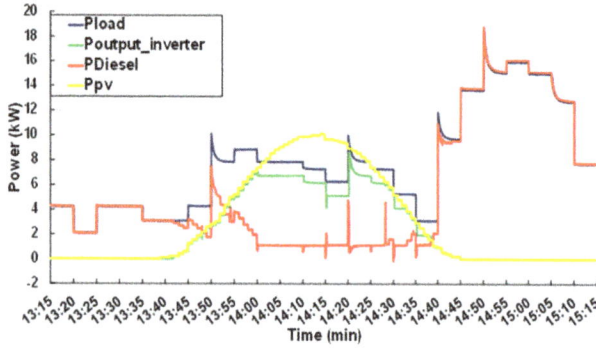

Figure 15. Experimentation of energy flow balance on sunny day under inverter active power control.

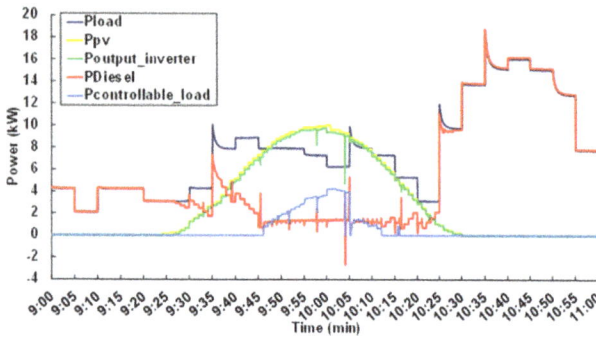

Figure 16. Experimentation of energy flow balance on sunny day under control with controllable loads.

5.2.2. Voltage and Frequency Fluctuation from Experimental Analysis

The integration of a PV system into a microgrid like hybrid PV/Diesel system including a diesel generator can have significant impacts on power quality in general and especially on voltage and frequency as investigated in [8]. In the following sub-section, the energy system's frequency and voltage under the energy management strategies proposed are presented and discussed.

i.   Inverter active power control strategy
• Frequency fluctuation

Figures 17 and 18 display the frequency fluctuation in the microgrid respectively on the cloudy and sunny day. The frequency fluctuation can be observed during the operation of the energy system. It is clear that, the variation of diesel generator output power influences the system's frequency value. In fact, when the diesel output power increased, the frequency dropped slightly and vice versa. The frequency as well as on the cloudy and sunny days with this control strategy ranged between 49.3 and 50.2 Hz. It was also found from the tests that, the frequency has a good dynamic and did not deviate out of the acceptable range 50 ± 4.5 Hz as recommended in [13] for off-grid PV systems.

• Voltage fluctuation

The voltage values in the three-phase hybrid energy system under inverter active power control strategy in cloudy day and on the sunny day are respectively plotted in figures 19 and 20. Additionally on the cloudy day and sunny day under this strategy, the voltage value during the tests ranged from 225.5V to 231V. According to the standard EN 50160,

the variation of the voltage should be in the range of ±10% of the nominal value (230V) [8]. On the two days considered the same voltage fluctuation level in the order of 2% has been observed.

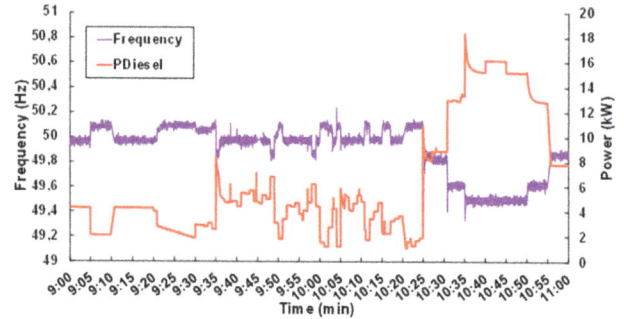

Figure 17. Frequency fluctuation on cloudy day under inverter active power control strategy (experimental).

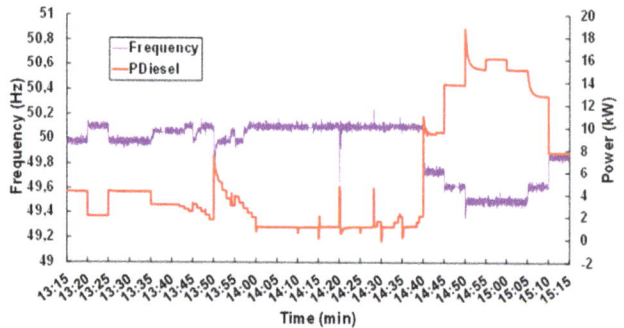

Figure 18. Frequency fluctuation on sunny day under inverter active power control strategy (experimental).

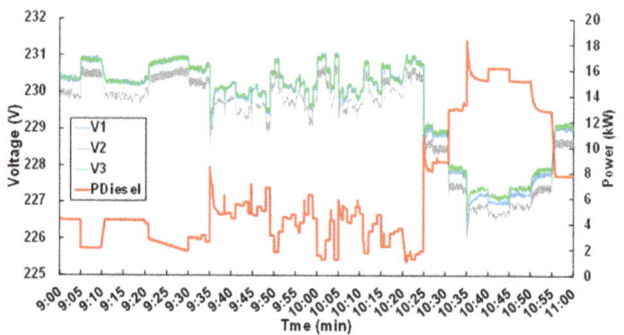

Figure 19. Voltage fluctuation on cloudy day under inverter active power control strategy (experimental).

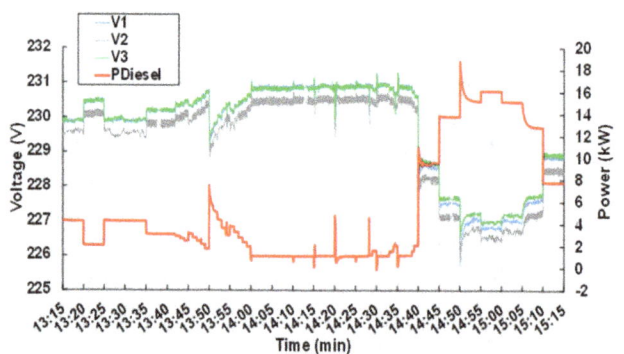

Figure 20. Voltage fluctuation on sunny day under inverter active power control strategy (experimental).

ii. Control strategy with controllable loads
• Frequency fluctuation

With controllable loads control strategy too, the frequency fluctuate according to the output power of the diesel generator as show in figure 21 for the sunny day. The controllable loads output power connected to the system has no impact on the frequency. This latter is around it nominal value (50Hz) during that time (from 09h45 to 10h15).

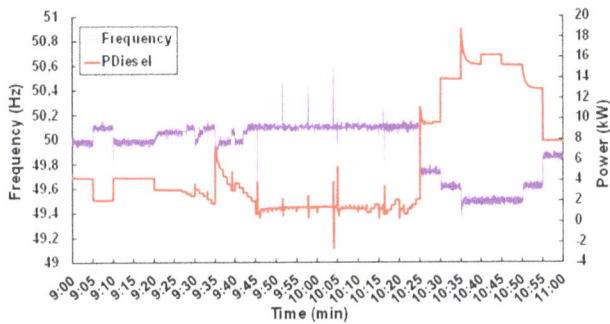

Figure 21. *Frequency fluctuation on sunny day under control strategy with controllable loads (experimental).*

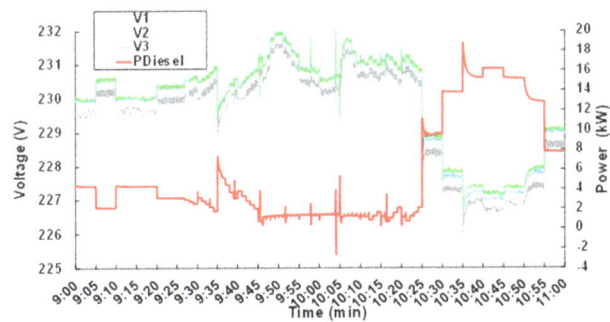

Figure 22. *Voltage fluctuation on sunny day under control strategy with controllable loads (experimental).*

• Voltage fluctuation

Although the frequency was stable when the controllable loads were connected to the system, some weak fluctuations on the system's voltage were observed. Indeed, as we can see in figure 22, the voltage has increased from its nominal value (230 V) to 231.8 V. this means that, the voltage has fluctuated just 1% in that period. This fluctuation level is too slight compared to the standard fluctuation level.

Based on the power quality analysis results, it can be summarized that the two control strategies based respectively on inverter active power control and on the controllable loads use does not deteriorate the system stability.

6. Conclusion

The present work presents a simulation and experimental analysis of energy management control strategies for a hybrid PV/Diesel energy system. The proposed control strategies based on inverter active power control and on the controllable loads use have been presented and applied to a typical rural loads profile under two different climate conditions. The simulation results compared to inverter

ON/OFF control strategy have shown a good performance of the strategies proposed especially when high PV generation and low load demand occur. From this experimental analysis, the effectiveness of the control strategies proposed has also been demonstrated in term of energy flow balance and of the system's stability. In future work, this control strategies approach for energy management will be applied to hybrid PV/Diesel energy systems including several inverters and several diesel generators.

Acknowledgments

Authors want to thank the European Union Commission through the Energy Facility II, the International Institute for Water and Environmental Engineering (2iE) and the National Institute for Solar Energy (INES)-France for their financial support of this study. The Authors are also grateful to Mrs Jessica Ouedraogo of the 2iE - Penn State Centre for the english polishing of this paper.

References

[1]   U. Suresh Kumar and P. S. Manoharan, "Economic analysis of hybrid power systems (PV/diesel) in different climatic zones of Tamil Nadu," *Energy Convers. Manag.*, vol. 80, pp. 469–476, 2014.

[2]   S. M. Shaahid and I. El-Amin, "Techno-economic evaluation of off-grid hybrid photovoltaic–diesel–battery power systems for rural electrification in Saudi Arabia—A way forward for sustainable development," *Renew. Sustain. Energy Rev.*, vol. 13, no. 3, pp. 625–633, Apr. 2009.

[3]   International Energy Agency, "World Energy Outlook (WEO)." 2013.

[4]   D. Tsuanyo, Y. Azoumah, D. Aussel, and P. Neveu, "Modeling and optimization of batteryless hybrid PV (photovoltaic)/Diesel systems for off-grid applications," *Energy*, pp. 1–12, 2015.

[5]   D. Yamegueu, Y. Azoumah, X. Py, and N. Zongo, "Experimental study of electricity generation by Solar PV/diesel hybrid systems without battery storage for off-grid areas," *Renew. Energy*, vol. 36, pp. 1780–1787, 2011.

[6]   B. I. Ouedraogo, S. Kouame, Y. Azoumah, and D. Yamegueu, "Incentives for rural off grid electrification in Burkina Faso using LCOE," *Renew. Energy*, vol. 78, pp. 573–582, 2015.

[7]   K. Y. Lau, M. F. M. Yousof, S. N. M. Arshad, M. Anwari, and a. H. M. Yatim, "Performance analysis of hybrid photovoltaic/diesel energy system under Malaysian conditions," *Energy*, vol. 35, no. 8, pp. 3245–3255, 2010.

[8]   D. Yamegueu, Y. Azoumah, X. Py, and H. Kottin, "Experimental analysis of a solar PV/diesel hybrid system without storage: Focus on its dynamic behavior," *Int. J. Electr. Power Energy Syst.*, vol. 44, pp. 267–274, 2013.

[9]   D. Alfonso, H. E. Ariza, J. C, a Correcher, G. Escriv, R. Roig, C. Rold, I. Segura, C. Vargas, E. Hurtado, and F. Ib, "Experimental veri fi cation of hybrid renewable systems as feasible energy sources n," vol. 86, pp. 384–391, 2016.

[10] A. Kaabeche, M. Belhamel, and R. Ibtiouen, "Techno-economic valuation and optimization of integrated photovoltaic/wind energy conversion system," *Sol. Energy*, vol. 85, pp. 2407–2420, 2011.

[11] M. S. Ismail, M. Moghavvemi, and T. M. I. Mahlia, "Techno-economic analysis of an optimized photovoltaic and diesel generator hybrid power system for remote houses in a tropical climate," *Energy Convers. Manag.*, vol. 69, pp. 163–173, 2013.

[12] N. A. Luu, "Control and management strategies for a microgrid", Thèse de doctorat, Université de Grenoble, 2015.

[13] SMA, "PV Inverters - Use and Settings of PV Inverters in Off-Grid Systems, http://www.sma.de/en/products/solarinverters/sunny-tripower-20000tl-25000tl.html#Downloads-108649," pp. 2–5.

24

Evaluation of Energy Efficiency, Energy Consumption and Energy Saving in the Production of Oil, Gas and Energy Sectors

Eduard A. Mikaelian[1], Saif A. Mouhammad[2, *]

[1]Department of Thermodynamics and Heat Engines, Gubkin Russian State University of Oil and Gas, Moscow, Russia
[2]Department of Physics, Taif University, Taif, Saudi Arabia

Email address:
Saifnet70@hotmail.com (S. A. Mouhammad)

Abstract: Energy efficiency, energy consumption and energy savings depend on the performance of energy-technological equipment characterising the operational suitability of it. The operational suitability of the equipment determines the quality of it. Based on energy consumption, fuel consumption and reliability indicators as the formalised quality indicators in energotechnological equipment it is possible to determine the energy efficiency, energy consumption and energy savings of the equipment and therefore select the right production volume indicator (PVI). The aim in determining the fuel and power parameters (FPP) of the oil and gas production units and power sector is to get dependence influencing the energy efficiency (EE), power consumption (PC) and energy saving (ES) of those production units. To solve this problem it was necessary to choose the right (PVI). For that purpose, the issues concerning justifying the choice of the volume indicator for the oil, gas and power sectors using the gas turbine units (GTU) were examined. The correct choice of (PVI) it was possible to determine the available efficient power of gas turbine units which should meet the power consumption requirements for the gas compression by gas pumping units (GPU) on the compressor station (CS) of gas transmission systems for a given mode [4,5].

Keywords: Energy Efficiency of Equipment, Energy Consumption of Equipment, Energy Saving of Equipment, Production Volume Indicator, Oil and Gas Industry, Gas Pumping Unit, Pipeline Power Consumption, Effective Power of Machines

1. Introduction

The choice of (PVI) is quite essential in terms of assigned tasks. Analysis of the methods used of the determination of the (PVI) revealed shortcomings in the process. In one case it was due to the gross nature of (PVI), in other words, it was taken into account how much gas was transported at a predetermined distance but it was not considered how much power was consumed in that case. In another case, the work expended was taken into account but without proper accounting for the total energy losses. The correct assessment of the gas compression effectiveness associated with a specific energy consumption in production of the oil, gas and power sectors provided basis for determining (PVI).

On the value of (PVI) at compressor stations of gas pipelines with a gas turbine drive is influenced by both external and internal factors. External factors are mainly related to the outdoor temperature at the inlet of a gas-turbine drive at the compressor stations which are taken into consideration in the article. Based on the thermodynamic characteristics of a gas turbine drive derived are the effects of the outdoor temperature on the gas turbine drive of the gas pumping units of compressor stations.

In this regard, the article discusses regularities of outdoor temperature influence on the effective power of the gas-turbine drive of gas compressor units [1].

2. Materials and Methods

To assess energy efficiency (EE), energy consumption (EC) and energy savings (ES) production of the oil and gas sector (OGS) and energy sector it is necessary, first of all, to choose the (PVI). The article assesses the (EE), (EC) and

(ES) based on the (PVI) [1, 2].

The assessment of (POI) used in the industry is carried out on the basis of the commodity transport activities (CTA). The commodity transport activities are used as an indicator to characterise the productions volume of the (OGS) and energy sector, for example, the gas transportation pipeline system (GTS) and is a conditional work on the displacement of a unit of the volume of gas transported per length unit of a gas pipeline (gas pipelines) [3], [6].

The (CTA) are of the bulk nature and originates from the Ministry of Railway Transport (ton-kilometer) concerning the movement of goods and cargoes; it does not take into account a change in the physical properties of cargoes during transportation [3]. The pipeline transport is taken as an example. In the pipeline transport the physical properties of a working medium change, and the temperature, pressure, density of the working medium in the energotechnological equipment (ETE) do change as well, which greatly influences the (EE), (EC) and (ES). In this regard, using (CTA) it is impossible to assess the (EE), (EC) and (ES) of the considered (ETE) which is represented by gas pumping units (GPU) at the compressor stations (CS) along the pipeline in the pipeline transport.

The proposed equivalent commodity transport activities are based on determining the polytropic compression without considering all the losses of energy, in particular, the external energy loss determined by the mechanical efficiency of (GPU), a gas centrifugal compressor (CC) of the gas pipeline compressor station and the power spent for mechanical losses during the transmission of mechanical energy from the drive to the centrifugal compressor [4], [7]. It is mainly the gas turbine unit (GTU) that is used as the drive in the gas turbine system. Gas turbine units are also used in the power sector, i. e. thermal power plants (TPP), heat and power plants (HPP) to drive electric generators. Later we will mainly focus on the (GTU) used in the (GPU) to drive gas (CC).

In this regard, it is proposed to use the efficient energy required to compress a predetermined volume of gas transported as the (POI) and, therefore the specific energy consumption in kW. h for the transport of the unit of a given volume of transported gas in million. m3 [5, 6]:

$$d = N_e / Q_k , \left[kw.h / million.m^3 \right], \qquad (1)$$

where N_e is the effective power of GPU of CS of GPS; Q_k $\left[million.m^3 / day \right]$ - volume (quantity) of the compressed gas of (GCU) of (CS) of (GPS).

3. Energy Efficiency of the Energy and Gas Pumping Units with the Electric Drive and Gas Turbine Drive

Untimely preventive maintenance to clean the pipeline from solids, crystalline hydrates and condensates as well as unreliable operation of the systems designed for gas cleaning, drying and stripping lead to overload (GPU) and increased (POI), i. e. an increase in the specific energy consumption for compressing. Based on the study of the modes of operation of gas transmission systems Urengoy - Chelyabinsk, NPTR (Northern Parts of the Tyumen Region) - Nizhnyaya Tura (N. Tura), N. Tura- Gorky in the first years of operation it was found that for gas pipelines in the normal state, the power consumption is up to 90-95 kWh / (million .m³. km) with an average operating pressure in a gas pipeline of 5.6 MPa and up to 115-125 kW. h / (million. m³. km) at 7.5 MPa. To prevent the formation of hydrates in gas and eliminate hydrate blocks, it is necessary to reduce the pressure of the gas transportation, raise its temperature, lower humidity and apply various inhibitors, condensate traps etc.

In turn, the poor technical condition of a gas-turbine gas-pumping unit leads to a fall in its productivity and ultimately a decrease in a certain amount of gas flow through the pipeline, or it causes an overload in the gas-turbine gas-pumping unit during the operation of the planned volume of gas transported. This causes an increase in specific energy consumption for transportation of gas, that is, an increase in the (POI).

An analysis of the mode of operation of the gas-turbine gas-compressor units of compressor stations of pipelines shows the impact from the technical state of the machinery and equipment on the (POI), deterioration of the overall performance, power performance of the gas-turbine gas-pumping unit and pipeline.

The research has shown that lowering the degree of compression in respect of (CS) is one of the most effective ways to reduce energy consumption including the fuel gas flow.

In the reporting practices of the gas pipeline systems (GPS) based on the recommendation of STO Gazprom 2-3.5-113-2007 [7], the energy efficiency of gas pumping units and that of compressor workshops, compressor stations and (GPS) are evaluated differently. If for (GPU) of (CS) with electric engines the efficiency is determined by the power consumption in specific values as $d_e = N_{ee} / Q_k$ or in absolute terms, the absolute energy consumption of (GPU) of (CS) of (GPS) - kW. h for the reporting period:

$$E_{ee} = d_e . Q_k , \qquad (2)$$

In the case of evaluation of efficiency of a gas-pumping unit with a gas-turbine drive compressor used is the fuel gas consumption of the gas pipeline system- in thousand. m³ (million. m³) for the reporting period. It should be noted that taking into account quality categories in the energotechnological equipment (ETE) fuel consumption can be characterised by the economical efficiency [8]. In this regard, to assess the energy efficiency of a gas-pumping unit with a gas-turbine drive, power consumption in calculations should be in specific terms $d_g = N_{eg} / Q_k$, or in absolute terms, i. e. the absolute power consumption of a (GPU) of (CS) of (GPS)- kW. h for the reporting period:

$$E_{eg} = d_g . Q_k, \qquad (3)$$

To assess the energy efficiency, energy consumption and energy saving of the production of (GPU) of (CS) of (GPS) according to [9, 10], it is proposed to divide the calculated flow rate of the fuel gas for the gas-turbine and gas-pumping unit (power consumption for the (GPU) with an electric engine) into perform useful work units. At the same time, the polytropic compression work is used as the useful work done by the gas pumping unit, compressor shop and compressor station. The proposed design ratio of the polytropic compression work with the work of gas-turbine and gas-pumping unit does not account for power losses that it is explained factors given below.

To the value of the (POI) at the (CS) of pipelines from the gas-turbine and gas-pumping unit is influenced by two factors affecting the energy loss, i. e. the external and internal factors that determine the energy losses of (CC). Internal energy losses are assessed by the efficiency of the process which occurs in our machine across the flow part of the machine, from the inlet to the outlet of the machine. The gas (CC) is related to the flow impeller machinery and according to the theory of flow turbochargers the internal energy losses are evaluated by the adiabatic efficiency. The external factor relates mainly to the outdoor temperature at the inlet to an axial compressor (T_a) of (GTU). Known is the sensitivity of the gas turbine drive of (GPU), for which (CC) is used. According to the thermodynamic theory [5], on average every degree of Ta change changes the effective power of the (GPU) drive on average by 0.8 - 1.2% in the normal variation interval of T_a from -15 to 45°C.

In this regard, examines the regularities of influence of T_a on an effective power of the gas-turbine drive of (GPU). To evaluate (EE), (EC) and (ES) in addressing mode-related issues concerning the compression of a predetermined amount of gas transported, it is necessary to determine the available efficient power of the (GPU) drive that should not be less than the required power of (GPU) for gas compression of (GPU) of (CS) for fulfilling the assigned regime:

$$N_{e(gtu)} \geq N_{e(gcu)}. \qquad (4)$$

Certain problems are related to the determination of power of the (GPU) drive ($N_{e(gtu)}$) in choosing the method to determine this power. The work [1] considers the sensitivity of the gas turbine unit to the change in temperature of the outside air Ta that enters at the inlet of the axial compressor (AC) of (GTU) and how in this case changes the available power efficient of drive of GPU- $N_{e(gtu)}$ using those or other methods. This subject is studied in a number of well-known works [1-3] etc.

Greater sensitivity of (GTU) to change in the temperature of air entering the compressor of (GTU) can be explained, firstly, by the fact that these units have constant passage sections of the gas turbine thus ruling out the possibility of regulating the air flow at constant parameters of the working medium and, secondly, the fact that modern gas turbine units are characterised by a high ratio of effective compression power of the axial compressor Nec of (GTU) and expansion of a gas turbine $\left(N_{et} \right)$, respectively, for compressor and turbine ($p_e = N_{ec}/N_{et}$ – the characteristics of the power ratio, respectively, for compressor and turbine).

According to the conducted research [9], [11, 12], it follows that at low degrees of compression (C) for (AC) the effective efficiencies and power of (GTU) are less sensitive to the outdoor temperature. For example, when C = 2 the relative change in the effective efficiency and power are respectively on average 0.25 and 0.43% per 1 degree of change in temperature T_a relative to the value of the data sheet T_{a0} = 288.2°K. Hereinafter the index '0' refers to the nominal values of the parameters of the data sheet. With an increase in C for (AC), the sensitivity of (GTU) to changes in air temperature at the inlet to (AC) dramatically increases. Thus, at C = 6, the relative change in the effective efficiency and the gas turbine power per 1 degree of T_a, on average reaches 0.60 and 0.82% respectively. Greater sensitivity of (GTU) to changes in the temperature of the ambient air T_a entering the inlet of the axial compressor (AC) of (GTU) creates the problem of determining the available power of (GTU) for performing process-related tasks, for example, in the gas-and-oil-pipeline industry for the transport of gas or oil. In the industry known is a method of determining the available effective power of (GTU) based on the application of the formula of VNIIGAS (a leading Scientific Research Institute in Russia for gas) presented in [1], [3] taking into account the correction factors (K_N, K_{ai}, K_{re}, K_T)- as the empirical method:

$$N_{ea} = N_{e0} K_N K_{ai} K_{re} \left[1 - K_T (T_a - T_{a0})/T_a \right] (p_a/p_{a0}), \qquad (5)$$

where K_N is technical status index of the gas-turbine and gas-pumping unit; K_{ai} is the factor taking into account the effects of the anti-icing system; K_{re} is a factor which takes into account the work of recycling systems; K_T is a factor which takes into account the effect of outside temperature. The factors K_N, K_{ai}, K_{re}, K_T are presented in [1]. In calculating the available power of (GTU) and making other calculations relating to the mode of operation and determination of the technical condition of units of (CS) instead of the nominal power units according to the datasheet (N_{e0}), the power value according to the industry standards can be taken as the basic power of units.

The temperature coefficient K_m influencing the value of available power of the gas-turbine and gas-pumping unit (N_{ea}) is determined taking into account [3] is as follows:

$$K_T = 1 - k_T (T_a - 288)/T_a, \qquad (6)$$

where T_a is the calculated temperature of air at the gas-turbine and gas-pumping unit; κ_T is the reduced temperature coefficient the values of which for various gas-turbine and gas- pumping units are presented in the annex [3] and are changed on average within the range of 1, 3 - 3, 7 (see Table 1). For estimations proposed is the value of $\kappa_T = 3.0$.

It should be noted that a change in the available power of (GTU) depending on the ambient temperature in the station conditions, determined can also be, along with the above methods, according to the formula proposed by Professor N. I. Belokon. A known method for determining the available power of a unit based on the well-known equation proposed by Professor N. I. Belokon [5]:

$$N_{ea} = N_{e0}\left\{\left[1-(T_a - T_{a0})/T_a\right]*\left[\lambda_e/(1-\lambda_e)\right]*(T_a - T_{a0})/T_a\right\}. \tag{7}$$

In calculating the mode of the gas-turbine and gas-pumping unit at (CS) on the condition of the example under consideration it is necessary that the effective power supplied for the (CC) operation does not exceed the available power of the gas turbine unit (4).

In this formula, the ratio of the power consumed by the axial compressor of (GTU) and produced by the gas turbine (λ) depends on the ambient temperature (T_a). The effective power of (GTU) becomes equal to (1- λ) relative the power of the axial compressor.

Table 1. The λ values depending on the temperature.

T_a [°C]	-15	0	15	30	45
λ	0.520	0.540	0.572	0.615	0.668

The results of calculations to determine the relative value of the available power of (GTU) N_{ea}/N_{e0} depending on the ambient temperature according to the formula of Professor N. I. Belokon and the formula of VNIIGAS at different values of the temperature coefficient (K_T) are given in Table 2.

Table 2. Relative available power of GTU of various types at: $K_N =1$; $K_{ai} =1$; $K_{re} =1$; $P_a = P_{a0}$; $T_{a0} =288\ °K$.

T_a [°K]	According to the formula of N. I. Belokon			According to the reduced formula of the Russian Research Institute for Natural Gases and Gas Technologies (VNIIGAS) at values K_T					
				1.3	2	2.8	3.2	3.4	3.7
	$\Delta T=T_a-T_{a0}$	$\lambda/(1-\lambda)$	N_{ea}/N_{e0}	GPU-C-6.3	GTK-10I	GPU-C-16	GTN-16	GTN-25	Other GTU
258	-30	1.083	1.22	1.15	1.23	1.32	1.37	1.39	1.43
273	-15	1.174	1.11	1.07	1.11	1.15	1.17	1.19	1.20
288	0	1.337	1.00	1.00	1.00	1.00	1.00	1.00	1.00
303	15	1.596	0.86	0.93	0.90	0.86	0.84	0.83	0.82
318	30	2.012	0.70	0.87	0.81	0.73	0.70	0.68	0.65
333	45	2.448	0.53	0.82	0.72	0.62	0.56	0.54	0.50

From the data it follows that the value of N_{ea}/N_{e0} by the formula of VNIIGAS varies within a wide range for various types of gas turbine units. This variation in the temperature range T_a from -15°C to 15°C comes to 18 - 20%, while for the other temperature values it reaches 35% or higher; with an increase in coefficient K_T the ratio N_{ea}/N_{e0} in the range of negative temperatures increases, and it decreases in the range of positive temperatures. The value N_{ea}/N_{e0} by the formula of N. I. Belokon is stable for all types of gas turbine units, and is approximately 1% in the range of positive temperatures T_a. This gives a considerable spread of power of the gas-turbine and gas-pumping units by the formula VNIIGAS. It should be noted that the method of determining the effective power of (GTU) by the formula of Professor N. I. Belokon is based on the analytical method and may indicate the most accurate results of calculations (see Table 2).

Calculation of the available power of units has a great importance in the study of modes of the gas-turbine and gas-compressor units of (CS) of pipelines makes it possible to determine the shortage of power of units in summer and the value of underutilising it in the winter time at (CS), the load factor of units etc. When calculating the load factor of the gas-turbine and gas-pumping units, the actual effective capacity of a unit is referred to as the effective base power, and at the same time, various values, i. e. nominal, nameplate power, regulatory or available power are taken as the latter one. It is difficult to achieve the nameplate power under operating conditions not only on the units that are in

operation for a long time but also on a number of new units. Therefore, various types of units in use, set the industry standard for the power of (GPU), which is far below the nominal one. On average, the decrease is 15%.

A number of methods of determining the effective power of (GPU) depending on the outdoor temperature (T_a) [5], [9]. In this regard, we consider the problem of determining the effective driving power of (GPU) depending on T_a varying ways for comparative evaluation. The effects of other factors on the value of the available power (use of various recycling systems, anti-icing systems, the technical condition of units etc.) thus treated the same for different ways depending on temperature.

Then, taken as an example is a unit of the type GTK-10-4 with the known characteristics according to the datasheet, and the temperature of the outside air at the inlet of (AC) of (GTU) is taken as $T_a = 0°C$ for the following calculations. In determining the available power of the unit is used in different ways as input a set of known reference values and specifications for the unit datasheet in doing sums.

3.1. Determination the Available Power of the Gas-Turbine Unit

To determine the available power of the unit type GPU-C-6, 3 at the nominal temperature before the high-pressure turbine with the following values, $N_{e0} = 6,300$ kW; $K_N = 0.95$; $K_{ai} = 1.0$; $K_{re} = 0.985$; $K_T = 1.3$; $T_a = 273°K$; $p_a = 0.099$ MPa.

Solution: The determination of the available power of N_{ea} is carried out by two ways.

At first according to the formula VNIIGAZ taking into account the corrective coefficients according to (5) we have:

N_{ea1} = 6300*0, 95*1, 0* 0,985[1 – 1, 3 (273-288)/273] (0,099/0,1013)= 6173 Kw.

We obtained N_{ea}<N_{e0} i. e. the available power of the unit by the condition of the example under consideration is understated relative to the nominal power of (GTU) on 127 kW, which is 2%.

Then, the determination of the available power N_{ea} is conducted using formula (7) taking into account the auxiliary coefficients by the formula (5), the values of which are adopted as in the above example:

N_{ea2} = N_{e0} K_N K_{ai} K_{re} {[1- (T_a -T_{a0})/T_a] * [λ_e /(1-λ_e)]* (T_a - T_{a0})/T_a}(p_a / p_{a0}).

After substituting the numerical values we obtain the final result:

N_{ea2}=6300*0,95*1,0*0,985{[1- (273 -288)/273] * [0.54/(1-0.54)]*(273-288)/273}(0,099/0,1013)=7135 Kw.

We obtained N_{e0} < N_{ea} i. e. the available power of the unit according to the condition of this example is greater than the nominal engine power of (GTU) on 835 Kw, which is 13,3%.

The obtained result of comparing two methods of determining the N_{ea} indicates that the method of determining the effective power of the (GTU) following Professor N. I. Belokon's formula is based on the analytical method and may indicate the most accurate results of calculations relative to the first empirical method.

3.2. The Problem of Choosing the Available Power of GTU–N_{ea}

On the basis of the technical documentation of the manufacturer concerning servicing (GTU) it is not allowed to exceed the maximum power of the engine by 10% higher than the rated power according to the datasheet, $N_{e(max)}$ = 1.10 N_{e0}. An analysis of the results of the calculations show that in the first and second cases, N_{ea} exceeds N_{e0} by 2% and 13.3%:

N_{ea1} = 1,02 N_{e0}; N_{ea2} = 1,133 N_{e0}.

The calculations according to the 1st empirical method allow running gas turbine units, and according to the 2nd analytical N_{ea2} exceeds N_{e0} by more than 10%. If we take the results of the 1st method, then (GTU) will operate overloaded as the result is incorrect and in fact overload will not be 2% but 13.3% relative to N_{e0} according to the analytical method of calculation by Professor N. I. Belokon. Therefore, the conditions faced with the task to reduce the overload of up to 10%. For this purpose, it will be necessary to develop a method for reducing the air temperature T_a at the (GTU) inlet [9].

4. Conclusions

The article describes the validity of choice for assessment of the volume of production in the oil and gas industry as well as the energy sector that use gas turbine units. As a production volume indicator in enterprises with gas turbine

units given is the specific energy consumption at compressor stations of gas pipelines for the transport of a given volume of gas transported in the gas pipeline system, allowing an objective evaluation of the efficient use of energy consumed.

Taking into account the influence of temperature of the ambient air on the available effective power of the drive, considered are various ways to determine this power when the temperature of outside air is changed in the range of 258°K to 333°K. In this context, the method of determining the available effective power of the gas turbine drive based on the equation proposed by Professor N. I. Belokon is proposed as the most suitable one.

References

[1]　V. A. Shurovsky, Yu. N. Sinitsyn and A. k, "Analiz sostoyaniya i perspektiv sokrasheniya zatrat prirodnogo gaza pri ekspluatacii gazoturbinnyh kompressornyh cehov", Moscow, Issue. 2, 1982.

[2]　Eduard A. Mikaelyan, "From the energy saving solutions to the resource-saving technology, from the energy audit to the resources audit", Gas industry, № 4, 2014, PP. 70–73.

[3]　Normy tehnologicheskogo proektirovanija magistral'nyh gazoprovodov: STO Gazprom 2-3.5-051- 2006.

[4]　Eduard A. Mikaelyan, "Harmonisation of the Characteristics of Gas-Turbine and Gas Centrifugal Compressors regarding the Mechanical Power losses in Transmission from the Drive", Quality Management in the Oil and Gas Industry, № 1, 2015, pp. 15-18.

[5]　Eduard A. Mikaelyan, "Maintenance of the power technology equipment, the gas-turbine gas pumping units of the gas collection and transport system", Methodology, research, analysis and practice, Moscow, p. 314, 2000.

[6]　Eduard A. Mikaelyan, "Environmental air temperature influence on available power of gas-turbine machines", Energy Security- Energy Saving, № 3, 2014, pp. 23-28.

[7]　Appraisal procedure of power efficiency of gas-transportation installations and systems: STO Gazprom 2-3.5-113- 2007.

[8]　Eduard A. Mikaelyan, "Improvement of quality, provision of reliability and safety of main gas oil pipes in order to improve the operational suitability", Ed. by Professor G. D. Margulov, Moscow, p. 640, 2001.

[9]　A. I. Vladimirov, V. Ya. Kershenbaum, "Industrial safety of gas-turbine-driven compressor plants", National Institute of Oil and Gas, Moscow, P. 551, 2008.

[10]　Gas pumping units driven with gas turbine. General specifications, GOST R 54404-2011.

[11]　Eduard A. Mikaelian, "Recourse-saving technologies of gas transportation", Quality Management in Oil and Gas Industry, № 2, 2012, pp. 12-13.

[12]　V. A. Shchurovskii, Yu. N. Sinitsyn, V. I. Korneev, A. V. Cheremin, G. S. Stepanov, "Methodological guidelines for thermotechnical and gasdynamic calculations during tests of turbine pumping units", PR 51-31323949-43-99, Moscow, 1999.

25

An Environmental Value Model to Examine Recycling and Green Development in the Traditional Chinese Medicine Industry

Zhang Haibo[1], Shen Yuan[2, *], Shen Junlong[1], Yuan Pan[1]

[1]Nanjing University of Chinese Medicine, Nanjing, China

[2]School of Applied Mathematics, Nanjing University of Finance & Economics, Nanjing, China

Email address:

ZHB305@126.com (Zhang Haibo), ocsiban@126.com (Shen Yuan)

*Corresponding author

Abstract: Green development is a new model of development for China's economy to create a new normal by transforming and upgrading the TCM industry. A green TCM industry must solve the problems of three wastes because wastes generated from processing Chinese medical material (CMM) resources seriously pollutes the ecological environment and puts pressure on the environment. From the environmental value model and green economy theory, this study explains the social value of the environment, clarifies the social costs of TCM resources that create negative externalities from waste generated in the TCM industry. This study on the internal mechanism of external effects and discussion of the green economy model and use of green technology to recycle TCM waste proposes an economic network model to promote the use of technology to reuse TCM waste in a green network of industries.

Keywords: Green Development, Environmental Value Model, Waste Reutilization of Tcm, Economic Network Model, Green Technology

1. Introduction

Green development represents a social change from traditional agriculture to industry, and finally to an ecological civilization. In 1989, Pierce in the *Green Economic Blue Book* first proposed the green economy concept, wherein western developed countries experienced eight serious events that shocked the world in the mid-20th century. Since the 21st century, issues around global warming have become more prominent, and ecological crises, environmental crises, resource crises, and climate crises occur frequently. To cope with the increasing environmental crises, the United Nations (UN) Environment program launched the *Global Green New Deal: Policy Brief* in October 2008. In 2012, the main theme at the Rio+20 was the "Green economy in the context of sustainable development and poverty eradication," at the same time; the green economy was also the theme of the UN Conference on Sustainable Development.

China's economic development is entering a new normal, and the "Made in China 2025" plan was formulated, the Chinese version of Industry 4.0. Green manufacturing is one of five major projects in the plan, which includes sustainably developing high efficiency, intensity, environmental protection, and resource recycling, all of which can help transform and upgrade China's economy. The Fifth Plenary Session of the Eighteenth Central Committee of the Communist Party of China emphasized, "Achieve the implementation of the '13th Five-Year' planned development goals, firmly establish and effectively implement the innovation, coordination, green, open, shared development philosophy. This is a profound change related to the overall development of our country." This plan intends to address the limitations of the current carrying capacity of China's ecological environment that have become a bottleneck to economic upgrade and transformation. Therefore, the Fifth Plenary Session stated, "to adhere to green development, we must adhere to the basic state policy of resource conservation and environmental protection, sustainable development, development of production, affluent lifestyles, and a path to a good society based on ecological development."

Most agricultural waste occurs as litter or is released into the environment at an annual growth rate of approximately 5%~10%, which creates significant environmental pollution and squanders resources [1]. TCM waste accounted for a large proportion, especially in manufacturing processes that generate waste residue, waste water, and waste gas caused by the negative external effect of wastes in the industrialization process for TCM resources. Transforming and upgrading the TCM resources industry requires effective governance of these negative externalities. Professor Duan Jinao [2] noted that wastes associated with traditional Chinese medicine have not been used in Chinese traditional medicine production, including waste tissues, unused available substances, waste exhaust gas, waste residue, and waste resources using appropriate methods or techniques to recycle waste or develop available resources from the waste products. This article will focus on Chinese herbal medicine resources in the process of industrialization of waste resource utilization research in the context of traditional Chinese medicine, the industrialization of Chinese traditional medicine resource waste into available or renewable resources, and the negative externalities associated with Chinese medicine wastes to encourage the industry to conserve resources, recycle, protect the ecological environment, the development of green industry. [3]

2. Model of Ecological Environment Value of TCM Resources

The World Bank in 1992 listed production capital, social capital, human capital, and ecological environment as the four major types of capital to promote the development of a social economy [4]. The United Nations Environment Programme (UNEP) believes: "the ecological environment under certain conditions can generate economic value, and improvement in

ecological environment factors can improve current and future human well-being." Ecological environment value theory is generally based on labor value theory, utility value theory, and existence value theory. Researchers previously only focused on use value rather than non-use value, leading to a global environmental crisis. Since the development of TCM resources that are affected by modern industrial civilization, pharmaceutical processes only extracted the small portion of the chemical composition currently considered useful and discarded the vast majority of chemical substances. The Chinese pharmaceutical manufacturing processes produced three wastes with a serious effect on the ecological environment, the value of which developed along with the value of TCM resources. Figure 1 illustrates the total value model of environmental resources to explain the value of green development of TCM resources. In the figure, Total Economic Value (TEV) is divided into Use Value (UV) or instrumental value and Non-Use Value (NUV), or intrinsic value [5]. UV is divided into Direct Use Value (DUV) and Indirect Use Value (IUV). NUV is divided into Existence Value (EV), Bequest Value (BV), Option Value (OV), and Quasi-Option Value (QOV), though OV and QOV are also sometimes attributed to UV.

The green transition in the TCM industry, compared to the pre-industrial black development that had a "high-efficiency, high pollution" model of development, must move beyond recycling (reduce, reuse, recycle) to promote ecological restoration and environmental improvement as a precondition of development. While the need for sustainable development in the Chinese medicine industry also requires recycling methods, the industry's development must be based on the concept of green development as a new economic norm guided by the green concept, with a green system and green technology as the driving force. This will also mean building a sustainable development model for a green TCM industry, which is feasible based on the characteristics of TCM resources because they include natural resources.

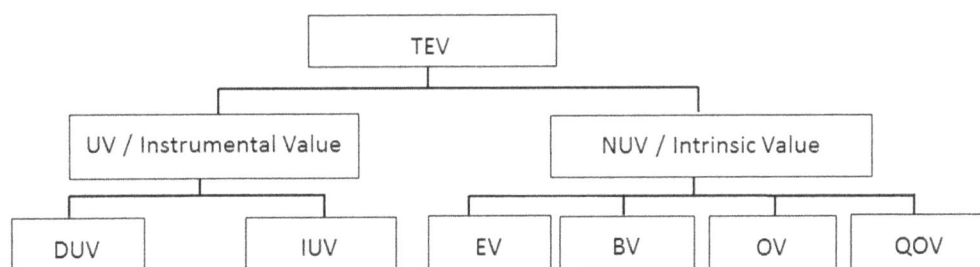

Fig. 1. Total Economic Value (TEV) structural model.

3. Green Development to Promote TCM Waste Reutilization

The green economy includes the population, resources, and environmental subsystems that create an organic combination

in a social-economic-ecology complex. In this complex, development is not only related to coordinating the interests of people, enterprises, and society, but also the coordination of humans with nature, social development, and ecological environmental protection. The green economy must build a green economic network system using the recycling industry to solve environmental problems. The TCM industry must use its resources to transform waste into treasure by formulating a new industry chain to conduct ecological production. We should learn from successful international experiences and adapt this according to the regional situation.

3.1. The Green Economic Point and the Internal Resource Recycling Mechanism

The green economic point is the basic level of the green economic network, which is the basic unit or node of the whole network of the green economy focused on green products produced from TCM resources and the transformation of TCM pharmaceutical enterprises based on the Du Pont Company model and the United States Acushnet company model. The DuPont Company model uses the concept of a circular economy in the chemical industry to develop a creative 3R principle and 3R manufacturing method. In the production process, the company abandoned some of the chemicals that are harmful to the environment and reduced the use of some chemicals, and found the new technology to recycle waste. By 1994, DuPont reduced plastic waste by 25% and air pollutants by 70%. The company simultaneously developed methods to recycle waste plastic and chemicals as well as durable ethylene and other new materials.

China can begin with a pilot in a TCM pharmaceutical enterprise. The government encourages enterprises to internally address external problems by means of tax cuts or subsidies to establish circular patterns and mechanisms for a circular enterprise model. Enterprises should adhere to the concept of cleaner production and the principle of "low consumption, low emission, high efficiency" in order to achieve internal material circulation between enterprises, control pollution in the production process, maximize the use of renewable resources, and achieve the purpose of reducing emissions or even emitting zero emissions to protect the environment [6, 7]. For example, waste residues from Chinese herb production can be made into biological fertilizer, biomass carbon (to adsorb cadmium in soil, heavy metals such as lead, and provide general soil improvement) or extract other chemicals, not only reducing the waste of renewable resources and protecting the ecological environment.

3.2. Green Economy Sheet and Spatial Economic Mechanism

The green economy sheet makes production in TCM and other related enterprises complementary to centralized production in a certain geographical area to recycle the resources and control pollution. The entire area thus becomes a subsystem of the green economy network. Local governments should establish TCM resources for a regional demonstration of a project focusing on recycling and the three wastes and use a green economic model to protect the ecological environment. We can refer to patterns at the Danish Karen Fort ecological industrial park, which is a typical representative of an ecological system. The park is within a refinery factory, power plant, pharmaceutical factory, and gypsum board factory as the four core factories with other small enterprises which trade using production process, which frequently create by-products and waste, for material and energy exchange and cycling. This has reduced wasted emissions and processing costs, not only significantly upgrades the economy by also greatly improves the ecological environment. Treatment transforms $9m^3$ of pharmaceutical factory wastewater into fresh water, reducing water use in refineries by 1.2×10^6 m^3. The refinery will supply gas to the power plant, reducing its annual coal and oil consumption by 3×10^4 t and 1.9×10^5 t, respectively. Organic residue from the pharmaceutical factory can become organic fertilizer, thus eliminating the need for a landfill. By replacing some of the coal and oil, the reduced sulfur dioxide and carbon dioxide emissions also reduce air pollution [8].

The local government should use the policy mechanism to encourage the establishment of a modern Chinese medicine ecological industrial park in line with the principles of green economic theory and industrial ecology, imitating the natural ecosystem model, similar to the food chain model in industry with a linear economic model of one-way flow from resources to products to change traditional wastes. Enterprises in the park will exchange surplus material, where waste from one business will serve as raw materials for another, thus achieving material interdependence, a closed cycle, multi-level energy use, and ultimately maximize the efficient use of resources and minimize waste emissions. The economy develops while protecting the ecological environment [9]. For example, the TCM GAP cultivation base, Herbal Pieces Processing Enterprises with TCM pharmaceutical enterprises, organic fertilizer manufacturing enterprises, and fine chemicals enterprises will be integrated by design. The waste materials produced in the process of planting and processing Chinese medicinal herbs will be transported for processing and use in the TCM pharmaceutical factory for different products, the waste residue produced by the pharmaceutical enterprises will become inputs in the organic fertilizer production plant, and ultimately supply fertilizer to the Chinese medicine planting base. Chinese medicine waste residue, waste water, and waste gas will supply useful material to the fine chemical enterprise, so that TCM resources are recycled and effectively improve the utilization efficiency in Chinese herbal medicine and make an entire ecological industrial park of green economic firms.

3.3. Green Economy Line and Supply Chain Management System

Green economy lines transforms and upgrades the TCM resources industry. By using green technology to optimize the combination of industries to enhance the green connotation of the entire industry, the entire industry becomes a subsystem of a green economic network. Manufacturing is different for Chinese herbal medicine and Western medicine. In Western medicine manufacturing, raw materials are a clear and transparent component, but TCM uses many ingredients, though the exact active ingredients need research and classification. However, despite the importance of the use of raw materials in medical manufacturing, the TCM industry should learn from Starbucks supply chain management and the collaborative model of French wine production. Chinese medicine manufacturers must be good at cooperating in terms of cultivation of germplasms, planting, collection processing, preparation technology, and integrating enterprises to form strategic economic cooperation between partners.

3.4. Green Economy Plane and Recycling-Oriented Social Model

The green economy plane is in the larger regional and even nationwide TCM resource companies that work with food companies, health care products companies, daily chemical enterprises, and environmental protection enterprises, among others at a deeper level of collaboration. This is a larger area to become a green subsystem. We can refer to the Japanese model. For example, in June 2000, the Japanese government announced the "Promotion of circular society fundamental law" as a basis, and then introduced the "Solid waste management and clean public law" and a policy "Promoting the effective use of resources" as a second-level synthesis. The third aspect relates to the legislative level and industry-specific products. In April 2001, Japan implemented the "Home appliances circulation method," a law stipulating that car manufacturers are obliged to recycle used cars and to recycle waste. At the end of May in the same year, Japan implemented the "Construction of recycling law," a 2005 provision mandating 100% recycling of construction site waste cement, asphalt, mud, and wood. The third level of legislation also includes "The promotion of container and packaging waste separation law," the "Food recycling law", the "Green purchasing law," and so on. Japan's circular society development model is reflected in the consumer, government, enterprise, social, and many other aspects. Consumers advocate green consumption and encourage consumers to purchase products with environment labels on products to promote green consumption, green production, and thus the development of a green economy. The Japanese government has stepped up publicity efforts to fundamentally change perceptions such that citizens do not despise garbage but rather regard it as a useful resource. Enterprises have been aware of energy conservation due to the scarcity of Chinese resources. In the vigorous development of the eco-industrial creed, an increasing number of enterprises have focused on clean production. According to the circular society concept, production processes are designed to promote recycling of raw materials and energy. The Japanese government urged companies to develop advanced new technology. In product design we should consider the issue of resource reuse. From the social aspect, Japan vigorously developed a green consumer market and promoted the development of industrial resource recycling and utilization.

The long chain of Chinese medical resources waste generated by planting, maintenance, acquisition, rough machining of some waste branches, leaf litter, flowers, fruit, and debris, the government and social organizations should support the establishment of a unified Chinese medicine resource recycling and processing system, using the social circulation model and the mechanism of ecological environment protection, akin to an "arteries, veins, recovery system" nationwide, with full recycling analogous to a venous system in terms of the resources used in Chinese herbal medicine. This will create new value and ensure the full and effective use of renewable resources, promote TCM resource waste recycling in the industry, establish a circular economic industrial chain, and balance economic and ecological development to protect China's environment.

4. Green Technology and Selection of Waste Reutilization in TCM

The core of green development is to save resources and protect the environment. Green development in the TCM industry must make full use of high technology to improve the utilization rate of TCM resources and help solve environmental pollution. The TCM industry is the most likely industry for green development because the use of innovative technology to grow the enterprise does not harm the environment and human health in the process of cultivation, production, circulation, and consumption. In the range of the carrying and regenerative capacities of TCM resources under non-reducing conditions, the ecological environment is conducive to healthy choices to balance people and nature through green technology. Green development needs green technology innovation, selection and use of environmentally friendly technologies, to ensure the protection of the ecological environment as the premise of innovative technology and multi-purpose development, maximum efficiency, and recycling of TCM resources.

In the past, due to the lack of scientific research and technical development in the process of harvesting and processing materials related to TCM, the selection of some part of the medicinal material is often used and the remaining non-medicinal parts are discarded. Manufacturing TCM products will produce a significant amount of waste gas, waste liquid, waste residue, and so on [10, 11]. For example, manufacturers only use angelica root, and discard the stems and leaves. For peonies, firms discard the fibrous root, root head, root bark, etc., thus wasting material. For licorice extract, only the licorice acids are use; the rest is waste, which further wastes resources in Chinese medicine manufacturing. China's medicine industry's average consumption of plant-based medicines for about a year is 7×10^5 t, while producing up to one million tons of residue [12]. This wastes resources and affects the ecological environment. Therefore, the medicine must use the appropriate technology for waste treatment resources, to convert it to available or renewable resources, and improve resource utilization and extend the TCM industrial chain. To save resources and encourage environmentally-friendly development in a circular economy, firms should promote the industrialization of TCM resources and sustainable development, which will lead to good health [13].

4.1. Raw Material Crushing Technology

Because most TCM wastes are typically solid cellulose material, it forms a relatively stable network structure that is difficult to use fully. Solid wastes thus require preprocessing by steam blasting, crushing, radiation, microwaving, and ultrasonic techniques to deconstruct the organic structure; acid hydrolysis, alkaline hydrolysis, ozone decomposition, ionic liquid with chemical technologies to break the crystalline cellulose structure; and fungi, enzymes, and other biological treatments to degrade the cellulose and lignin. These methods may also have to be used in combination [14]. Biochemical techniques will help transform solid TCM waste for all types of carbohydrates, saccharification to destroy the crystal structure, degrade the cellulosic material so the TCM residue can be used to produce fuel, chemicals, polymers, and other raw materials. The Jiangsu Baoying area is rich in lotus root, and the area has more than 60 lotus root production enterprises. Until the end of 2013 the annual export volume of lotus products was more than $5 * 10^4$ t, though this produces a lot of lotus root residue, with only a small fraction used as feed and most abandoned, thus wasting resources. Bao-yu CAI [15] found that crushing technology can degrade hemicellulose and lignin in lotus root residue, undermining its crystallinity to provide a source of dietary fiber. This shows that the techniques will be beneficial to Chinese traditional medicine in terms of recycling and reducing waste, and thus reduce ecological environment pollution.

4.2. Use of Biomass Energy Development Technology

From the sustainable development perspective, TCM waste is a clean renewable energy resource, so conserving TCM residue biomass energy to become an important method to industrialize TCM waste resources and create an industrial chain. We can use curing and compression molding and other physical techniques, to transform TCM resource recovery processing of waste stem leaf and residues as fuel directly; by anaerobic fermentation and biochemical technology, transform TCM solid organic waste using bacteria to convert part of the mass to methane, ethanol, and other energy sources; by thermochemical conversion technology to, transform solid wastes into clean solid, liquid, and gaseous fuels, such as of Salvia *miltiorrhiza* from the waste residue of catalytic pyrolysis to create bio oil [16, 17]. Shangdong BaichuanTongchuang Energy Co.ltd An energy company could use biomass thermos-chemical reaction mechanisms to convert residue into combustible gas or a high quality steam for conversion into electrical energy. These processes help create clean energy, which is both a valuable product but also helps solve issues with fuel shortages and provides an important source of biomass energy development. By August 2014, Shandong BaiChuang built more than 400 biomass gas demonstration projects in 13 provinces and cities.

4.3. Fermentation Technology

TCM waste contains many high molecular weight types of organic matter, such as protein, cellulose, and so on. Using fermentation technology can extract a variety of high value-added products. For example, TCM residue fermented with some microorganisms can create biological organic fertilizer and help improve soil via composting technology. Fermentation technology can also convert TCM residue into protein feed for poultry, livestock, and fish. Herbal residues can provide a fermentation substrate for edible fungi and a cultivation matrix for seedlings [18, 19, 20]. The current fermentation technology used in TCM residue conversion to study organic manure exists in a demonstration project run by Guangdong Yili Pharmaceutical Group to develop a "Chinese medicine extraction waste and environmental protection project for the national recycling economy," and addresses key demonstration projects. The project adopts high temperature fermentation technology to transform TCM waste into organic fertilizer, back to the planting base of Chinese medicine, the annual amount will be $1.5 * 10^5$ t Chinese medicine waste resources of equal amounts of organic fertilizer matrix, the formation of the cultivation of Chinese Medicine - Chinese medicine extraction - organic fertilizer manufacturing green circular economy industrial chain in the Twelfth Five Year Plan.

4.4. Extractive Techniques and Enrichment Technology

TCM residue resource materials are mainly flavonoids, polysaccharides, alkaloids, and triterpenoid glycosides, which can therefore benefit from extraction and enrichment techniques to significantly protect the ecological environment [21]. It is possible to recycle some solid Chinese medicine wastes through a solid-liquid phase equilibrium extraction, chromatographic separation, and enrichment techniques, to extract tanshinone IIA from Salvia *miltiorrhiza*. Liquid wastes treated with adsorption, membrane filtration, and biological treatment technology not only recycles water resources, but also recovers proteins, polysaccharides, and other TCM ingredients. Gaseous wastes, through condensation, adsorption, and biological purification technology can yield useful resources such as ageratum and mint aromatics[22,23]. The present applications of extraction and enrichment technology are mature in terms of transforming TCM waste. Gegen Ankang City as advantageous resources, and will meet the considerable demand for puerarin, which will generate large amounts of waste liquid containing medically valuable components such as isoflavones, which first squandered Ankang City's resource advantage. Yasuko Yoshiro, a biological resource firm, applies water extraction and membrane separation technology to Pueraria waste to enrich isoflavonesdaidzeinyuan to develop drugs and health foods, extending the chain of Puerarialobata industry development in Ankang City, and thus promoting recycling.

5. Conclusion

Green development requires the establishment of a green system, development of a green economic model, and achievement of the "Green New Deal" that has green resource technology innovation at its core. Green technology is required to solve the external problems of wastes generated in Chinese medical material (CMM) resource industrialization to promote green technology innovation, protect the environment, and efficiently use TCM resources on the path to green development. It is an effective way for the TCM industry to transform TCM waste into renewable resources, which is one of China's advantages in terms of planting, breeding, quality identification, and extraction and processing technology. TCM waste reutilization is not only conducive to sustainable development and ecological resource protection, but takes full advantage of TCM resources to develop China's TCM industry and promote sustainable development in China's economy.

Acknowledgments

The study was sponsored by "12th Five-Year" National Science and technology support program (2010BAI04B 03)& Natural Science in Colleges and universities in Jiang Su province level project (13KJD110002). We thank Prof Shen yuan for contributions to study design, study coordination; Shen junlong for contributions to study design and study management; Yuan pan for checking the Article's language and style.

References

[1] Wu Qun. Practical significance and Countermeasures of agricultural waste resource utilization [J]. Modern Economic Research, 2013(10): 50-52.

[2] Duan Jing-ao, Su Shu-lan, Guo Sheng etc. Production of castoff from process in Chinese materia medica resources industrialization as well as resource utilization strategies and modes [J]. Chinese Traditional and Herbal Drugs, 2013, 44(20): 2787-2797.

[3] Wu Xue-Ming, Xu Ting-Ting, He Bing-Fang, etc. Establishment of non-aqueous biotransformation system and its application in castoff from Chinese materia medica industrilization [J]. Chinese Traditional and Herbal Drugs, 2015, 46(3): 313-319.

[4] Costanza R.d Arge R.de Groot R. et.al. The value of the world's ecosystem services and natural Capital [J]. Nature, 1997, 387:1-260

[5] Davis RK. Recreation planning as an economic problem [J]. Natural Resources Journal, 1963, (3): 239-249.

[6] Pan Dong, Liu Dong-Huang, Tong Qun-Wang. Path Choice of China's industrial enterprises circular economy [J]. Lanzhou Academic Journal, 2014(10): 163-166.

[7] Liu Liang. Learn from the Models of Developed Countries' Circular Economy and Design the Models of China's Circular Economy [D]. Changchun: Northeast Normal University, 2008

[8] Chi Xing-Yun, Research on Mode and Stability Of Regional Circular Economy [D]. Jinan: Shandong University, 2009.

[9] Zhang Jing-bo. Research on the Resources utilization Modes of Industrial waste Based On Recycling Economy [D]. Hefei. HeFei University of Technology, 2007.

[10] Shi Jian-Yong. Chinese medicine industry economy and development [M]. Shanghai. Shanghai science and Technology Press, 2002.

[11] Yuan Pan, Shen Jun-Long. Causes and optimized strategies of price fluctuations for genuine traditional Chinese medicinal materials [J]. Chinese Traditional and Herbal Drugs, 2014, 45(23): 3503-3508.

[12] Waste extraction of traditional Chinese Medicine can be recycling environmental treatment [J]. Guiding Journal of Traditional Chinese Medicine and Pharmacy, 2012, 18(5): 119.

[13] Duan Jing-ao, Su Shu-lan, Guo Sheng etc. Research practices of conversion efficiency of resources utilization model of castoff from Chinese material medica industrialization [J]. China Journal of Chinese Materia Medica.2013, 40(23): 3991-3996.

[14] Pang F, Xue S L, Yu S S, et al. Effects of combination of steam explosion and microwave irradiation (SE-MI) pretreatment on enzymatic hydrolysis, sugar yields and structural properties of corn stover [J]. Ind Crops Prod, 2013, 42: 402-408.

[15] Cai Bao-Yu, Wang Lei, Tao Guan-Jun, etc. The preparation procedure of fiber in lotus root residue [J]. The Food Industry, 2004(3): 14-15.

[16] Fu Peng. Study on Biomass gasification gas phase product release characteristics and Coke structure evolution behavior, [D].Wu Hang. Huazhong University of Science and Technology, 2010

[17] Wang-Pan, Zhan Si-Hui, Yu Hong-Bing, Study On the catalytic pyrolysis of herb residue for bio-oil, etc. [J]. Environmental Pollution and Control, 2010, 32(5): 14-19.

[18] Wang Jian-Fei, Yu Qun-Ying, Chen Shi-Yong, Environmental hazards of Agricultural organic solid waste and the prospect of composting technology, etc. [J]. Journal of Anhui Agricultural Sciences, 2006, 34(18): 4720-4722.

[19] Huang Xiao-Guang, Qiu Zhe-Shi. Application of Chinese herb residue as feed additive [J]. Guangdong feed, 2007, 16(6): 32-33.

[20] Wu Yan-Xin, Ji Yan-Xi, Ren Ang, Review of the Research of Edible and Medicinal Fungi Cultivated by Chinese Herbal Medicine Residue [J]. Edible Fungi of China, 2011, 30(4): 3-6.

[21] Duan Jin-Ao. Resources Utilization of Chinese Herbal Medicine Wastes [M]. Beijing, Chemical Industry Press, 2013.

[22] Shi Ling, Hong Hao, Zhang Yan, etc. Separation and purification of tanshinoneÒA from Danshen dregs [J]. Journal of Dalian University of Light Industry, 2010, 29(2): 106-108.

[23] Chen Shi-Lin, The introduction of sustainable utilization of Chinese medicine resources [M]. Beijing, China Medical Science Press, 2006.

26

Thermal Analysis and Modelling of Thermal Storage in Solar Water Heating Systems

Samuel Sami[1], Jorge Zatarain[2]

[1]Research Center for Renewable Energy, Catholic University of Cuenca, Cuenca, Ecuador
[2]Faculty of Energy Engineering, Universidad Politécnica de Sinaloa, Sinaloa, México

Email address:

dr.ssami@transpacenergy.com (S. Sami)

Abstract: A numerical mathematical model has been developed to predict the thermal behavior of phase change material during thermal storage in a thermal tank. The model is based upon energy conservation equations and includes fusion of the phase change material. The thermal behavior of the phase change material during charging and discharging have been studied numerically, and analyzed under different conditions. Comparisons were made against experimental data for validation purposes of the predictive model. The model fairly predicted experimental data obtained at various inlet conditions of the phase change material.

Keywords: Phase Change Material, Thermal Storage, Thermal Tank, Numerical Model, Simulation, Experimental Validation

1. Introduction

Solar collectors convert solar irradiation energy to thermal energy that is transferred to working fluid in solar thermal applications. The heat carried by the working fluid can be used to either provide domestic hot water/heating, or to charge phase change material in a thermal storage tank for heat supply during non-solar periods.

Excellent thermal properties such as high thermal storage capacity, good heat transfer rate between the heat storage material, heat transfer fluid and good stability to avoid chemical and mechanical degradation, are the key factors to viable solar thermal energy storage system [1-8].

Phase change materials (PCMs) can store/release significant amount of heat during melting/solidification phase change processes. PCM during sensible heat storage experiences a large temperature rise/drop when storing/releasing thermal energy. However, latent heat storage occurs at nearly isothermal way. This makes latent heat storage desirable and favourable for thermal storage applications. Paraffin waxes are widely used as heat of fusion storage materials due to their availability to store thermal energy in a large temperature range [7].

A comparison of five commercial paraffin waxes as latent heat storage materials have been presented by Ukraincyk et al. [8]. The temperatures, heat capacities, of solid, and liquid paraffin waxes were measured by differential scanning calorimetry DCS. The thermal diffusivity was determined utilizing transient method.

An experimental investigation was carried out by Khot [10] to understand the improvement of thermal storage system using water-PCM in comparison with water. The experiments were conducted for the same heat input during accelerated conditions in both Water-PCM and also water inside thermal storage tank system.

Lin et al. [11] reported on an experimental investigation of solar water heater with thermal storage. In the study, an outdoor integrated phase change material flat plat solar collector was carried out with paraffin wax. The PCM gave the highest performance when considering the day and night time efficiency compared to the case without PCM.

Another study to investigate the characteristics of thermal energy storage in solar system using phase change paraffin wax in a thermal storage tank was presented by Kanimozhi [12]. The tank was instrumented to measure inlet and outlet water temperature as well as water mass flow rate. The solar energy was absorbed and stored in PCM storage unit as latent heat.

Recently, another investigation has been presented by Kuolkarni and Deshmukh [13] to undertake a study of the feasibility of storing solar energy using phase change materials PCMs to heat water for domestic purposes during the night times. The experiment showed that with using PCM, cooling rate during the night decrease as well as efficiency and heat storage capacity increases. Hot water storage tank systems store sensible energy; however, this requires added weight. However, the study showed that using phase change materials PCMs solves this problem and can satisfy the night and day-time energy demand.

Milisic [14] in his thesis described the state of the art progress in applying PCM materials for energy storage in thermal tanks, and opportunities of their future applications. In addition, he presented a mathematical model consisted of charging process and discharging process. Furthermore, Muhmud et al. [15] presented a theoretical model to investigate the thermal and physical properties of a phase change material consisted of paraffin wax and 5% aluminum powder in solar air heater. The results showed that the air temperature gained was due to energy discharge process decease with increasing the air mass flow rate.

A mathematical model for describing the heat stored in thermal storage tank and phase change material behaviour in thermal solar tanks is presented hereby. The model was established after the energy conservation coupled with the heat transfer equations. In the following sections, simulations of a thermal solar tank using paraffin wax will be presented and analyzed. In addition, several experimental data are used in order to validate the mathematical model. The mathematical model, in particular, was used to study of the effect of operating conditions such as solar radiation, working fluid flow rates, initial working fluid and paraffin wax temperatures on the phase change material behaviour and system performance as well as the thermal conversion efficiency.

2. Mathematical Model

A schematic of the thermal solar system under study is depicted in Figure.1. The system consists of a thermal solar panel collector, thermal tank, and paraffin wax, piping and pump as well as control valves. The phase change material; paraffin wax was placed in the thermal tank where a single tube heat exchanger was placed in the tank to charge and discharge the paraffin (C.F. Figure.1 and 2). The thermal tank with the single tube heat exchanger was numerically divided into different elements as shown in Figure.2 to permit writing the energy and heat transfer equations in finite-difference format for the heat transfer fluid HTF and paraffin wax PCM. The model is based on the following assumptions; PCM is homogeneous and isotropic, HTF is incompressible and it can be considered as a Newtonian fluid, inlet velocity and inlet temperature of the HTF are constant, PCM is in the solid phase for melting or in the liquid phase for solidification, thermophysical properties of the HTF and the PCM are constant. The phase change material experiences during

charging and discharging processes: three phase; solid, liquid and mushy. The solid and liquid phases have sensible heat and the mushy one has latent heat.

The conservation equations and heat transfer equations were written for each element as follows for each of the phases; solid, mushy and liquid phases;

Energy conservation and heat transfer equations:
The heat released by the heat transfer fluid HTF can be written as follows, [14],

$$\rho_{PCM} V_{PCM} C p_{PCM} \frac{\Delta T_{PCM}}{\Delta t} = Q = m_w C p_w \Delta T_w \qquad (1)$$

ΔT_w: the heat transfer fluid temperature difference
ΔT_{PCM}: the phase change material temperature difference.
The heat balance for the heat exchanger tube in the tank can be as follows [15];

$$(T_{in} - T_{out}) C p_w m_w = 2\pi R l h (T_{in} - T_{sfc}) \qquad (2)$$

Where the heat transfer coefficient is approximated as [15];

$$h = \frac{K_w}{D_H} b_2 Re^n \qquad (3)$$

$$Re = \frac{m_w D_H}{\mu A_f} \qquad (4)$$

Charging phase:
During the charging phase the water (HTF) mass flow rate can be calculated from the heat released by the solar radiation,
Mass flow rate of water:

$$m_w = \frac{G A_{Panel}}{1000 \times C p_w \Delta T_w} \qquad (5)$$

Equation (2) with the finite difference formulation of the time derivative can be written for the solid phase as follows (Figure. 2);
Solid phase:

$$T_{PCM_{m+1}} = T_{PCM_m} + \frac{m_w C p_w \Delta T_w}{\rho_s V_{PCM} C p_s} \Delta t \qquad (6)$$

Where: T_{PCM_m}, Temperature of PCM at m element ($°C$)
m_w, Water mass flow rate $\left(\frac{kg}{s}\right)$
$C p_w$, Specific heat of water $\left(\frac{kJ}{kg\ K}\right)$
V_{PCM}, PCM volume (m^3)
$C p_s$, PCM specific heat at solid phase $\left(\frac{kJ}{kg\ K}\right)$
ρ_s, Density of PCM at solid phase $\left(\frac{kg}{m^3}\right)$
G, Radiation $\left(\frac{W}{m^2}\right)$
R, Tube radius (m)
l, Tube length (m)
h, Heat transfer coefficient
b_2 & n, Constants equal to 0.3 and 0.6 respectively
D_H, Hydraulic diameter (m)
μ, Water viscosity $\left(\frac{m^2}{s}\right)$
K_w, Thermal conductivity of water $\left(\frac{kJ}{ms°C}\right)$

A_{Panel}, Area of solar panel (m^2)

A_f, Flow area (m^2)

Δt: time interval in the finite difference formulation,

Re: Reynolds number

Same finite difference formulation of the time derivative can be applied to the mushy and liquid phases as follows;

Mushy phase:

The liquid fraction is calculated from the heat balance at the mushy phase region as follows;

$$\gamma_{m+1} = \gamma_m + \left(\frac{m_w Cp_w \Delta T_{w,mushy}}{\rho_L V_{PCM} h_L}\right)\Delta t \tag{7}$$

Where the heat absorbed during the mushy phase change is given by;

$$Q = \rho_L V_{PCM} h_L \gamma \Delta t \tag{8}$$

Where, γ_m, Liquid fraction at m element (%)

h_L, PCM latent heat $\left(\frac{kJ}{kg}\right)$

V_{PCM}, Volume of PCM (m^3)

ρ_L, Density of PCM at liquid phase $\left(\frac{kg}{m^3}\right)$

Q, Heat (kJ)

Liquid phase:

With the finite difference formulation of the time derivative, the PCM liquid temperature can be calculated as;

$$T_{PCM_{m+1}} = T_{PCM_m} + \frac{m_w Cp_w \Delta T_w}{\rho_L V_{PCM} Cp_L}\Delta t \tag{9}$$

Therefore, the total heat absorbed during the charging process by the phase change material during solid, mushy and liquid phases is;

$$Q_{Charge} = m_{PCM}(Cp_s \Delta T_s + h_L + Cp_L \Delta T_L) \tag{10}$$

Discharge phase:

During the discharge process, phase change material experiences phase change from liquid to mushy and solid while yielding heat absorbed during the charging process. The water mass flow rate of heat transfer fluid during the discharge process can be calculated by;

$$m_w = \frac{Q_{Charge}}{Cp_w \Delta T_w} \tag{11}$$

Where;

ΔT_w: heat transfer fluid temperature difference

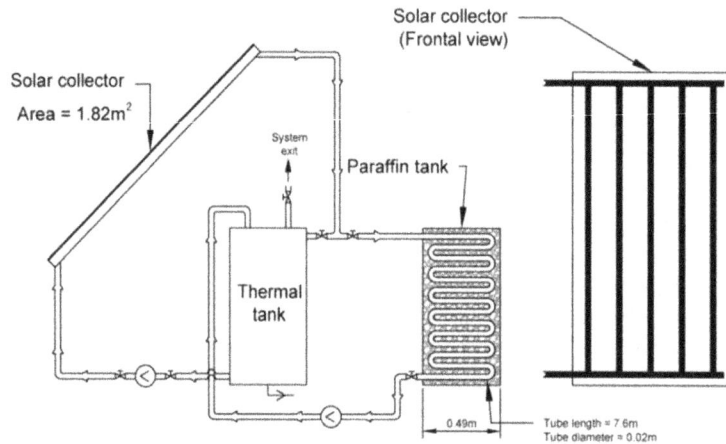

Figure 1. Schematic diagram of thermal solar panel and thermal tank system with paraffin.

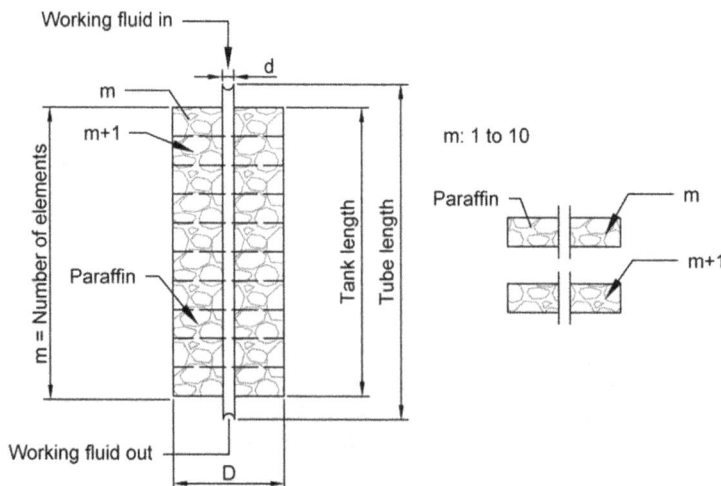

Figure 2. Detailed formulation of finite-difference scheme for the thermal solar tank with paraffin.

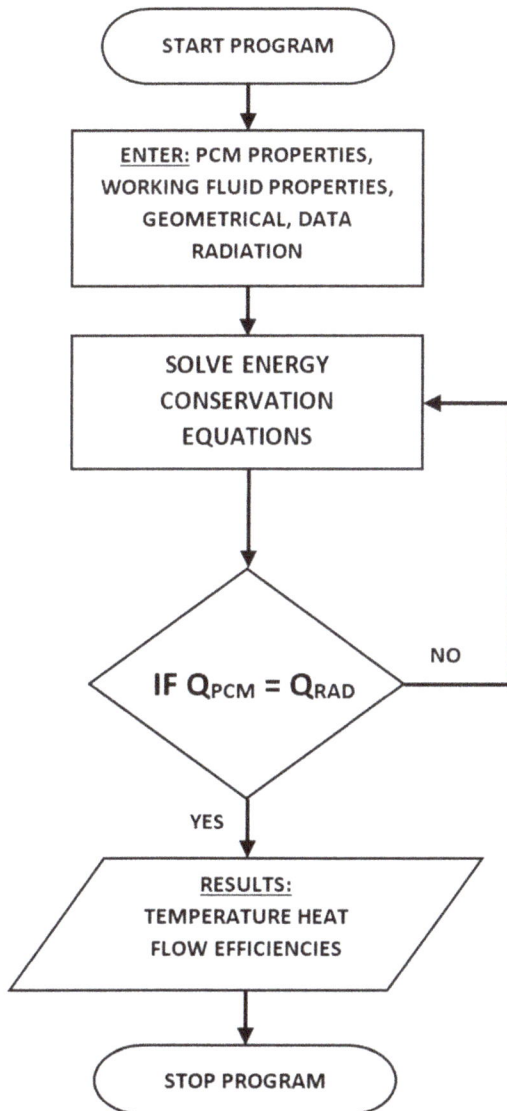

Figure 3. Logical flow diagram for finite difference scheme.

3. Numerical Procedure

The energy conversion and heat transfer taking place during charging and discharging periods for phase change material paraffin wax placed in the thermal solar tank have been outlined in equations (1) through (11). The aforementioned equations have been solved as per the logical flow diagram shown in Figure.3, where the input independent parameters are defined and other dependent parameters were calculated and integrated in the finite-difference formulations. Iterations were performed until a solution is reached with acceptable iteration error. The numerical procedure starts with the use of the solar radiation to calculate the mass flow water circulating in the solar panel. This follows by predicting the temperature profile of the phase change material and heat transfer fluid during the charging and discharging periods for the three phases; solid, mushy and liquid using the finite difference formulation.

4. Discussion and Analysis

The aforementioned system of equations (1) through (11) in finite-difference forms have been integrated and numerically solved and samples of the predicted results are plotted in Figures 5 through 13, at different inlet conditions.

The thermal tank has diameter of 0.49 meter and height 0.99 meter with capacity of 100 liters and one path tube heat exchanger of 0.025 meter and 7.5-meter length. The heat transfer fluid; water flows inside the aforementioned one path tube heat exchanger as shown in Figures. 1 and 2. Three PCMs paraffin waxes are considered in the present simulation with their thermal and thermophysical properties as presented in Table. 1. Paraffin waxes were selected because they are the most commonly used commercial organic heat storage PCM due to their large latent heat and moderate thermal energy storage density, low vapor pressure, good thermal and chemical stability, lack of thermal separation and environmentally sound [9].

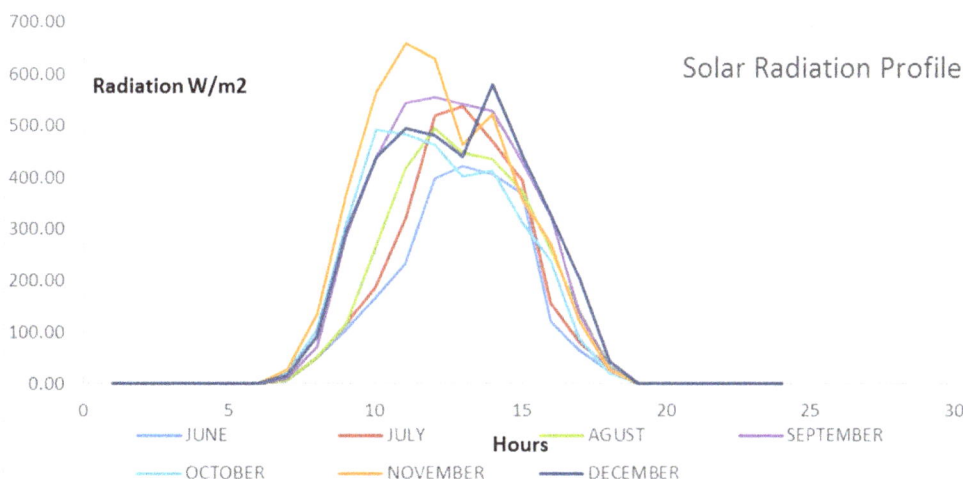

Figure 4. Time variation of solar intensity.

In general, it is quite clear from figures (5) through (13) that initially the phase change material (paraffin wax) temperature increases during the solid phase until reaches the fusion temperature (melting point) where the temperature stays constant until the onset of the liquid phase where the temperatures increases until reaches the charging temperature that is determined by the heat transfer fluid conditions. Therefore, during this process, the phase change material experiences sensible heat addition during the solid and liquid phases and latent heat addition during the mushy phase. In the following sections, we present simulation results of different paraffin wax phase change material at different conditions during the charging and discharging process. Thermodynamic and thermophysical properties of the paraffin wax materials are presented in Table. 1.

In particular, Figure. 4 presents the time variation of solar insolation (W/m2) measured at the site and employed in the simulation. It is quite clear that the intensity of radiations depends upon the hour of the day and the month of the year.

Figure 5 displays the predicted behavior of phase change material paraffin wax [5] during the charging process for a cycle of 8 minutes under constant solar radiation of 550 W/m^2. This figure presents the temperature profile of the heat transfer fluid and consequent temperature increase of the paraffin wax. Such a behavior was presented during the solid phase and fusion (mushy) phase. It can be observed that the heat transferred from the HTF to paraffin wax caused the paraffin wax's temperature to rise and the HTF's temperature to decrease. Furthermore, Also Figures. 5 through 7 show the same behavior for three paraffin waxes [9] under investigation at different solar radiations. As can be seen from the figures, the charging time can vary depending upon the properties of each paraffin wax, namely, the melting temperature, and specific heat among other properties as well as the conditions of the heat transfer fluid properties.

In particular, Figure. 6 and 7 display the temperature time variation of paraffin 1, 2 and 3 under different solar intensity radiations. It is quite clear from the results presented in these figures that the initial paraffin temperature increases during the charging process until reached the melting temperature and the threshold of the mushy phase, where the paraffin temperature remains constant until the end of the mushy/latent heat phase. As shown in the figures, the paraffin temperature starts to increase until the end of the charging phase where the paraffin is fully converted into liquid. It is worthwhile noting that this behavior has been observed for the three paraffin waxes.

On the other hand, figures 8 and 9 present the time variation of the heat absorbed during the charging process for three paraffin waxes under different solar radiation. The results clearly show that sensible heats were observed during the solid and liquid phase where heat was increased similar to the temperature time variation presented in the previous figures. Latent heat was constant in the mushy phase where the melting temperature is constant. Furthermore, the results also showed that the higher the solar intensity radiation the

higher the heat absorbed. Similar behavior has been observed for the other paraffin waxes.

Figure.10 illustrates the temperature time variation of the paraffin wax and the heat transfer fluid where the liquid phase material is converted into mushy and solid at the end of the discharging process. The paraffin temperature decreases during the discharging process due to the heat transferred to the heat transfer fluid. The amount of the heat transferred depends upon the final condition of the paraffin wax by the end of the charging process and the heat transfer fluid thermal and thermophysical properties.

Figures 11and 12 present the temperature profiles during discharging process for paraffin waxes PCM2 under different solar intensity radiation. Obviously, the amount of heat transferred to the heat transfer fluid is dependent upon the thermophysical and thermodynamic properties of the paraffin wax [Table.1]. As discussed previously, and similar to the behavior observed of the paraffin waxes during the charging process, the discharging time has a functional dependence upon the type of paraffin wax. On the other hand, Figure.12 illustrates that the discharging characteristics of paraffin wax depend also upon the solar radiation intensity used during the charging process and the higher the radiation the lower charging time and consequently the shorter discharging time. Clearly, the required energy to raise the water temperature is influenced by solar radiation intensity.

Furthermore, Figure.13 shows that the amount of heat released from the paraffin waxes to the heat transfer fluid is impacted by thermal and thermophysical properties of the phase change material. As shown in Figure 13, the amount of heat released is dependent upon the final condition by the end of the charging process and the heat transfer fluid conditions. Similar behavior was observed under different solar radiation intensity.

Figure.14 and 15 have been constructed to show the impact of heat transfer fluid conditions on the heat released by paraffin during charging and discharging process at different water flow rates and constant solar radian intensity. It is evident from results displayed in these figures that higher water flow rate yields to higher heat released and consequently higher paraffin and water temperatures during the charging and discharging processes, respectively. In addition, similar behavior was observed at other solar isolations.

From the aforementioned discussion, it is worthwhile noting that in sensible heat storage, thermal energy transferred from or to the heat transfer fluid is stored by raising the temperature of a solid or liquid. The solar and phase change material system, in question, utilizes the heat capacity and the change in temperature of the material during the process of charging and discharging. However, the latent heat storage is based on the heat absorption or release when a storage phase change material undergoes a phase change from solid to liquid or liquid to solid.

In order to validate the proposed model to predict the behavior of phase change material in a solar thermal tank, experimental data presented by Fazilati, and Alemrajaki, [16] have been simulated during the charging and discharging process have been presented in Figures 16 and 17. It is quite clear from the results presented in Figure.16 that the proposed model fairly predicted the data of time variation of the paraffin wax temperatures during the charging process and the sensible heat periods. The slight difference between the data and model prediction suggest that there some heat losses during the latent heat region which was not taken into account fully by the model formulation.

However, Figure 17 presents the comparison between the model's prediction and the paraffin data during the discharging process. It appears from this figure that the model over predicted the data in the solid phase during the discharging process. This was attributed to heat losses that were not taken into account in the model.

Finally, Figure 18 has been constructed to compare the thermal storage energy conversion efficiencies of different paraffin waxes. In this figure, the efficiencies of seven paraffin waxes were presented; paraffin-1 [17], paraffin-2 [18], paraffin-3 [19], paraffin-4 through-7 [20]. The energy conversion efficiency is defined as the ratio between the discharge heat to the charging heat of paraffin wax. As discussed in the aforementioned sections, the energy conversion thermal storage efficiency mainly has a functional dependent upon the thermodynamic and thermophysical properties of paraffin wax. The results presented in this figure also show that the higher the efficiency the lower the melting point and the ability to discharge more heat to the heat transfer fluid. In addition, it is worthwhile mentioning that higher melting temperature requires higher thermal energy during the charging process. However, in some special applications, higher melting temperatures are preferred if higher discharging temperature is needed and energy conversion efficiency is compromised.

Table 1. Paraffin wax properties.

Paraffin Wax 1 [5]	
Melting point	46.7°C
Specific heat (solid)	2.89 kJ/kg°K
Specific heat (liquid)	2.89 kJ/kg°K
Density (solid)	947 kg/m³
Density (liquid)	750 kg/m³
Latent heat	209 kJ/kg
Paraffin Wax 2 [9]	
Melting point	41°C
Specific heat (solid)	2.48 kJ/kg°K
Specific heat (liquid)	2.76 kJ/kg°K
Density (Solid)	829 kg/m³
Density (Liquid)	765 kg/m³
Latent heat	288 kJ/kg
Paraffin Wax 3 [9]	
Melting point	37°C
Specific heat (solid)	1.82 kJ/kg°K
Specific heat (liquid)	2.17 kJ/kg°K
Density (solid)	911 kg/m³
Density (liquid)	799 kg/m³
Latent heat	201 kJ/kg

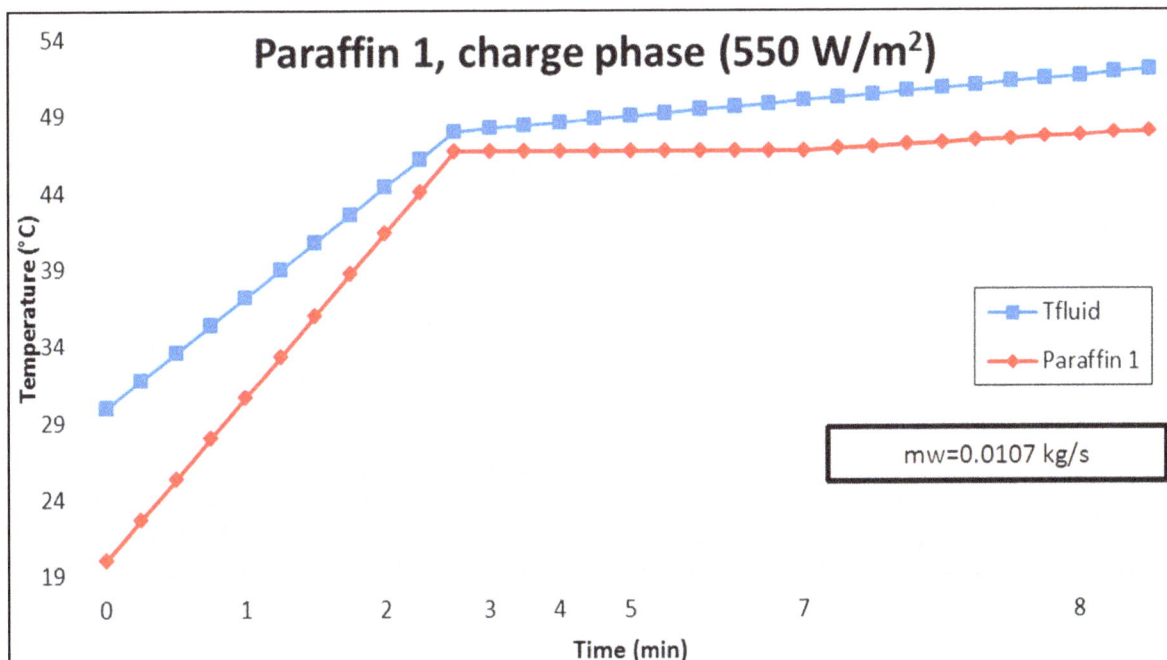

Figure 5. Temperature profile during charging phase under solar radiation 550 W/m².

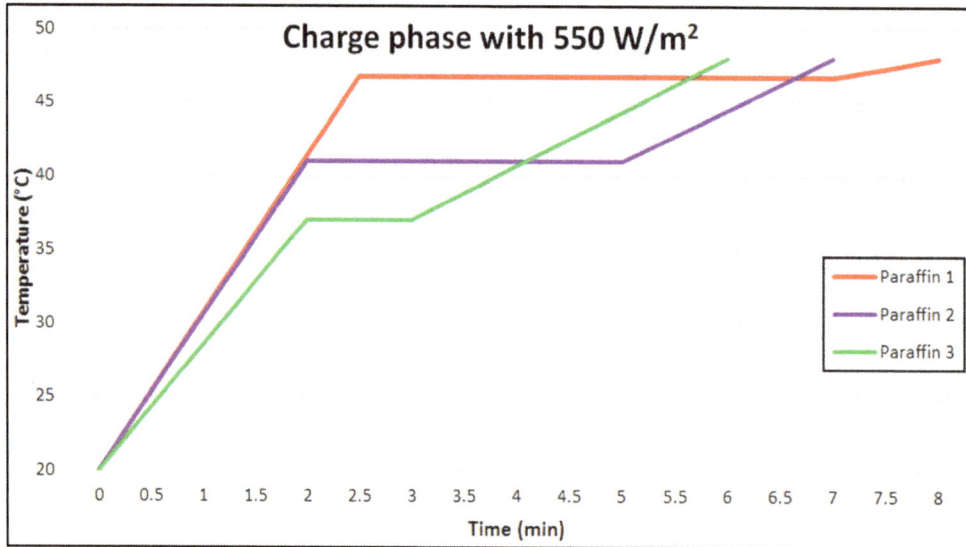

Figure 6. *Comparison of different paraffin wax [Table.1] under different solar radiations.*

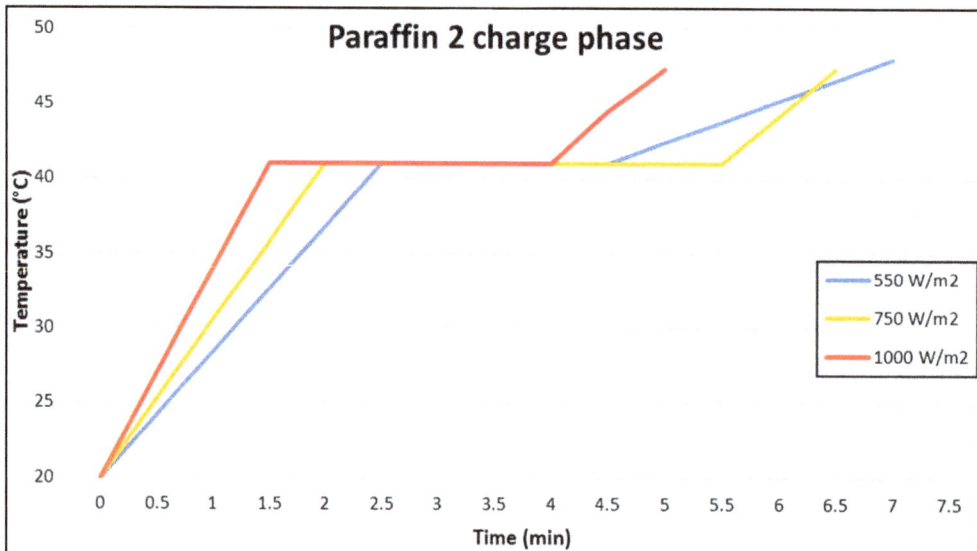

Figure 7. *Behaviour of paraffin 2 [Table.1] under different solar radiations.*

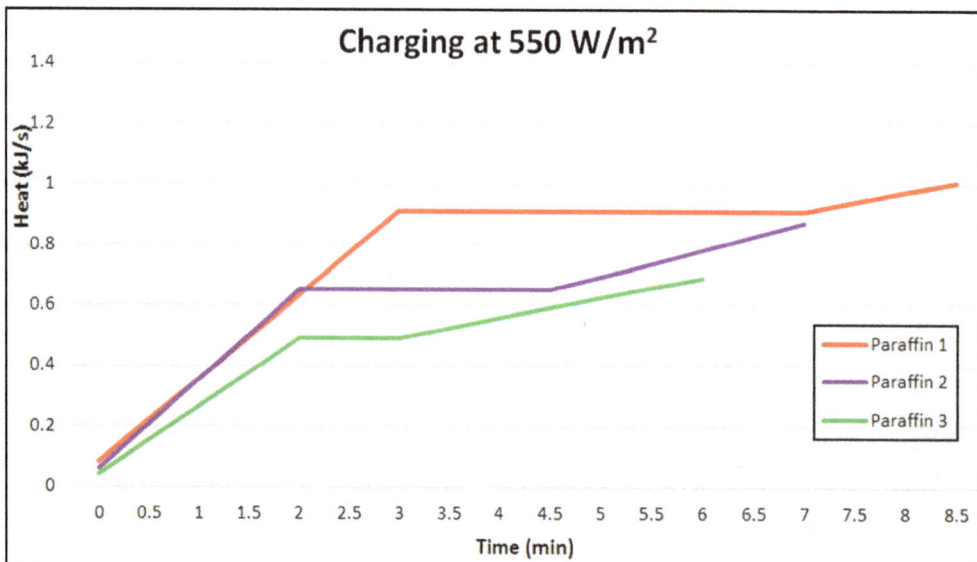

Figure 8. *Heat absorbed by different paraffin wax [Table.1] under constant solar radiation.*

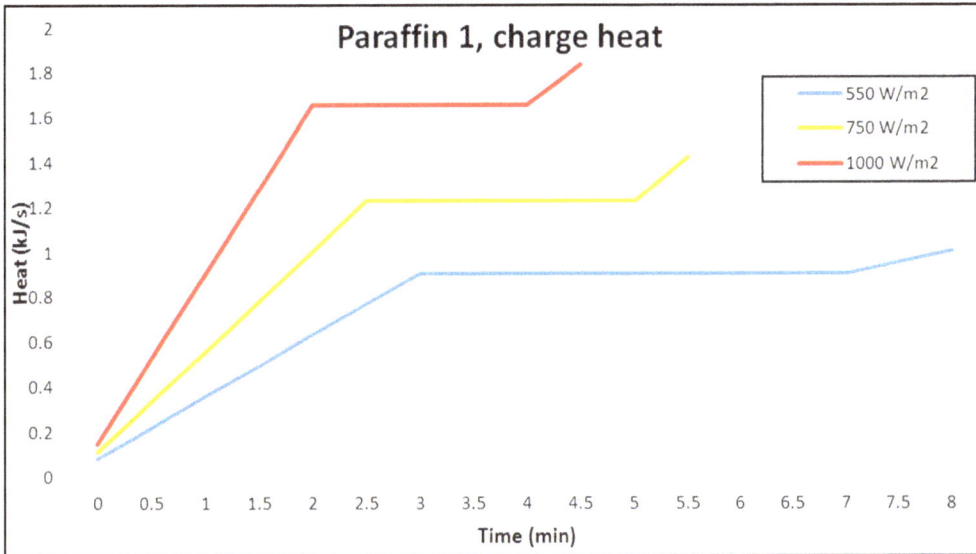

Figure 9. *Heat absorbed by paraffin 1 [Table.1] under two different solar radiations.*

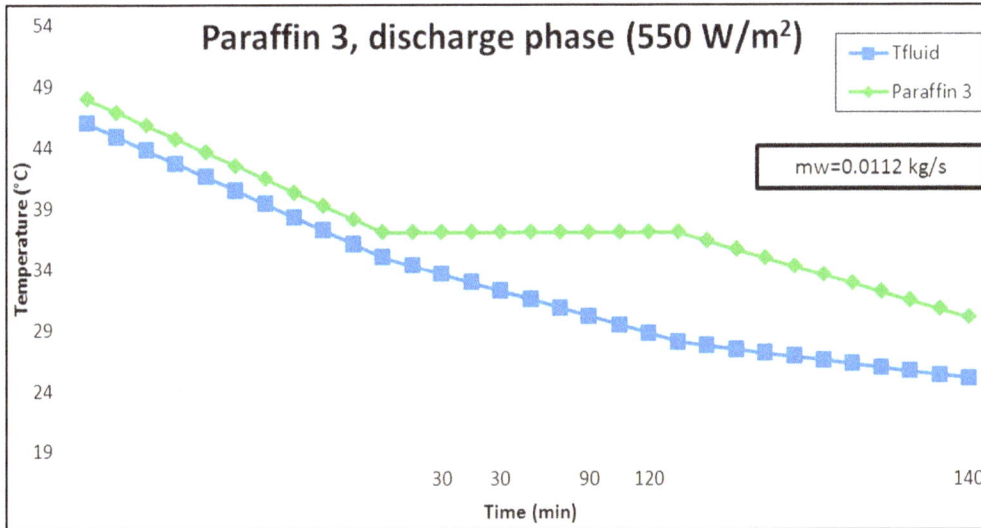

Figure 10. *Behaviour of paraffin wax [Table.1] during discharging process.*

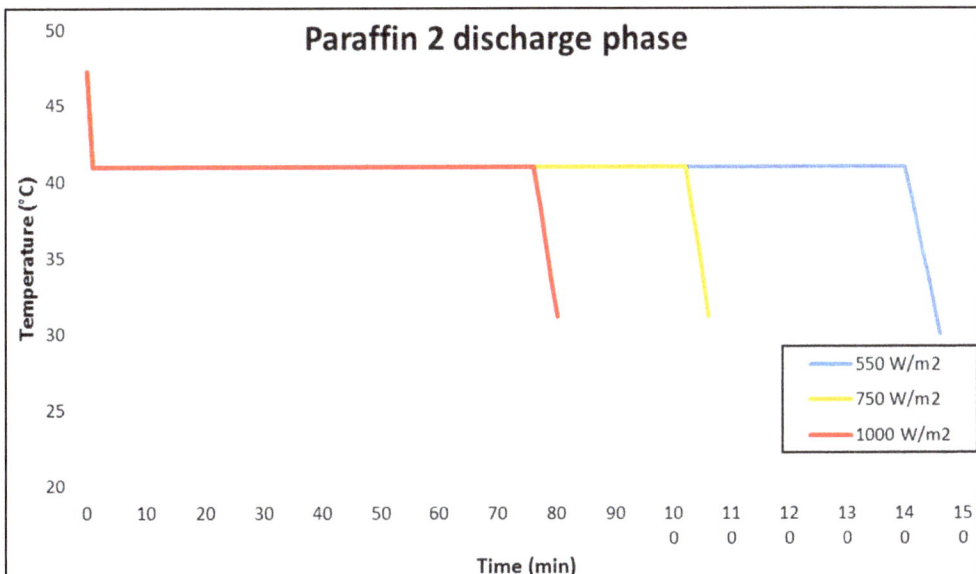

Figure 11. *Paraffin wax 2 [Table.1] during discharging process while charging under different solar radiations.*

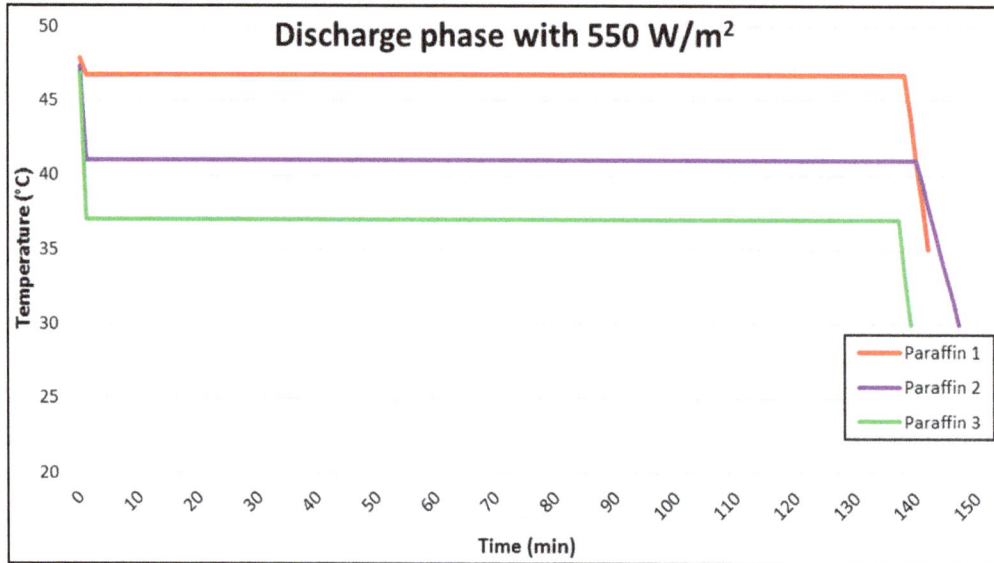

Figure 12. Temperature time variation during discharging by different paraffin wax [Table.1] under solar radiation 550 W/m2.

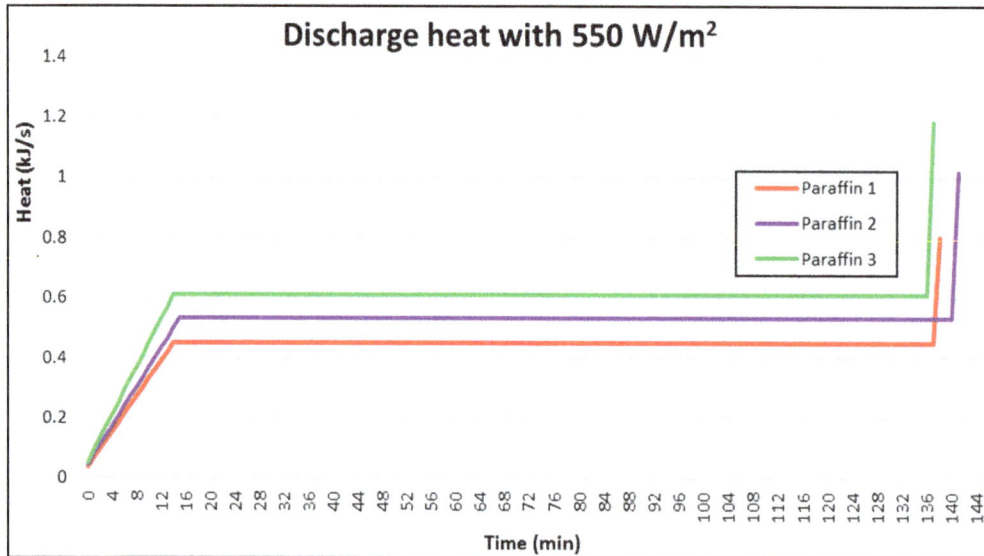

Figure 13. Heat released during discharging by different paraffin wax [Table.1] under solar radiation 550 W/m².

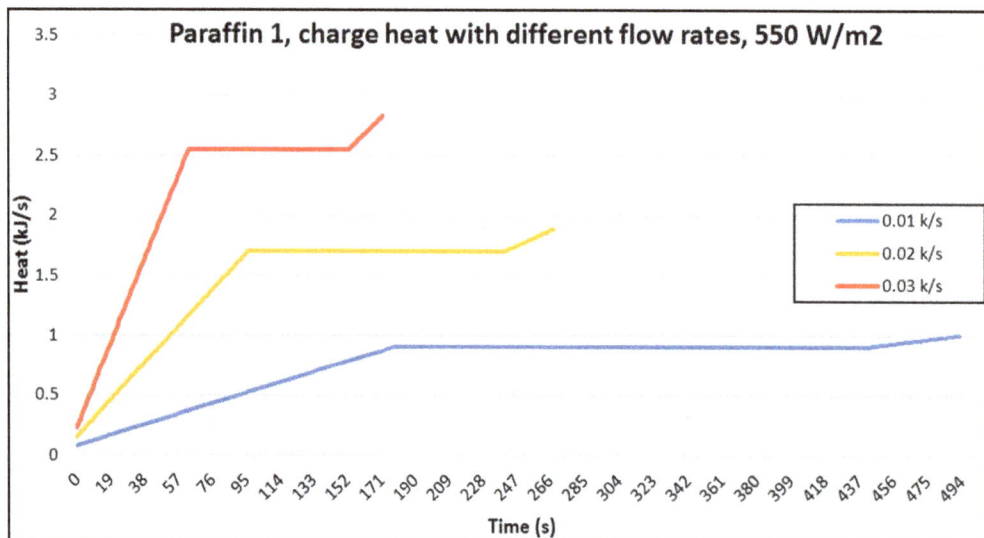

Figure 14. Heat released by paraffin wax 1 [Table. 1] at different HTF mass flow rates during discharging process.

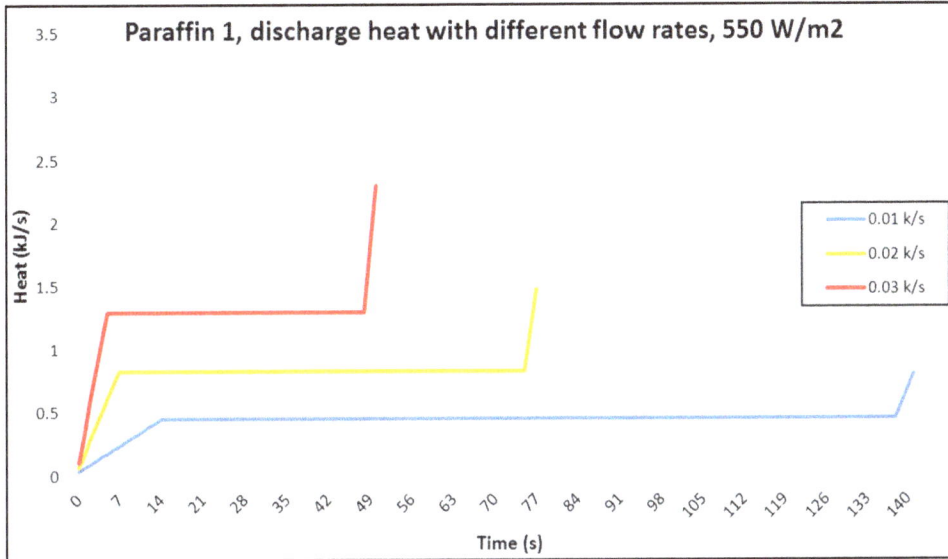

Figure 15. *Heat released by paraffin wax 1 [Table. 1] at different HTF mass flow rates during discharging process.*

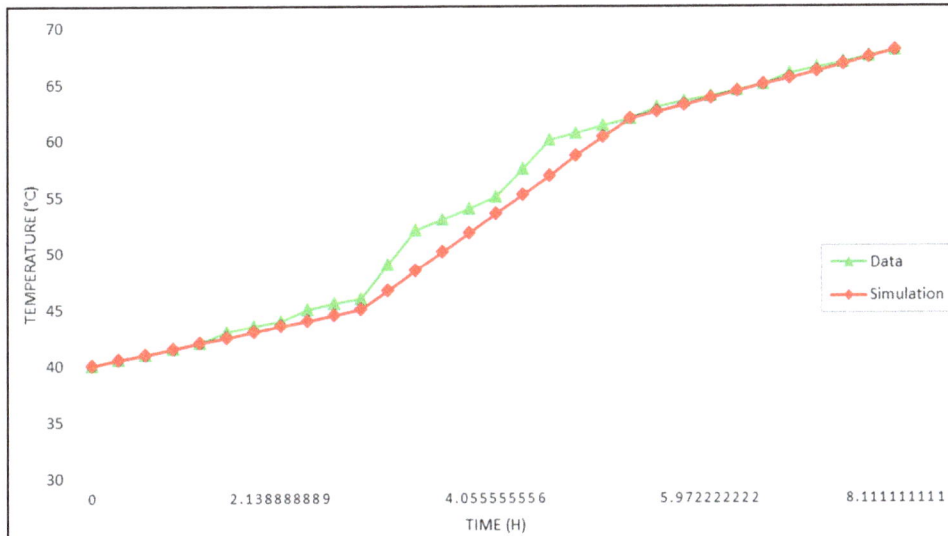

Figure 16. *Comparison between model prediction of charging paraffin wax and data [16].*

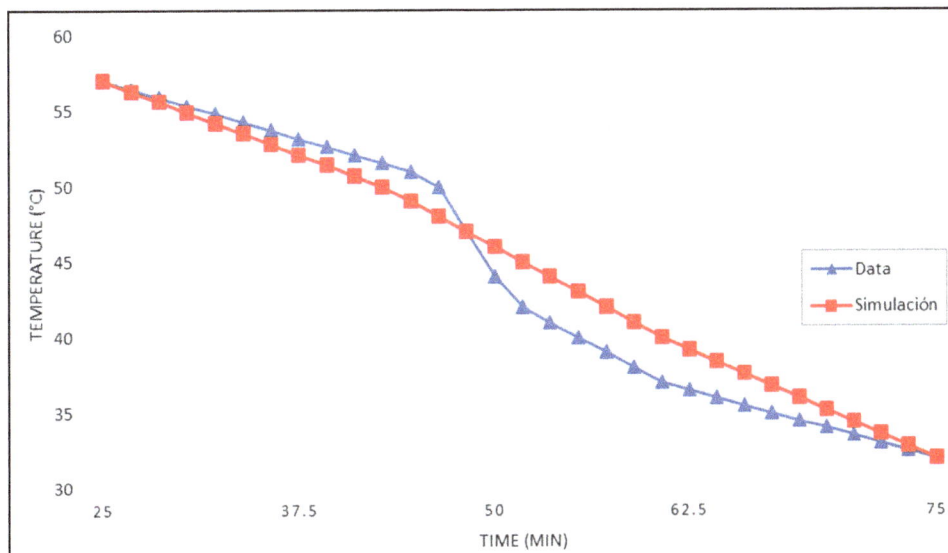

Figure 17. *Comparison between model prediction and data [16] during discharging process.*

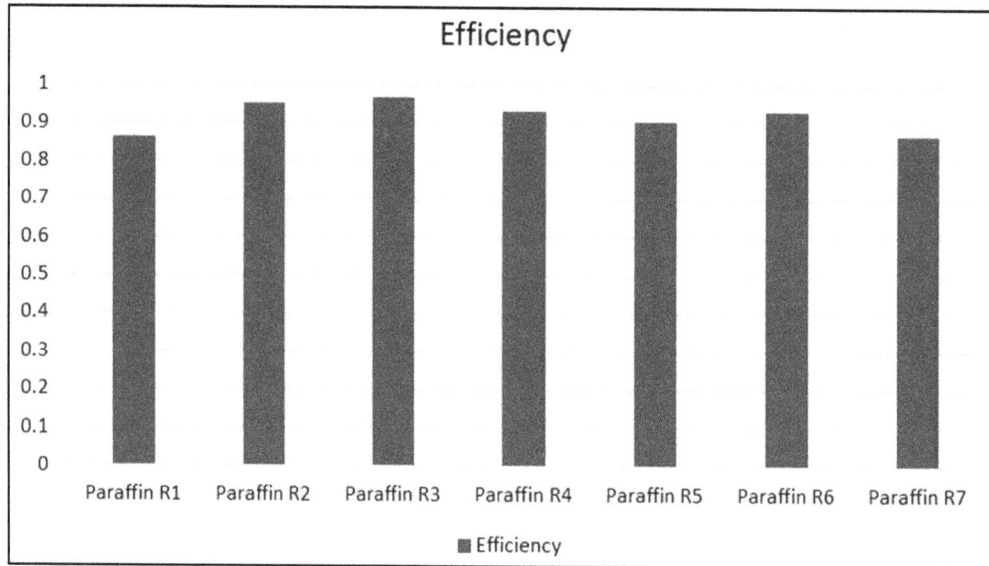

Figure 18. *Energy Conversion Efficiency comparison between different paraffin waxes.*

5. Conclusions

During the course of this study, the phase change material characteristics during the charging and discharging processes of paraffin wax in solar thermal tank have been modeled, presented and analyzed. The model was established after the energy conservation equations coupled with the heat transfer equations. In general, the presented numerical model fairly predicted the phase change material heat transfer characteristics during charging and discharging processes and compared well with the experimental data.

Nomenclature

A_{Panel}, Area of solar panel (m^2)

A_{f}, Flow area (m^2)

C_{ps}, PCM specific heat at solid phase (kJ/(kg K))

Cps, PCM specific heat at solid phase (kJ/(kg K))

C_{pw}, Specific heat of water (kJ/(kg K))

D_{H}, Hydraulic diameter (m)

G, Radiation (W/m^2)

h, Heat transfer coefficient

h_L, PCM latent heat (kJ/kg)

K_w, Thermal conductivity of water (kJ/(ms°C))

l, Tube length (m)

m_w, Water mass flow rate (kg/s)

N: number finite different element (N: 1-12)

Q_{tub}, Heat (kJ)

R, Tube radius (m)

$T_{PCM, m}$, Temperature of PCM at "m" element (°C)

V_{PCM}, PCM volume (m^3)

V_{PCM}, Volume of PCM (m^3)

Greek

ρ_s, Density of PCM at solid phase (kg/m^3)

ρ_L, Density of PCM at liquid phase (kg/m^3)

μ, Water viscosity (m^2/s)

γ_m, Liquid fraction at m element (%)

Subscripts

H: Hydraulic

L: Liquid

PCM: Phase change material

S: Solid

Tub: tube

W: Water

Acknowledgement

The research work presented in this paper was made possible through the support of the Catholic University of Cuenca.

References

[1] Tian, Y, and Zhao, CY A review of solar collectors and thermal energy storage in solar thermal applications. Applied Energy 104 (2013): 538–553.

[2] Farid, M. M, Khudhair, A. M., Razack, S. A. and Al-Hallaj S. A review on phase change energy storage material. Energy Conversion and Management, 2004; 45: 1597.

[3] Elawadhi, E. M. Phase change process with free convection in a circular enclosure; numerical simulation. Computer & Fluids, 2005; (33), 1335-148.

[4] Tardy, F. and Sami, S. (2008), An experimental study determining behaviour of heat pipes in thermal storage, International Journal of Ambient Energy, 2008; 29, (3).

[5] Razali, T, Hamdani, Irwasnsyah, Zaini, (2004), Investigation of performance of solar water heater system using paraffin wax, ARPN Journal of Engineering and Applied Sciences, 2014; 9, (10).

[6] Thirugnanm, C, and Marimuthu, P (2013), Experimental analysis of latent heat thermal energy storage using paraffin wax as phase change material, International Journal of Engineering and Innovative Technology (IJEIT), 2013; 3, (2).

[7] Hale DV, Hoover MJ, O'Neill MJ. (1971), Phase change materials hand book. Alabaa: Marshal Space Flight Center.

[8] Sami, S., and Tardy, F. (2015), Numerical prediction of thermal storage using phase change material" IJIRE, Volume 3, No 4.

[9] Ukrainczyk, N., Kurajica, S., and Sipusic, J, (2010), Thermophysical comparison of five commercial paraffin waxes as latent heat storage materials, Chem. Biochem. Eng. Q 24, (2), 129-137.

[10] Khot, S. A., (2014), Enhancement of thermal storage system using phase change material, Energy Procedia, 54, 142-151, 2014.

[11] Lin, S. C, Al-Kayiem, H. H. and Bin Aris, M., S., (2012), Experimental investigation on performance enhancement of integrated PCM-Flat solar collector, Journal of Applied Sciences, 12, (23), 2390-2396, 2012.

[12] Kanimozhi, B, and Bapu, R., (2012), Experimental study of thermal energy storage in solar system using PCM, Transaction on Control and Mechanical Systems, vol. 1,No.2, Pp. 87-92, 2012.

[13] Kulkarni, M. V. and Deshmulkh, D. S., (2014), Improving efficiency of solar water heater using phase change materials, IJSSBT, vol. 3, No. 1, Dec. 2014.

[14] Milisic, E, (2013), Modeling of energy storage using phase-change materials (PCM materials), Master thesis, Norwegian University of Science and Technology, July 2013.

[15] Mahmus, A, Sopian, K, Alghoul, M. A. and Sohif, M, (2009), Using a paraffin wax-aluminum cpound thermal storage material in solar air heater, ARPN Journal of Engineering and Applied Sciences, Vol.4, No.10, pp. 74-77, Dec 2009.

[16] Fazilati, M. A. and Alemrajaki, A. (2013), Phase change material for enhancing solar water heater, an experimental approach, ECM Energy Conversion and Management, Vol. 71, pp.138-145, 2013.

[17] Razali, T. et al. (2014). Investigation performance of solar water heater system using paraffin wax. ARPN Journal of Engineering and Applied Sciences, Vol. 9, 1749-1752.

[18] Novak, P. et al. (1996). Thermal storage of solar energy in the wall for building ventilation. Renewable Energy, Vol. 8, Issue 1-4, 268-271.

[19] Fortunato, B. et al. (2012). Simple Mathematical Model of a Thermal Storage with PCM. ELSEVIER, Procedia 2, 241-248.

[20] Ukrainczyk, N., et al. (2010). Thermophysical Comparison of Five Commercial Paraffin Waxes as Latent Heat Storage Materials. ChemBioChem, Vol. 24, 129-138.

27

Experimental Investigation of SAHs Solar Dryers with Zigzag Aluminum Cans

Mustafa Adil, Osama Ibrahim, Zainalabdeen Hussein, Kaleid Waleed

Renewable Energy Research Center, University of Anbar, Ramadi, Iraq

Email address:

mnmaab@leeds.ac.uk (M. Adil), osama_eng21@yahoo.com (O. Ibrahim), zeen_huss2002@yahoo.com (Z. Hussein),
kaleidwaleed@yahoo.com (K. Waleed)

Abstract: This experimental study investigates the thermal performance of two different solar-air collector designs for Ramadi climate conditions. Two types of absorber plate are fabricated and tested. Type (I) uses an absorber plate without cans, whereas Type (II) uses one with cans, these cans are arranged in a zigzag pattern. These collectors are a single-duct double-pass type. Air first enters through the inlet and then passes over the absorber plate before returning underneath the absorber and moving toward the outlet duct. Moreover, the plate is covered with 4 mm thick glass. An axial fan is used for air circulation. As a result, the increase in temperature difference is approximately 3 °C to 10.5 °C when using aluminum cans with a zigzag array. The increase in thermal efficiency between Types I and II is approximately 20%. Additionally, at an average mass flow rate of 0.075 kg/s, the difference between the practical and theoretical thermal efficiencies for the two models is approximately 3%.

Keywords: SAHs, Solar Dryer, Aluminum Cans, Zigzag, Thermal Performance

1. Introduction

Solar radiation is one of the more diverse, abundant and obtainable renewable energy resources. Such energy can be captured and used both directly and indirectly. Moreover, solar power can significantly contribute to reducing carbon emissions from fossil fuels. Solar solutions offer additional opportunities to meet the requirements of planning policies and building regulations [1].

Solar radiation can be directly converted into heat, numerous types of equipment are available to achieve such conversion. A flat plate collector is a device used to achieve this aim. This collector consists of three main parts that govern the design requirements. A glass cover is fixed above the absorber plate, and the system is insulated thermally from the back and sides. Solar air heaters (SAHs) are simple in design and easy to maintain. Corrosion and leakage problems are less severe than those in liquid heater solar systems. Flat plate collectors have been in service for a long time without any significant changes in their design and operational principles [2].

The main drawback of a SAH is that the heat-transfer coefficient between the absorber plate and the air stream is low, which results in low thermal efficiency. If the suggested modifications are implemented, this could improve in the heat-transfer coefficient between the absorber plate and air [3].

Numerous numerical and experimental investigations have been conducted to enhance the thermal performance of flat plate solar collectors. Ion V. ION and Jorge G. MARTINS [4] showed that the performance of air-solar collectors can be enhanced in different ways. These methods include using good thermal insulation, a cover with high transmittance, low absorbance and thermal conductivity of the material as well as using a low-cost absorber with high absorption and thermal conductivity. Also constructing a flow duct with low-pressure losses and using a fan with an appropriate power-flow rate characteristic. Hikmet Esen [5] experimentally analyzed a novel flat plate SAH with several obstacles and another without obstacles. Four types of double-flow solar air collectors under a wide range of operating conditions were studied to evaluate energetic and exergetic efficiencies. Esen showed that the flat plate collector with obstacles had the highest efficiency. Filiz Ozgen et al. [6] experimentally investigated the effect of inserting an absorbing plate made of aluminum cans into the double-pass channel in a flat-plate SAH on thermal performance. Three different absorber plates were designed and tested for the experimental study. They found that the optimal collector efficiency was achieved for zigzag cans at a 0.05 kg/s mass flow rate. K. Sopian et al. [7] concluded that the addition of porous media to the second channel of the double-pass solar air collector enhanced collector performance. As a result, the typical thermal efficiency of the double-pass solar collector with porous media reached approximately 60 - 70%. They also found close agreement between the theoretical simulation and experimental data. Ali Zomorodian and Maryam Zamanian [8] experimentally investigated a flat plate solar air collector under direct solar radiation to enhance the thermal efficiency with a slatted glass cover and two different absorber plates with two thicknesses under different air mass flow rates. They found that a maximum thermal efficiency of 0.88 was achieved for the thickest, more porous absorption plate at the highest air mass flow rate.

In this study, a solar air heater with aluminum cans in a zigzag arrangement was constructed to determine the thermal performance of modified SAHs experimentally. The performance was then compared with that of conventional flat plate solar heaters.

The applications of SAHs include the drying or treatment of agricultural goods, space heating, regeneration of dehumidifying agents, seasoning of timber and curing industrial products such as plastics.

2. Experimental Procedure

In this experimental work, the performance of conventional absorber plate is compared with the newly proposed absorber plate having aluminum cans.

2.1. Model I: SAH Collector Without Cans

Model I, is a solar air collector using an absorber plate without aluminum cans, as shown in Fig. 1(a). Model I has an outdoor flow loop which comprises a conventional absorber plate a single-duct double-pass solar air collector with entrance and exit sections, an air blower, a vane-type anemometer, a pyrometer, and a thermocouple for temperature measurement. A schematic illustration of the SAH collector of Model I is shown in and Fig. 2(a).

a) Model I   b) Model II

Figure 1. *Photographs of SAH collector Models I and II.*

2.2. Model II: SAH Collector with Cans

Model II, is a solar air collector using an absorber plate with aluminum cans in a zigzag arrangement, as shown in Fig. 1(b). Model II has an outdoor flow loop comprising aluminum cans as obstacles, a single-duct double-pass solar air collector with entrance and exit sections, an air blower, a vane-type anemometer, a pyranometer, and a thermocouple for temperature measurement. A schematic illustration of the SAH collector of Model II is shown in Fig. 2(b).

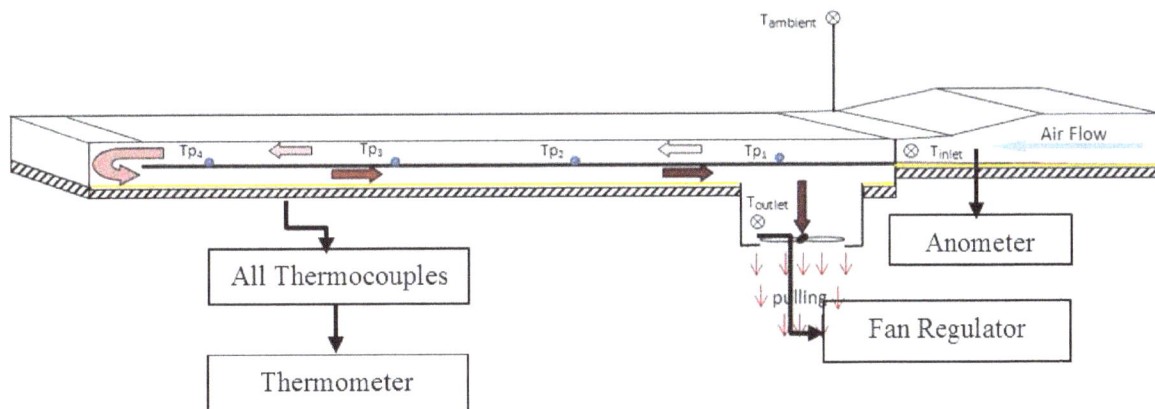

a) Model I (Flat plate collector without Al cans)

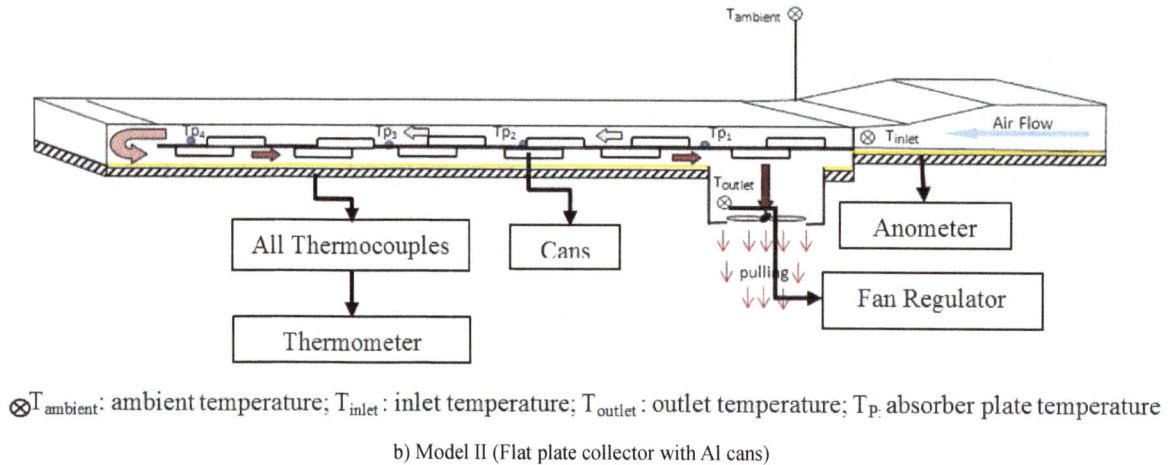

$\otimes$T$_{ambient}$: ambient temperature; T$_{inlet}$: inlet temperature; T$_{outlet}$: outlet temperature; T$_{P}$ absorber plate temperature

b) Model II (Flat plate collector with Al cans)

Figure 2. Schematic illustration of the SAH collectors: a) Model I and b) Model II.

2.3. Materials

The collector plates were 1 mm thick and made of aluminum sheet with the size 2.20 m × 0.84 m having a 32° angle of inclination. The absorber plate was then painted with solar thermal black paint (selective coating). The copper sheet was 0.1 mm thick and was installed on the bottom duct. The sheet had a specific heat capacity of 0.385 J/g·°C, thermal conductivity of 386 W/m·K, the absorbance of 0.94, and permeability of 0.03. The aluminum cans served as obstacles and were placed on the top and bottom of absorber plate. A single glass cover 4 mm thick was used as glazing for the two collectors. The frame of the collectors was constructed using wood. Therefore, the thermal losses from the back of collectors attributed to conduction, convection, and radiation are assumed negligible. All the specifications of these collectors are shown in Table 1.

Table 1. Specifications of SAH collector.

Collector tilt angle (degree)	32 [9]
Collector length (cm)	240
Collector width (cm)	88
Overall height (cm)	15
Upper duct height (cm)	8
Lower duct height (cm)	7
Inlet Area (m^2)	0.0869
Outlet Area (m^2)	0.049
Exposed Area (m^2)	1.8 * 0.81
Plate type	Flat plate
Cover material	Commercial clear glass T=0.86 [10]
Number of covers	1

2.4. The Measurement

An auto-controller was used to vary the speed of the air blower. An anemometer was used to measure air velocity. Calibrated copper–constantan thermocouples were used to measure the temperature of the inlet and outlet air. The heated absorber plates were placed in different positions, and temperatures were measured by using a digital thermometer.

All components were checked, and the instruments were calibrated. The axial fan was then switched "on" and joints were checked for leakages. All experiments were conducted under the climatic conditions of Al-Ramadi City (longitude: 33.25° N; latitude: 43.18° E; facing south) for two selected clear days on 11 March and 7 April 2014. The average mass flow rate was set at 0.075 kg/s. The following parameters were measured: solar intensity, absorber plate temperature, inlet air temperature, outlet air temperature and air velocity.

3. Thermal Performance Calculation

To calculate the heat gain (qu) and thermal efficiency (η_{th}) of the solar air collector, the following must be calculated: the average of inlet and outlet temperature Ti and To, respectively; specific heat capacity of air Cp; density ρ; air velocity v$_{inlet}$, and the inclination of the collector [9]. The thermal gain produced by the SAH can be calculated as in Eq. (1):

$$q_u = \dot{m} \times cp\,(T_o - T_i) \tag{1}$$

Where, ($\dot{m}$) is a mass air flow rate (kg sec^{-1}), which can be calculated as in Eq. (2):

$$\dot{m} = \rho \times v_{inlet} \times A_{inlet} \tag{2}$$

Where A$_{inlet}$, is the area of the inlet. The thermal efficiency of the solar air collectors (η_{th}) is defined as the ratio between the energy gain and the solar radiation incident on the collector plane and can be evaluated as in Eq. (3):

$$\eta_{th} = \frac{q_u}{I_{tilt} \times A_{exposed}} \tag{3}$$

where A$_{exposed}$ is the area covered by the absorber plate (m^2), and I$_{tilt}$ is the intensity of radiation (W·m^{-2}).

4. Estimation of Solar Radiation

In this work, the flowchart shown in figure 3 is employed to estimate global solar radiation data by using commonly available meteorological equations through a mutable function, as shown below. Many empirical equations and coefficients have been used to estimate global solar radiation.

These equations include the location of tests, number of days, the inclination of the collector, as well as longitudinal, latitudinal and atmospheric attenuation coefficients [11, 12].

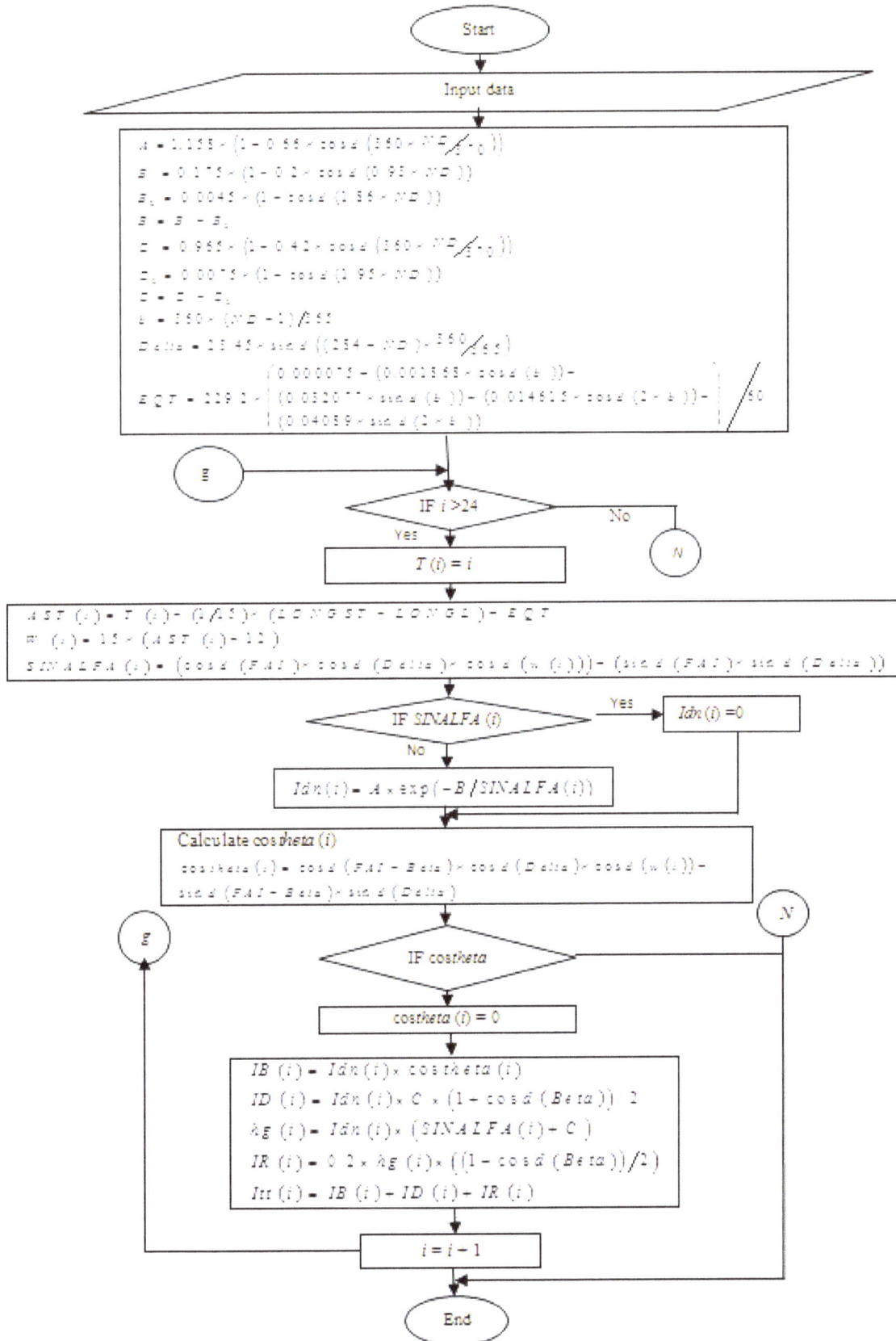

Figure 3. Flowchart used to estimate global solar radiation.

5. Results and Discussion

The practical results acquired in Al-Ramadi City are presented and discussed. All experiments for both collectors were conducted at an average air flow rate of 0.075kg/s.

5.1. Model I: SAH Collector Without Cans

The values of intensity solar radiation for practical and theoretical conditions which is taken in Al-Ramadi City for the two selected days (11 March and 7 April 2014) are shown in Figs. 4 and 5, respectively. These measurements is also carried out for slope and horizontal plan. The values of incident solar radiation increased gradually to the peak value at midday and then decrease steadily from sunrise to sunset. The practical and theoretical data of the figures 4 & 5 have been used to evaluate the hourly and daily variation of a number of parameters such as thermal efficiency. In the other words, these parameters were evaluated based on the experimental data and the estimated values of intensity solar radiation.

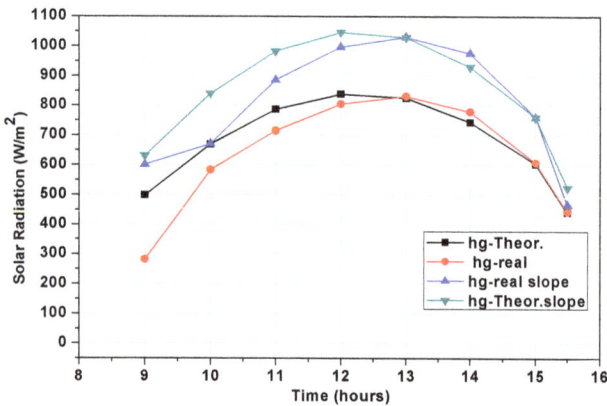

Figure 4. *Practical and theoretical values of solar radiation on a clear day on 11 March 2014.*

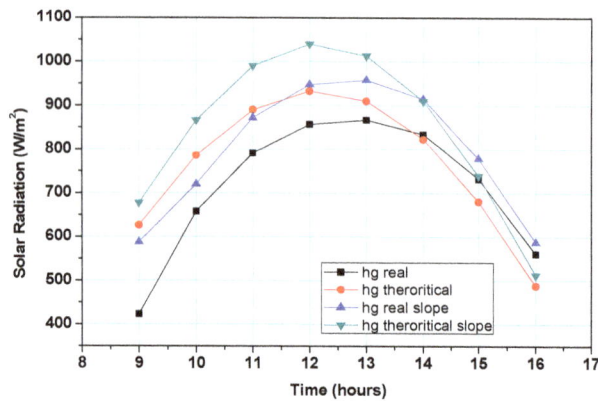

Figure 5. *Practical and theoretical values of solar radiation on a clear day on 7 April 2014.*

Figure 6 shows the temperature distribution for Model I (without cans) on a clear day on 11 March 2014. For every half hour during daytime; the normal behavior of inlet, outlet, and surface temperatures might be observed for two essential reasons; firstly, the air flow rate fluctuation which could be caused by a number of reasons for instance, the effect of variation of wind speed that could change the air inlet velocity. This variation also affected on wind heat transfer coefficient [10] as shown below:

$$hw = 5.7 + 3.8Vw \qquad (4)$$

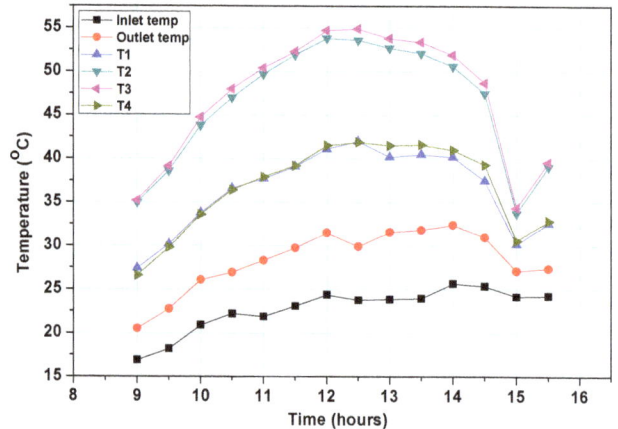

Figure 6. *Temperature variation of the inlet, outlet, and surface for Model I on a clear day on 11 March 2014.*

The increasing of this coefficient will increase the overall heat transfer losses. In addition, the absence of sun rays for a short time.

The average surface, inlet, and outlet air temperature variations on a clear day on 11 March 2014 are shown in Fig. 7. The maximum difference between inlet and outlet air temperatures was approximately 7°C at mid-daytime. Besides, the minimum air temperature difference was about 3.2°C at 15:00. This attributed for the causes aforementioned above.

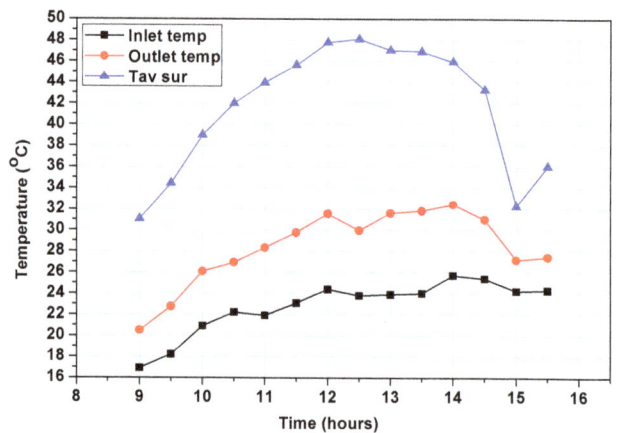

Figure 7. *Average surface, inlet, and outlet air temperature variation on a clear day on 11 March 2014.*

Fig. 8 shows the hourly practical and theoretical thermal efficiencies for Model I (without cans) on 11 March 2014. The averages of practical and theoretical thermal efficiencies were 49.47% and 46.5%, respectively.

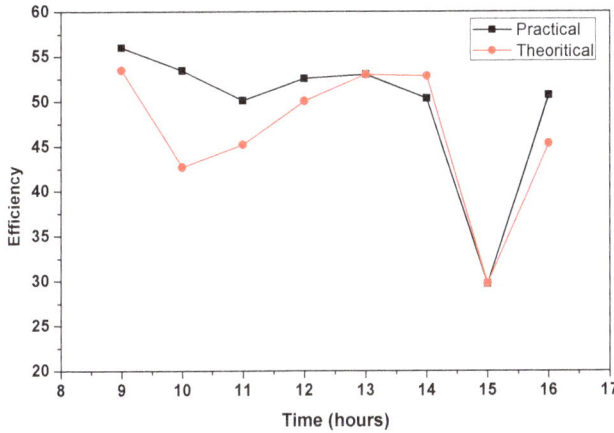

Figure 8. Practical and theoretical thermal efficiencies for Model I on a clear day on 11 March 2014.

5.2. Model II: SAH Collector with Cans

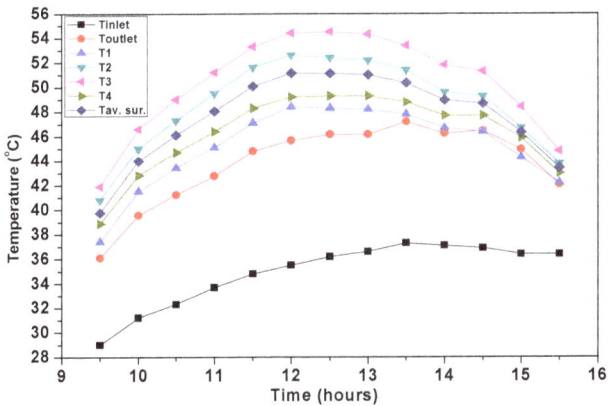

Figure 9. Temperature variation of the inlet, outlet, and surface for Model II on a clear day on 7 April 2014.

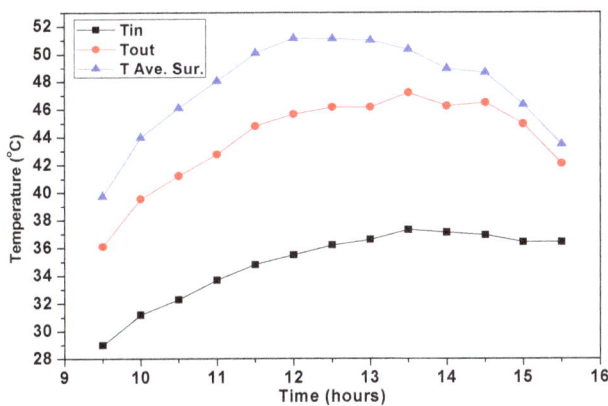

Figure 10. Average surface, inlet, and outlet air temperature variation for Model II on a clear day on 7 April 2014.

Figure 9 shows the temperature distribution for Model II (with cans) on a clear day on 7 April 2014. For every half hour during the daytime, the fairy normal behavior of inlet, outlet, and surface temperatures was observed during the daytime. The average surface, inlet, and outlet air temperature variations on a clear day on 7 April 2014 are shown in figure 10. The maximum difference between inlet and outlet air temperatures was approximately 10 °C at 13:30.

This value is attributed to presence of aluminum cans that act as extended surfaces. Also, the air temperature difference values decreased to less than 2°C at 15:30 as minimum value owing to the value of intensity solar radiation is quite small which in turn reducing the heat absorbed from aluminum plate.

Fig. 11 shows the hourly practical and theoretical thermal efficiencies for Model II (with cans) on a clear day on 7 April 2014. The averages of practical and theoretical thermal efficiencies were 68.5% and 65.2%, respectively.

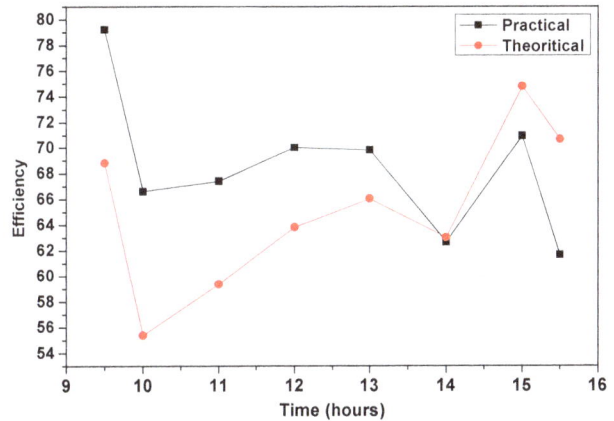

Figure 11. Practical and theoretical thermal efficiencies for Model II on a clear day on 7 April 2014.

5.3. Comparison Between Model I and Model II

Figs. 12 and 13 show the increase in the temperature difference between Models I and II for two selected clear days on 11 March and 7 April 2014, respectively. The increase in temperature differences is approximately 3.2 °C to 10.5 °C. This increase as result of the using aluminum cans with a zigzag array. The aluminum cans serve as fins that increase the capability of the absorber plate to absorb energy, consequently increasing the heat transfer coefficients, as well as contributing to the breakage of the boundary layer and reducing its growth.

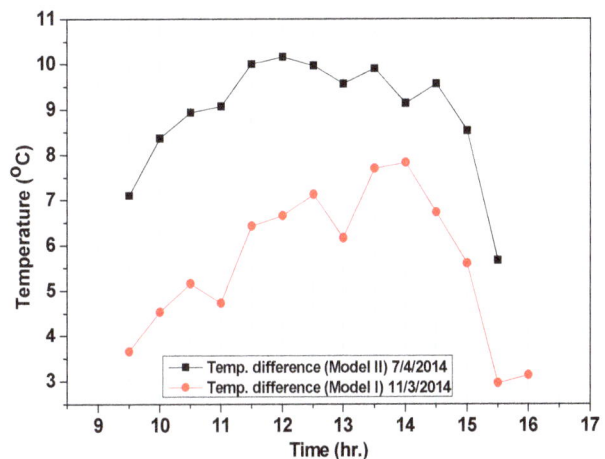

Figure 12. Inlet and outlet temperature difference for Models I and II for the two selected clear days on 11 March and 7 April 2014.

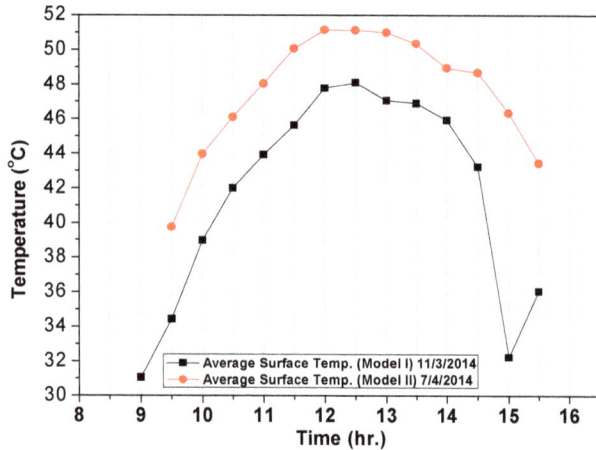

Figure 13. *The average surface temperature for Models I and II for two selected clear days on 11 March and 7 April 2014.*

Fig. 14 and Table 2 show the experimental and theoretical thermal efficiencies for Models I and Model II for the two selected days. We observe that the difference in the practical and theoretical thermal efficiencies between Models I and II is approximately 20%. In addition, the difference between practical and theoretical thermal efficiencies for the two models is approximately 3% at an average mass flow rate of

0.075 kg/s. The enhanced efficiency of the solar air collector in Model II is attributed to the following reasons: using the cans as extended surfaces (fins) and placing the can array in a zigzag pattern increases the heat transfer coefficient by increasing air turbulent flow and reducing the growth of the boundary layer.

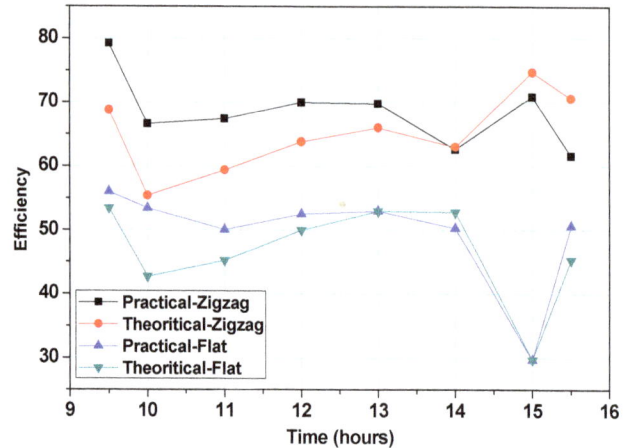

Figure 14. *Variation of hourly practical and theoretical thermal efficiencies for Models I and II.*

Table 2. *Practical and theoretical thermal efficiencies of Models I and II for two selected sunny days on 11 March and 7 April 2014.*

Time	$\eta_{Theoretical}$	$\eta_{Practical}$	e	$\eta_{Theoretical}$	$\eta_{Practical}$	e
	Model I			**Model II**		
9:00	53.458	56	1.797465438	68.803	79.217	7.363810019
10:00	42.687	53.43	7.59644815	55.380	66.610	7.940809153
11:00	45.209	50.1	3.458459267	59.375	67.409	5.68089588
12:00	50.042	52.55	1.773423807	63.814	70.006	4.378405189
13:00	52.974	53	0.018384776	66.027	69.818	2.680641807
14:00	52.825	50.33	1.764231419	63.008	62.664	0.243244733
15:00	29.740	29.7	0.028284271	74.763	70.924	2.714582933
16:00	45.321	50.7	3.803527376	70.666	61.650	6.375274739
Ave.	46.5	49.476	2.081899141	65.229	68.537	2.338932455

The degree of agreement between the theoretical and experimental results have been calculated by using a statistical analysis [10], as shown below:

$$e = \sqrt{\frac{\sum_{i=1}^{N}(e_i)^2}{N}} \qquad (5)$$

$$e_i = \frac{X_i - Y_i}{X_i} * 100 \qquad (6)$$

Where Xi and Yi are the theoretical and the experimental results of SAH collector, respectively, (e) the root mean square of percentage deviation.

This difference between experimental and theoretical results could be caused by few reasons. One of them, the use practical and theoretical intensity solar radiation values. As it is mentioned before, the percentage error between these solar radiation values around 3 %. This percentage affected directly on the calculation of the thermal efficiencies. Another reason, the values of practical intensity solar radiation vary based on the weather conditions such as dust, humidity etc., some

values therefore behave convergent at 14:00 and behave divergent at 9:00.

6. Conclusions

An experimental investigation was conducted by the Renewable Energy Research Center in Al-Ramadi City to evaluate the thermal performance of two different models of single-pass double-duct SAHs (solar dryers) with and without cans (obstacles) arranged in a zigzag pattern as fins. The thermal performance of the single-pass double-duct type SAH, in which air flows over and returns under the absorber plate, is efficient because the flowing air collects and absorbs most of the supplied energy. The results show an increase in thermal efficiency of Model II by approximately 20%. Model II was more efficient and had a maximum temperature difference that was approximately 10.5 °C higher than that of Model I. Moreover, the theoretical values of solar radiation used were satisfactory, and the error in the results was approximately 3%.

Acknowledgements

This work is supported by the University of Anbar-Iraq/Renewable Energy Research Center with Grant No. RERC-PP34.

References

[1] Rawlings R and Butcher K. Capturing solar energy. Chartered Institution of Building Services Engineers 2009.

[2] Raj Thundil Karuppa R, Pavan P, and Reddy Rajeev D., "Experimental Investigation of a New Solar Flat Plate Collector", Research Journal of Engineering Sciences, 1(4), pp. 1–8, 2012.

[3] Mohamad AA., "High efficiency solar air heater", Solar Energy, 60(2), pp. 6–71, 1997.

[4] Ion IV and Martins JG., "Design, developing and testing of a solar air collector", Research Journal of Engineering Sciences, 1(4), pp. 1–8, 2012.

[5] Esen H., "Experimental energy and exergy analysis of a double-flow solar air heater having different obstacles on absorber plates", Building and Environment, 43, pp. 1046–1054., 2008.

[6] Ozgen F, Esen M, and Esen H., "Experimental Investigation of Thermal Performance of a Double-flow Solar Air Heater Having Aluminium Cans", Renewable Energy, 34, pp. 2391–2398, 2009.

[7] Sopian K, Alghoul MA, Alfegi EM, Sulaiman MY, and Musa EA., "Evaluation of thermal efficiency of double-pass solar collector with porous–nonporous media", Renewable Energy, 34, pp. 640–645, 2009.

[8] Zomorodian A and Zamanian MA., "Designing and Evaluating an Innovative Solar Air Collector with Transpired Absorber and Cover", International Scholarly Research Network ISRN Renewable Energy; 2012 (Article ID 282538), 5 pages, 2012.

[9] ASHRAE Applications Handbook, American Society of Heating, Solar Energy Use (ASHRAE). Atlanta, GA, 1999, Chapter 32.

[10] Amori KE and Abd-AlRaheem MA., "Field study of various air based photovoltaic/thermal hybrid solar collectors", Renewable Energy, 63, pp. 402–414, 2014.

[11] Duffie JA and Beckman WA., Solar Engineering of Thermal Processes, John Wiley & Sons, Inc 2006.

[12] Kalogirou S., Solar Energy Engineering: Processes and Systems, Cyprus University of Technology, Elsevier's Science and Technology, USA 2009.

28

Extended P21-Based Benchmarking

Zhiguang Cheng[1], Behzad Forghani[2], Tao Liu[1], Yana Fan[1], Lanrong Liu[1]

[1]Institute of Power Transmission and Transformation Technology, Baobian Electric Co., Ltd, Baoding, China
[2]Infolytica Corporation, Place du Parc, Montreal, Canada

Email address:

emlabzcheng@yahoo.com (Zhiguang Cheng), forghani@infolytica.com (B. Forghani)

Abstract: This paper highlights two important aspects of the electromagnetic field modeling and simulation when used for industrial applications, namely the application based benchmarking activities and the magnetic material modeling. It emphasizes the relationship between the two, and briefly reviews the recent progress in extending the TEAM (Testing Electromagnetic Analysis Methods) Problem 21 Family (P21) and the related modeling results, and proposes a new benchmarking project which includes the upgraded benchmark models that can handle extreme excitations, i.e. current sources with a DC bias, as well as multiple harmonics.

Keywords: Extended Benchmarking, Extreme Excitation, Finite Element, Industrial Application, Magnetic Flux, Magnetic Loss, Problem 21 Family (P21), Working Magnetic Property Modeling

1. Introduction

The effectiveness of the numerical modeling and simulation is dependent on the electromagnetic analysis method, computational software being used, and access to sufficient material property data. Consequently, the development and validation of both the numerical computation method and the material property modeling, under working conditions, are of great interest.

In order to validate the numerical modeling methods, since 1985, the international COMPUMAG (the biennial conference on the computation of the electromagnetic fields) society (ICS) has paid great attention to organizing the TEAM activities worldwide, in order to test and compare the electromagnetic analysis methods, and has established a series of benchmark problems that are now widely used in the computational electromagnetics community [1]. Meanwhile, the IEEE Standard for validation of computational electromagnetics computer modeling and simulations has also been issued [2]. The authors have devoted a lot of their work to the engineering-oriented TEAM activities for many years [3-15], have proposed an engineering-oriented benchmark family of problems, Problem 21[1,3,4], and have updated it three times since 1993, which has been of interest to many scientists and engineers up to now.

On the other hand, more advanced material modeling

techniques have been investigated systematically, involving extreme magnetization conditions [16]. As a result, there has been significant progress in the efficient design of electromagnetic devices [17-20]. However, so far, what is widely used in industrial applications is the standard one-dimensional B-H curves, obtained from either the Epstein frame or the single sheet tester (SST). Of course, magnetic properties change according to the working conditions, e.g. they can vary with the frequency, the temperature, and the stress action. Therefore, it is necessary to validate the magnetic property data under the same working conditions that the device is subjected to.

In very large electromagnetic devices, for example a EHV (extra high voltage) power transformer, the reduction of the stray-field loss, produced by the leakage flux from the transformer winding and heavy current leads, and the protection against unallowable loss concentrations, and then the resulting local overheating have become more and more significant[21-27]. Various types of power frequency shields are widely utilized to effectively save energy and ensure a reliable operation. In addition, the various shields can change and control the global distribution of the 3-D electromagnetic field inside a large electromagnetic device. It is important to accurately model and estimate the multi-shielding effects and optimize the shielding configurations at the electromagnetic design stage [14, 22, 23].

The purpose of this paper is to focus on the engineering-oriented benchmarking and the application-based magnetic material modeling in electromagnetic devices, to examine the effect of the variation in the different B-H representations used in different solvers on the iron loss and flux in GO (grain-oriented) silicon steel sheets. In addition, the modeling, simulation and validation, under extreme excitations, as a new benchmarking project, is also proposed in this paper [28].

2. Problem 21 Family and Selected Benchmarking Results

The electromagnetic and thermal field problems in large electromagnetic devices are usually very complicated, involving multi-physic field coupling, multi-scale (very thin sheet/penetration depth and very large bulk) configurations, and multi-materials subjected to varying working conditions. In order to obtain an effective solution, the strict validation of the analysis method and software, to be used for solving such complex field problems, is certainly needed. However, it is impossible to do that via a large real electromagnetic device. Therefore, the verification based on the engineering-oriented benchmark models becomes the best and most practical way.

2.1. TEAM Problem 21 Family

To investigate the stray-field loss problems in electromagnetic devices, a benchmark family, TEAM Problem 21, consisting of 16 benchmark models, has been well established (the definition of Problem 21, v.2009, can be found at www.compumag.org/TEAM)[1]. See Table 1.

Table 1. TEAM Problem 21 Family (V.2009).

Member	Models	Electromagnetic features	Industry background	Proposed at
P21^0	P21^0-A P21^0-B	3-D nonlinear eddy current and hysteresis model with multiply connected regions.	Solid magnetic components models, e.g., power transformer tank or other solid parts.	TEAM-Miami, USA, 1993.
P21^a	P21^a-0 P21^a-1 P21^a-2 P21^a-3	3-D linear eddy current model with multiply connected regions.	Slotted solid plate models, e.g., core tie-plates in power transformers.	TEAM-Yichang, China, 1996.
P21^b	P21^b-MN P21^b-2M P21^b-2N	3-D nonlinear eddy current and hysteresis model with magnetic or/and non-magnetic steel plates separately placed.	Hybrid steel structure models with magnetic and non-magnetic material aiming to reduce stray-field loss, also can be seen in power transformer tank.	IEE CEM, Bournemouth, UK, 2002.
	P21^b-MNM P21^b-NMN	3-D nonlinear eddy current and hysteresis model with magnetic and non-magnetic steel plates welded together.		ACES, Miami, USA, 2006.
P21^c	P21^c-M1 P21^c-M2 P21^c-EM1 P21^c-EM2	Magnetic shielding and electromagnetic shielding models: 3-D nonlinear eddy current and hysteresis model with anisotropic lamination.	Magnetic shunt and electromagnetic barriers models under low frequency, Widely used in power transformers.	Compumag-Shenyang, China, 2005.
P21^d	P21^d-M	3-D nonlinear eddy current and hysteresis model with anisotropic lamination without solid magnetic steel.	Laminated frame models for detailing the electromagnetic behaviour.	IEEE CEFC-Athens, Greece, 2008.

All the member models of Problem 21 Family come from typical structures found in large power transformers, each presenting a different electromagnetic behavior. The engineering-oriented benchmarking activities have the following goals:

(a) Test electromagnetic analysis methods

According to the original motivation of benchmarking (TEAM), Problem 21 works for testing and comparing the electromagnetic analysis methods and the developed computation software, as well as the commercial software being used.

(b) Verify computation models

It is important to build a correct numerical computation model that can take into account the nonlinearity, the electric and/or magnetic anisotropy of the material, the skin effect and loss concentration in components, allowing for a reasonable simplification for reducing the computational cost in large scale electromagnetic analysis and design.

(c) Detail the field behavior of typical product structure

Problem 21 is engineering-oriented, including power transformer tank, core-plate, and shielding models. The detailed modeling of the stray field loss generated in different components and the electromagnetic field distributions are helpful in improving the product design.

Table 2. Brief summary of P21-based modeling.

P21^0	In the iron-loss calculation, the hysteresis loss component and the nonlinearity of the magnetic steel must be accounted (P21^0 includes Model A and Model B, i.e., P21^0-A and P21^0-B). The practical calculation method of hysteresis loss and the measured hysteresis loss curves are available [1, 7].
P21^a	The satisfied results of eddy current losses in non-magnetic steel can be achieved using different 3-D analysis methods based on different potential sets, even with coarse mesh in non-magnetic components, but 2-D results is not available[4,9,33].
P21^b	The detailed examination and comparison of both the total loss and the loss concentration in the hybrid steel plate structure are of importance to improve the product design [10].
P21^c	The evaluations of the power loss and magnetic flux inside both electromagnetic and magnetic shields, as well as that of the separation-type shields, are given [12, 13].
P21^d	Both the iron loss and magnetic field inside the GO silicon steel lamination with different excitation patterns, and the additional iron loss induced by normal magnetic flux are detailed [4-6].

(d) Benefit to large-scale numerical modeling

A benchmark model is different from a large

electromagnetic device, yet the benchmarking results are useful for solving the large-scale field problems, by simplifying the real problem, the treatment of material properties, and the choice of the solvers.

Table 2 shows a P21-based benchmarking note, which is expected to be helpful for the numerical modeling and computation in electromagnetic devices.

2.2. Selected Benchmarking Results

A number of P21-based benchmarking results have been presented by the authors and other researchers worldwide [4]. A summarized loss calculation results of Problem 21 Family is shown in Fig. 1 [9].

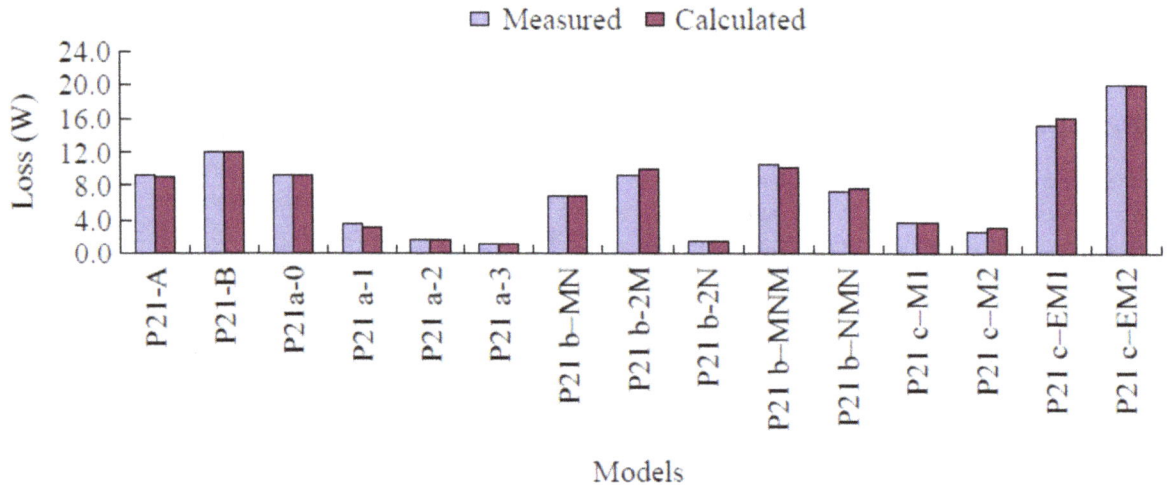

Figure 1. Loss spectrum of Problem 21 Family (with rated excitation condition of 10A, 50Hz).

In version 2009 of Problem 21 Family, the rated exciting current of 10 A (rms, 50 Hz) was upgraded to 50A. As a typical benchmarking result, the measured and calculated results of both the iron loss and magnetic flux inside the magnetic plate for $P21^0$-B are shown in Table 3 and Table 4 [1, 4].

Table 3 and Table 4 show that the 3-D electromagnetic solver based on potential set A_r-V-A_r [29, 33], developed by the authors, provides good results for both the iron loss and flux under different exciting currents, agreeing well with the measured results.

Table 5 shows that the 3-D eddy current analysis for $P21^a$, using different methods based on A_r-V-A_r and T-φ-φ potential sets [29-33], also provides good results, but the 2-D calculated results are not available.

Table 3. Iron loss inside magnetic steel ($P21^0$-B).

Exciting currents (A,rms,50Hz)	Measured(W)	Calculated (by A_r-V-A_r) $P_{calc.}$(W)	$(P_{calc.} - P_{meas.})/ P_{meas.}$ (%)
5	3.30	3.30	0.0
10	11.97	12.04	0.6
15	26.89	27.12	0.9
20	49.59	50.92	2.7
25	82.39	84.78	2.9
30	123.70	128.67	4.0
35	179.10	183.15	2.3
40	248.00	250.45	1.0
45	330.00	330.91	0.3
50	423.00	425.07	0.5

Table 5. Eddy currents losses in non-magnetic steel ($P21^a$).

Models	Meas. (W)	3-D (W)			2-D (W)
		T-φ-φ	A_r-V-A_r	A-V	
$P21^a$-0	9.17	9.50	9.31	9.22	14.75
$P21^a$-1	3.40	3.37	3.34	3.35	6.23
$P21^a$-2	1.68	1.67	1.66	1.68	3.07
$P21^a$-3	1.25	1.15	1.14	1.15	1.86

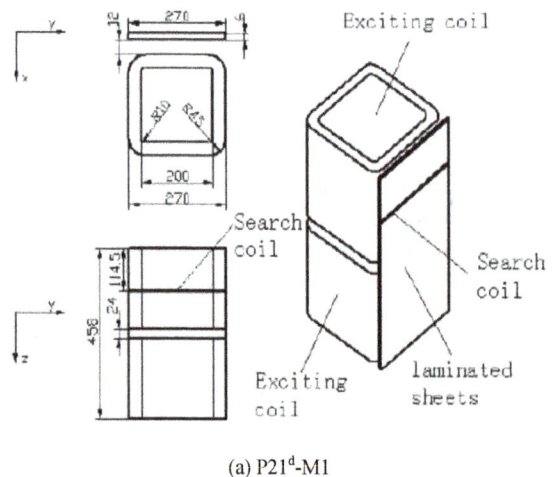

Table 4. Flux inside magnetic steel ($P21^0$-B).

Exciting currents (A,rms,50Hz)	Measured $\Phi_{meas.}$(mWb)	Calculated (by A_r-V-A_r) $\Phi_{calc.}$ (mWb)	$(\Phi_{calc.} - \Phi_{meas.})/ \Phi_{meas.}$ (%)
5	0.158	0.151	-4.12
10	0.318	0.306	-3.86
15	0.478	0.458	-4.13
20	0.618	0.605	-1.98
25	0.770	0.750	-2.66
30	0.936	0.890	-4.90
35	1.064	1.024	-3.76
40	1.206	1.152	-4.48
45	1.357	1.276	-5.97
50	1.486	1.396	-6.06

(a) $P21^d$-M1

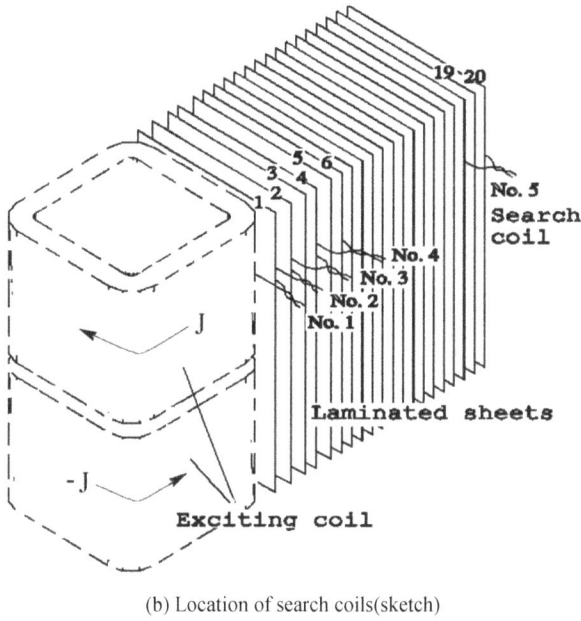

(b) Location of search coils(sketch)

Figure 2. Measurement of flux densities inside sheet.

In order to investigate the iron loss and magnetic flux densities in the laminated sheets, a very simplified model, P21^d-M of Problem 21 Family, as shown in Fig. 2(a), has been proposed, which enables us to accurately calculate and measure the loss and the flux within one sheet [11-13], as shown in Fig. 2(b).

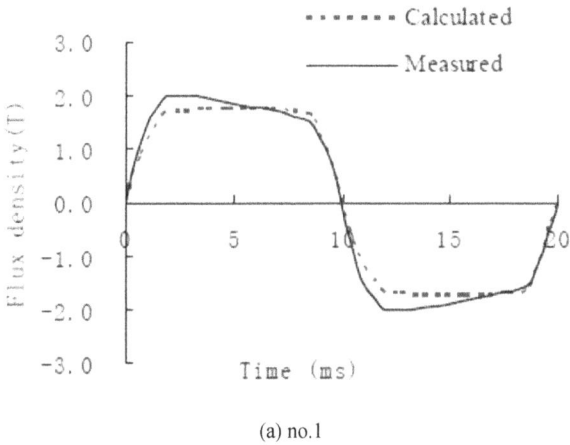

(a) no.1

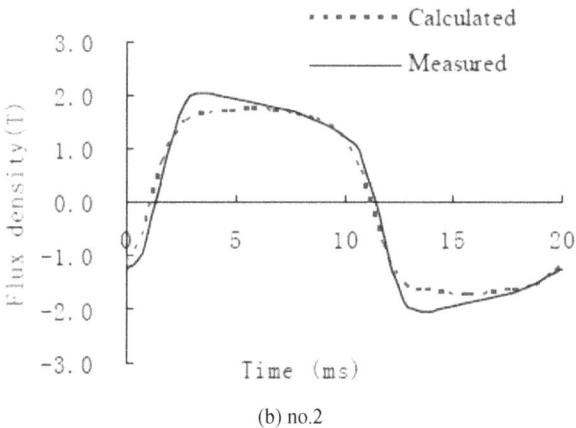

(b) no.2

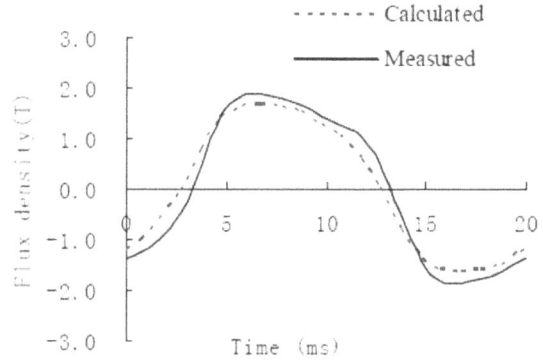

(c) no.3

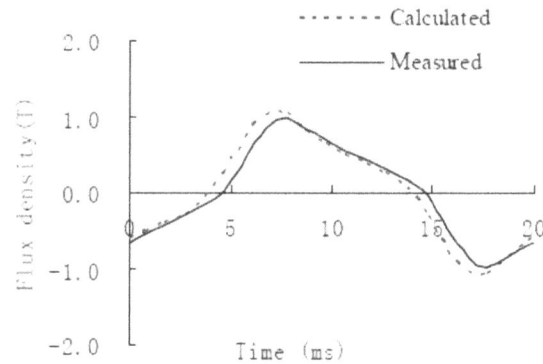

(d) no.4

Figure 3. Waveforms of flux densities inside lamination -layes(exciting currents: 25A, rm, 50Hz).

The measured and calculated waveforms of the flux densities inside different sheet-layers (from no.1 to no.4) at the exciting current of 25A(rms, 50 Hz) are shown in Fig. 3. The waveforms of flux densities inside the laminated sheets are distorted at different levels and the measured and calculated results agree well.

Note that in the current version of Problem 21 (V.2009, posted on the ICS website, www.compumag.org), the exciting currents have a sinusoidal waveform. The latest extension to the Problem 21 Family can handle the extreme excitation condition, i.e., the electromagnetic components of the member models of Problem 21 are excited by a DC-biased AC supply which may contain a number of harmonics [28]. Table 6 shows two newly proposed models.

Table 6. New upgraded models with extreme excitations.

Proposed models	Electromagnetic features	Industry background	Remarks
P21^0-B$^+$	3-D nonlinear transient field in solid magnetic plate under multi-harmonic and/or DC-biasing excitations.	Magnetic loss in solid magnetic components under extreme conditions.	Upgraded P21^0-B with magnetic flux compensation.
P21^c-M1$^+$	3-D nonlinear transient field in laminated sheets under multi-harmonic and/or DC-biasing excitations.	Magnetic loss in laminated magnetic components under extreme conditions.	Ungraded P21^c-M1 with different types of laminated sheets.

3. P21-Based Benchmarking Notes

3.1. Finite Element Model for Lamination Configuration

In order to reduce the electromagnetic computation costs and obtain the solutions very efficiently, a number of homogenization methods for the lamination structures, such as the transformer core and the magnetic shields, have been proposed [34-38]. The following benchmarking model aims to deal with the standard iron loss and additional iron loss based on a simple model [12].

The simplified finite element model of the laminated GO silicon sheets has the following characteristics:

1) Treatment of electric anisotropy

Pattern 1: Modeling the first few laminations individually and modeling the rest as bulk; the 3-D eddy currents flow in the individual laminations and the 2-D eddy currents are limited in each lamination in the bulk region where the anisotropic conductivity is used, see Fig. 4(a).

Pattern 2: Fine meshing within a thin surface layer, and coarse meshing inside the bulk. In the entire conducting domain, the anisotropic conductivity is assumed. See Fig.4 (b).

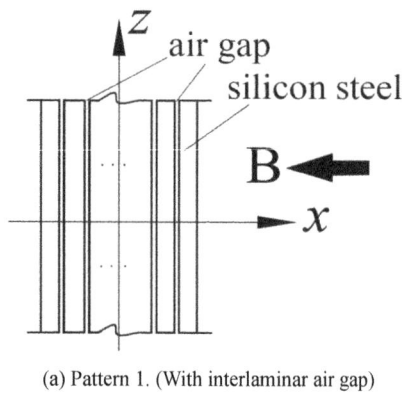

(a) Pattern 1. (With interlaminar air gap)

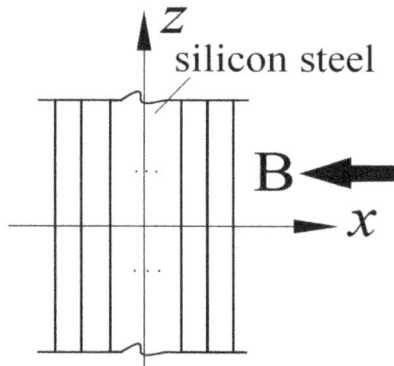

(b) Pattern 2. (With electric anisotropy)

Figure 4. *Simplification of laminated sheets.*

2) Treatment of magnetic anisotropy

The resulting magnetic field inside the laminated sheet is almost in one direction (along the z-axis), making it a weak magnetic anisotropy problem, and the orthogonal anisotropic permeability is assigned to all the laminations.

3.2. Eddy Current Analysis

The well-established eddy current solvers, based on various potential sets, have been developed and applied in the computational electromagnetics research and industrial applications [29-33]. In the method based on the T-Ω potential set, used by MagNet, the software used for the work in this paper, the magnetic field is represented as the sum of two parts, i.e., the gradient of a scalar potential Ω and T. In the conductors, an additional vector field is represented with vector-edge elements. As a result, the solution vector consists of the magnetic scalar potential at the nodes plus edge-degrees of freedom associated with the current flow in solid conductors. The T-Ω based solver does not run into the convergence and instability issues associated with other formulations [32]. The governing equation in the eddy current region, and in the presence of anisotropic and nonlinear materials, is given by (1),

$$\nabla \times ([\sigma]^{-1} \nabla \times T) + [\mu]\frac{\partial (T - \nabla \Omega)}{\partial t} = 0 \qquad (1)$$

The anisotropic and nonlinear permeability $[\mu]$ in (1) can be represented by (2)

$$[\mu] = \begin{bmatrix} \mu_0 / (1 - c_p) & & \\ & C_p \mu_y & \\ & & C_p \mu_z \end{bmatrix} \qquad (2)$$

where C_p is the packing factor.

The anisotropic conductivity $[\sigma]$ of the sheets can be dealt with as

$$[\sigma] = \begin{bmatrix} c_p \sigma_x & & \\ & c_p \sigma_y & \\ & & c_p \sigma_z \end{bmatrix} \qquad (3)$$

where $\sigma_y = \sigma_z = \sigma_{yz}$ in the sheets, while σ_x is expressed as

$$\sigma_x = \begin{cases} \approx 0 \, (\textit{2D eddy current region}) \\ assigned \, (\textit{3D eddy current region}) \end{cases} \qquad (4)$$

The iron loss and flux generated in the GO laminations are computed based on the field results, as part of the post processing operation. The additional iron loss P_a, caused by the flux entering normal to the laminated sheets, is not included in the measured total iron loss (by using standard magnetic property measurement methods), however, it cannot be neglected. As a practical solution, the total iron loss P_t referred to as standard iron loss, can be divided into two parts, i.e.,

$$P_t = P_s + P_a \qquad (5)$$

where P_s can be numerically computed based on the 3-D field solution (B_m^e), the measured standard loss curve B_m-W_t, and the elements volumes, and then summed up, as shown in (6),

$$P_s = \sum_e P_s^e(W_t, B_m^e) \cdot V^e \qquad (6)$$

where P_s^e and V^e are the total specific loss and volume of each element in the lamination-layers.

While P_a can be calculated based on the induced eddy current field solution J flowing in the lamination-layers, with anisotropic conductivity σ according to equation (7),

$$P_a = \int \frac{J \cdot J}{\sigma} dv \qquad (7)$$

3.3. On Magnetic Property Modeling

Both, keeping track of advanced material modeling technologies and promoting large-scale applications, using the existing material property data, from the point of view of industrial application, are really important.

(a) A bottleneck problem of industrial application

The material property modeling is one of the key topics of the engineering electromagnetics. In the author's opinion, it is still a bottleneck-problem of industrial applications. This is because the measurement conditions used in measuring the material properties are standard, using standard equipment, such as the Epstein frame, SST, or other equipment used for obtaining vector magnetic properties [17, 20]. However, the working conditions that the components in a device are subjected to are not standard.

Up to now, the material property data provided by the material manufacturers are one-dimensional, but the field problems in the real products are three-dimensional. So another problem to figure out is how to use the existing property data when solving the real problems?

(b) Combination of material modeling and numerical computation

The majority of the current electromagnetic-thermal analysis software can access the one-dimensional or the so-called orthogonal-anisotropic property data. The software must be upgraded if the vector property data of the material is to be used.

(c) Improvement of magnetic property modeling technology

For many years, the standard testing equipment have been used for the measurement of the magnetic material property. There is a need for an extension to the measurement function and upgrades of the measurement values, for example, using the Epstein frame to measure the different types of B-H curves (B_m-H_m and B_m-H_b), or when the mean path length of the Epstein depends on many factors and is not a constant value [39,40].

3.4. Effect of B-H Properties on Iron Loss and Flux

3.4.1. Different B-H Curves

Two kinds of B-H curves, namely B_m-H_m and B_m-H_b, are currently used in electromagnetic field computation [13]. The B_m-H_m curve takes the maximum values of both the flux density (B_m) and the magnetic field strength (H_m) within a cycle. Generally, B_m and H_m cannot achieve the maximum

value at the same time inside the magnetic steel due to the eddy current, especially at low flux density, as shown in Fig.5. Thus there is another magnetic field strength H_b corresponding to the maximum value of the flux density B_m, i.e. B_m-H_b curve, see Fig.5. The eddy current becomes zero at the instant when the flux becomes the maximum, therefore, B_m-H_b curve can be referred to as a dc B-H curve.

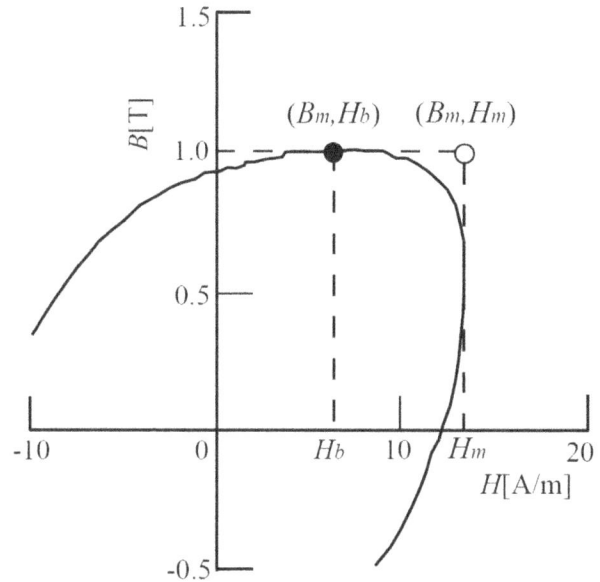

(a) B_m=1.0T

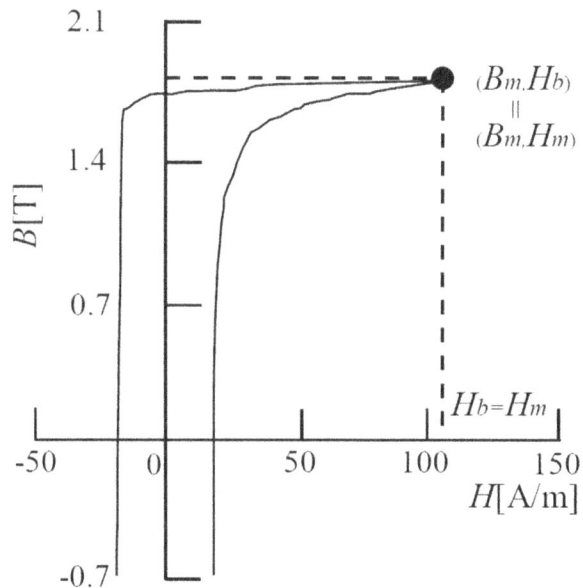

(b) B_m=1.8T

Figure 5. Definition of B_m, H_m and H_b (30P105, f=50Hz).

Fig.6 shows the examples of forming B_m-H_m and B_m-H_b curves based on hysteresis loops (30P105), and Fig.7 demonstrates the measured B-H curve family (SST with H coil) at different frequencies. Both Fig.6 and Fig.7 indicate that the B_m-H_b curves at around a commercial frequency (e.g., 50Hz) are similar to the dc B-H curve at low frequency (0.01Hz), but B_m-H_m curves are different from the quasi dc B-H curve.

3.4.2. Different Sampling of GO Silicon Steel Sheets

The B-H curves (B_m-H_m and B_m-H_b) and the specific loss curve (W_t-B_m), measured at different sampling angles to the rolling direction of the GO steel sheet (30P120) and at different frequencies using the Epstein frame, are shown in Fig. 8 [13].

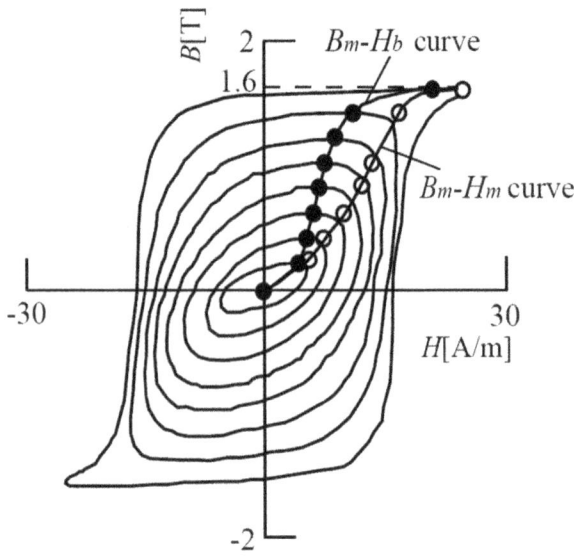

(a) f=50Hz

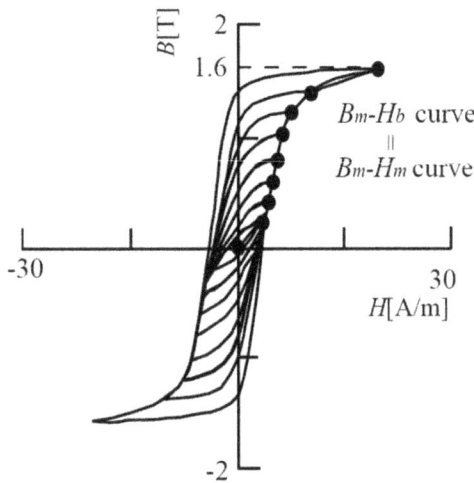

(b) f=0.01Hz

Figure 6. *Hysteresis loops at different frequencies (30P105).*

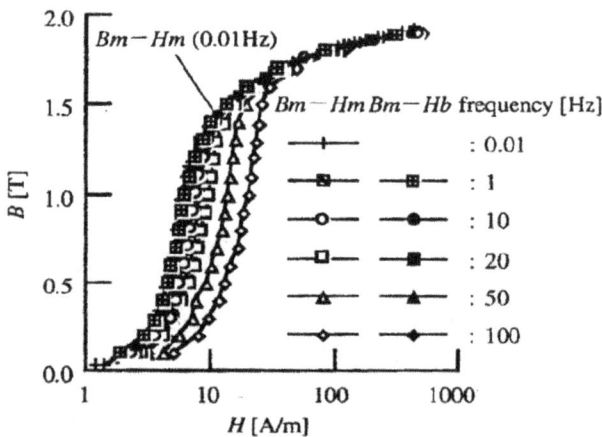

Figure 7. *Comparison of B-H curves (30P105).*

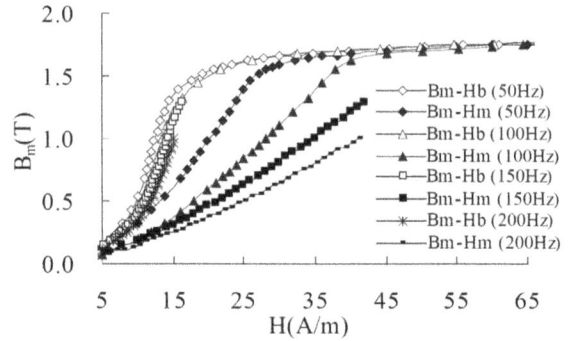

(a) B-H curves (0° to rolling direction).

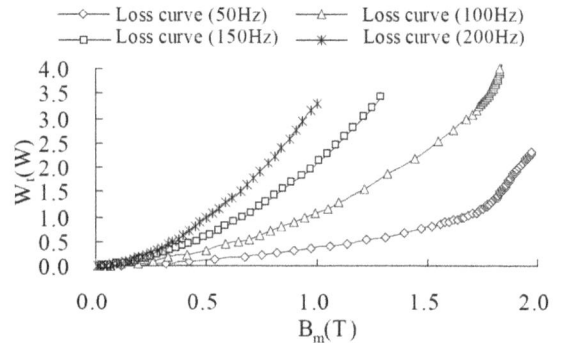

(b)B_m-W_t curves (0° to rolling direction).

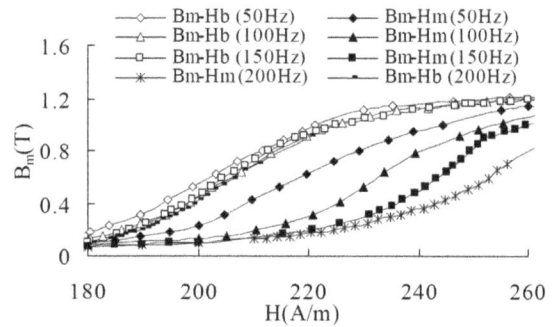

(c) B-H curves (90° to rolling direction).

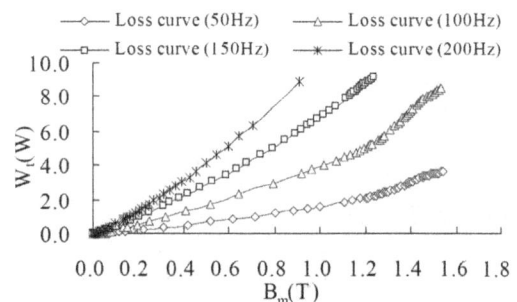

(d) W_t-B_m curves (90° to rolling direction).

Figure 8. *B-H and B_m-W curves of 30P120.*

A comparison between the B-H curves, i.e., B_m-H_m and B_m-H_b, measured by using Epstein frame and SST respectively, is given. See Fig. 9.

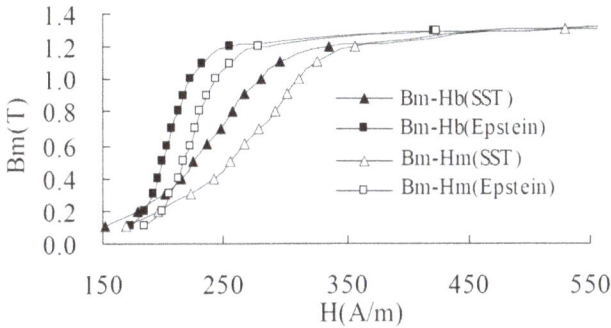

Figure 9. *B-H curves (90° to rolling direction) measured using SST and Epstein frame (30P120).*

3.4.3. Test Models and Results

Table 7. *Specification of test models.*

Models	Magnetic plate of 10 mm thick	GO silicon sheets	Sheet-coil distance (mm)	Sheet size (mm)
T1	Without	30P120	15.0	500×500
T2	Without	30P120	12.0	80×458 (3 sets)
P21ᶜ-M1	With	30RGH120	12.0	270×458
P21ᶜ-M2	With	30RGH120	12.0	80×458 (3 sets)

To examine the effects of the different B-H representations on the iron loss and flux inside the lamination, two test models have been proposed, i.e. Model T1 and Model T2, which are derived from the benchmark Model P21ᶜ-M1 and P21ᶜ-M2 of Problem 21 Family, respectively. A brief comparison among the benchmark models and the newly proposed Problem 21-based test models is shown in Table 7.

In Model T1, only six silicon steel sheets of 500×500mm (30P120) are driven by a twin AC source (50 to 200Hz), a 3-D excitation, see Fig. 10 (a). The purpose of Model T2 is to show the effect of the division of the wide sheets on the reduction of iron loss, see Fig. 10 (b).

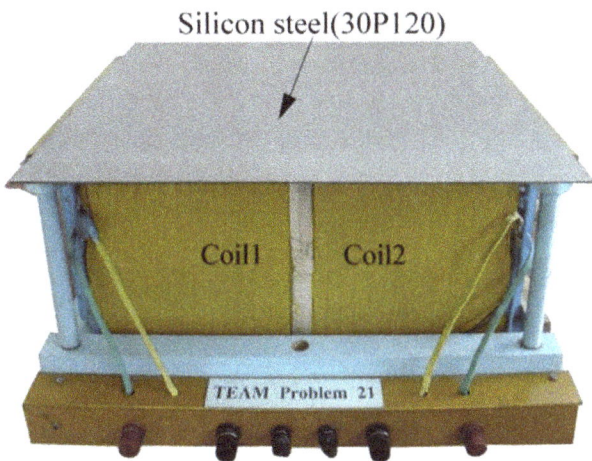

(a) Model T1.

Silicon steel(30P120)

Coil1 Coil2

TEAM Problem 21

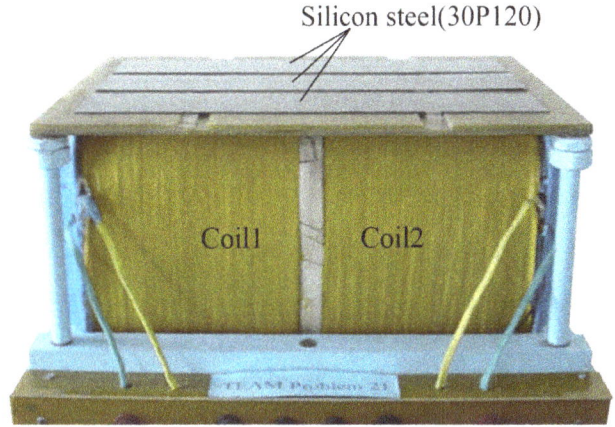

(b) Model T2.

Figure 10. *Models T1 and T2.*

3.4.4. Discussion

According to the pre-measurement results obtained by the authors, the saturation level of the laminated sheets is not so high, especially at lower excitations. Both the time harmonic (TH) and the time stepping (TS) solvers of the T-Ω-based MagNet, Infolytica, are used to solve the 3-D eddy current problem. Table 8 shows the calculated and measured results of the total iron loss P_t of Model T1.

All the calculated results, using different B-H curves and different solvers, indicate that the use of TH solver and B_m-H_m curve can offer better results when compared to measurement. This is, because the element-permeability is dependent on the quasi-maximum values of B and H within a cycle in the TH solver. On the contrary, in the case of the TS solver, B_m-H_b curve can offer better results. This is, because in the TS solver the element-permeability is determined according to the instantaneous values of B and H at an instant. As a result, a more precise analysis is possible by the time stepping method using the B_m-H_b curve, which is almost the same as the dc B-H curve.

Table 8. *Total iron loss results (Model T1).*

Current (A, rms, 50Hz)	Calc. (W)				Meas. (W)
	TH solver/Pattern 1		TS solver/Pattern 2		
	B_m-H_b	B_m-H_m	B_m-H_b	B_m-H_m	
10	2.61	2.74	2.54	2.35	2.52
15	7.26	7.89	6.66	6.24	7.12
20	12.74	14.14	13.37	12.65	13.7
25	20.31	22.47	23.68	24.01	23.8

Table 9 shows the contributions of additional iron loss P_a and standard specific iron loss P_s to the total iron loss P_t, using different solvers and/or different kinds of B-H curves. Table 9 also indicates that P_s calculated using B_m-H_b curve is larger than that using the B_m-H_m curve. This is, because the permeability taken from B_m-H_b curve is higher than that taken from B_m-H_m curve. See Fig.11.

However, from the calculated results, P_a has a different tendency, using different B-H curve and/or different (TH or TS) solver.

Table 9. Components of iron loss (Model T1).

Current (A, rms, 50Hz)	TH solver (W)				TS solver (W)			
	B_m-H_b		B_m-H_m		B_m-H_b		B_m-H_m	
	P_a	P_s	P_a	P_s	P_a	P_s	P_a	P_s
10	1.53	1.08	1.81	0.93	1.73	0.81	1.72	0.63
15	5.50	1.76	6.35	1.54	5.20	1.46	5.12	1.12
20	10.44	2.31	12.07	2.07	11.15	2.22	10.95	1.70
25	17.19	3.12	19.89	2.58	21.22	2.46	21.59	2.42

The calculated results by both the TH and TS solvers show that the iron loss is mainly concentrated in a few layers on the side facing the exciting source, and goes down with the increase of layer-number going from 1 to 6, at 20A(50Hz) for Model T1, as shown in Fig. 9.

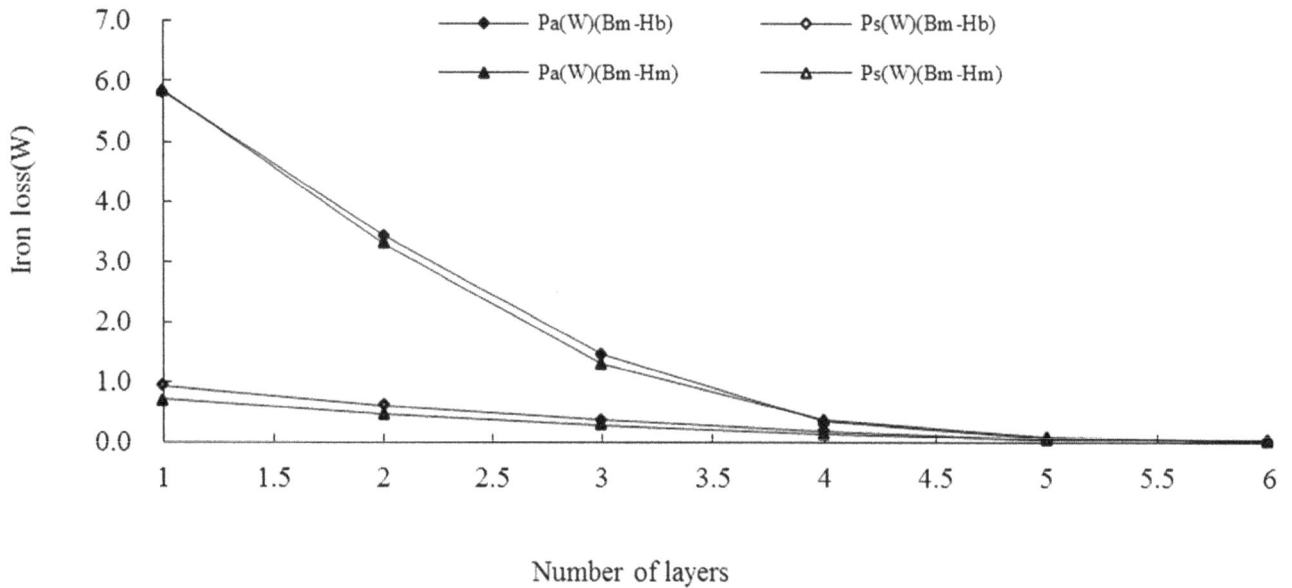

Figure 11. Iron loss distribution in layers (calculated by TS solver, Model T1, at 50Hz).

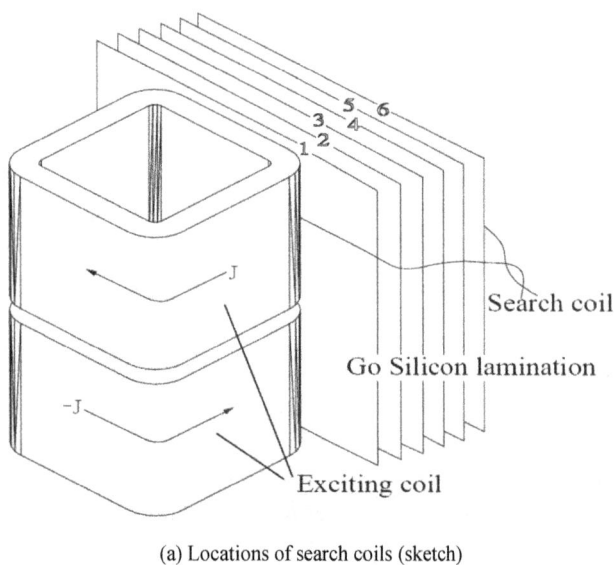

(a) Locations of search coils (sketch)

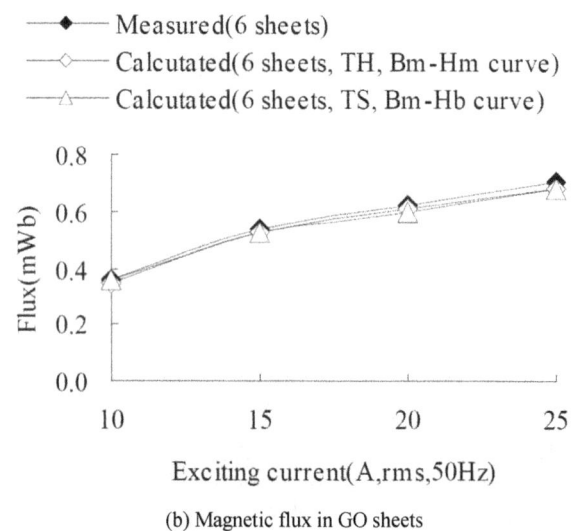

(b) Magnetic flux in GO sheets

Figure 12. Magnetic flux inside GO laminated sheets (Model T1, 50Hz).

To determine the magnetic flux inside the laminated sheets, search coils are set up in Model T1; see Fig. 12(a). The magnetic fluxes, at the prescribed positions of the laminations under the different exciting currents, are calculated. Either TH solver and B_m-H_m curve or TS solver and B_m-H_b curve are applied. The calculated results agree well with the measured ones. See Fig. 12(b).

In Model T2, the iron loss produced in the three sets of the narrow GO silicon steel sheets of 80×458mm, which are placed in parallel, is considerably lower compared to that of Model T1 for the same exciting currents.

Fig. 13 shows the calculated and the measured results of iron loss of Model T2 at 10A, under a frequency range from 50 to 200Hz. The results also indicate that the iron loss increases with the exciting frequency.

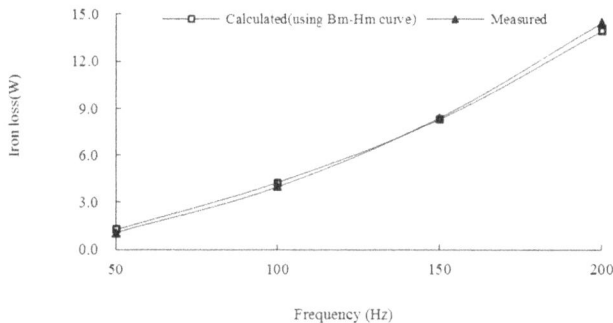

Figure 13. Iron loss varying with frequency (calculated by TH solver using B_m-H_m curve, Model T2).

Following is a very brief summary:

1) The examination of the effect of the different B-H curves (obtained by different means and data access modes and at different frequencies) on the iron loss and flux in GO silicon steel laminations is carried out. The numerical modeling results of the iron loss and flux based on the test models are in practical agreement with the measured ones.

2) All the numerical modeling results suggest that the B_m-H_b curve is desirable for the use in the transient solver, but the combination of the B_m-H_m curve and the time harmonic solver is also available for the problem with lower saturation levels.

3) The additional iron loss (P_a), due to the normal flux, exponentially drops from the surface facing the exciting source to the opposite side of the laminated sheets. On the other hand, the specific iron loss (P_s) generated by the parallel flux drops slowly compared to P_a.

4. Concluding Remarks

The progress of TEAM Problem 21 Family and long-term P21-based benchmarking are briefly reviewed, and a selected set of benchmarking results are presented.

The magnetic loss and flux inside both the solid and laminated components are numerically computed using different solvers based on different potential sets and measured using well-established experimental setups.

The effect of the variation in the B-H representations used in different solvers on the iron loss and flux in laminated sheets are examined in detail.

As the future research project, the Problem 21 Family is being extended to cover the modeling and computation issues that a device experiences when under extreme excitation conditions.

Acknowledgement

The authors would like to thank the heads and all the colleagues of R & D Center, Baobian Electric Group, for their cooperation, support and help in the long-term P21-based experiment and numerical modeling, and thank the ICS for all the approved updates of TEAM P21. Specially, here and now the authors would express their grief to the late Prof. Norio Takahashi for his co-research for many years.

References

[1] TEAM Benchmark Problems. [Online]. available: www.compumag.org/TEAM.

[2] IEEE Std 1597.1™-2008: IEEE Standard for validation of computational electromagnetics computer modeling and simulations.

[3] Z. Cheng, Q. Hu, S. Gao, Z. Liu, C. Ye, M. Wu, J. Wang, and Z.Hu, "An engineering-oriented loss model (Problem 21)," Proc. of the international TEAM Workshop, Miami, pp.137-143, 1993.

[4] Z. Cheng, N. Takahashi, B. Forghani, X. Wang, et al, "Extended progress in TEAM Problem 21 family," COMPEL, 33, 1/2, pp.234-244, 2014.

[5] Z. Cheng, N. Takahashi, B. Forghani, and Y. Wang, "Engineering-oriented benchmarking and application-based magnetic material modeling in transformer research", Presented at the International Colloquium Transformer Research and Asset Management(invited), Dubrovnik, Croatia, May 16 – 18, 2012.

[6] Z. Cheng, Q. Hu, N. Takahashi, and B. Forghani, "Stray-field loss modeling in transformers," *International Colloquium Transformer Research and Asset Management*, Cavtat, Croatia, Nov.12-14, 2009.

[7] N. Takahashi, T. Sakura and Z. Cheng, "Nonlinear analysis of eddy current and hysteresis losses of 3-D stray field loss model (Problem 21)," *IEEE Trans. Magn.*, vol.37, no.5, pp.3672-3675, 2001.

[8] Z. Cheng, R. Hao, N. Takahashi, Q. Hu, and C. Fan, "Engineering-oriented benchmarking of Problem 21 family and experimental verification," *IEEE Trans. Magn.*, vol. 40, no.2, pp.1394-1397, 2004.

[9] Z. Cheng, N. Takahashi, S. Yang, T. Asano, Q. Hu, S. Gao, X. Ren, H. Yang, L. Liu, and L. Gou, "Loss spectrum and electromagnetic behavior of Problem 21 family", *IEEE Trans. Magn.*, vol.42, no.4, pp.1467-1470, 2006.

[10] Z. Cheng, N. Takahashi, S. Yang, C. Fan, M. Guo, L. Liu, J. Zhang, and S. Gao, "Eddy current and loss analysis of multi-steel configuration and validation," *IEEE Trans. Magn.*, vol.43, no.4, pp.1737-1740, 2007.

[11] Z. Cheng, N. Takahashi, B. Forghani, G. Gilbert, J. Zhang, L. Liu, Y. Fan, X. Zhang, Y. Du, J. Wang, and C. Jiao, "Analysis and measurements of iron loss and flux inside silicon steel laminations," *IEEE Trans. Magn.*,vol.45, no.3, pp.1222-1225, 2009.

[12] Z. Cheng, N. Takahashi, B. Forghani, et al, "Effect of excitation patterns on both iron loss and flux in solid and laminated steel configurations," *IEEE Trans. Magn.*, vol.46, no.8, pp.3185-3188, 2010.

[13] Z. Cheng, N. Takahashi, B. Forghani, et al, "Effect of variation of B-H properties on loss and flux inside silicon steel lamination," *IEEE Trans. Magn.*, vol.47, no.5, pp.1346-1349, 2011.

[14] Z. Cheng, N. Takahashi, B. Forghani, L. Liu, Y. Fan, T. Liu, J. Zhang, and X. Wang, "3-D finite element modeling and validation of power frequency multi-shielding effect," *IEEE Trans. Magn.*, vol.48, no.2, pp.243-246, 2012.

[15] Z. Cheng, N. Takahashi, B. Forghani, et al, *"Electromagnetic and Thermal Field Modeling and Application in Electrical Engineering,"* Science Press (in Chinese), ISBN 978-7-03-023561-9, Beijing, 2009.

[16] A. J. Moses, "Characterisation and performance of electrical steels for power transformers operating under extremes of magnetisation conditions," *International Colloquium Transformer Research and Asset Management*, Cavtat, Croatia, Nov.12-14, 2009.

[17] M. Enokizono, H. Shimoji, A. Ikariga, et al, "Vector magnetic characteristic analysis of electrical machines," *IEEE Trans. Magn.*, vol.41, no.5, pp.2032-2035, 2005.

[18] K. Fujiwara, T. Adachi, and N. Takahashi, "A proposal of finite-element analysis considering two-dimensional magnetic properties," *IEEE Trans. Magn.*, vol.38, no.2, pp.889-892, 2002.

[19] H. Nishimoto, M. Nakano. K. Fujiwara, and N. Takahashi, "Effect of frequency on magnetic properties," *Papers of Technical Meeting on Magnetics, IEE Japan*, MAG-98-56, 1998 (in Japanese).

[20] J. Zhu, J. J. Zhong, Z. W. lin, et al, "Measurement of magnetic properties under 3-D magnetic excitations," *IEEE Trans. Magn.*, vol.39, no.5, pp. 3429-3431, 2003.

[21] J. Turowski, M. Turowski, and M. Kopec, "Method of three-dimensional network solution of leakage field of three-phase transformers," *IEEE Trans. Magn.*, vol. 26, no. 5, pp. 2911-2919, 1990.

[22] N. Takahashi, S. Nakazaki, and D. Miyagi, "Optimization of electromagnetic and magnetic shielding using ON/OFF method," *IEEE Trans. Magn.*, vol.46, no.8, pp.3153-3156, 2010.

[23] M. Horii, N. Takahashi, and J. Takehara, "3-D optimization of design variables in x-, y-, and z-directions of transformer tank shield model," *IEEE Trans. Magn.*, vol.37, no.5, pp.3631-3634, 2001.

[24] J. Turowski, X. M. Lopez-Fernandez, A. Soto, and D. Souto, "Stray losses control in core- and shell-type transformers," *Advanced Research Workshop on Transformers*, Baiona, Spain, 29-31 Oct., 2007.

[25] K. V. Namjosji and P. P. Biringer, "Efficiency of eddy current shielding of structural steel surrounding large currents: a circuit approach," *IEEE Trans. Magn.*, vol.27, no.6, pp.5417-5419, 1991.

[26] R. Tang, Y. Li, F. Lin, and L. Tian, "Resultant magnetic fields due to both windings and heavy current leads in large power transformers," *IEEE Trans. Magn.*, vol.32, no.3, pp.1641-1644, 1996.

[27] R. M. D. Vecchio, "Eddy current losses in a conducting plate due to a collection of bus bars carrying currents of different magnitudes and phases," *IEEE Trans. Magn.*, vol.39, no.1, pp.549-552, 2003.

[28] Z. Cheng, B. Forghani, Y. Liu, Y. Fan, T. Liu, and Z. Zhao, "Magnetic Loss inside Solid and Laminated Components under Extreme Excitations," to be published in the Special Issue (no.164022) of *International Journal of Energy and Power Engineering*.

[29] O. Biro and K. Preis, "Finite element analysis of 3-D eddy currents," *IEEE Trans. Magn.*, vol.26, no.2, pp.418-423, 1990.

[30] O. Biro, K. Preis, and K. R. Richter, "Various FEM formulation for the calculation of transient 3D eddy currents in nonlinear media," *IEEE Trans. Magn.*, vol.31, no.3, pp.1307-1312, 1995.

[31] O. Biro, K. Preis, U. Baumgartner, and G. Leber, "Numerical modeling of transformer losses," presented at *International Colloquium Transformer Research and Asset Management*, Cavtat, Croatia, Nov.12-14, 2009.

[32] J. P. Webb and B. Forghani, "T-Omega method using hierarchal edge elements," *IEE Proc.-Sci.Meas. Technol.*, vol.142, no.2, 1995, pp.133-141.

[33] Z. Cheng, S. Gao, and L. Li, *"Eddy Current Analysis and Validation in Electrical Engineering"*, Higher Education Press (in Chinese), ISBN 7-04-009888-1, Beijing, 2001.

[34] H. Kaimori, A. Kameari, and K. Fujiwara, "FEM computation of magnetic field and iron loss in laminated iron core using homogenization method," *IEEE Trans. Magn.*, vol.43, no.4, pp.1405-1408, 2007.

[35] K. Preis, O. Biro, and I. Ticar, "FEM analysis of eddy current losses in nonlinear laminated iron cores," *IEEE Trans. Magn.*, vol.41, no.5, pp.1412-1415, 2005.

[36] T. Kohsaka, N. Takahashi, S. Nogawa, and M. Kuwata, "Analysis of Magnetic characteristics of three-phase reactor model of grain-oriented silicon steel," *IEEE Trans. Magn.*, vol.36, no.4, pp.1894-1897, 2000.

[37] H. Igarashi, K. Watanabe, and A. Kost, "A reduced model for finite element analysis of steel laminations," *IEEE Trans. Magn.*, vol.42, no.4, pp.739-742, 2006.

[38] W. Zheng, and Z. Cheng, "Efficient finite element simulation for GO silicon steel laminations using inner-constrained laminar separation," *IEEE Trans. Magn.*, vol.48, no.8, pp.2277-2283, 2012.

[39] P. Marketos, S. Zurek, and A. Moses, "A method for defining the mean path length of Epstein," *IEEE Trans. Magn.*, vol.43, no.6, pp.2755-2757, 2007.

[40] Z. Cheng, N. Takahashi, B. Forghani, A. Moses, P. Anderson, Y. Fan, T. Liu, X. Wang, Z. Zhao, and L. Liu, "Modeling of magnetic properties of GO electrical steel based on Epstein combination and loss data weighted processing," *IEEE Trans. Magn.*, vol.50, no.1, 6300209, 2014.

Permissions

All chapters in this book were first published in IJEPE, by Science Publishing Group; hereby published with permission under the Creative Commons Attribution License or equivalent. Every chapter published in this book has been scrutinized by our experts. Their significance has been extensively debated. The topics covered herein carry significant findings which will fuel the growth of the discipline. They may even be implemented as practical applications or may be referred to as a beginning point for another development.

The contributors of this book come from diverse backgrounds, making this book a truly international effort. This book will bring forth new frontiers with its revolutionizing research information and detailed analysis of the nascent developments around the world.

We would like to thank all the contributing authors for lending their expertise to make the book truly unique. They have played a crucial role in the development of this book. Without their invaluable contributions this book wouldn't have been possible. They have made vital efforts to compile up to date information on the varied aspects of this subject to make this book a valuable addition to the collection of many professionals and students.

This book was conceptualized with the vision of imparting up-to-date information and advanced data in this field. To ensure the same, a matchless editorial board was set up. Every individual on the board went through rigorous rounds of assessment to prove their worth. After which they invested a large part of their time researching and compiling the most relevant data for our readers.

The editorial board has been involved in producing this book since its inception. They have spent rigorous hours researching and exploring the diverse topics which have resulted in the successful publishing of this book. They have passed on their knowledge of decades through this book. To expedite this challenging task, the publisher supported the team at every step. A small team of assistant editors was also appointed to further simplify the editing procedure and attain best results for the readers.

Apart from the editorial board, the designing team has also invested a significant amount of their time in understanding the subject and creating the most relevant covers. They scrutinized every image to scout for the most suitable representation of the subject and create an appropriate cover for the book.

The publishing team has been an ardent support to the editorial, designing and production team. Their endless efforts to recruit the best for this project, has resulted in the accomplishment of this book. They are a veteran in the field of academics and their pool of knowledge is as vast as their experience in printing. Their expertise and guidance has proved useful at every step. Their uncompromising quality standards have made this book an exceptional effort. Their encouragement from time to time has been an inspiration for everyone.

The publisher and the editorial board hope that this book will prove to be a valuable piece of knowledge for researchers, students, practitioners and scholars across the globe.

List of Contributors

Huang Wen-bo
Institute of International Law, Wuhan University, Wuhan, China

Moses Peter Musau, Nicodemus Odero Abungu and Cyrus Wabuge Wekesa
Department of Electrical and Information Engineering, School of Engineering, The University of Nairobi, Nairobi, Kenya

Rahim Jassim
Saudi Electric Services Polytechnic (SESP), Baish, Jazan Province, Kingdom of Saudi Arabia

Galal Zaki, Badr Habeebullah and Majed Alhazmy
Mechanical Engineering, King Abdulaziz University, Jeddah, Saudi Arabia

M. Magesh Kumar
ME power system engineering, Chandy college of engineering, Tuticorin, India

R. Sundareswaran
AP/EEE Department, Chandy college of engineering, Tuticorin, India

Oluwatosin M. Dada, Ilesanmi A. Daniyan and Temitayo M. Azeez
Department of Mechanical and Mechatronics Engineering, Afe Babalola University, Ado Ekiti, Nigeria

Olalekan O. Adaramola
Department of Computer Engineering, Afe Babalola University, Ado Ekiti, Nigeria

L. Mogaka and M. J. Saulo
Electrical and Electronics Department, Technical University of Mombasa, Mombasa, Kenya

D. K. Murage
Electrical and Electronics Department, Jomo Kenyatta University of Agriculture and Technology, Nairobi, Kenya

Musa H. and Ibrahim S. B.
Department of Electrical Engineering, Bayero University, Kano, Nigeria

Singhal M. K. and Arun Kumar
AHEC, Indian Institute of Technology, Roorkee, India

Halima Ibrahim ElSaeedy and Karam Fathy Abd El-Rahman
Department of Physics, Faculty of Science for Girls, King Khalid University, Abha, Saudi Arabia

Maryam Ayidh Saad Al Shahrani
Department of Physics, Faculty of Science for Girls, Bisha University, Abha, Saudi Arabia

Sayed Taha Mohamed Hassan
Department of Physics, Faculty of Science, King Khalid University, Abha, Saudi Arabia

Kitheka Joel Mwithui
Jomo Kenyatta University of Agriculture and Technology, Nairobi, Kenya

Michael Juma Saulo
Department of Electrical and Electronic Engineering/ Faculty of Engineering and Technology, Technical University of Mombasa, Mombasa, Kenya

David Murage
Department of Electrical and Electronic Engineering, Jomo Kenyatta University of Agriculture and Technology, Nairobi, Kenya

Fátima Aparecida de Morais Lino and Kamal Abdel Radi Ismail
Energy Department, Faculty of Mechanical Engineering, State University of Campinas, Barão Geraldo, Campinas, Brazil

K. P. V. B. Kobbekaduwa and N. D. Subasinghe
National Institute of Fundamental Studies, Hanthana Road, Kandy, Sri Lanka

Samuel Sami, Edwin Marin and Jorge Rivera
Research Center for Renewable Energy, Catholic University of Cuenca, Cuenca, Ecuador

Pham Thanh Tuan, Vu Ngoc Hai and Seoyong Shin
Department of Information and Communication, Myongji University, Yongin City, Republic of Korea

Xue Kang, Yiping Wang and Xusheng Shi
School of Chemical Engineering and Technology, Tianjin University, Tianjin, China

Ganchao Xin
Toppley (Zhuhai) Chemicals. Co., LTD. Guangdong, China

Brahim Menacer and Mostefa Bouchetara
Aeronautics and Systems Propelling Laboratory, Department of Mechanical Engineering, University of Sciences and the Technology of Oran, L.P 1505 El -Menaouer, Oran, Algeria

Abdalla Gomaa
Refrigeration and Air Conditioning Technology Department, Faculty of Industrial Education, Helwan University, Cairo, Egypt

Motoo Fumizawa, Naoya Uchiyama and Takahiro Nakayama
Department of Mechanical Engineering, Shonan Institute of Technology, Fujisawa, Kanagawa, Japan

Mahamat Barka
Faculté Des Sciences Exactes Et Appliquées de l'Université De Ndjaména, Département De Technologie, Ndjaména, Tchad

Abakar Mahamat Tahir
Faculté Des Sciences Exactes Et Appliquées de l'Université De Ndjaména, Département De Technologie, Ndjaména, Tchad
Laboratoire Ampère Insa-Lyon 20, Avenue Al. Einstein, Villeurbanne Cedex, France

Amir Moungache
Faculté Des Sciences Exactes Et Appliquées de l'Université De Ndjaména, Département De Technologie, Ndjaména, Tchad
Laboratoire LT2C (ex DIOM 2), Rue Du Dr Paul Michelon, Etienne Cedex, France

Dominique Ligot and Pascal Bevilacqua
Laboratoire Ampère Insa-Lyon 20, Avenue Al. Einstein, Villeurbanne Cedex, France

Jean-Jacques Rousseau
Laboratoire LT2C (ex DIOM 2), Rue Du Dr Paul Michelon, Etienne Cedex, France

Motoo Fumizawa, Yoshiharu Saito, Naoya Uchiyama and Takahiro Nakayama
Department of Mechanical Engineering, Shonan Institute of Technology Fujisawa, Kanagawa, Japan

Ngoc Hai Vu and Seoyong Shin
Department of Information and Communication Engineering, Myongji University, Yongin, South Korea

Sayeda Fahmida Ferdousi and Raju Chowdhury
Department of Natural Science, Stamford University Bangladesh, Dhaka, Bangladesh
Department of Mathematics, Bangladesh University of Engineering & Technology, Dhaka, Bangladesh

Md. Abdul Alim
Department of Mathematics, Bangladesh University of Engineering & Technology, Dhaka, Bangladesh

Gabin Koucoi, Daniel Yamegueu and Yézouma Couliblay
Laboratory for Solar Energy and Energy Savings (LESEE), International Institute for Water and Environmental Engineering (2IE), Ouagadougou, Burkina Faso

Quoc-Tuan Tran and Hervé Buttin
Smart Grid Laboratory (LSEI), National Solar Energy Institute (CEA/INES), Bourget Du Lac, France

Eduard A. Mikaelian
Department of Thermodynamics and Heat Engines, Gubkin Russian State University of Oil and Gas, Moscow, Russia

Saif A. Mouhammad
Department of Physics, Taif University, Taif, Saudi Arabia

Zhang Haibo, Shen Junlong and Yuan Pan
Nanjing University of Chinese Medicine, Nanjing, China

Shen Yuan
School of Applied Mathematics, Nanjing University of Finance & Economics, Nanjing, China

Samuel Sami
Research Center for Renewable Energy, Catholic University of Cuenca, Cuenca, Ecuador

Jorge Zatarain
Faculty of Energy Engineering, Universidad Politécnica de Sinaloa, Sinaloa, México

Mustafa Adil, Osama Ibrahim, Zainalabdeen Hussein and Kaleid Waleed
Renewable Energy Research Center, University of Anbar, Ramadi, Iraq

Zhiguang Cheng, Tao Liu, Yana Fan and Lanrong Liu
Institute of Power Transmission and Transformation Technology, Baobian Electric Co., Ltd, Baoding, China

Index

www.ingramcontent.com/pod-product-compliance
Lightning Source LLC
Chambersburg PA
CBHW080511200326
41458CB00012B/4162